W0042314

THIRD EUROPEAN RHEOLOGY CONFERENCE

and Golden Jubilee Meeting of the British Society of Rheology

Proceedings of the Golden Jubilee Meeting of the British Society of Rheology and Third European Rheology Conference, Edinburgh, UK, 3–7 September 1990.

ORGANISING COMMITTEE

C.J.S. PETRIE (*Chairman*)
J. FERGUSON
N. HUDSON
C. MOULES
D.R. OLIVER
S. ALDHOUSE

THIRD EUROPEAN RHEOLOGY CONFERENCE

and Golden Jubilee Meeting of the British Society of Rheology

Edited by

D.R. OLIVER

School of Chemical Engineering,
University of Birmingham, UK

ELSEVIER APPLIED SCIENCE
LONDON AND NEW YORK

ELSEVIER SCIENCE PUBLISHERS LTD
Crown House, Linton Road, Barking, Essex IG11 8JU, England

Sole Distributor in the USA and Canada
ELSEVIER SCIENCE PUBLISHING CO., INC.
655 Avenue of the Americas, New York, NY 10010, USA

WITH 33 TABLES AND 272 ILLUSTRATIONS

© 1990 ELSEVIER SCIENCE PUBLISHERS LTD
© 1990 CROWN COPYRIGHT — pp. 94–97

British Library Cataloguing in Publication Data
European Rheology Conference (3rd; 1990; Edinburgh, Scotland)
    Third European rheology conference: and golden jubilee
    meeting of the British Society of Rheology.
    1. Rheology
    I. Title   II. British Society of Rheology, Golden Jubilee
    Meeting   III. Oliver, D.R.
    531.11

Library of Congress CIP data applied for

ISBN-13: 978-94-010-6838-3   e-ISBN-13: 978-94-009-0781-2
DOI: 10.1007/978-94-009-0781-2

Softcover reprint of the hardcover 1st edition 1990

No responsibility is assumed by the Publisher for any injury and/or damage to persons or property as
a matter of products liability, negligence or otherwise, or from any use or operation of any methods,
products, instructions or ideas contained in the material herein.

Special regulations for readers in the USA

This publication has been registered with the Copyright Clearance Center Inc. (CCC), Salem,
Massachusetts, Information can be obtained from the CCC about conditions under which photocopies
of parts of this publication may be made in the USA. All other copyright questions, including
photocopying outside the USA, should be referred to the publisher.

All rights reserved. No part of this publication may be reproduced, stored in a retrieval system, or
transmitted in any form or by any means, electronic, mechanical, photocopying, recording, or otherwise, without the prior written permission of the publisher.

EDITORIAL

The British Society of Rheology

The British Rheologists' Club was founded in one of the darkest years of the war — 1940. The founder members were H.R. Lang, V.G.W. Harrison and the redoubtable G.W. Scott-Blair, with E.W.J. Mardles an early member and L.R.G. Treloar joining in 1942. The first President of the club was G.I. Taylor. Another leading British rheologist of the period was J. Pryce-Jones, the self-styled 'stormy petrel of rheology'. The club was pre-dated by the American Society of Rheology, correctly termed 'Society of Rheology', which was founded by E.C. Bingham in 1929.

The club grew rapidly and changed its name to 'The British Society of Rheology' (1950). An emblem or logo was designed in 1966 and first appears on the lower left side of *Rheology Abstracts* (issue number one) of the following year. The emblem combined the simple concepts of flow and elasticity in a circular symbol.

By 1951, Scott-Blair was describing efforts to form an International Union of Rheology, but the International Committee on Rheology was not formed until 1953, during the second International Congress on Rheology, which was held at Oxford. At that time, the national societies were those of the USA, Britain, Germany, Japan and The Netherlands, with France's society expected in the near future. Other countries represented on the first committee were Brazil, Israel, Sweden and Yugoslavia.

The Society has always had strong links with our friends in Australia; around 1960 the New South Wales and Victorian Branches became part of the BSR, and not until 1984 was the decision taken (equally well received by both parties) to form an independent Australian Society of Rheology. This, of course, led to the organisation of a most successful World Congress on Rheology in Sydney in 1988.

By a happy chance, the Golden Jubilee of the British Society of Rheology coincides with one of the four-yearly dates allocated for European conferences, giving the present meeting its double title. We hope it will be as successful as its predecessors.

The Third European Rheology Conference

This conference is the third major European conference, following the 'First Conference of European Rheologists' held at Graz (Austria) in 1982 and the 'Second Conference of European Rheologists' held in Prague in 1986. The adoption of the title 'Third European Rheology Conference' reflects the nature of these conferences as being truly international and, while they take place in Europe, the participants have come from all over the world. Reports on the previous two meetings, plenary lectures and short versions of papers presented are contained in 'Progress and Trends in Rheology I' and 'Progress and Trends in Rheology II'. These were published by Dr Dietrich Steinkopff Verlag of Darmstadt, in 1982 as a special issue of *Rheologica Acta* and in 1988 (for the 1986 conference) as a 500-page supplement to *Rheologica Acta*. The present conference is being recorded for posterity both in this volume of short papers (extended abstracts) and in 'Progress and Trends in Rheology III' which will be a special issue of *Rheologica Acta* due to appear at the end of 1990. This should contain the three plenary lectures and eight keynote papers (which are represented here by abstracts).

An account of the history of these conferences may be found in the introduction to 'Progress and Trends in Rheology I' (*Rheol. Acta* **21**, 1982, 355–356). At the time of writing, no decision has been taken about the fourth European Rheology Conference; the European delegates will meet in Edinburgh to discuss this. It seems likely from the success of the first two meetings and the excellent response to this meeting that there will be a fourth European conference in 1994, two years after Belgium hosts the eleventh International Congress on Rheology.

It may also be useful to record that the series of newsletters entitled *European Rheological Activity* continues to appear from time to time. These are not sent to individuals but to secretaries of societies of rheology and to delegates to the International Committee on Rheology. A copy is always sent to the secretary of the ICR, currently Professor David James of the University of Toronto. The intention of promoting European activities has never been to do so at the expense of world-wide cooperation. Dr C.J.S. Petrie, whose help is acknowledged in the preparation of this information, is the European Secretary.

D.R. OLIVER

CONTENTS

Contributed Papers

xii

.

xvii

CARRIED ALONG ON A PATH LINE IN MODELLING CONSTITUTIVE EQUATIONS OF VISCOELASTIC FLUIDS

HANS WALTER GIESEKUS
Department of Chemical Engineering, University of Dortmund
P.O. Box 50 05 00, D-4600 Dortmund 50, FRG

A retrospect is given on the "path line" whereon the author was carried along in search for structural constitutive models which predict visco-elastic properties. In the first section of this path an analysis was carried out of a dilute solution of rigid dumbbells, and subsequenty of linear-elastic dumbbells with finite equilibrium length. In doing so, the associated constitutive equation for the special case of Hookean dumb-bells could be identified with that of an Oldroyd B fluid. The second section was devoted to the indication of conditions under which second normal-stress differences would appear. This was shown to be the case for solutions of rigid dumbbells with spheroidal beads without and with inclusion of hydrodynamic interaction and also for solutions of rigid spheroids. In a third section of this path line attention was focused on the dependence of rheological properties of suspensions with rigid par-ticles of different shape on orientation and motion subject to different types of flow. Eventually, in a last section an approach was attempted to comprehend more concentrated solutions in these investigations. This was based on the concept of configuration-dependent tensorial mobility as a function of the mean configuration tensor. Already a linear relation between these tensors resulted in a model which predicted most features of polymer fluids. When substituting instead a slightly more complicated relation of relaxation type, these predictions could be substantially improved. Finally, inclusion of a quadratic and an exponential term, respectively, in the first mentioned linear relation allowed for a description of flow hardening phenomena.

In conclusion, a new attack is pushed on shear hardening. It starts from a result of a paper, written about 30 years ago, in which the modifica-tion of the effective flow field owing to the action of particles is cor-related with the extra stresses resulting therefrom. In the special case of macroscopic shear flow the effective flow field is no longer a plane "weak" flow but a spatial "strong" flow in which elastic particles, e.g. polymer coils, can be oriented and continuously stretched in a fixed position if they are slender enough. If counteracting influences, as are, e.g., described in terms of a configuration-dependent mobility, are allowed to be overcome by this effect, shear thinning will transit into shear thickening and, at the same time, the second normal-stress dif-ference will become positive. This may cause instability phenomena even in straight flows.

THE RHEOLOGY OF POWDERS

SIR SAM EDWARDS
University of Cambridge
Department of Physics
Madingley Road, Cambridge CB3 OHE, UK

Powders can be treated in a manner analogous to thermal systems by the introduction of a new variable, which I call the compactivity, $\partial V/\partial S$, which replaces the temperature $\partial E/\partial S$, and permits a precise way of describing the variable density possible with powders.

The normal equations of visco-elasticity containing ρ, v and T are replaced by equations for ρ, v and X. Several examples will be offered; a particularly interesting case is the transition to plug flow for the flow of powder in a pipe.

SOME INVERSE PROBLEMS IN RHEOLOGY LEADING TO INTEGRAL EQUATIONS

ALEXANDER MALKIN
Research Institute of Plastics,
35, Perovskii Proezd, Moscow 111112, USSR

In various applications we meet the necessity to solve inverse problems, that means to find the properties of the matter (expressed by some constants or functions) from the experimentally observed macro-behaviour of a sample. Such inverse problems can be differential (the unknown constants being the coefficients of differential equations) or integral. The last case consists in existence of relationships between inherent properties of the matter and gross properties of a body, expressed by integral equations or functions. In this lecture two typical rheological themes leading to inverse integral problems are discussed. The first one represents rheology as a method of reflection of molecular composition of a matter. It is flow curve - molecular weight-distribution correlation. It is shown that this problem has the exact solution, but this solution is unstable in principle and it means that the inverse problem is incorrect by its nature. The second inverse problem, discussed in the lecture, lies in its proper field of rheology. It is the calculations of creep from relaxation functions and vice versa. This problem has a correct solution but the possibility to find such a solution depends upon the method of approximation of an experimental curve.

THE RHEOLOGY OF FIBRE SUSPENSIONS

D V BOGER, D U HUR and †C J S PETRIE
Department of Chemical Engineering, University of Melbourne
Parkville, Victoria 3052, Australia
†*Permanent address*:   Department of Engineering Mathematics, The University,
Newcastle upon Tyne, NE1 7RU, UK

ABSTRACT

Theoretical and experimental work on the rheology of suspensions of slender fibres is reviewed. This is an area of enormous importance in applications such as the manufacture of fibre reinforced polymeric materials and the fabrication of articles out of reinforced plastic. The particular features of suspension rheology when the suspended particles are long and slender, notably the response in extensional flows, are also of fundamental scientific interest in their own right and the issues we shall be surveying are of interest to both theorists and experimenters.

The published literature relevant to this topic is vast; one of us has listed well over 300 references [1]. In the rheological literature, attention is drawn to useful sections in a number of books [2, 3, 4]. Note that for suspensions of fibres and for solutions of long rigid polymers the theoretical approach, and indeed many theoretical results, are identical. Review articles which can be recommended include the comprehensive theoretical study by Brenner [5] and a review by Batchelor [6] which focus on the fundamental aspects of the topic. This contrasts with the more pragmatic approach of Jinescu [7]. A valuable contribution to bridging the gap between theory and practice was produced by Jeffrey and Acrivos [8]. More specifically devoted to fibre suspensions are the reviews by Maschmeyer and Hill [9] and by Ganani and Powell [10]. The behaviour of fibre suspensions in extensional flow has been recently reviewed in [11].

The starting point for theoretical work has to be the dynamics of a single particle suspended in a flowing fluid. From there we trace the steps necessary for the construction of a model from which the macroscopic behaviour of a suspension may be predicted. The influence of non-Newtonian behaviour of the suspending fluid has been studied to a limited extent. The main practical drawback of dilute suspension theory is that, for long slender fibres, diluteness means that fibres have to be far apart on a scale based on the length of the fibres rather than their diameter, and so dilute means *extremely* dilute in terms of the volume fraction of solids in the suspension. Consideration of the hydrodynamic interaction between particles is obviously necessary. In the extreme case of suspensions that are so concentrated that, in effect, the fibres have no choice but to lie parallel to one another we find the regime of liquid crystalline behaviour. The issue of wall effects, i.e. hydrodynamic or direct mechanical interaction between the suspended fibres and the walls of the container within which the suspension is flowing, is extremely important.

There is much published experimental work on suspensions both in Newtonian liquids, which is of fundamental scientific interest, and in molten polymers, with obvious relevance to the processing of fibre-reinforced plastics. Suspensions of fibres in well-characterized non-Newtonian liquids have perhaps been studied less until recently and we discuss this area critically. We shall present in tabular form references to papers on both these areas and attempt to make clear the rheological content of the reports. Studies on fibre orientation and attempts at flow visualization are also of considerable interest and we offer a further tabulation of work in this area.

The contribution of C J S Petrie was made possible by a Visiting Research Fellowship of the University of Melbourne and by support from the Departments of Mathematics and Chemical Engineering of the University of Melbourne. This support and the granting of a period of study leave by the University of Newcastle upon Tyne are gratefully acknowledged. Research in non-Newtonian fluid mechanics at the University of Melbourne is supported by the Australian Research Council.

REFERENCES

1. D U Hur, **Flow of semidilute glass fibre suspensions in tubular entry flows** Ph D thesis, University of Melbourne, 1987.

2. R B Bird, C F Curtiss, R C Armstrong and O Hassager, **Dynamics of Polymeric Liquids, Second Edition, Volume 2: Kinetic Theory**, Wiley, New York, 1987.

3. M Doi and S F Edwards, **The Theory of Polymer Dynamics**, Clarendon Press, Oxford, 1987.

4. W R Schowalter, **Mechanics of Non-Newtonian Fluids**, Pergamon, Oxford, 1978.

5. H Brenner, *Int. J. Multiphase Flow*, **1** (1974) 195-341.

6. G K Batchelor, *Ann. Rev. Fluid Mech.*, **6** (1974) 227-255.

7. V V Jinescu, *Intern. Chem. Eng.*, **14** (1974) 397-420.

8. D J Jeffrey and A Acrivos, *AIChE J.*, **22** (1976) 417-432.

9. R O Maschmeyer and C T Hill, *ACS Adv. Chem.*, **134** (1974) 95-105.

10. E Ganani and R L Powell, *J. Composite Mater.*, **19** (1985) 194-215.

11. J Mewis and C J S Petrie, **Hydrodynamics of spinning polymers in Encyclopedia of Fluid Mechanics**, ed. N P Cheremisinoff, Volume 6, pp 111-139, Gulf Publishing, Houston, 1987.

CHALLENGES IN PROCESS RHEOMETRY

JOHN M. DEALY
Department of Chemical Engineering
McGill University
3480 University Street/Montreal, Canada H3A 2L7

ABSTRACT

Process rheometers are useful as process sensors for on-line
quality control as well as for process control. Unlike passive
sensors for equilibrium properties such as temperature and
pressure, a rheometer must subject the material of interest to
a controlled deformation while monitoring the resulting
stress. To accomplish this in a compact, robust device
suitable for use in a manufacturing environment is not
straightforward, and considerable ingenuity has been applied
to the development of the several types of instrument now
offered commercially.

However, there are many potential applications for which
presently available units are not suitable. For example,
problems arise when a very rapid instrument response is
required, when the fluid is non-Newtonian, and when suspended
solids are present. In addition, in biotechnology
applications, it may be necessary that all exposed surfaces be
autoclavable. For a fast response, an in-line unit is
necessary, i.e., one that does not draw a side stream from the
process flow. The challenge here is to provide for an adequate
sample renewal rate without interfering with the operation of
the rheometer. For non-Newtonian fluids, the problem is the
generation of a uniform shear rate.

Solving these problems poses major problems to the rheometer
developer, and some interesting advances have recently been
made. In particular, the problem of sample renewal has been
addressed in one laboratory by incorporating a mechanism for
periodic forced flow to clear the shearing gap of "old"
material. In another laboratory, the problem posed by non-
Newtonian fluids has been solved by use of a novel shear
stress transducer. These developments are described, and some
remaining challenges are outlined.

THE RHEOLOGICAL PROPERTIES OF POLYACRYLAMIDE SOLUTIONS IN SHEAR AND EXTENSION

J Ferguson*, K Walters**, and C Wolff***
*Department of Pure & Applied Chemistry,University of Strathclyde, Glasgow; ** Department of Mathematics, UCW, Aberystwyth;*** ENSITM, University of Haute Alsace, Mulhouse.

As part of an EEC Science Stimulation programme on extensional viscosity, two major conferences were organised on the subject. The second of these [1] was devoted to the results obtained on a standard fluid, M1. The data obtained in shear flow was remarkably consistent from laboratory to laboratory. Extensional flow results presented quite a different picture. Using a series of nonequilibrium techniques such as the spinline rheometer, opposing jet, falling drop and converging flow, extensional viscosity results were obtained which differed by as much as 2 to 3 orders of magnitude. Nevertheless, it was apparent that consistency did exist within similar techniques. It is in the context of this information that the measurements described below have been made .

The shear and extensional flow properties of partially ionised polyacrylamide in solution at concentrations ranging from 5ppm to 500ppm were measured. The method of solution preparation was found to have a profound effect on the behaviour of the solutions in shear flow . The influence of salt concentration and pH was investigated and is discussed in the context of molecular shape in solution.

Extensional flow measurements, using the spinline rheometer, showed that the solutions are strongly strain thickening even at concentrations as low as 5ppm. As change in salt concentration and pH altered the degree of ionisation of the carboxylic acid groups in the polyacrylamide chain, the macromolecular chain altered its dimensions in solution. Coil expansion was found to lead to an increase in extensional viscosity over all ranges of strain rate examined. These results are considered in the light of polymer entanglement and association in the strong flow field.

REFERENCE

1. Journal of Non Newtonian Fluid Mechanics, <u>35</u>, 2&3 (1990)

INFRARED POLARIMETRY STUDIES FOR MULTI COMPONENT POLYMER MELTS

GERALD G. FULLER
CAROLINE M. YLITALO
Department of Chemical Engineering
Stanford University
Stanford, CA 94305-5025

A rheo-optical technique to measure dynamic infrared dichroism was used with deuterium labeling to study component relaxation in a variety of multi component polymer melts. For each sample overall relaxation was observed simultaneously using birefringence measurements.

Nearly monodisperse poly(ethylene propylene) samples were used to construct three different sets of bidisperse blends. For each sample, results for step strain relaxation and oscillatory response of each component and of the total sample are presented and discussed. Effects of intermolecular orientational coupling were observed and a coupling strength of 0.45 was measured. The results of these experiments were compared to a reptation based constraint release model, and a qualitative agreement was found.

Using nearly monodisperse polybutadiene block copolymers, selected segment relaxations following step strains were examined. It was found that segments located at chain ends relax faster than segments located at chain centers, and that both the Rouse model and the tube model of Doi and Edwards correctly predict the qualitative features of segmental relaxation. The tube model however, gave exact quantitative agreement with experiment when the effects of orientational coupling interactions of strength 0.4 were incorporated into the model.

Additional studies on orientation coupling effects in polymer melts were conducted using a series of monodisperse polybutadiene oligomers dissolved in a polybutadiene polymer matrix. The oligomers had molecular weights ranging from well below the entanglement molecular weight, M_e, to several times M_e. The results indicate that orientational coupling interactions are significantly reduced for oligomers with molecular weights above M_e.

This technique can be further applied to study component dynamics in blends of two different polymers where the relaxation of each polymer in the melt can be observed separately.

PROGRESS AND CHALLENGES IN COMPUTATIONAL RHEOLOGY

Roland Keunings

Unité de Mécanique Appliquée
Université Catholique de Louvain
Place du Levant, 2
1348 Louvain-la-Neuve, Belgium

Abstract

The mathematical models and numerical techniques that provide the theoretical foundation of classical Computational Fluid Dynamics have a range of validity limited to gases or low molecular weight Newtonian liquids. The provocative flow phenomena observed with polymeric fluids, for example, cannot at all be predicted by the Navier-Stokes equations. Non-Newtonian behavior has many facets, the most striking being memory effects associated with the elasticity of the material. The theoretical challenge is to translate the complex rheological behavior of non-Newtonian fluids into suitable constitutive equations, and to develop accurate numerical procedures for use of these models in the analysis of complex flows.

The aim of this lecture is to review the field of large-scale non-Newtonian flow simulations as it stands in 1990, with a special emphasis put on the modeling of viscoelastic effects. We begin with a discussion of the mathematical models that are of current use in numerical work. These models combine ellipticity and hyperbolicity in a rather subtle way. We then introduce a classification of numerical approaches which is used to describe the entire spectrum of available numerical techniques. Some of the published simulations that predict significant viscoelastic effects are reviewed. Finally, we focus on the considerable mathematical, numerical and algorithmic challenges associated with the prediction of viscoelastic flows.

RHEOLOGY OF RODLIKE POLYMERS IN THE NEMATIC PHASE

G MARRUCCI

Department of Chemical Engineering, University of Naples
Piazzale Tecchio, I-80125 Napoli, Italy

ABSTRACT

The theory of rodlike polymers in the nematic phase is now sufficiently well developed to allow predictions of rheological behaviour which compare favourably with many experimental observations. These include peculiar effects such as negative normal stresses, damped oscillations in start-up flows and oscillation frequencies inversely proportional to shear rate. The basic points of the theory are presented, together with a more direct interpretation of the above-mentioned phenomena.

PHYSIOLOGY MECHANICS AND RHEOLOGY

A. SILBERBERG
Department of Polymer Research
Weizmann Institute of Science, Rehovot 76100, Israel

ABSTRACT

Most interactions between living systems and the surrounding inanimate world are mechanical. From the act of self-reproduction, to defence against predators, to the search for food and its ingestion, the physiological processes involved are fundamentally mechanical in nature. Even the cells liberate energy and generate forces with the purpose, mainly, of the performance of mechanical work for physiological effects. Biomechanics deals with this aspect of living systems. How in a given biological situation, i.e. with a given structure, a given set of moving parts and with materials of certain properties, the biological system performs its set task. In Biomechanics one does not design, but try to understand. One tries to mathematize a certain situation and describe its mechanical essentials, without being faced with costs, or other extraneous considerations, which could influence efficiency. The design one is trying to analyse and understand may be presumed, moreover, to have had the benefit of evolutionary development and thus to represent a highly tested optimum. The challenge is to understand the nature of this optimum. Not only is mechanical adequacy, indeed perfection, achieved, often also an economy in energy is attained, where this kind of savings can be undertaken with safety. The required biomechanical effects must, however, be achieved with the biological materials at hand. Biorheology deals with these aspects, with the characterisation and matching of the response of these materials, in deformation and flow, to the physiological functions to be performed. As rheologists it is here that we find the most interesting features of biological materials, the most interesting examples of how molecular properties are tuned to achieve rheological effects. Though nature has a relatively limited chemical repertoire from which to build up its materials it is well able to make macromolecules and can thus derive full use of the great variety of properties achievable with changes in macromolecular architecture. With only a limited number of building blocks it has been able to create materials uniquely adapted to their tasks. The rheologists and materials engineer has much to learn in this context.

By way of example the clearance of impinged particles (dust, bacteria, pollen etc.) which are breathed into the lungs will be discussed. Transport of these particle loads occurs by propulsion by cilia. Cilia like bacterial flagella, are rapidly beating flail-like cellular protusions which densely cover the airway epithelia. For propulsion a layer of mucus has to be present over the cilia where it acts as coupler between the load and the bed of cilia which move in union The role of mucus in this is rheological and a fairly sharp optimum in the storage modulus is associated with this function. We know that the function is rheological since one can substitute mucus by systems which are chemically rather different, but rheologically equivalent. A model will be presented which describes mucus deformation, stress development and relaxation in this role. The discussion provides an example for biorheological matching.

RHEOLOGY OF POLYMER GELS

Horst Henning Winter, University of Massachusetts, Department of Chemical
Engineering, Amherst, MA 01003

ABSTRACT

A network polymer at its gel point is in a critical state between liquid and solid. Linear
viscoelasticity reduces to a simple behavior which is described by the **gel equation** for the
stress (J. Rheol., **30**, 367 (1986) and **31**, 683 (1987)):

$$\tau(t) = \int_{-\infty}^{t} S \ (t-t')^{-n} \ \dot{\gamma}(t') \ dt$$

in which the strain history is defined by the rate of deformation tensor $\dot{\gamma}(t')$. Molecular
parameters determine the gel strength, S, and the relaxation exponent, n. The above power
law relaxation behavior seems to be a universal property at the gel point. It has been found
with a large variety of chemically or physically crosslinking polymers, i.e. with all
crosslinking polymers which we studied above the glass transition temperature. Initially,
we seemed to find a universal relaxation exponent of 0.5 for all gels as it might be expected
for a critical state. However, measured values between 0.15 and 0.8 indicate that the
relaxation exponent may cover the entire range between 0 and 1. The front factor S
depends on the magnitude of the relaxation exponent, and for classes of network polymers
it is found to follow the relation $S = G_0 \lambda_0^n$, where G_0 and λ_0 are characteristic properties
which are constant within each class of network polymers (to appear in Macromolecules,
1990). G_0 is of the order of the plateau modulus of the network and λ_0 is believed to be the
relaxation time of a network strand.

The universality of the power law relaxation behavior of polymers at the gel point
allows a most accurate determination of the instant of gelation (Polym. Eng. Sci., **27**, 1698
(1987)). The gel point is reached when the complex rheological behavior reduces to power
law relaxation in the terminal frequency range (fractal behavior, mechanical selfsimilarity).
The simple technique of Fourier Transform Mechanical Spectroscopy (J. Non-Newt. Fluid
Mech., **27**, 17 (1988)) not only allows the direct determination of the gel point. It also
allows, by extrapolation, to predict the distance from the gel point in a sol. Recent
advances of the technique and a set of applications will be given.

When replacing the rate of deformation tensor in the gel equation with a suitable
strain measure, it can predict the stress in critical gels up to about 2 strain units (for
PDMS). Beyond this, a large strain will irreversibly rupture the molecular network.

Acknowledgement: This work is supported by the National Science Foundation
grant MSM 8601595 and by the Center of the University of Massachusetts for Industrial
Research on Polymers (CUMIRP). The main coworkers on this project were Francois
Chambon, Sundar Venkataraman, and Jim Scanlan.

Closure Approximations for Three-Dimensional Orientation Structure Tensors

by

Suresh G. Advani
Department of Mechanical Engineering
Center for Composite Materials
University of Delaware, Newark, DE   19716

and

Charles L. Tucker III
Department of Mechanical and Industrial Engineering
University of Illinois at Urbana- Champaign, Urbana, IL 61801

Abstract

A tensor description of the orientation structure in a fiber suspension provides an efficient way to compute flow-induced fiber orientation, but the scheme requires an accurate closure approximation for the higher-order moments of the orientation distribution function. This presentation evaluates a number of different closure approximations, comparing transient orientation calculations using the tensor equations to a full calculation of the distribution function.

Closure approximations are often derived by requiring the approximate form of a_4 to have some of the properties of the actual fourth-order tensor. The exact fourth-order tensor is symmetric with respect to any pair or indices, and the contraction of any two indices produces the second-order tensor (i.e., $a_{ijjk} = a_{ik}$). However, experience suggests that imposing *fewer* conditions tends to produce *better* closure approximations.   A linear closure approximation for a_4 may be formed by combining products of a_2 and the unit tensor δ. Hand [1] required his expression to be symmetric with respect to any pair of indices, and to meet the normalization condition. The quadratic closure approximation is formed by taking the dyadic product of the second-order tensor with itself. The linear closure approximation is exact for an isotropic (random) distribution of orientation while the quadratic approximation is exact for perfectly aligned fibers. Other closure approximations attempt to improve on these by being correct in both limits. Hinch and Leal [2] derived a number of different closure approximations in their study of suspensions of fiber-like particles with Brownian motion. They begin by writing simple forms that are exact for either isotropic orientation or for perfectly aligned orientation. Then they form a composite of these formulas that is exact in both limits. Hinch and Leal also developed a second set of approximations by deriving asymptotically correct closure approximations based on the steady-state solutions. Their second approximations are not only correct for perfectly aligned and isotropic orientations, but are also correct for distributions approaching the two extremes. Hinch and Leal do not give explicit formulas for a_{ijkl}; rather they derive approximations for the product $a_{ijkl}\,\dot{\gamma}_{kl}$. We have extracted the corresponding closure approximations for a_{ijkl}. To create a closure approximation that is accurate over the entire range of orientation, one can also mix the quadratic and linear forms according to some scalar measure of orientation, f. The scalar measure of orientation f must be independent of the coordinate system and free from assumptions about the distribution function. We introduced this in our previous paper for planar orientation field [3]. Our conclusions about closure approximations for planar orientations cannot automatically be extended to three-dimensional orientation. The three-dimensional case offers more freedom for both the flow field and the fiber orientation, making the task of finding a good closure approximation more difficult. Here we report an improved hybrid closure approximation for three-dimensional orientation, formed by modifying the scalar measure of fiber alignment. The best closure approximation will be the one that most accurately approximates a_4 in the equations of change.

The examples presented concern orientation and deformation fields that are spatially uniform. Hence, the convective terms in the material derivative of the equation of change are zero, the rate-of-deformation and vorticity tensors are constant, and the orientation tensor components are functions of time only. Solution methods include solving numerically for the complete distribution function in the equation of change for the orientation distribution function and integrating it at each time step to obtain the second order orientation tensors. This provides the standard for comparison: We call these "exact" results as no closure approximations were made and the only errors were discretization and round-off errors in the numerical solution. Then, we solve for second order orientation tensors directly by introducing the various closure approximation for the fourth order tensor in the equation of change for the second order tensor. We compare the "exact" results with our new hybrid closure, the commonly employed quadratic closure and the more general Hinch and Leal composite closures in a variety of three dimensional homogeneous flow fields. The behavior of the different closure approximations was examined in simple shear flow, pure shear flow, uniaxial and biaxial elongational flows and in a combination shearing/stretching flow.

None of these closure approximations provide accurate solutions for all the flow and orientation fields. The quadratic closure exhibits stable dynamic behavior, but predict neither the correct transient behavior nor accurate steady-state values, especially for nearly random to intermediately aligned orientations and rotational flow fields. Hinch and Leal's closures work quite well for low to intermediate alignment, but one form displays artificial oscillations in simple shear flow for strong alignment. The other composite makes up for this deficiency in simple shear flow, but it is consistently less accurate than the former one in other flows and gives physically impossible values in biaxial elongation. Our new hybrid closure is always well behaved. In fact, it is the only approximation other than the quadratic closure that never exhibits artificial oscillations or pathological behavior. Its steady-state predictions are slightly better than the quadratic form in shearing flows and performs best for combined shearing/stretching flow over a wide range of orientations.

References

1.   Hinch E. J. and L. G. Leal, *J. Fluid Mech.*, **76**, 187 (1976).
2.   Hand, G. L., *J. Fluid Mech.*, **13**, 33 (1962).
3.   Advani, S. G. and C. L. Tucker , *J. Rheol.*, **31**, 751 (1987).

COMPRESSION CREEP OF MOLTEN POLYMERS

J.V. Aleman
Instituto de Ciencia y Tecnologia de Polimeros
Juan de la Cierva 3. 28006 Madrid. SPAIN

INTRODUCTION

Bulk compression creep of molten polymers has received almost no attention to-date (1) because of the commonly accepted notions that no volume changes of importance take place in polymers during actual practice, and the inherent practical difficulties related to the precise measurement of volume deformations. For this purpose an Instron Capillary Rheometer with a steel plug instead of the capillary has been used (Figure 1-a). Actual data look as these shown in Figure 1-b.

To have results representative of a broad spectra of polymer structures, the rigidity ($-\Delta F$) of the chain was used as the leading parameter. The following polymers were accordingly chosen. Rigid chain: a) poly(butylene terephthalate) (PBT) (1) with planar conformation and aromatic rings in the backbone, b) polystyrene (PS) (2) with helicoidal conformation and side-groups. Flexible chains: c) low density polyethylene (LDPE) (3) planar, non-polar, long-branches, d) poly (ethylene oxide) (PEO) (4) helicoidal, polar, linear chains.

EXPERIMENTAL RESULTS

Volume deformations ($k = \Delta V/V_o$) (Figure 2) increase as pressure (P) increases and temperature (T) decreases (PS, LDPE) or increases (PEO). Plots of the bulk compliance ($B(t) = k(t)/P$) versus time (t) (Figure 2) may be shifted to provide master curves. As the pressure and temperature increase, the pressure shift factors (b_p, a_p) increase non-linearly, whilst the temperature shift factors (b_T^p, a_T^p) decrease. The steady state creep compliances (B^s) allow to describe the recoverable storage of elastic energy (Be), and seem to be related to the extrusion die-swell (B_c^s/B_d^s) (Figure 3). Volume viscosity (η_K) decreases with increasing stress and decreasing (PS, PBT) and increasing (LDPE, PEO) temperature (Figure 3).

REFERENCES

1) Sanchez, M., Aleman, J.V., Rheologica Acta, 1988, 27, 634-638.
2) Aleman, J.V., Angew. Makrom. Chem., 1989, accepted.
3) Aleman, J.V., J. Polym. Sci. (Physics Ed.), 1990, submitted.
4) Aleman, J.V., 1990, unpublished results.

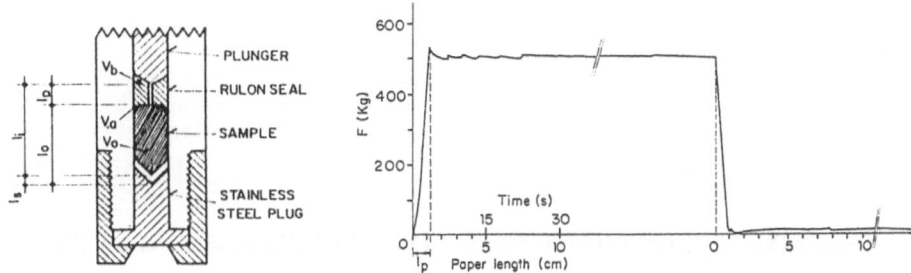

Figure 1.- Compression chamber, and force (F) versus time (t) recording.

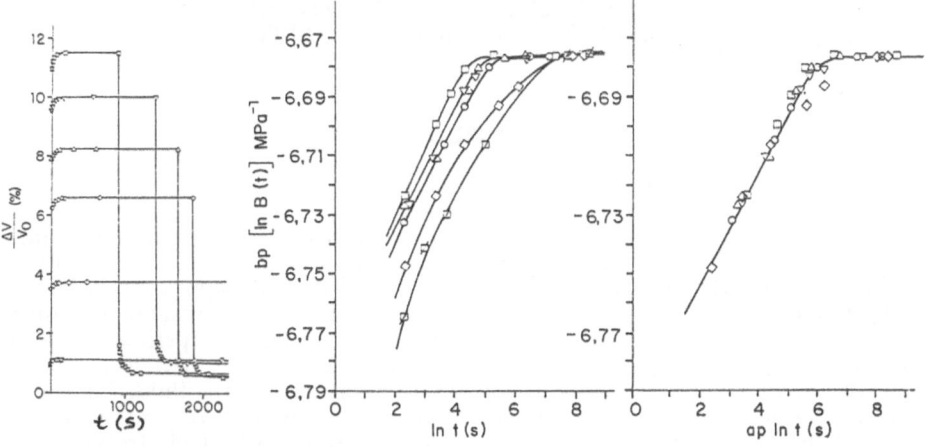

Figure 2.- Volume deformation (k%) and creep compliance (B(t)).

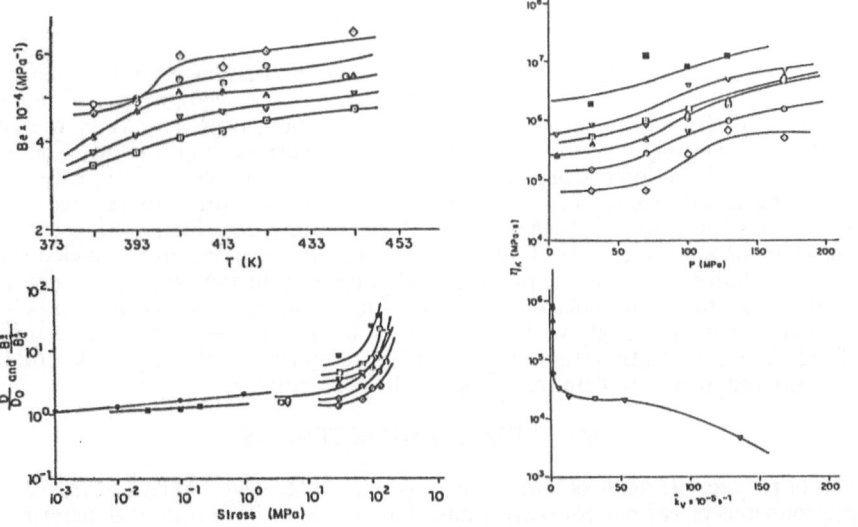

Figure 3.- Steady-state creep compliance B(t), and volume viscosity (η_K).

RHEOMETRY OF DILUTE SOLUTIONS OF POLYELECTROLYTES

DIRK ASCHOFF   PAUL SCHÜMMER
Institut für Verfahrenstechnik, RWTH Aachen
Turmstraße 46, D–5100 Aachen, FRG

ABSTRACT

Dilute solutions of polyacrylic acids have been investigated in steady and oscillatory shear experiments. The relative number of ionic groups on the polymer chain was varied by addition of different amounts of alkali. Further variables were the concentration and the average molecular weight of the polymer. With increasing degree of dissociation of the polyacid the viscosity of the solutions shows the well–known rapid increase. The shear thinning becomes more and more pronounced. Also the dynamic viscosity η' increases strongly with the degree of ionization as the oscillatory experiments showed. On the other hand, there is only a slight influence on the imaginary part η'' of the complex viscosity, which is associated with the elastic properties of the fluid. The electroviscous effects increase with decreasing concentration and with increasing average molecular weight. They may be suppressed by addition of simple strong electrolytes.

INTRODUCTION

Polyelectrolytes carry a large number of groups that can be ionized in solution. For instance, a part of the carboxylic groups of polyacids will dissociate after addition of sufficient alkali to the solution. Due to the thermal motion, a great part of the counter ions will leave the polymer coil and diffuse into the surrounding solvent. The electrostatic repulsion of the ionic groups that are bonded to the polymer chain causes an extension of the coiled macromolecule. With increase of the degree of ionization the repulsion overcomes the coiling tendency of the Brownian motion. Thus during the process of ionization the molecule takes all forms from a nearly spherical, highly coiled shape to an extended filament /1/. As the properties of polymer solutions are largely determined by the shape of the solute molecules, it is not surprising that the rheological behaviour of the solution varies strongly with the degree of ionization. Besides those intramolecular effects the electrostatic interaction between the molecules of the polyelectrolyte has to be considered for a complete treatment of the problem /2/.

MATERIALS AND METHODS

Samples of polyacrylic acids of various average molecular weights have been used to prepare solutions of various concentrations. The degree of ionization was determined by regulation of the pH–values. In the pure aquaeous solution the degree of dissociation is rather low. In the investigated range of concentrations the pH–values vary from 3 to 4.3. By basifying the solution with sodiumhydroxid a certain number of carboxylic groups can be dissociated. The titration curves of the solutions show that at pH–values about 9 the polyacrylic acid is completely dissociated. This limit is nearly independent of the concentration as well as of the molecular weight of the samples we investigated. The steady shear experiments at moderate shear rates and the oscillatory experiments were carried out on a Haake CV–100 Couette rheometer. The viscosities at low shear rates were measured with a Contraves LS 200 rheometer.

RESULTS AND DISCUSSION

Figs. 1 and 2 show the rapid increase of the shear viscosity with increasing ionization, due to the enlargement of the coil. This effect was described by several authors /3,4/.

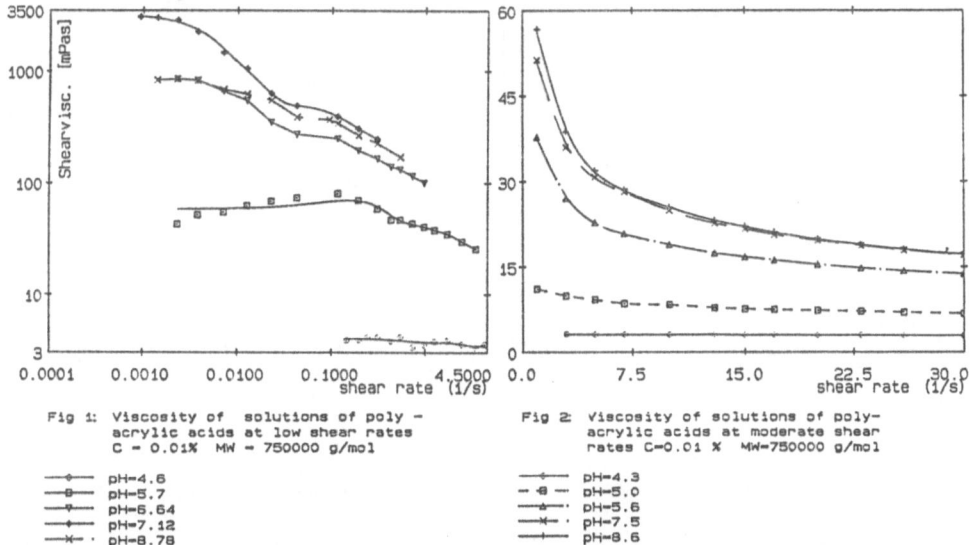

Fig 1: Viscosity of solutions of poly – acrylic acids at low shear rates C – 0.01% MW – 750000 g/mol

———◇——— pH–4.6
———□——— pH–5.7
———▽——— pH–6.64
———+——— pH–7.12
——×— · pH–8.78

Fig 2: Viscosity of solutions of poly- acrylic acids at moderate shear rates C–0.01 % MW–750000 g/mol

———+——— pH–4.3
– □ – pH–5.0
———△——· pH–5.6
———×— · pH–7.5
———+——— pH–8.6

The zero shear rate viscosity of a solution of a polymer at a high degree of ionization may exceed that of the non–ionized polymer by several hundred times (compare plots for pH=4.6 and 7.12 in Fig. 1). Especially at lower pH–values the increase is very strong (see plots for pH=4.3 to 5.6 in figs. 1 and 2). At higher pH–values the polymer chains are largely extended. A further dissociation is less significant. On the contrary, the ionic groups of the polymer are screened by the ions brought to the solution by the redundant sodiumhydroxide. A contraction of the coil results and a drop of the shear viscosity will occur (see plot for pH=8.8 in Fig. 1). Adding further alkali or other strong electrolytes the rheological behaviour approaches that of solutions of neutral polymers /3/.
With increasing degree of ionization the shape of the macromolecule approaches that of an extended filament. The stronger alignment in flow of those molecules gives rise to an increase of the shear thinning efgfect (see figs. 1 and 2).
The electroviscous effects are stronger in solutions of low polymer concentrations /4/ because there is more free volume to which the counter–ions can migrate.The net charge of the polymer coil increases.
On the right side of fig. 3 the drop of the real part of the complex viscosity with increasing frequency of the oscillation is shown. Correspondingly, the imaginary parts increase (left–hand side of fig. 3). At higher pH–values (see pH=8.6 in fig. 3) nonlinearities occur.The complex viscosity is a function of the strain amplitudes. Higher order harmonics have to be taken into account. Corresponding to the shear vis

cosity, η' the real part of the complex viscosity which is a measure for the viscous contribution, strongly increases with the degree of ionization. The increase of the imaginary part of the complex viscosity that is associated with the elastic properties of the fluid is much smaller. On the one hand the extension of the macromolecule by the electrostatic repulsion yields a higher elasticity caused by the alignment of the molecules during a flow. On the other hand the rigidity of the rodlike molecules prevents elasticity effects caused by a deformation of the coil. These effects may compensate each other. As a result the relation of imaginary and real part of the complex viscosity decrease with increasing ionization (see fig 4).

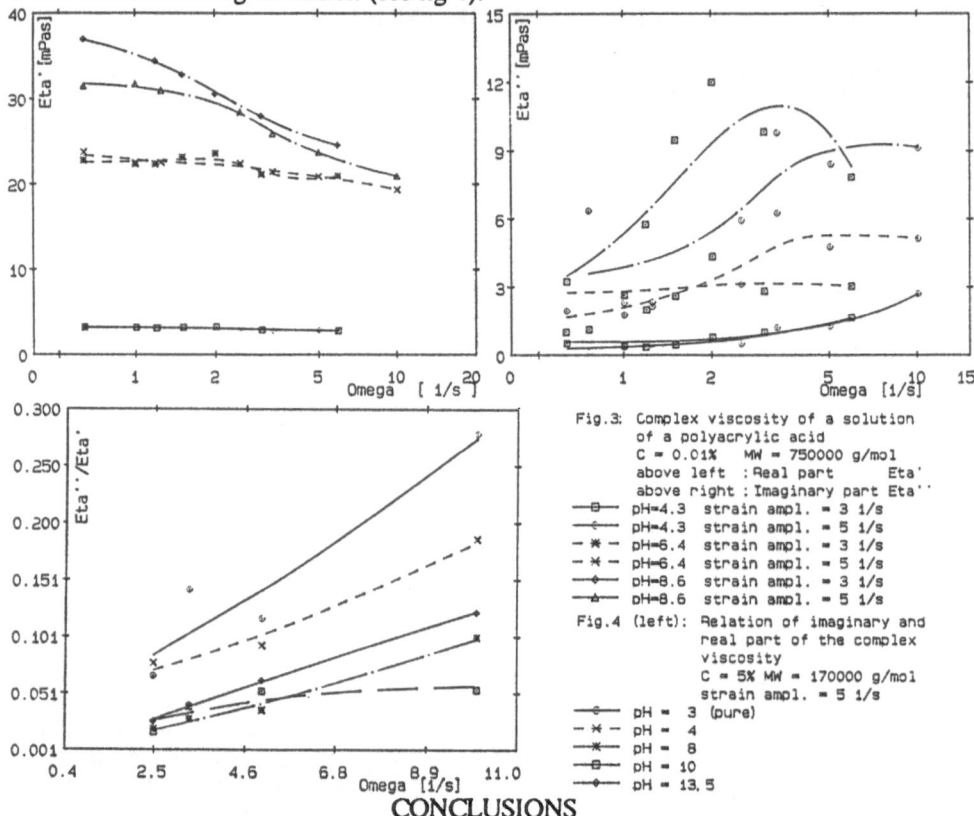

Fig.3: Complex viscosity of a solution of a polyacrylic acid
C = 0.01%   MW = 750000 g/mol
above left : Real part   Eta'
above right : Imaginary part Eta''
—□— pH=4.3 strain ampl. = 3 1/s
—◊— pH=4.3 strain ampl. = 5 1/s
— * — pH=6.4 strain ampl. = 3 1/s
— ✻ — pH=6.4 strain ampl. = 5 1/s
—◆— pH=8.6 strain ampl. = 3 1/s
—▲— pH=8.6 strain ampl. = 5 1/s

Fig.4 (left): Relation of imaginary and real part of the complex viscosity
C = 5% MW = 170000 g/mol
strain ampl. = 5 1/s
—◊— pH = 3 (pure)
— ✻ — pH = 4
—✻— pH = 8
—□— pH = 10
—◆— pH = 13.5

CONCLUSIONS

The presence of ionic substituent groups on a macromolecule causes an extension of the polymer coil. Due to the increase of the effective hydrodynamic volume of the molecule the shear viscosity and the real part of the complex viscosity increase strongly. The increase of the elastic effects caused by the alignment of the filaments is partly compensated by the loss of flexibility of the chain.

REFERENCES

/1/ Kuhn, Künzle, Katchalsky   Helv.Chim.Act., Vol 31, 1948 and
                                  J.Polym Sci , Vol. V, 1949;
/2/ Conway and Dobry—Duclaux   in Eirich, Rheology, Academic Press 1960
/3/ Fuoss et al.   Polyelectrolytes I—IX J.Polym.Sci
                                  Vol. 2–6, 1946–1950;
/4/ Yamanaka et al.   J.Am.Chem.Soc., 1990, 112

MODELLING OF SLIT DEVOLATILIZATION OF POLYMERS

GIANNI ASTARITA and PIER LUCA MAFFETTONE

Dipartimento di Ingegneria Chimica, Universita' di Napoli

Piazzale Tecchio, 80125 Napoli Italy

ABSTRACT

A new technique of polymer devolatilization, where the polymer-volatile solution simply flows through a heated slit, is modelled. We assume that the mass transfer is linear in the driving force, and that lubrication theory [1] can be applied to the gas phase; a strong hypothesis on the pressure gradient is also made. For some operating conditions a good agreement is found with experimental results. An initial tentative argument about cyclic behaviour, as observed experimentally, is presented.

INTRODUCTION

A new static devolatilization technique, which overcomes the problem of long residence times of the polymer at high temperature, is recently being developed. In this process the separation takes place during the passage of the polymer solution through a heated slit. Fig.1 shows the scheme of the system considered and the coordinates chosen.

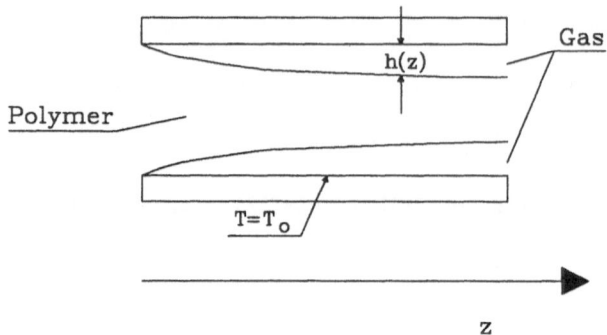

Figure 1 - A sketch of the scheme of the system considered

A ZERO ORDER APPROACH TO MODELLING

The basic assumption of our model is that the evaporated gas forms two channels, of thickness h(z), which separate the polymeric phase from the slit walls.

If the lubrication approximation holds for the gas phase, the momentum balance is:

$$-dp/dz = (12\mu G)/(\rho_g h^3), \qquad\qquad z=0 \quad p=p_0 \qquad\qquad (1)$$

where p is pressure, G the gas mass flow rate per unit width, ρ_g the density and μ the viscosity of the gas.

Since the heat diffusion time in the polymer is significantly less than the residence time in the slit, at a zero order level of approximation it is reasonable to assume that the polymer temperature T is, at the slit entrance itself, equal to the wall temperature T_0. Since heat transfer is by conduction through the gas layer, the heat balance is:

$$dT/dz = M (T_0 - T)/ h - N dG/dz, \qquad z=0, \quad T=T_0 \qquad\qquad (2)$$

where $M=K_g/(c_{pol}G_{pol})$ and $N=\lambda/(c_{pol}G_{pol})$ with K_g the gas conductivity, c_{pol} and G_{pol} the polymer specific heat and mass flowrate per unit of slit width, respectively, and λ the volatile latent heat.

The mass transport is assumed to be linear in the driving force, so it can be written as:

$$dG/dz = A \ [p^o(T,C) - p], \quad z=0 \quad G=0 \tag{3}$$

in which p^o is the monomer vapor pressure at temperature T and volatile concentration C, and the unknown mechanism of mass transfer is lumped into the value of A, which is the only adjustable parameter of the model. The mass balance is:

$$C = C_o - G/(G+G_{pol}) \tag{4}$$

In order to close the modelling the momentum balance for the polymeric phase ought to be written. This very difficult task can be avoided by making the strong hypothesis that the pressure gradient is constant. With this assumption, the system (1-4) can be solved numerically, choosing arbitrary values for A. The results obtained with this zero order model, although capable of predicting qualitatively some of the trends observed experimentally, do not predict well the shape of the experimental temperature profiles shown in Fig. 2 [2].

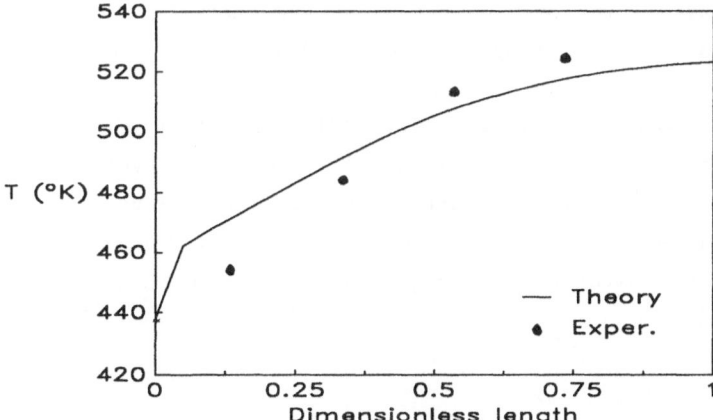

Figure 2 - Experimental temperature profile for a system styrene-polystyrene compared to theoretical results obtained at second level of approximation for A=5 10^{-8} [s/cm].

REFINEMENTS OF THE MODEL

At a second level of approximation, the hypothesis that T(0)=To can be removed, and the heat balance can be written considering also heat conduction in the polymeric phase (although only in a linearized form, which is justified since heat conduction in the polymer plays a role only near the slit entrance). This only changes the form of Eq.2. Fig.2 shows the results obtained at this second level with A=5 10^{-8} (s/cm). Good agreement is found with experimental results [2] obtained for small flow rates of the polymer solution.

A CONCEPTUAL DRAWBACK OF THE MODEL. FUTURE WORK

It's worthwhile noting that the model suffers of a conceptual drawback: the assumption of a finite pressure gradient is in contrast with the consideration of the formation of two gas layers. In fact, under steady state conditions the pressure gradient should be balanced by some shear force acting on the polymer, and no such force can exist if the polymer is sandwiched between two gas layers. One could now consider the possibility that a gas layer forms only on one side, and that a shear force is then exerted on the other side by the solid wall. However, the polymer would tend to become hotter on the side touching the wall, thus favouring formation of a gas layer on that side. This line of thought would suggest that the process is intrinsically unstable, and that cyclic behaviour is expected. It is interesting to note that cyclic behaviour has indeed been observed experimentally [2], with both temperature and pressure undergoing sustained cycles with periods comparable to the residence time of the polymer.

REFERENCES

1. G.K. Batchelor, "An introduction to Fluid Dynamics", Cambridge Univ. Press, 1983, Cambridge.
2. Montedipe private communications.

MAGNETIZED POLYMER SOLUTION FLOW IN POROUS MEDIA

SAYAVVUR I. BAKHTIYAROV
Department of Theoretical Mechanics,
Azerbaijan Inst. of Oil and Chem., 20 Lenin
Avenue, Baku -10, 370601, U.S.S.R.

ABSTRACT

A theoretical analysis indicates that any model for the flow
of magnetized polymer solutions in porous media must include
expansion and contraction regions,the changes in the
rheological behavior of the fluid and the intensity of
magnetization of fluid. The interacting effects of polymer
properties and intensity of magnetization on flow performance
are considered.Experiments were conducted with the polyacrylamides with molecular weights ranging from 3000000 to over
25000000.The porous matrix consisted of a flow cell packed
with glass balls.The polymer solutions were characterized by
shear and normal stress measurements.Under certain conditions,
high flow resistance was observed.An analysis indicates that
this phenomenon is a function of flow velocity,molecular
weight of polymer and magnetic induction.It was found that
there is an optimum magnetic induction which yields a maximum
value of the flow resistance for the polyacrilamide solutions.

INTRODUCTION

Polymer flooding is one of the more popular recovery methods.
Several laboratory studies of polymer flooding have shed
light on some aspects of polymer solution flow behavior(1,2).
This paper concerns magnetized polymers that are being tested
widely as an additive for injection water: it is partially

hydrolyzed polyacrylamide.The present investigation surveys
some ways in which magnetized polymer solution flows are
affected by fluid properties,intensity of magnetization and
flow rate.

MATERIALS AND METHODS

Samples of three different polyacrilamids were examined.They
were differed mainly in molecular weight and may be described
as follows:
   PAA - 1: molecular weight $3x10^6$.
   PAA - 2: molecular weight $10x10^6$.
   PAA - 3: molecular weight $(20-25)x10^6$.

   All polymer solutions were prepared carefully to avoid
mechanical degradation of the polymer molecules. The solution
was protected from bacteria with a low concentration of
formaldehyde.Polymer solutions were premagnetized at different
magnetic induction from 0 to 2800 Gs.The porous matrix
consisted of a flow cell (copper tube with internal diameter
1.498 cm and length 10.166 cm) packed with glass balls
(diameters 0.9 - 1.1684 mm).

EXPERIMENTAL PROCEDURE

The permeability of the porous media was first determined.
The cell was saturated by evacuating and forcing distilled
water into the cell.Distilled water was then pumped into the
cell at a constant rate and the pressure drop measured.The
permeability was then calculated from Darcy's equation.
Premagnetized polymer solution was then injected into the
flow cell at a constant rate,displacing the distilled water.
After full polymer concentration in the effluent had been
reached and the pressure drop stabilized,the flow rate and
pressure drop were recorded.The flow rate was then changed,
and measurements of flow rate and pressure drop again were
recorded.

RESULTS

Under certain conditions the mobility of the magnetized
polymer solutions in porous media was found to be lower
than non-magnetized polymer solutions.The magnetized polymer
solutions appeared not to plug the matrix, and the low
mobility is explained upon an anomalous viscosity effect.
Viscosities and the elastic properties of the solutions were
measured in a RMS - 800 viscometer (cone - plate).Viscosity
measurements show that the magnetized polymer solutions
exhibit typical pseudoplastic behavior, i.e.,a decreasing
viscosity with increasing shear rates.
   The mobility reduction is used to describe the flow of
magnetized polymer solutions.It is defined as the ratio of
initial water mobility M_i (darcy velocity divided by pressure
gradient) to the following magnetized polymer solution
mobility M_p :

$$MR = M_i \, / \, M_p$$

The behavior of polymers magnetized at magnetic inductions 0, 1600 and 2800 Gs is depicted in Figure for flow through porous media.The non-magnetized polymer is not much more effective than water since the mobility reduction never becomes greater than 2. The magnetized polymers show a greater mobility reduction. At any given rate of advance,the higher the magnetic induction, the greater the mobility reduction. The shapes of the curves are qualitatively the same for all three polymers: the curves tend to reach a minimum at low velocities and reach a maximum at high velocities.

Experiments show that polymer effectivness in reducing mobility is greatest with polymers having highest average molecular weight (PAA-3). The mobility of magnetized polymer solutions in porous media decreases as the flow rate increases.

CONCLUSIONS

1.The mobility of polymer solutions in porous media decreases markedly as the magnetic induction increases.

2.Polymer effectiveness in reducing mobility is greatest with polymers having highest averege molecular weight.

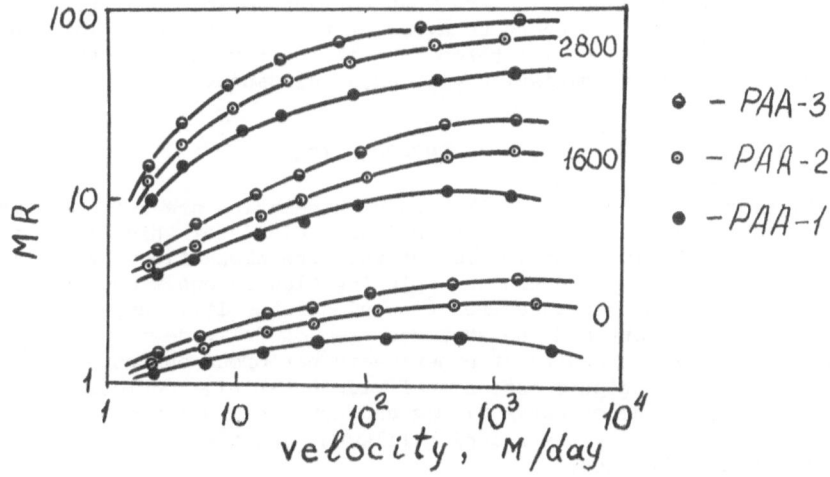

REFERENCES

1. Mirzadjanzade, A.Kh.,et al. Improvement of Oil and Gas Well Cementing Quality, Nedra,Moscou, 1975,pp. 44-70.

2. Mirzadjanzade, aA. Kh. Aspects of Hydrodynamics of Visco-Plastic and Viscous Fluids in Oil Production,Azerneftneshr Baku, 1959, 412 p.

FIBER SPINNING EXPERIMENTS ON POLYOX AND POLYACRYLAMIDE SOLUTIONS

H. BAO, O. MANERO*, R.K. GUPTA
State University of New York at Buffalo
Buffalo, N.Y. 24260
*Instituto de Investigaciones en Materiales, UNAM.
Apartado Postal 70-360, Coyoacán 04510, México, D.F.

ABSTRACT

Spinning experiments were performed using two different polymer solutions. The systems included solutions of Polyox(molecular weight of $4x10^6$) and Alcomer (copolymer of sodium acrylate and acrylamide). Extensional flow data were obtained by using the Gupta-Sridhar extensional viscometer, with results that show different behavior of the two systems. The experimental measurements show a coil-stretch transition if we examine the apparent spinning viscosity variation with axial distance.

INTRODUCTION

Fiber spinning is a type of flow whose kinematics are sufficiently close to those of elongational flows. From the analysis of this flow, it is possible to obtain an estimation of the true elongational viscosity of polymeric fluids. However, the spinning flow is one where the strain rate varies with axial distance from the spinneret, it is dependent on the preshearing history and the stresses are not fully developed. In this work, we present results of experiments performed on two polymer solutions using the Gupta-Sridhar extensional viscometer [1]. Calculation of the kinematics is made by curve fitting and numerical differentiation of the velocity field along the spinning filament and the stresses are calculated by a momentum balance.

MATERIALS AND EXPERIMENTS

The test fluids were two polymer systems: a 3000 ppm solutions of polyoxiethylene ($4x10^6$ molecular weight) in water and a 200 ppm aqueous solution of a copolymer of sodium acrylate and acrylamide. The apparatus used and data acquisition have been described elsewhere [1]. The filament profile was discretized and the stresses were calculated from the following momentum balance (neglecting gravity, surface tension and air drag):

$$(N_1)_{n+1} = v_{n+1}[(N_1)_n/v_n + \rho (v_{n+1} - v_n)] \qquad (1)$$

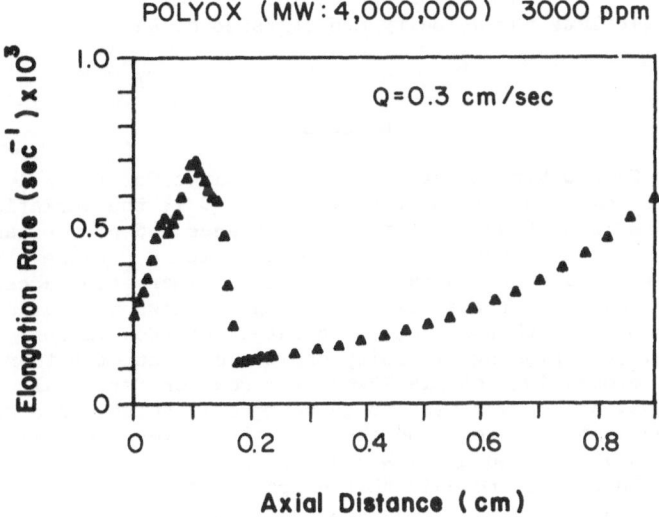

Figure 1. Variation of the elongation rate with axial
distance. 3000 ppm Polyox solution.

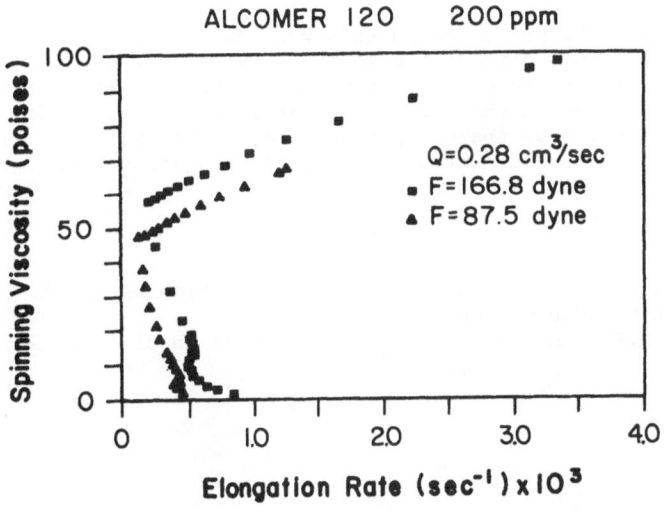

Figure 2. Variation of the spinning viscosity with
elongation rate, for two values of the
force. 200 ppm Polyacrylamide solution.

where $N_1 = \tau_{zz} - \tau_{rr}$ is the stress difference, v is the velocity and ρ is the fluid density. Given the initial values, Eq. (1) provides a method to calculate the velocity and stresses at each point along the filament.

RESULTS

Figures 1 and 2 show experimental results obtained for the Polyox and Polyacrylamide solutions, respectively. In Fig. 1, the variation of the elongation rate is plotted against axial distance, depicting an unusual maximum close to the spinneret region. There is a range where the elongation rate is almost constant, followed by increasing acceleration up to the lower extreme of the filament. Fig. 2 shows the variation of the spinning viscosity with the elongation rate, for two values of the applied force. The spinning viscosity shows a coil-stretch transition, and the critical elongation rate is lower for smaller force values. At high extension rates, the curves tend to an asymptotic value whose magnitude is larger for increasing force. The shape of the curves is similar to those of elastic liquids studied by Mackay and Petrie [2].
Analysis of both figures reveals that entrance effects, such as preshearing, may have an influence on the elongation rate close to the spinneret and hence, it may also affect the critical elongation rate for sudden extension. On the other hand, at the lower extreme of the filament, it is possible that end effects and the air drag produced by the suction device may induce an apparent increasing acceleration.

CONCLUSIONS

Fiber spinning data presented in this work have shown that changes in the applied force induce different asymptotic values for the spinning viscosity at high extension rates and also they affect the critical elongation rate for sudden increase in the viscosity. Entrance effects, such as preshearing, influence the elongation rate profile which in turn modifies the values of the above mentioned critical elongation rate.

REFERENCES

1. Sridhar, T., Gupta, R.K., Boger, D.V. and Binnington, R. (1986): J. Non-Newt. Fluid Mech. 21: 115.
2. Mackay, M.E. and Petrie, Ch. J.S. (1989): Rheol. Acta 28:281.

DETECTION AND QUANTIFICATION OF BARUS EFFECT IN VISCOELASTIC FLUID MIXTURES

M.BARRACÓ-SERRA ; MªA.ADRIÁ-CASAS
Departament de Mecànica de Fluids. Escola Tècnica Superior
d´Enginyers Industrials.Universitat Politècnica de Catalunya.
Avinguda Diagonal 647
08028 BARCELONA (Spain)

ABSTRACT

Barus (or Merrigton) effect was studied and analysed in mixtures binary and ternary fluid systems.

INTRODUCTION

Viscoelastic products structure

The viscoelastic products have a wide behaviour range:from viscous liquids with a light elasticity (Maxwell Type) to solids with some viscous properties (Kelvin type).
The macromolecular elastic deformation is the mean reason of the products viscoelasticity.Often,in the viscoelastic products there is not a full recuperation of the initial stress values. Its elastic tridimensional structure is partially destroyed by the stress effect.How much stronger and more complex would be the structural net,higher would be the elastic behaviour of the product.A new macromolecular structure is developed when the initial net has been broken down.This is one of the more significant aspects of the viscoelastic behaviour at macromolecular level.
On modelling the viscoelastic behaviour two different models can be used.The first one is based on the dispersion of the elastic particles in a solute (viscoelastic fluids).The second one uses a tridimensional elastic network with more or less tendence to degradate (viscoelastic solids)

Analitical development

Equilibrium equation:

$$\sigma_{11}\,dA + A\,d\sigma_{11} - \rho A u_1\,du_1 + \rho g A\,dx_1 = 0$$

$$\left(\frac{d\sigma_{11}}{dx_1} + \sigma_{11}\frac{dA}{dx_1}\right) + \left(\frac{\rho u_1^2}{A}\frac{dA}{dx}\right) + \rho g = 0$$

Continuity equation:

$$\frac{d(\rho u A)}{dt} = 0$$

Normal stresses appears into the fluid,and corresponding shears are the cause of Barus effect.The main viscoelastic function is the difference between principal normal stresses:

$$N_1 = \sigma_{11} - \sigma_{22}$$

Relations of N_1, γ_r (partial recuperation of viscoelastic material) and J_e^o (flow compliance) used usually are:

$$\gamma_r = (\sigma_{11} - \sigma_{22})/2\,\sigma_{21} \quad ; \quad J_e^o = (\sigma_{11} - \sigma_{22})/2\sigma_{21}^2 = \gamma_r/\sigma_{21}$$

The equation proposed is based in no lineal viscoelastic constitutive expression:

$$N_1 = 2\sigma_{21}\left[2\left(d\gamma/D\right)^6 - 2\right]^{1/2} \cdot K$$

with:K =f(concentration,temperature,pressure..).This equation
(Tanner equation)with particular K values for each polymer has
given us a very good results in all tested fluids.

MATERIAL AND METHODS

Tested fluid mixtures

Detection and measurement is done with surfactant solution wi-
th:a)water,oleic ethoxilated phosphoric alcoholic esther,syn-
tetic oil (18 cSt of kinematic viscosity)and oleic ethoxilated
alcohol. b)water and oleic ethoxilated phosphoric alcoholic es
ther.

Test conditions

The fluid were placed into a piston and preassured with a cons
tant speed.The diameter modification by post extrusion effect
was measured by video-photografic methods.The temperature was
20ºC.

RESULTS AND CONCLUSIONS

-Barus effect was detected in a little wide interval of con-
 centrations.For high concentrations of oil or alcohol (solvents)this interval is reduced.
-The Figure 1 shows with a phase diagram the different zones
with Barus effect or without Barus effect and the limit of both zones.
-The Barus effect measurement by experimental methods is according with values obtained by analitical development.Howe-
 ver,these results that we have obtained are only for one particular test.We must do exhaustive test series in order to
 get wider use of values.

KNOWLEDGEMENTS

To:Ana Oliva Brañas;Sergi Grau Torrent;Toni Trias Torres for
To:Pere Surià Lladó for his valious suggeriments and aids.
To:PULCRA,S.A

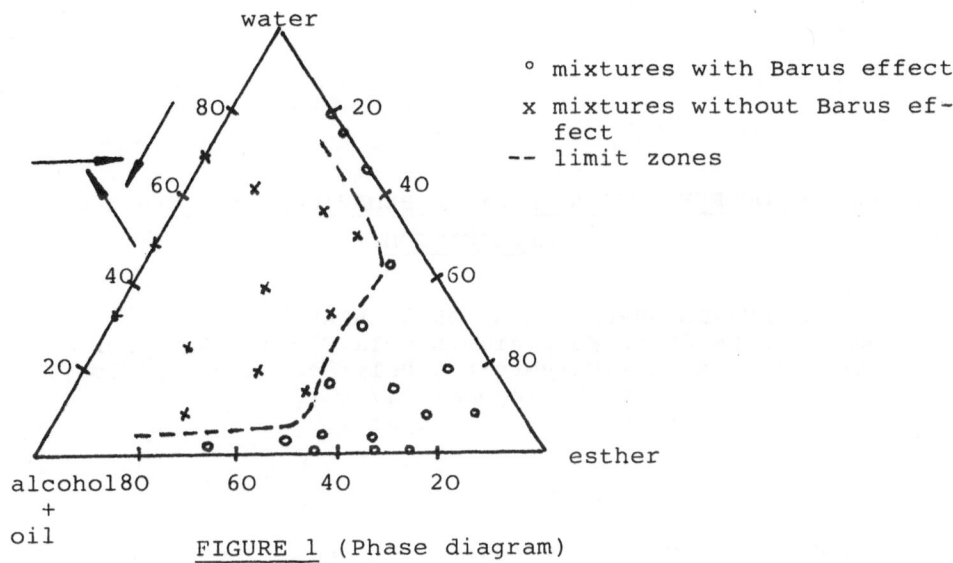

FIGURE 1 (Phase diagram)

MODIFICATION OF RHEOLOGICAL POLYMER BEHAVIOUR BY ULTRA-SOUND APPLICATION

M.BARRACÓ-SERRA ; MªA.ADRIÁ-CASAS
Departament de Mecànica de Fluids.Escola Tècnica Superior
d'Enginyers Industrials.Universitat Politècnica de Catalunya.
Avinguda Diagonal 647
08028 BARCELONA (Spain)

ABSTRACT

In this paper we present the modifications in rheological properties of some polymers by ultra-sound application.

INTRODUCTION

Rheological properties of solution polymers can be modified
by application of an ultra-sound process.
The modifications in the polymer samples was carried out in order to industrial application.
The effects of ultra-sound application consists in a no des-
tructive structural process of macromolecules,but in an orientation in the same direction of the stresses applied.
For this reason,viscosity was modified in function of ultra-
sound time,centrifugation time,ultra-sound power,molecular
weight and all environmental conditions.

MATERIAL AND METHODS

The basic fluid considerated in this study consists in Breox
polyethylene glycol solutions.
Breox polyethylene glycols are polymers of ethylene oxide with
the generalised formula:

$$HOCH_2(CH_2CH_2O)_n CH_2OH$$

Breox polyethilene glycols are not apprecially corrosive,are
a very low toxicity and present no undue hazards to health under normal condition of industrial use.They are stable solvents for many pharmaceuticals,cosmetics,dyestuffs and resines,
and are used like lubricants.
Molecular weight range tested were 1000,1500,4000 and 6000.
The concentrations tested were 30% and 50%.

Ultra-sound method
-Equipment description

The Labsonic U consist of a generator (with repeating duty cycle),a six foot line cord,a transducer,a standard probe tip (3/4"-19 mm) and two spanner wrenches for changing probe tips.

Principle of operation

the Labsonic U generator produces output power at a frequency of 20 KHz.A piezoelectric transducer in the probe assembly convers the output of the generator into vertical mechanical motion.A solid titanium probe tip submits the fluid to extremely high acoustic pressures and is responsible for the production of a phenomenon called cavitation.
Cavitation is produced by the sudden formation and collapse of vapour bubbles fluid by preassure.
Extremely high shearing pressures,temperatures and processes as microdisruption,emulsification and the production of suspensions are the original cause of this cavitation.

Rheological measures

The rheological measures were realised in a Ferranti-Shirley cone-plate viscometer,using the following experimental conditions:a)shear stress vs.shear rate rheograms:maximum speed 50 r.p.m;scanning time 240 s. b)temperature:20ºC.

Test conditions

The ultra-sound application time was in 90 seconds periods,with a pause of 45 seconds for the 6000 and 1500 molecular weight polymers.For the rest of samples the sound application periods were 180 seconds with pauses of 90 seconds.
Powers ultra-sounds were 50 W and 30 W.
A group of polymers which are treated with ultra-sounds,were centrifugated at 8000 r.p.m during 20 min.

RESULTS AND CONCLUSIONS

-Tested solutions of polymers were newtonian under experimental conditions.
-Ultra-sound application time is not very important for the obtained results at own experimental level.
-The centrifugation process is not a outstanding factor in the modification of rheological properties.
-The relatioship viscosity-molecular weight is showed in Figure 1.

REFERENCES

A.J.Giacomin et alt."The use of the piezoelectric composite oscillator technique for measuring the viscoelasticity of liquid cristalls".Mat.Res.Soc.Symp.Proc.Vol.152 (1989)

KNOWLEDGEMENTS

To:Lluís Prat-Camós;Jordi Dachs-Marginet;Eulàlia Planas-Cuchí for your practical collaboration.
To:Pere Surià Lladó for his practical consults and valious suggeriments.
To:Dra.Blanca Madariaga of Piochemical Dept.Univ.of Barcelona.

(1)High concentration.No ultra-sound treated samples

$\hat{\mu}=2.70.10^{-4}(MW)^{1,1684}$

(2)High concentration ultra-sound treated samples

$\hat{\mu}=2.76.10^{-4}(MW)^{1.1441}$

(3)Low concentration.No ultra-sound treated samples

$\hat{\mu}=9.32.10^{-5}(MW)^{1.0927}$

(4)Low concentration ultra-sound treated samples

$\hat{\mu}=1.75.10^{-5}(MW)^{1.2791}$

($\hat{\mu}$ is a reduced dynamical viscosity)

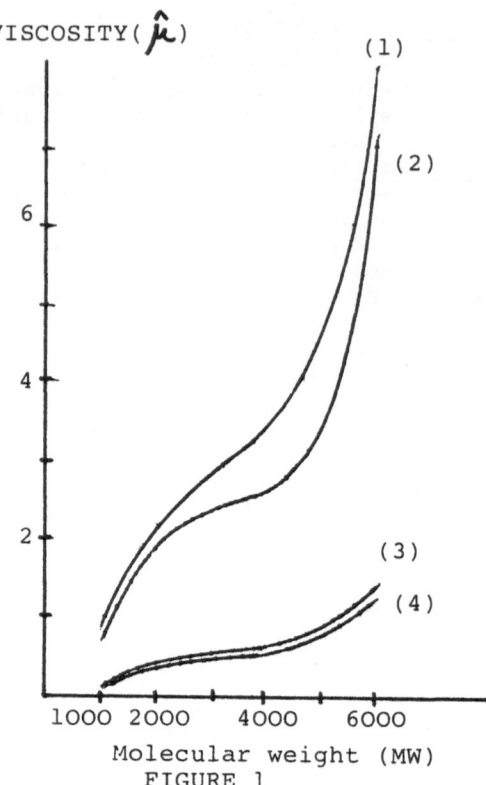

FIGURE 1

RHEOLOGICAL PROPERTIES OF A MICELLE SYSTEM IN SOLUTION TO BE USED AS REFERENCE LIQUID WITH VISCOPLASTIC BEHAVIOUR

HARRO BAUER and NORBERT BOESE

Physikalisch-Technische Bundesanstalt, Labor 3.23

Bundesallee 100, D-3300 Braunschweig, West Germany

ABSTRACT

Solutions of a di-block copolymer in hydrocarbons form under suitable condition a micelle system. These solutions exhibit a viscoplastic behaviour with a yield value. Measurements were carried out with a controlled stress rheometer and with several controlled shear apparatus in order to test the rheological behaviour of the solutions under different experimental conditions in a wide stress range.

INTRODUCTION

The exact determination of the yield values and the rheological behaviour of viscoplastic liquids is quite difficult due to thixotropic characteristics, sedimentation and other problems [1,2]. The operation of laboratory rheometers may be problematic especially selecting a suitable measuring system. Often the vane method is selected, but in this case it is difficult to obtain representative values for the shear rate and the stress.

We have selected a commercial di-block copolymer consisting of polystyrene and polyisoprene with an averaged molar mass of 100 000 g/mol and a very narrow molar mass distribution. The solvent used was paraffinic hydrocarbon with addition of cyclohexane. This solution forms micelles with slightly solved polystyrene cores. The polyisoprene ends of the copolymer molecules may be entangled with other micelles. This polymer system has been found to have a good longterm stability and to be stable against shear stress and temperature [3,4]. Therefore it is preferred to act as reference material for comparison measurements.

In a shear rate region from 1 s^{-1} up to 4000 s^{-1} the samples can be described in a good approximation as Bingham fluids, as shown in ref. [3]. Under special solvent composition, they may exhibit a sudden shear thickening in the range of about 100 s^{-1} [5]. Measurements in the stress region below and arround the yield value have been done with different measuring techniques using sensitive rheometer equipment.

EXPERIMENTAL

Measurements were carried out with a controlled stress rheometer (CS) from Bohlin Reologi AB, Sweden and a modified shear controlled rheometer CV100 from Haake, West Germany. The CS has been used with concentric cylinder measuring systems (Searle type) with 14 and 25 mm of inner radius according to DIN 53019 and a 25 mm system with rippled cylindrical part of both cup and bob in order to check wall slip effects. The torque range applied by the instrument is up to 10 mNm with a resolution of $2 \cdot 10^{-4}$ mNm. The rotational speed may be in the range up to 60 rad/s with a resolution of $2 \cdot 10^{-5}$ radians. The CV100 rheometer has been used with a DIN measuring system, too. Additionally some measurements were carried out with two systems with Mooney-Ewart geometry with different gap widths. Both systems have been shielded with a guard ring. This prevents from an end effect from the upper end of the bob. In order to obtain low shear rates, the instrument has been modified. A geared down stepper motor and an optical angle encoder with a maximal resolution of $2 \cdot 10^{-4}$ radians have been added. The resolution of the torque detection is 10^{-4} mNm.

Measurements were carried out in different ways. Time dependent effects were tested by sweeping the stress or shear rate up and down. In most cases the sample was allowed to come to a steady state before the measured value was taken. In some cases the measuring time was fixed to a small varied value in order to obtain relaxation effects. In addition creep measurements were carried out.

Two different samples were found to be suitable for these investigations with respect to a possible use as reference liquid for comparison measurements. Sample A consists of 7% copolymer in n-paraffin I (mainly consisting of $n-C_{11}/n-C_{12}$) with 20% cyclohexane. Sample B consists of 6% copolymer in n-octane with also 20% cyclohexane. The measuring temperature for all investigations has been held constant to 20 °C unless otherwise noted.

RESULTS AND DISCUSSION

The following figures carry on the main ordinate at the left side the rotational speed ω in radians per second. This is due to the fact that those plastic samples in the yield region don't exhibit flow through the whole gap. The really flowing area is only a small cylinder of the gap. The calculation of the shear rate on the right side of the figures is done under the assumption that flow over the whole gap takes place and it therefore may be incorrect. All measurements are corrected with the end effect correction and are reduced to the stress at the inner radius.

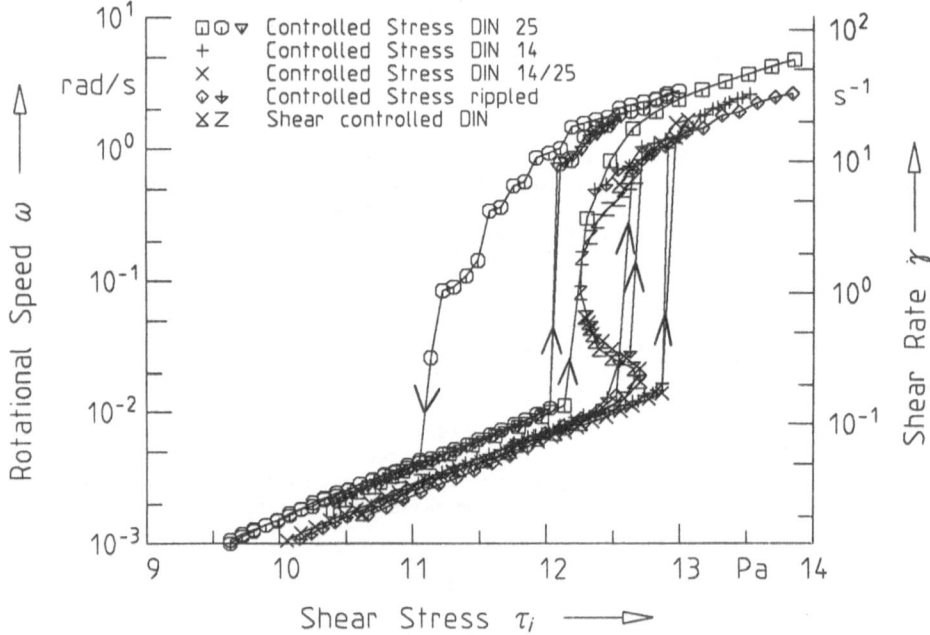

FIGURE 1: *Measurements of sample B in different measuring systems and rheometers at 20 °C*

Figure 1 shows a semilogarithmic plot of measurements of sample B with different measuring sytems and instruments. First, one can obtain from the figure that the sample exhibits a 'creep-flow' in the stress region below the real yield value. This leads to the question for the mechanism of this flow. A measuring system similar to the 25 mm DIN system but with vertical ripples both in cup and bob was used with the CS rheometer in order to test if wall slipping is a problem in this region. These measurements were also drawn in fig. 1 and show no difference, so wall slipping has been excluded. Furthermore, a combination of the bob 14 mm and the cup 25 mm has been used with the CS rheometer and has been treated as a 14 mm measuring system. The measurement is presented in fig. 1 with the diagonal crosses and shows an exact conformity with that measurement using the complete 14 mm system (vertical crosses). This leads to the above mentioned assumption that flow takes place in a small inner region of the gap only. The first significant deviation to the 14 mm system is observed beyond $\omega = 7$ rad/s. All these facts lead to the result, that the 'creep flow' below the yield value bases on a dynamic motion of the entangled micelles due to a disentanglement and reentanglement of the polyisoprene ends under the influence of shear stress. In all measurements elasticity has not been observed, only a relaxation to rearrange the micelle network has been detected by sweeping the stress down (open octagons in fig. 1).

The solutions seem to have a '1. yield value' which may be seen as the starting point of the 'creep flow'. This has been tested in the CV100 rheometer and additionally in a RV2 measuring unit from Haake with a very weak spring by switching off the shear and observing the remaining torque over a long time. The asymptotic value of the appropiate stress has been found to be about 6 Pa for sample A and about 3 Pa for sample B at 20 °C.

It can be obtained from figure 1 that the flow curves measured with the shear controlled CV100 differ from those obtained by the CS rheometer. The CV100 detects due to its principle of operation a break point when the stress reaches the real yield value. Increasing the rate of shear leads to a decreasing stress. This point can be seen as the point of destruction of the threedimensional network of the entangled micelles. Exceeding this point the micelles slip to eachother resulting in a decresing stress. Using the CS rheometer with the same sample the flow curve shows an incontinuity at the point of destructing the micelle network since the stress is held constant and the measuring system starts speeding up until a new equilibrium is reached.

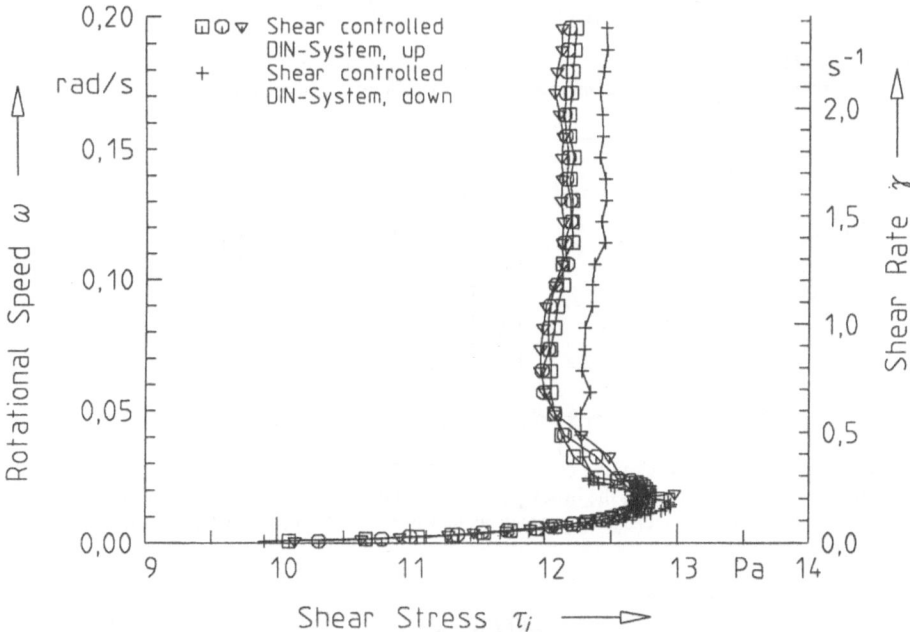

FIGURE 2: *Measurements of sample B with the CV100 rheometer sweeping the shear rate up and down*

Figure 2 shows four measurements with the CV100 and sample B in a linear plot. The first three measurements are of the sweep up type with a constant measuring time of 2, 20 and 60 seconds for each point. They coincide with each other in the range of measurement accuracy, therefore a time dependency can be neglected. The fourth measurement has been recorded as sweep down with 60 seconds measuring time per point (vertical crosses in fig. 2). As it can be seen, the reservibility is good in the lower region of the flow curve. Beyond 0.04 rad/s the down sweep shows a deviation to the up sweep curves due to a relaxation needed to rebuild the network of the micelle system.

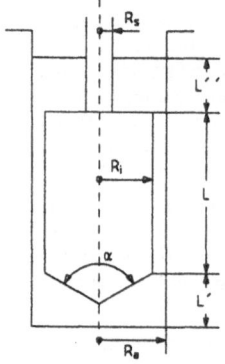

It seems to be evident that the stress at the yield value has to be calculated for the inner cylinder. Also one has to take into account an end effect correction c_L which is greater than the value for Newtonian fluids [4]. Calculating this end effect correction one has to summarize the different torque values originating from the several parts of the bob of a DIN measuring system (fig. 3) to a total torque M. In the following equation the four additive terms in the square brackets originate (from left to right) from the cylinder, the upper surface of the cylinder, the cone and the shaft.

$$M = 2\pi\tau_e \cdot [R_i^2 L + (R_i^3 - R_s^3)/3 + R_i^3/(3 \cos(90°-\alpha/2)) + R_s^2 L'']$$

Setting the preferable values stated in DIN 53019 ($R_s = 0.3R_i$; $\alpha = 120°$; $L' = L'' = R_i$; $L = 3R_i$) this equation evaluates to

$$M = 2\pi\tau_e \cdot R_i^2 \cdot (L + 0.799 \cdot R_i)$$

FIGURE 3: *Dimensions of a measuring system according to DIN 53019*

With $c_L = 1+(\Delta L/L)$ this gives $c_L = 1 + 0.799 \cdot R_i / 3 \cdot R_i = 1.266$. This value is in very good agreement with the experimental value found in ref. [4]. The values for the representative stress τ_{rep} have to be corrected using this value for c_L in the following equation giving τ_i for the inner cylinder:

$$\tau_i = 2\delta^2 / (1+\delta^2) \cdot c_{L,Newtonian} / 1.266 \cdot \tau_{rep} \qquad \text{(with } \delta = R_a/R_i)$$

This correction has been done with all measurements presented in this paper. Table 1 shows the stress value for the break point τ_b for both samples A and B with different measuring systems.

Measuring system	Rheometer	τ_b sample A	τ_b sample B
DIN 25 mm	CS	18.35 Pa	12.04 Pa
DIN 14 mm	CS	-	12.91 Pa
25 mm rippled	CS	17.77 Pa	12.53 Pa
Mooney-Ewart	CV100	19.46 Pa	12.86 Pa
Mooney-Ewart narrow gap	CV100	18.89 Pa	13.25 Pa
DIN	CV100	18.89 Pa	12.78 Pa

TABLE 1: *Corrected stress values for the break point of the micelle network*

The averaged value for the break point of sample A is 18.67 ± 0.9 Pa, for sample B it is 12.73 ± 0.7 Pa. The accuracy for the measurement of the break point was found to be within ± 5 %. This value covers a lot of measurements with different measuring geometries and instruments and seems to be satisfactory for this complex polymer system.

REFERENCES

[1] Cheng D.C-H.; Yield stress: A time-dependent property and how to measure it; Rheol Acta **25** (1986) 542-54

[2] Nguyen Q.D., Boger D.V.; Characterization of yield stress fluids with concentric cylinder viscometers; Rheol Acta **26** (1987) 508-15

[3] Bauer H., Meerlender G., Stern P.; Viscoplastic flow of ternary polymer solutions: application as a non-Newtonian standard material; Progr and Trends in Rheol II, Suppl. Rheol Acta **26** (1988) 181-83

[4] Bauer H., Boese N.; Non-Newtonian Reference Liquids; Proc of the Xth Int. Congr. on Rheology 14.-19.Aug. 1988, Sydney, Australia

[5] Bauer H., Meerlender G.; Polymer solutions with threshold values of shear stress and shear rate in the flow curve; Progr Colloid & Polymer Sci **72** (1986) 106-11

LIQUID FILAMENT MICRORHEOMETER AND SOME OF ITS APPLICATIONS

A.V.BAZILEVSKY, V.M.ENTOV, A.N.ROZHKOV
Institute for Problems in Mechanics, USSR Academy
of Sciences, pr.Vernadskogo 101, 117526 Moscow,
USSR

ABSTRACT

Liquid filament microrheometer (LFM) is a device using the
ability of dilute and semidilute polymeric solutions to form
very thin stable filaments. It allows to measure relaxation
times of order of 0.01-1.0 sec using very small quantities
of polymer solution (one drop or 0.01 cu. cm approximately)
and rather quickly (a few seconds per measurement). It makes
it possible to follow time variation of properties of polymer
solutions, polymer degradation processes etc. The LFM apparatus and a number of its applications are discussed in some
detail.

INTRODUCTION

The ability of dilute, semidilute and concentrated polymer
solutions to form very thin stable filaments is both striking
physical phenomenon and perspective source of new rheological
techniques. Its theory and some preliminary experiments are
reviewed briefly in (1). A close-up of liquid filament of
(poly)ethylenoxide 500 ppm solution in water is shown in
Fig.1 (left). Such a filament lives as long as 1 sec while
thinning due to action of capillary forces so it is quite observable. The theory predicts that the thinning kinetics is
governed by the fluid rheology. In particular for Newtonian
viscous and Maxwellian elastoviscous fluids one has respectively (2)

$$a = a_o - \frac{\alpha}{6\eta}t, \quad a = a_o \exp(-t/(3\theta))$$

Here $a(t)$ is the filament radius, $a_o = a(0)$, α is the
surface tension, η is the fluid viscosity and θ is the
fluid relaxation time. So by monitoring of the filament radius history one can both discriminate between the fluid models and determine numerical values of rheological constants

such as η or θ.

MEASURING DEVICE AND SOME EXPERIMENTAL DATA

This idea is realized in a number of LFM devices which use
for measurements a drop of liquid which is placed in a gap
between two discs. The discs then pulled apart quickly so
that a thin liquid filament is generated and its subsequent
thinning is monitored. It is necessary only that the liquid
develops observable filaments. This requirement is met readily by a number of sufficiently viscous and elastic (polymeric) liquids. Fig.1 (right) shows some characteristic experimental data for a viscous (mineral oil) and an elastic
(a(poly)ethylenoxide solutions) liquids.

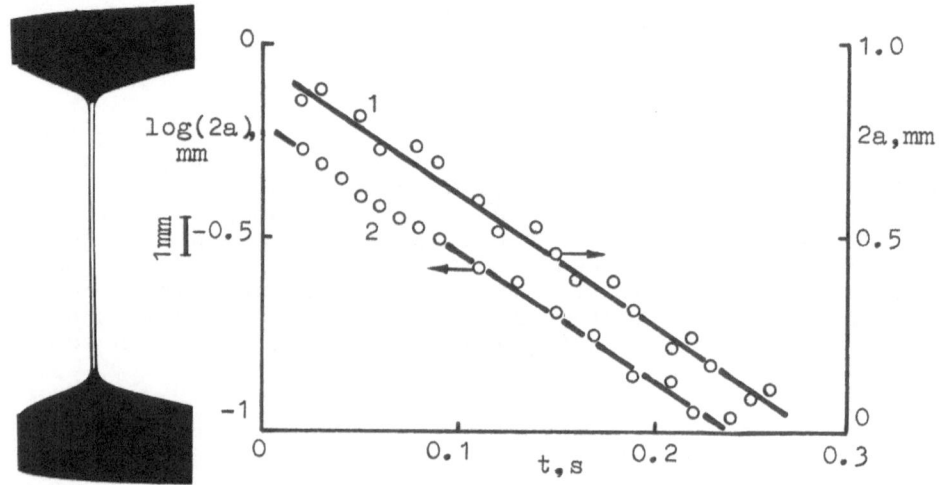

Figure 1. Liquid filament of 500 ppm water solution of PEO
          WSR-301 (left). Liquid filament radius versus time
          (experimental data) for Newtonian viscous liquid
          (a mineral oil, $\eta \simeq 30$ P) - 1, and for an elasto-
          viscous liquid (500 ppm water solution of PEO
          WSR-301) - 2 (right)

The data conform rather well to predictions and can be
processed to find η and θ. The measurements and the data
processing are automized readily. By repeating the measurements one can follow any time-dependent behaviour of the
fluid properties (say, that of relaxation time) and use its
as an indication of structural changes at the polymeric system such as polymer crosslinking or degradation. Some examples are shown in Fig.2. Rather small "specimen" volume (a
drop or 0.01 cu. cm of liquid), high time resolution (~ 1
msec), short measuring period (less that 1 sec), basic simplicity and versatility of LFM devices make it rather advanta-

geous in a number of applications both industrial and medi-
cal (3).

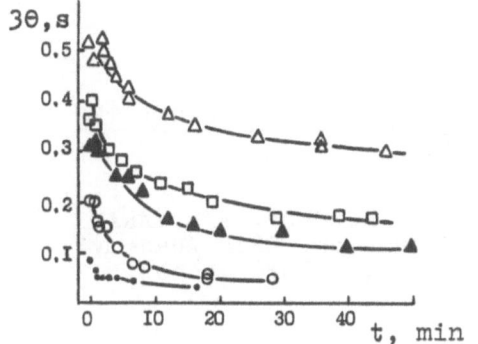 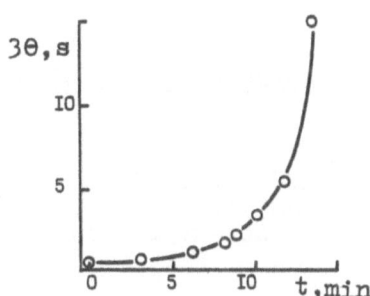

Figure 2. Time dependence of relaxation time θ of polymer
solutions (PAA in water) in the course of mechanical degradation due to mixing (left): • - 50 ppm,
o - 100 ppm, ▲ - 200 ppm, □ - 300 ppm, △ -500ppm.
Time dependence of relaxation time of PAA in water
solution (c=2000 ppm) during gelation due to addition of a crosslinking agent (right).

REFERENCES

1. Bazilevsky,A.V., Entov, V.M., Rozhkov, A.N. and Yarin,
   A.L., Polymeric jets beads-on-string breakup and related
   phenomena (this volume).

2. Entov, V.M., Elastic effects in flows of polymer solutions. In Heat- and Masstransfer in Polymeric Systems and
   Suspensions, ITMO AN BSSR, Minsk, 1984, pt 1, pp.15-30.

3. Dobrich, V.A., Bazilevsky, A.V. and Rozhkov, A.N., Study
   of viscoelastic properties of liquid contents of airways
   of lungs by thinning filament method. Laboratornoie Delo,
   1988, No 7, 26-27 (in Russian).

POLYMERIC JETS BEADS-ON-STRING BREAKUP AND RELATED PHENOMENA

A.V.BAZILEVSKY, V.M.ENTOV, A.N.ROZHKOV, A.L.YARIN
Institute for Problems in Mechanics, USSR Academy
of Sciences, pr.Vernadskogo 101, 117526 Moscow,
USSR

ABSTRACT

The phenomenon of beads-on-string capillary breakup of thin
jets of polymeric solutions is investigated both theoretically and experimentally. Basic physical mechanism, namely large
elastic extension of polymer molecules and overall hydrodynamic model are discussed.

INTRODUCTION

The phenomenon of beads-on-string breakup of thin jets of dilute polymer solutions was discovered by Goldin et al (1).
Its essence is that at later stages of capillary breakup
"necks" between forming drops cease to thin and transform
into thin liquid filaments gradually thinning without apparent change of shape of drops (Fig.1) (2-5).

0.2cm

Figure 1. Regular beads-on-string breakup of a jet of polymer solution.

THEORY AND EXPERIMENT

A number of questions arise relative to this phenomenon. One must explain physical mechanism of necks thinning slow-down, unusual shape of evolving jet, observed time dependence of liquid filaments radius and their hyperstability and finally the filaments breakup mechanism.

It was shown both theoretically and experimentally that the initial growth of disturbances over a polymeric jet does not differ drastically from that over a water jet, but at later stages polymeric molecules are stretched by strong elongational flow in necks so that the liquid becomes effectively stronger. The forming filaments between drops are effectively in the state of mechanical equilibrium which corresponds to zero axial stress, $\sigma_{11}=0$, $\sigma_{22} = \sigma_{33} =-q_\alpha$. Here $q_\alpha =+ \alpha/a$ is the capillary pressure, α is surface tension, a is the filament radius. So the thinning of filaments due to action of lateral capillary pressure can be predicted provided that the fluid rheology is prescribed. The experimental data correspond to effectively shear-thickening behaviour (6) and to that of "a perfect Maxwellian elastoviscous fluid". This last model leads to exponential filament radius decrease with time, the time constant being proportional to the liquid relaxation time (5). This theoretical prediction is well supported by experimental data. A "jet-rheometer" was developed both by Schummer and Tebel (7) and by present authors (2) for dilute and semidilute polymer solutions. It proved however that the jet rheometer is less suitable than the liquid filament microrheometer described in some detail in (8) for most of applications. Only in a range of rather low polymer concentrations the jet rheometer has some advantages. The key issue of the total approach, namely the assumption of zero axial stress, has been resolved conclusively by measuring total axial force in the liquid filaments which proved to be just equal to surface tension force $2\pi a\alpha \sim (1\div5)$ dynes in accordance with the theory. To this a vibrating was used. A momentary photograph of vibrating jet (Fig.2) allows one to evaluate axial force in the filaments with good accuracy.

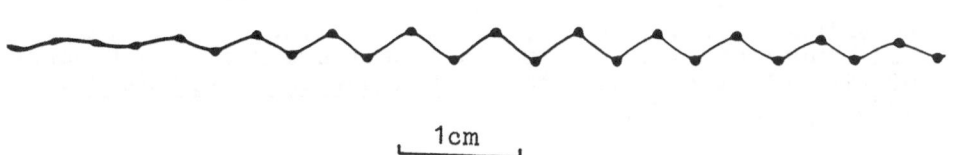

1cm

Figure 2. Beads-on-string breakup of a vibrating jet of a polymer solution (100 ppm PEO WSR-301 in water).

Predicted by theory highly oriented state of strain of polymeric liquid in filaments explains their hyperstability (4, 5).

The total process of breakup starting from the stage of small initial disturbances and up to final beads-on-string quasi-steady-state was described in the framework of quasi-unidimensional theory of jets dynamics in (9) and later using two-dimensional numerical simulation in (10).

REFERENCES

1. Goldin, M., Yerushalmi, J., Pfeffer, R. and Shinnar, R., Break up a laminar capillary jet of viscoelastic fluid. J. Fluid Mech., 1969, 38, No 4, 689-711.

2. Bazilevsky, A.V., Voronkov, S.I., Entov, V.M. and Rozhkov, A.N., Orientational effects during breakup of jets and filaments of dilute polymer solutions. Soviet Phys.Doklady, 1981, 257, No 2, 336-339.

3. Rozhkov, A.N., Dynamics of filaments of polymer solutions. J. Eng. Phys., 1983, 45, No 1, 72-80.

4. Entov, V.M., Elastic effects in flows of polymer solutions. In Heat- and Masstransfer in Polymeric Systems and Suspensions, ITMO AN BSSR, Minsk, 1984, pt 1, 15-30 (in Russian)

5. Entov, V.M., Elastic effects in flows of dilute polymer solutions. In Progress and Trends in Rheology II. Proc. 2nd Conf. Europ. Rheol., Darmstadt, 1988, pp.260-262.

6. Entov, V.M., Kordonsky, V.I., Kuzmin, V.A., Shulman, Z.P. and Yarin, A.L., Investigation of break-up of jets of rheologically complex fluids. J. Appl. Mech. & Techn. Phys., 1980, No 3, 90-105.

7. Schümmer, P., Tebel, K.H., Instability of jets of non-Newtonian fluids. In Rheology, 2, Fluids, Plenum Press, N.Y., 1980, pp.87-92.

8. Bazilevsky, A.V., Entov, V.M. and Rozhkov, A.N., Liquid filament microrheometer and some of its applications (this volume).

9. Entov, V.M. and Yarin, A.L., On the influence of elastic stresses on capillary breakup of jets of dilute polymer solutions. Fluid Dynamics, 1984, No 1, 27-35.

10. Bousfield, D.W., Keuning, R., Marrucci, G. and Denn, M.M., Non-linear analysis of the surface tension driven breakup of viscoelastic filaments. J. Non-Newton. Fluid Mech., 1986, 21, No 1, 79-97.

NUMERICAL INVESTIGATION OF BREAK–UP OF VISCOELASTIC JETS

CHRISTIAN BECKER PAUL SCHÜMMER

Institut für Verfahrenstechnik, RWTH Aachen

Turmstraße 46, D–5100 Aachen, FRG

ABSTRACT

On the basis of one–dimensional equations derived from the three-dimensional theory
the dynamics of controlled break–up of harmonically disturbed viscoelastic liquid jets
are studied numerically using a finite-difference technique. Controlled jet break–up can
be investigated considering the temporal instability of spatially harmonic disturbances
of an infinite jet. This procedure is approximately equivalent to the use of a frame of
reference moving along the jet axis with the mean velocity of the jet. A more realistic
point of view is to look at the problem as a spatial instability of time harmonic disturbances of a semi-infinite jet. The latter method is employed here and some results are
presented.

INTRODUCTION

Controlled break–up of liquid free jets is of interest in several technical applications
e.g. ink-jet printing or the manufacture of spherules of ion exchange resin. It can be
realized experimentally by applying a harmonic disturbance at the nozzle. Due to the
slenderness of jets a theory developed by Green et al. [2–4] is useful. To describe the
motion of slender bodies they proposed a method to derive a set of one-dimensional
equations from the three-dimensional theory. With the help of this method we studied
break-up of viscoelastic jets by means of a finite difference technique.

EQUATIONS AND METHODS

In our study of viscoelastic jets we took the Jeffreys model as constitutive law. We used the above mentioned method developed by Green et al. to obtain its one-dimensional equivalent. The complete set of equations is then given in dimensionless form by

$$R_t^2 + (v\,R^2)_z = 0 \tag{1}$$

$$v_t + vv_z - \tfrac{1}{2}R\,R_z\,(v_{zt} + vv_{zz} - \tfrac{1}{2}v_z^2 - \tfrac{1}{8}R^2(v_{zzt} + vv_{zzz}) =$$

$$\frac{1}{We}\left[\frac{R_z}{R^2(1+R_z^2)^{1/2}} + \frac{R_z\,R_{zz}}{R(1+R_z^2)^{1/2}} + \frac{R_{zzz}}{(1+R_z^2)^{3/2}} - \frac{3\,R_z R_{zz}^2}{(1+R_z^2)^{5/2}}\right] \tag{2}$$

$$+ \frac{1}{Re_N}\left[\frac{6R_z v_z}{R} + v_{zz}(3 - \tfrac{3}{2}R_z^2 - \tfrac{1}{2}RR_{zz}) - RR_z\,v_{zzz} - \frac{R^2}{8}v_{zzzz}\right]$$

$$+ \frac{1}{R^2}\left[n_{M^z} - q_{M^z} + p_{M^z} - \frac{R_z}{R}p_{M^{zz}} - (\frac{R_{zz}}{R} - \frac{R_z^2}{R})p_{M^z}\right]$$

$$n_M + De_1(n_{M^t} + vn_{M^z} - v_z n_M) = \frac{2}{Re_M}\,Re^2\,v_z \tag{3}$$

$$q_M + De_1(q_{M^t} + vq_{M^z} + 2v_z q_M) = \frac{1}{Re_M}\,R^2(\tfrac{1}{8}R\,R_z v_{zz} - v_z) \tag{4}$$

$$p_M + De_1(p_{M^t} + v\,p_{M^z} + v_z p_M) = -\frac{1}{8}\,\frac{1}{Re_M}\,R^4\,v_{zz} \tag{5}$$

with the balance of mass (1) and momentum (2) and in eqs. (3), (4) and (5) the one-dimensional form of the rheological equation of state. The Jeffreys stress has been split into a Newtonian and a Maxwellian part. In the above equations R and v are the shape and velocity functions. n, p and q are one-dimensional stress resultants, Re is the Reynolds number, De the Deborah number and We the Weber number. For further details see [5]. The equations were discretized using finite differences proposed by Bogy et al. [1] for the numerical solution of non-viscous and Newtonian jet break-up employing the same boundary conditions for the shape function and the velocity. The stress quantities n, p and q were taken to be zero on all boundaries.

49

RESULTS

Controlled break-up of viscoelastic jets can be investigated by the procedure outlined above. The usual difficulties in obtaining results at high Deborah or Weissenberg numbers are much less important here. Figure 1 shows an example of the calculated jet shape for $We=180$, $Re=24$, $De=40$ and $De_2=10$. However, for these high Deborah numbers very small time steps have to be taken to obtain good results. In the given example the grid width was $\Delta z=0.01$ and the time step $\Delta t=0.01$.

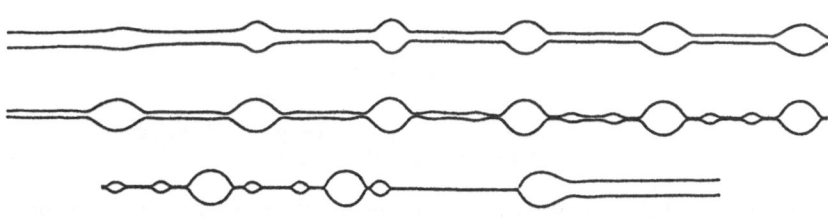

Figure 1: Radius calculated for jet break-up (parameters see text)

ACKNOWLEDGEMENT

We are grateful to the Deutsche Forschungsgemeinschaft for supporting this study.

REFERENCES

1. Bogy, D.B. , Shine, S.J. , Talke, F.E. , J. Comp. Phys. 38, 3(1980) 294–326
2. Green, A.E. , Laws, N. , Naghdi, P.M. , Proc. Camb. Phil. Soc. 64, (1968), 895–913
3. Green, A.E. , Naghdi, P.M. , Int. J. Solids Structure 6 (1970) 209–244
4. Green, A.E. , Naghdi, P.M. , Wenner, M.L. , Proc. R. Soc. London A 337 (1974) 451–507
5. Schümmer, P. , Thelen, H.G. , Rheol. Acta 27 (1988) 39–43

THE INFLUENCE OF LIQUID PHASE RHEOLOGY ON PARTICULATE PASTE EXTRUSION

J J BENBOW, T A LAWSON, E W OXLEY, J BRIDGWATER
School of Chemical Engineering
University of Birmingham
Edgbaston, Birmingham, B15 2TT, UK

ABSTRACT

Rheological results are presented for a variety of liquids which can be added to ceramic powders to make extrudable pastes.

A comparison of liquid properties to those of model pastes prepared by mixing the same powder with each liquid shows how liquid phase rheology affects pressure in the die-land.

INTRODUCTION

The liquid phase of a paste must enable powder and liquid to be well mixed on a macroscopic scale and then to be consolidated on a microscopic scale by high shear. The liquid phase has to lubricate the movement of particles over each other and it must also stabilise the paste so that liquid is not itself exuded by the pressure difference applied to it during extrusion (1). Finally the liquid must not produce surface imperfections; it must ensure that the extrudate retains its shape and that the extrudate can be cut cleanly without distortion and end tearing. These needs make the choice of the liquids demanding. (1).

The pressures necessary to produce extrusion are considered here. From research into the flow of pastes through square-entry dies having a variety of sizes we have shown that the relation between applied pressure P and extrudate velocity V can be described by the equation

$$P = 2(\sigma_o + \alpha V^m) \ln D_o/D + 4(\tau_o + \beta V^n) L/D \qquad (1)$$

The terms σ, τ, α, β, m and n are material properties which are characteristic of the paste formulation.

This equation is based on observations that plastic flow occurs in the die-entry followed by plug flow along the die-land with shearing confined to a thin liquid layer at the die wall. Equation (1) is the most general form. In some instances m and n are unity thus reducing the number of extrusion parameters to four. This is often a satisfactory approximation at practical extrusion velocities. When the liquid phase is a clay suspension, αV may also be negligibly small, there then being just three extrusion parameters. During flow through "square-entry" dies a continuous function of the velocity can be used to describe the flow without explicitly calculating the extrusion parameters (2). The notation employed is given in figure (1).

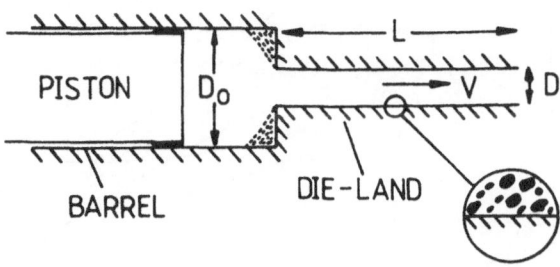

Figure 1
Terminology and notation for extrusion through a cylindrical square entry die.

EXPERIMENTAL

All the pastes described here contain the same blend of non-porous α-alumina particles. The complete size distribution extended from a diameter of $3\mu m$ to a maximum of $50\mu m$ with a mean value of $10\mu m$.

Three types of liquid were examined. These were suspensions of Wyoming Bentonite and starch in water, aqueous solutions of type HPM450P Celocol, and two mixtures of Bentonite in glucose solutions. Thus they included a Bingham fluid with a high yield value, pseudo-plastic liquids having no yield value and two liquids having yield stresses but which are Newtonian at high shear-rates. The rheological properties, measured in a Carrmied Type CSL rheometer, are shown as plots of apparent viscosity against shear-rate in figure (2). The low stress properties were measured but are not included here.

Pastes P, Q, R and S were prepared by mixing 190ml of the different liquids with the 1kg of the α-alumina powder in a Hobart mixer fitted with a paddle and were then subjected to high-shear mixing (pugging) by passing them twice through the mincer attachment of the Hobart machine. Equal amounts of liquid were added so that the thickness of the liquid layers surrounding the particles were approximately the same in all the pastes.
The rheological properties of the pastes were then measured in an instrumented ram extruder previously described (3). Paste compositions are listed in table (1).

PASTE	1. ALUMINA	2. CLAY	3. CELACOL	4. STARCH	5. GLUCOSE	WATER
	gm	gm	gm	gm	ml	ml
P	1000	38	–	–	127	63
Q	1000	38	–	–	95	95
R	1000	–	14.3	–	–	190
S	1000	38	–	38	–	190

1. WHITE BAUXILITE
2. WYOMING BENTONITE
3. CELACOL TYPE HPM 450P
4. POTATO STARCH
5. GLUCOSE SYRUP

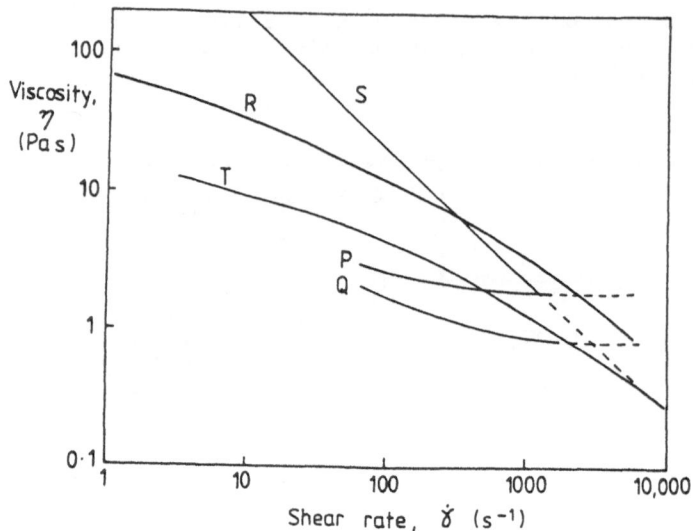

Figure 2
Viscosities of liquid phases versus shear rate.

DISCUSSION

From figure 2 it is evident that all the liquids have similar viscosities at high shear-rates 10^3 - $10^4 s^{-1}$. Estimation of the effective shear rate at the die-land wall is affected by the particle size distribution near the die wall. The maximum shear-rate occurs in the narrowest gap between the particles and the die wall. There is, however, a distribution of shear rates with values less than that in the minimum gap. Moreover, the regions of highest shear rate occupy only a small fraction of the total die-land perimeter, so that the effective shear-rate near the wall is necessarily less than the maximum. The effective average shear-rate can be estimated from the value of η for a liquid which is Newtonian at high shear-rates, ie those containing glucose solution pastes. If t is the effective thickness of the liquid between the paste bulk and the die-wall, it follows that $\eta/B = t$. For the paste P this gives a value of t = 1.4μm. For a typical extrudate velocity of 10mm/s, this value of t indicates that the effective shear-rate lies in the region of 6 x $10^3 s^{-1}$. The viscosities of all these liquids tend to converge at these shear-rates.

In figure 3(a) the values of τ, where $\tau = \frac{P_2}{4D}L$ are plotted against the extrudate velocity V. Viscometric results for the liquids employed to make these pastes are presented as curves of shear stress (σ) versus shear rate (γ) in figure 3(b).

Comparing the two sets of curves, it can be seen that the Newtonian viscosities of liquid P and Q at high shear rates is reflected in the steep, nearly linear, dependence of τ on V at high velocities. The high yield value of liquid R is reflected in the small dependence of pressure on velocity for paste R.

Finally, the pseudoplastic behaviour of liquid S is also found in the curvature of paste S, but the values of τ are lower than expected from the liquid properties. This could be due to the differences in the particle plus liquid structure or to differences during flow in the space between the particles and the die-wall.

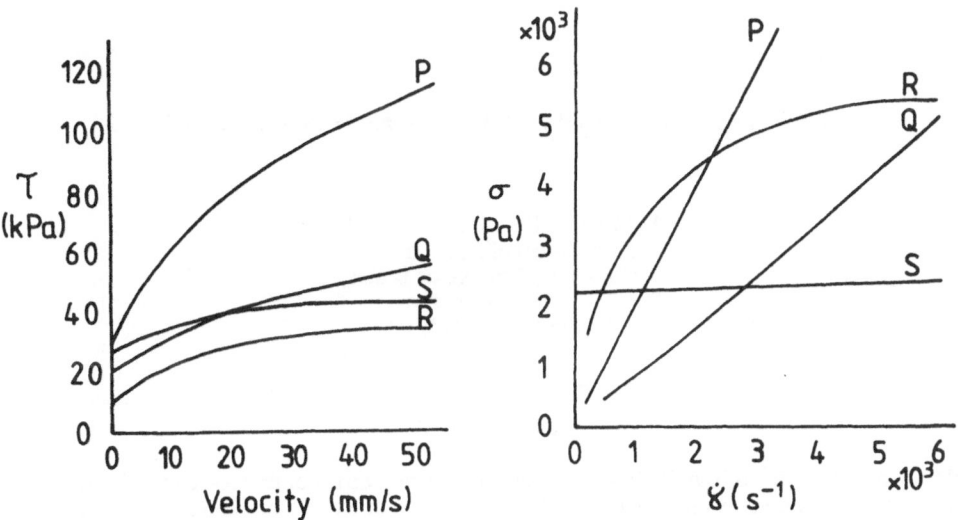

Figure 3 (a)
Curves of wall shear stress τ versus extrudate velocity V.

Figure 3 (b)
Curves of shear-stress σ versus shear rate ($\dot{\gamma}$) for the liquids used to make the pastes shown in figure 3(a).

CONCLUSIONS

Rheological measurements have been made, under conditions relating to flow through die-lands of several types of liquid. By studying the ram extrusion behaviour of model ceramic pastes made using these liquids some qualitative correlations have been indicated.

REFERENCES

1. Benbow, J.J., Jazayeri, S.M. and Bridgwater, J., Proc 1st International Conference on Ceramic Powder, Processing and Science 11 B 624-634 (1988).

2. Benbow, J.J., Lawson, T.A., Oxley, E.W., Bridgwater, J. Am. Ceram. Bull 1989 (In Press).

3. Benbow, J.J., Oxley, E.W. and Bridgwater, J. Chem.Eng.,Sci. 42 2151-2163 (1987).

ENTRY AND EXIT FLOWS OF NON-NEWTONIAN FLUIDS

D M BINDING and P R JONES
Department of Mathematics,
University College of Wales,
Aberystwyth
Dyfed SY23 3BZ, UK.

ABSTRACT

We consider the flow of mobile fluids through an axisymmetric contraction and a similar expansion. Of interest is the experimental determination of excess entry and exit pressures. The following types of fluid were tested: constant viscosity, highly tension thickening (Boger-type) fluids; shear thinning solutions of rigid-rod molecules (Xanthan gum solutions). It is shown how shear viscosity, extensional viscosity and elasticity all contribute to the observed behaviour in contraction and expansion flows.

INTRODUCTION

The entry flow problem has received considerable attention recently [1]. By contrast, relatively little effort has been devoted to exit flows. There have been very few attempts at measuring excess exit pressures, particularly for mobile polymer solutions (see, for example [2]).

EXPERIMENTAL PROCEDURE

The flow geometry is shown schematically in Figure 1. Fluid flows under pressure through a large circular (77 mm diameter) cylinder into a 6.36 mm diameter capillary by means of an abrupt contraction. Finally, the fluid exits into a cylinder of the same diameter as the entry. Normal stresses T_0, T_1, T_2 are measured using small diameter strain gauge type transducers (Druck Ltd, UK). Flow rates are measured using a catch and measure technique.

THEORETICAL CONSIDERATIONS

The approach adopted here is similar to that used by Vrentas and Vrentas [3]. It is assumed that the entry, central and exit regions are all sufficiently long for fully developed flows to be established at the sections "0", "1", "2" (see Figure 1) where normal stresses at the walls are measured. The pressure drop P between any two cross sections of the flow is written as W=PQ where W is the resultant rate of working of the relevant surface forces and Q is the volumetric flow rate. Making use of the fully developed flow conditions at "0", "1", "2" the following equations are obtained for the excess entry and excess exit pressures:

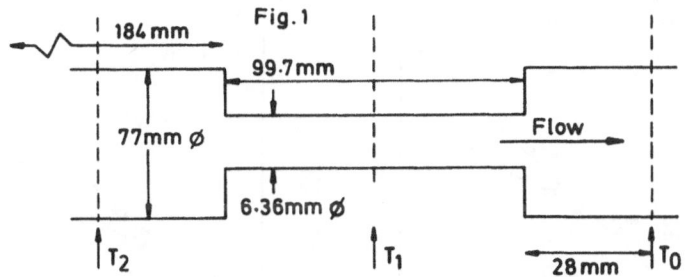

Figure 1. Schematic diagram of flow geometry.

$$P_{ent} = (T_2-T_1)+(2\pi/Q)\int_{S_1} N_1 v_z ds-(2\pi/Q)\int_{S_2} N_1 v_z ds-P_{2fd}-P_{1fd} \qquad (1)$$

$$P_{exit} = (T_1-T_0)+(2/Q)\int_{S_0} N_1 v_z ds-(2\pi/Q)\int_{S_1} N_1 v_x ds-P_{1fd}-P_{0fd} \qquad (2)$$

where N_1 is the first normal stress difference in shear, v_z is the fully developed axial velocity distribution and P_{1fd} is the fully developed pressure drop in region i. In deriving these equations, the second normal stress difference has been assumed to be negligible.

EXPERIMENTAL RESULTS

Characterisations of the fluids were obtained using a Weissenberg Rheogoniometer (model R16) and a Controlled Stress Rheometer (both Carrimed, UK). Shear viscosity and first normal stress difference data for all the fluids are shown in Figure 2.

Experimental results are shown in Figures 3-5. Each figure contains the measured upstream and downstream stress differences T_2-T_1 and T_1-T_0, respectively, as well as the

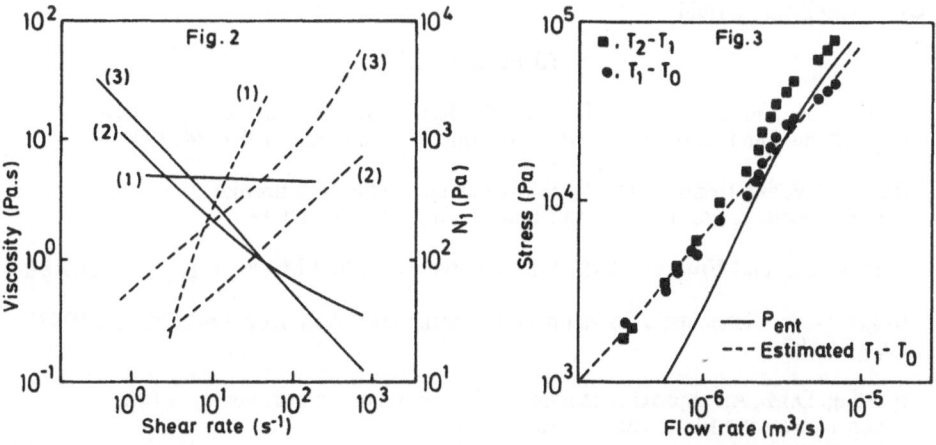

Figure 2. Shear viscosity (—) and 1st normal stress difference (---) for the Boger (1), Xanthan (2) and Polyacrylamide (3) solutions.
Figure 3. Experimental results for the Boger solution.

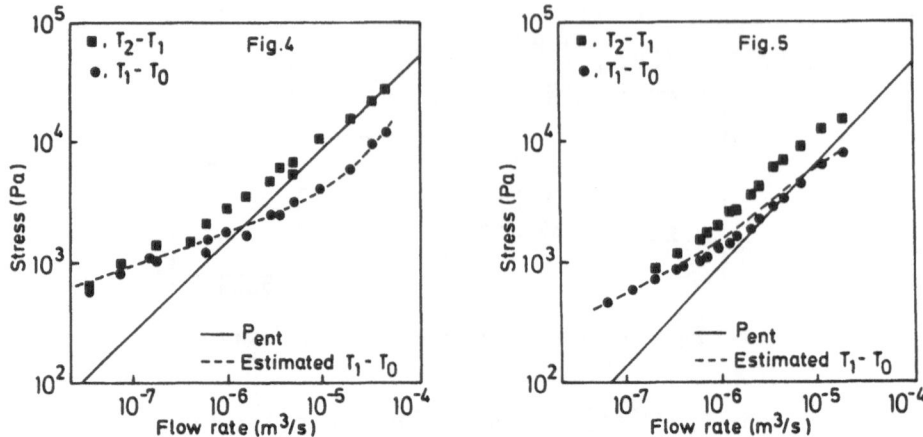

Figure 4. Experimental results for the polyacrylamide solution.
Figure 5. Experimental results for the Xanthan solution.

calculated excess entry pressure loss according to Equation (1). In addition, we show the estimated downstream stress difference based on Equation (2) with the predicted (see, for example [4]) excess exit pressure loss for an inelastic fluid with the same viscosity function.

CONCLUSIONS

It has been suggested [5,6] that excess entry pressures may be interpreted in terms of the fluids' shear and extensional viscosities. On the other hand, the experimental determination of excess pressures is clearly influenced by the elastic properties through N1. For the Xanthan solution the contribution of N1 to the excess entry pressure is small for the full range of flow rates achieved. For the more elastic fluids the contribution of N1 becomes significant. However, in all cases the effect of N1 is to produce small values for the excess exit pressure. In fact, Pexit is seen to be very close to the excess entry pressure expected for an equivalent inelastic fluid.

REFERENCES

1. White, S.A., Gotsis, A.D. and Baird, D.G., Review of the entry flow problem: Experimental and numerical, J. non-Newtonian Fluid Mech., 1987, 24, 121-160.

2. Boger, D.V. and Denn, M.M., Capillary and slit methods of normal stress measurements, J. non-Newtonian Fluid Mech., 1980, 6, 163-185.

3. Vrentas, J.S. and Vrentas, C.M., J. non-Newtonian Fluid Mech., 1983, 12, 211-224.

4. Boger, D.V., Viscoelastic flows through contractions, Ann. Rev. Fluid Mech., 1987, 19, 157-182.

5. Binding, D.M., An approximate analysis for contraction and converging flows, J. non-Newtonian Fluid Mech., 1988, 27, 173-189.

6. Binding, D.M. and Walters, K., On the use of flow through a contraction in estimating the extensional viscosity of mibile polymer solutions, J. non-Newtonian Fluid Mech., 1988, 30 233-250.

DETERMINATION OF THE RHEOLOGICAL PROPERTIES OF SOLIDS USING PULSE PROPAGATION METHODS

R.H. BLANC
Laboratoire de Mécanique et d'Acoustique du C.N.R.S.
31, Chemin Joseph-Aiguier F-13402 MARSEILLE cedex 9 France

ABSTRACT

The viscoelastic properties of solids are deduced from the change in the shape of a transient wave propagating within a slender bar of the medium under investigation. For this purpose, the shape of the wave is recorded after travelling two distances along the bar. If the waves can be separated, the solution is expressed simply in terms of the transfer function of the two wave recordings. One advantage of this technique is that it can be used with a non-contacting gauge. If the waves cannot be separated, a novel solution is proposed using the superimposition of the wave reflections produced by the ends of the bar. One advantage of this method is that it can be used with shorter bars and high damping materials. These methods entail no heating of the medium. They yield results on one to two decades within the audio frequency range. They therefore fill a gap between the available vibratory and ultrasonic methods.

INTRODUCTION

It is proposed to deduce the mechanical characteristics of a linear viscoelastic medium from the change in shape of a transient wave propagating along a bar of the medium under study. A homogeneous viscoelastic medium can be characterized in one-dimensional tension and compression in terms of the phase velocity $c(\omega)$ and the damping coefficient $\alpha(\omega)$ of the longitudinal waves. It is then possible to deduce the other equivalent viscoelastic functions, particularly the complex modulus $E^*(i\omega)$.

THEORY

In order to determine the functions $c(\omega)$ and $\alpha(\omega)$, an axial shock is applied to one end of a slender bar of the medium under investigation. This gives rise to a wave which propagates along the bar. After travelling a certain distance x, the wave can be represented by a Fourier integral involving the pulse at the origin, distance x and functions c(ω) and α (ω). Let us assume that it is possible to experimentally determine f(x,t) as a function of time after travelling two distances x_1 and x_2. It is proposed to then deduce functions c(ω) and α(ω). Three solutions to this problem are proposed. First, expressions are established for c(ω) and α (ω) in terms of the Fourier transforms of the wave at x_1 and x_2 ; this is the general solution /1/. It can be expressed very simply in terms of the transfer functions of the two waves. Secondly, the pulses can be applied to selective frequency filters /2/ and the results obtained by eliminating the filter characteristics. Here it is proposed to generalize the latter solution. Thirdly, we propose fast, approximate solutions, the advantage of which is that they provide expressions for c(ω) and α (ω) which are directly related to experimentally observable time functions /1/. It sometimes occurs however that the successive pulses cannot be separated : this is so in the case of short bars of materials with a high phase velocity dispersion and high damping. Two solutions are proposed in this case. First, a wave front method, which is another fast method /3/, and secondly, the method recently developed by B. Lundberg and the present author /4/, which focuses on the superimposition of the successive wave reflections produced by the ends of the bar.

EXPERIMENTAL SET-UP. RESULTS.

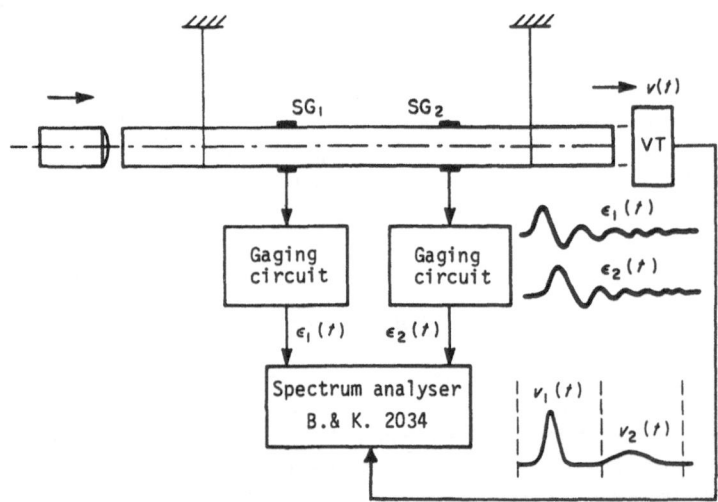

Figure 1. Diagram of the experimental set-up. Separate waves : the particle velocity is measured at the end of the bar. Superimposed waves : the strain is measured at two points.

These methods were applied to a wide range of materials such as polymers, elastomers, composites, etc. Excellent agreement has been obtained from one method to another, as well as with other existing methods.

CONCLUDING REMARKS

Wave methods have been developed for determining the viscoelastic properties of solids. For this purpose, the shape of a transient wave is recorded after travelling two distances along a slender bar of the medium under investigation. If the waves are separable, three solutions are proposed which have the advantage of requiring only a no-contact gauge. The general, exact solution is expressed quite simply in terms of the transfer function of the two wave measurements, which can be provided directly by a spectrum analyser. For cases where the waves are not separable, another solution is proposed which involves the multiple echo of the pulse produced in the bar. The advantage with this method is that it can be used with shorter bars and materials with extremely high damping. The two types of method are complementary. They entail no heating of the medium and they yield results on about one and a half decades within the audio frequency range, namely 20 to 20.000 Hz. They therefore fill a gap between the available vibratory and ultrasonic methods.

REFERENCES

1. BLANC R.H., Propriétés viscoélastiques des solides déterminées par des méthodes de propagation d'ondes transitoires. Cahiers Gr. Fr. Rhéol., C.R. 18e Coll., Paris, 1983, 6(4), 107-134.

2. BLANC R.H., Spectre instantané d'une impulsion dans un barreau viscoélastique. Rheol. Acta, 1974, 13(2), 228-232.

3. BLANC R.H. and CHAMPOMIER F.P., A wave-front method for determining the dynamic properties of high damping materials. J. Sound Vibr., 1976, 49(1), 37-44.

4. LUNDBERG B. and BLANC R.H., Determination of mechanical material properties from the two-point response of an impacted linearly viscoelastic rod specimen. J. Sound Vibr., 1988, 126(1), 97-108.

EXPERIMENTS AND COMPUTER SIMULATIONS ON THE FLOW OF CONCENTRATED DISPERSIONS

WILLEM H. BOERSMA, JOZUA LAVEN AND HANS N. STEIN
Laboratory for Colloid Chemistry
University of Technology Eindhoven
P.O. Box 513, 5600 MB Eindhoven, The Netherlands

ABSTRACT

The shear thickening transition often found in concentrated dispersions can be modelled by means of a force balance between stabilizing electrostatic and destabilizing hydrodynamic forces. The critical shear rate that results from a force balance between these two forces compares well with experiments. When these experiments are compared with Stokesian dynamics computer simulations on monolayers there is good qualitative agreement. In the simulations an order disorder transition and a rise in dispersion viscosity can be observed at the same critical shear rate as indicated by the theoretical model and the experiments.

INTRODUCTION

When studying the rheology of concentrated dispersions a phenomenon often observed with stabilized dispersions is shear thickening [1]. We recently deduced a criterion for the occurence of shear thickening [2] based on the assumption that shear thickening starts when the destabilizing, cluster formation inducing, hydrodynamic forces overrule the stabilizing forces. With electrostatically stabilized dispersions and the approximation of small interparticle distances and thus high volume fractions this leads to a dimensionless number N_d. Shear thickening occurs when N_d becomes larger than one.

$$N_d = \frac{6\pi\eta_0 \dot\gamma a}{2\pi\varepsilon_0 \varepsilon_r \psi_0^2} \frac{2}{\kappa} \frac{a}{h} \tag{1}$$

where $\varepsilon_o \varepsilon_r$ is the dielectric constant of the medium; ψ_0 is the surface potential; η_0 is the medium viscosity; a is the particle radius; h is the interparticle distance; and $1/\kappa$ is the Debye double layer thickness. h/a is a function of the volume fraction ϕ [2]. A comparison of this theory with experiments led to good results [2].

Computer simulations using the Stokesian dynamics method [3] showed that simulation results on the rheological behavior of dispersions compare well with experimental results [4]. The method takes into account many-body hydrodynamic interactions as well as interparticle forces as e.g. electrostatic stabilization.

RESULTS AND DISCUSSION

In fig. 1 experimental and simulation viscosities are plotted as a function of N_d. In both the experiments and the simulations $(\kappa h)/(2a)$ was nearly equal to $1/a$, altough the experimental volume fractions and thus h/a and κ were different from respectively the simulated volume fractions, h/a and κ.

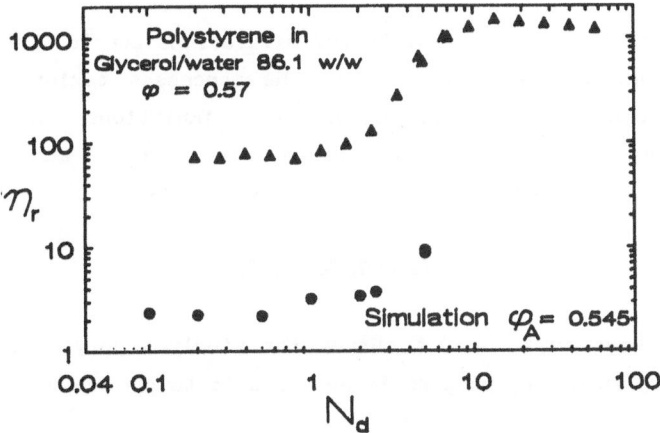

Figure 1. Relative viscosities as a function of N_d.

The simulations were performed with the Stokesian dynamics method in which constant surface potential electrostatic stabilization was taken into account. A monolayer of 25 particles was simulated.

It can be seen that the transition in both cases takes place around $N_d=1$, as was predicted form eq. 1. Only the viscosity level is different, because of the difference in volume (or areal) fraction. When looking at the particle configurations that result from the simulations it can be seen that there is a transition from an ordered statement of layers to a disordered statement with particle clusters. This is shown in fig. 2.

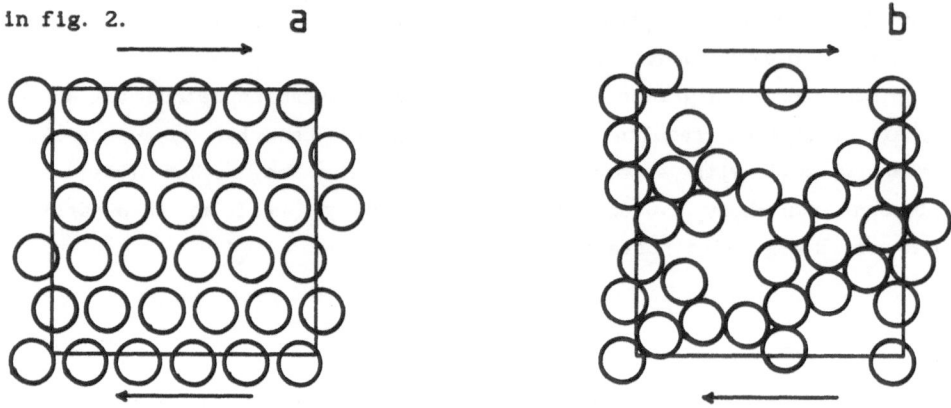

Figure 2. Snap shot of configuration at N_d=0.2 (a) and N_d=5 (b).

CONCLUSIONS

The results of theory, experiments and simulations are in agreement with each other and give good insight in the processes taking place in a stabilized concentrated dispersion. The simulations show that an order-disorder transition takes place around N_d=1 and that this transition is responsible for the rise in viscosity.

ACKNOWLEDGEMENTS

The authors wish to thank J.F. Brady for kindly supplying a Stokesian dynamics program and helping us in adapting it to our needs.

REFERENCES

1. Barnes, H.A., J. Rheol., 1989, 33, 329.
2. Boersma, W.H., Laven, J., and Stein H.N., AIChE J., 1990, 36, 321.
3. Durlofsky, L., Brady, J.F., and Bossis, G., J. Fluid Mech., 1987, 180, 21.
4. Brady, J.F., and Bossis, G., J. Fluid Mech., 1985, 155, 105.

Peristaltic Flow of Viscoelastic Liquids

G. Böhme, D. Nikolakis

Institut für Strömungslehre und Strömungsmaschinen,
Universität der Bundeswehr Hamburg, Holstenhofweg 85, D–2000 Hamburg 70,
Federal Republic of Germany

Abstract

Peristaltic pumping is investigated, generated by means of an infinite train of waves travelling along the wall of a cylindrical tube. The theory is based on the general second order integral constitutive equation for viscoelastic (simple) liquids. Analytical and closed form solutions are presented for the first order and the stationary part of the second order flow field approximations with respect to the amplitude ratio. It turns out that the zero–shear viscosity η_0 and the complex viscosity $\eta^*(\omega)$ are the only relevant fluid properties.

Outline of the theory

The main restrictions are: (i) the radius of the tube varies cosinusoidally along the tube axis, $r_0(x,z) = r_0 + \epsilon\, r_0 \cos\left[\omega(t - z/c)\right]$, (ii) the mean axial pressure gradient is zero, (iii) the ratio of the wave amplitude to the tube radius is small, (iv) the flow field has axial symmetry. The Reynolds number as well as the wave length are assumed to be arbitrary. The tube wall oscillates in radial direction so that the liquid particles adjacent to it possess zero axial velocity. At the tube axis the radial velocity vanishes and the axial one has a finite value. In case of small oscillation amplitude the momentum and continuity equation as well as the boundary condition can be expanded as power series of the wave amplitude parameter ϵ. The series have to satisfy the conditions of orthogonality and completeness. The perturbation method leads to linear boundary value problems for the first and the second order flow field. The relevant parameters for the primary flow field are the density ρ, the wave frequency ω, the mean tube radius r_0, the complex viscosity $\eta^*(\omega) = \eta'(\omega) - i\eta''(\omega)$ and the wave length l.

Cylindrical coordinates r, φ, z and the dimensionless parameters $St := \rho \, \omega \, r_0^2 \, / \, \eta'$, $H := 2\pi \, r_0 \, / \, l$, $V := \eta'' / \eta'$, $x = r/r_0$ have been introduced for further calculations. The solution of the primary velocity is as follows:

$$\mathbf{v}_1(r,z,t) \; = \; c \, r_0 \, \mathrm{curl} \, (A(r,z) \, \mathbf{e}_\varphi) \, e^{i\omega t} \; ,$$

$$A(r,z) \; = \; [B \, I_1(\xi) + C \, I_1(\zeta)] \, e^{-i\omega \, z/c} \; ,$$

where: $\quad \xi := H \, x \; , \quad \zeta := \sqrt{H^2 + k^2} \, x \; , \quad k^2 = \dfrac{St(i - V)}{1 + V^2} \; .$

$I_1(\xi)$, $I_1(\zeta)$ are modified first order Bessel functions. The constants B and C are determined by the boundary conditions. The RHS of the second order differential equation is a non–linear function of $\mathbf{v}_1$. The second order flow field consists of a stationary part and of a part which oscillates with double frequency 2ω. In view to the flow rate, only the stationary part has to be analysed. Here additional dimensionless parameters are introduced, i.e. $W := \eta' / \eta_0$; $K := \kappa(\omega) \, \omega / \eta''$; $\psi(x) := \psi_2^{st}(r)/c \, r_0^2$, where η_0 is the zero–shear viscosity, $\kappa(\omega)$ a frequency dependent normal stress coefficient and ψ_2^{st} the stream function of the stationary part. It is proved that ψ_2^{st} depends on r only. The solution of the stationary part has the following structure:

$$\mathbf{v}_2^{st} = \; \mathrm{curl} \left[\frac{\psi_2^{st}}{r} \, \mathbf{e}_\varphi \right] \; = \; v_z^{st}(r) \, \mathbf{e}_z \; ,$$

$$\psi(x; St, V, H, W, K) \; = \; \frac{\psi_2^{st}}{c \, r_0^2} \; = \; \psi_1(x; St, V, H) + W \cdot \psi_2(x; St, V, H) + W \cdot K \cdot \psi_3(x; St, V, H) \; .$$

Note that this field has a velocity component in the axial direction only. The mean flow rate $\dot{V}$ can be determined by integrating the axial velocity over the cross–section of the tube to give

$$\dot{V} \; = \; 2\pi \, \epsilon^2 \, c r_0^2 \, \psi(x{=}1; \, St, V, H, W, K) \; .$$

After extensive analytical calculations we are able to show that $\psi_1(x{=}1) = \psi_3(x{=}1) \equiv 0$. Therefore ψ_2 is the only relevant function to determine the flow rate,

$$\psi(x{=}1; \, St, V, H, W, K) \; = \; W \, \psi_2(x{=}1; \, St, V, H) \; .$$

Thus, the flow rate is independent of the normal stress parameter K, but is influenced by the complex viscosity $\eta*(\omega)$ and the zero–shear viscosity η_0 of the fluid.

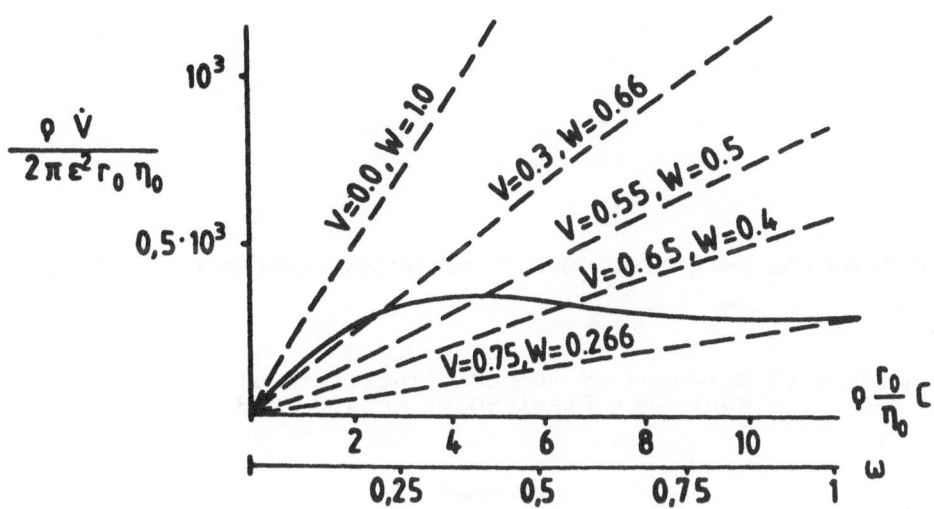

Figure 1: — — — Normalized flow rate as function of normalized wave velocity. V and W
are parameters of a real liquid. ——— Normalized flow rate as function of frequency
ω for a real liquid and fixed geometry. Note the existance of an optimal frequency.

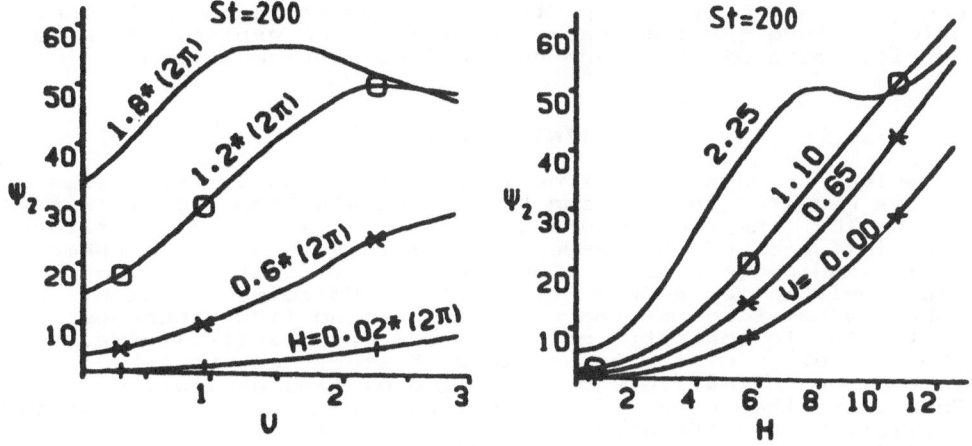

Figure 2: $\psi_2(x=1)$ as function of V and H

References

Böhme, G. and R. Friedrich: Peristaltic flow of viscoelastic liquids,
        J. Fluid Mech., 128 (1983), 109–122
Böhme, G.: Rheologisches Stoffgesetz für oszillierende Strömungen
        einfacher Flüssigkeiten, ZAMM 70 (1990), T360–362
Burns, J.C. and T. Parkes: Peristaltic motion, J. Fluid Mech. 29 (1967), 731–743
Shapiro, A.H., M.Y. Jaffrin and S.L. Weinberg: Peristaltic pumping with long wavelengths
        at low Reynolds number, J. Fluid Mech. 37 (1969), 799–825

RHEOLOGICAL RATE TYPE MODELS INCLUDING TEMPERATURE HISTORY

H.BRAUN

Academy of Sciences of the GDR, Institute of Mechanics,
Karl-Marx-Stadt 9010, PF 408, GDR

SUMMARY

For the calculation of nonisothermal flows of viscoelastic
fluids thermodynamically consistent relations (temperature
equation (TE), special rate type rheological constitutive
equation (RCE), temperature dependence of material parameters)
are derived for the limiting cases of entirely entropy and
energy elasticity, respectively. Some nonisothermal models,
which include the temperature history in an explicit form, are
special cases of the given generalized representation.

INTRODUCTION

Numerical simulation of nonisothermal flows of viscoelastic
fluids requires in addition to the RCE also the consideration
of the TE, the temperature dependence of material parameters,
the influence of the temperature history as well as thermodynamic requirements (1st, 2nd law). Consistent relations are
used only in few papers (e.g. [1]). Moreover, most of all
known simulations of thermally influenced flow processes are
restricted to inelastic or entire entropy-elastic fluids.
The influence of the temperature history can be taken into
account approximatively by inserting of time derivatives [2].
Sometimes the rate of temperature change is incorporated
implicitly by the glass temperature [3,4] or explicitly by
additional terms in the RCE [5-7].
In the present paper, RCE's of a special group (rate type
models) are generalized for nonisothermal conditions and thermodynamically consistent relations (RCE, TE and temperature
dependence of parameters) are derived for the limiting cases
of entirely entropy- and energy-elastic fluids.

THEORY

The considered group of RCE's (monomodale models) can be
described on global level (i.e. with observable variables) by:

$$\overset{\triangledown}{\tau} + \frac{1}{\Theta} \varphi(\tau) = 0 \qquad \text{with} \qquad \operatorname{tr} \tau \neq 0 \qquad (1a)$$

$$\varphi(\tau) = \beta_1 \tau^2 + \beta_2 \tau + \beta_3 \mathfrak{G} \qquad (1b)$$

Here τ is an extra stress tensor ($\tau=\delta-p\delta$, δ – Cauchy stress tensor, p – pressure, δ – unit tensor) with the equilibrium value $\tau_0=G\delta$, $G=\eta/\theta$ is the modul, η and θ are the characteristic viscosity and the relaxation time of the fluid, β_i are coefficients that can generally depend on the invariants of τ and $\triangledown$ denotes the upper convected time derivative.
On the level of hidden variables one obtains for the relations (1a,b):

$$\overset{\triangledown}{B_e} + B_e \cdot D_p + D_p \cdot B_e = 0 \qquad\qquad (2a)$$

$$\tau :\equiv \tau_e = 2 \left.\frac{\partial W}{\partial I}\right|_T B_e := G B_e \qquad\qquad (2b)$$

$$W = \frac{\eta}{2\theta} (I-3) \qquad\qquad \text{with} \qquad\qquad I = tr\ B_e \qquad\qquad (2c)$$

The basic idea [8,9] consists in an additive decomposition of the Finger tensor B ($B=F\cdot F^T$, F – deformation gradient), the deformation rate tensor D and the extra stress tensor into an elastic and inelastic part (indices e and p). The potential function in (2) is denoted as W, and T is the temperatur. It should be mentioned that an exact ansatz for $D_p(\tau)$ is required for the description according to (2) and that the expressions (1) and (2) can be transformed into each other by

$$\varphi = \theta (\ \tau \cdot D_p + D_p \cdot \tau\ ) \qquad\qquad (3)$$

Assuming $s=s(u,B_e)$ for the entropy (u – internal energy), the thermodynamic relation (4) can be derived:

$$2\ \varrho\ \left.\frac{\partial u}{\partial B_e}\right|_T \cdot B_e = \tau - T \left.\frac{\partial \tau}{\partial T}\right|_{B_e} \qquad\qquad (4)$$

In nonisothermal cases on principle $G=G(T)$ must be taken into account on the level of hidden variables. After transformation (3) one obtains an additional term in the global RCE (1a). Thus the so-called global nonisothermal RCE reads:

$$\overset{\triangledown}{\tau} + \frac{1}{\theta}\ \varphi(\tau) - \tau\ (\ln G) = 0 \qquad\qquad (5)$$

The derivation of the required **temperature equation** is possible from the entropy balance [9]. Thereby the dissipative source term Q occurs:

$$\varrho\ c\ \overset{\cdot}{T} = \nabla \cdot (\lambda\ \nabla T) + Q \qquad\qquad Q = \tau : D_p + D_e : T \left.\frac{\partial \tau}{\partial T}\right|_{B_e} \qquad\qquad (6)$$

In the following the relations (4) through (6) are specified for the limiting cases:

entropy elasticity   **energy elasticity**

i.e. $\left.\dfrac{\partial u}{\partial B_e}\right|_T = 0$ (7)   i.e. $\left.\dfrac{\partial s}{\partial B_e}\right|_T = 0$ (11)

$$(4): \quad \tau = T \left.\frac{\partial \tau}{\partial T}\right|_{B_e} \quad \longrightarrow \qquad\qquad \tau = 2 \varrho \left.\frac{\partial u}{\partial B_e}\right|_T \cdot B_e \quad \longrightarrow$$

$$G(T) = \frac{T}{T_0} G_0 \qquad (8) \qquad\qquad G = \frac{\eta(T)}{\theta(T)} \neq G(T) \qquad (12)$$

$$(5): \quad \overset{\triangledown}{\tau} + \frac{1}{\theta}\varphi - \tau(\overset{\cdot}{\ln T}) = 0 \quad (9) \qquad\qquad \overset{\triangledown}{\tau} + \frac{1}{\theta}\varphi = 0 \qquad (13)$$

$$(6): \quad Q = \tau:D = \phi \qquad (10) \qquad\qquad Q = \tau:D_p = \delta_m \qquad (14)$$

Beside the different temperature dependence of the modul (8),(12) and the forms of the RCE (9),(13), it is interesting that the dissipative source term coincides to the stress power ϕ (10) in the case of entropy-elastic fluids and to the mechanical dissipation rate δ_m (14) for energy-elastic fluids.

DISCUSSION

Relations (8)-(10) are thermodynamically consistent equations for the limiting case of entropy-elastic and (12)-(14) for entirely energy-elastic fluids. Some nonisothermal RCE's [5,6], which are derived on molecular level and include the temperature history explicitly by the rate of temperature change, are obtainable as special cases from (9). For entropy-elastic fluids, the RCE (9) shows the general way to modify isothermal RCE's of the considered group (e.g. the Leonov [10] and Giesekus [11] model) with respect to nonisothermal conditions by supplementing an additional term.
The differences in the dissipatively influenced flow behaviour of an entropy- or energy-elastic Leonov fluid are analysed in [12] for the plane Couette flow. In future special nonisothermal rheological experiments are necessary to measure the contributions of entropy and internal energy during elastic storage processes.

REFERENCES

1. Sugeng, F., Phan-Thien, N., Tanner, R.I., J. Rheol., 1987, 31, 37-58
2. Eringen, A.C., Mechanics of Continua, R.E. Krieger Publishing Company, Huntington NY, 1980
3. Matsumoto, T., Bogue, D.C., Trans. Soc. Rheol., 1977, 21, 453-468
4. Carey, D.A., Wust, C.J., Bogue, D.C., J.Appl. Polym. Sci., 1980, 25, 575-588
5. Marrucci, G., Trans. Soc. Rheol., 1972, 16, 321-330
6. Phan-Thien, N., J. Rheol., 1978, 22, 259-283
7. Gupta, R.K., Metzner, A.B., J. Rheol., 1982, 26, 181-198
8. Eckart, C., Phys. Rev., 1948, 73, 373-382
9. Stickforth, J., Rheol. Acta, 1986, 25, 447-458
10. Leonov, A.I., Rheol. Acta, 1976, 15, 85-98
11. Giesekus, H., J. Non-Newt. Fluid Mech., 1982, 11, 69-109
12. Braun, H., Friedrich, C., J. Non-Newt. Fluid Mech., 1989, 33, 39-51

THE CHARACTERISATION OF CONCENTRATED PARTICULATE SUSPENSIONS THAT EXHIBIT PLASTICITY

Brian J Briscoe* , Morad Kamyab* and Michael J Adams+
* Department of Chemical Engineering , Imperial College, London SW7 2BY, UK.
+ Unilever Research , Port Sunlight Laboratory, Merseyside, L63 3JW, UK.

ABSTRACT

The application of a simple engineering plasticity technique to the compression of a model soft particulate solid (Plasticine) showing rigid-plastic behaviour is described. The technique allows the constitutive equation and wall boundary conditions of the system to be obtained.

INTRODUCTION

There are a number of food products such as French mustard, cheeses, butter and chocolate that comprise of concentrated particulate suspensions. A characteristic feature of these soft solids is that they exhibit a yield stress. Food rheologists have been concerned with determining their constitutive relationships as a basis for understanding their tactile and masticatory properties. A review on the subject has been published by Sherman (1). We are all familiar with the difference in texture of Cheddar and German Loaf cheese but, in practice, quantifying these differences in a fundamental way is problematic. The appropriate techniques for studying rigid-plastic solids have been most fully developed in metalworking. To exemplify this approach we will consider an analysis for the compression of cylinders of Plasticine which involves similar experimental procedures to those described by Sherman for cheese (1).

EXPERIMENTAL AND DATA ANALYSIS

The compression of cylindrical specimens has been studied in considerable detail as a means of characterising the deformation of metals; the procedure is commonly termed as 'upsetting'. Here, we will consider the results of an equilibrium stress analysis due to Siebel (2). This analysis is based on two main assumptions: the material is a rigid-plastic solid and the deformation is homogeneous (i.e. the geometric shape of radial volume elements is preserved). In addition, explicit wall boundary conditions are required which are not established for pastes. We will introduce a general friction coefficient ψ which may be of the Tresca type, ($\psi = m = \tau' / k : \tau'$ is a constant wall shear stress; k is the bulk shear yield stress; m is a constant friction factor.), or Coulombic ($\psi = \mu \sqrt{3} = \tau''/ \sigma_z : \tau''$ is a radially varying wall shear stress ; σ_z is the corresponding local normal stress; and μ is the conventional coefficient of friction.).The upper limits on ψ ($0 < m < 1, 0 < \mu < 0.577$) correspond to 'no slip' boundary conditions. The mean normal wall stresse $\overline{\sigma}_z$ is then given by

$$\overline{\sigma}_z = \sigma_0 [1 + (\psi \, do / ho \, 3 \, \sqrt{3}) (1 - \varepsilon)^{- 3/2}] \qquad (1)$$

where σ_0 is the uniaxial flow stress, do and ho are the initial diameter and height of the specimen and ε is the engineering strain .

Figure 1a shows $\overline{\sigma}_z$ as a function of the aspect ratio do/ho for a number of strains, obtained by compressing Plasticine discs lubricated with 200cs silicone oil (PDMS). The measurements were performed at 21.5 °C and a constant uniaxial strain rate of $0.1s^{-1}$. At the lower aspect ratios and strains, the data conform to equation 1. The deviations arise from the breakdown of the lubricant film. The corresponding data for unlubricated specimens are shown in figure 1b and the deviations are more pronounced. The flow stresses obtained by extrapolating to a zero do/ho ratio are plotted for both the lubricated and the unlubricated conditions as a function of strain in figure 2a. The data corresponds to rigid-plastic behaviour with strain hardening. The corresponding computed ψ values are plotted in figure 2b. For strains of >35%, the ψ values can be regarded as extrapolation parameters since in this region the deformation was non-homogeneous which was observed as "barrelling" due to stick, rather than slip, at the platens.

DISCUSSION AND CONCLUSION

The influences of friction and aspect ratio on the compression of discs composed of a soft particulate solid are described. A simple engineering plasticity method can quantitatively account for these effects provided that the deformation is homogeneous and the boundary conditions are prescribed.

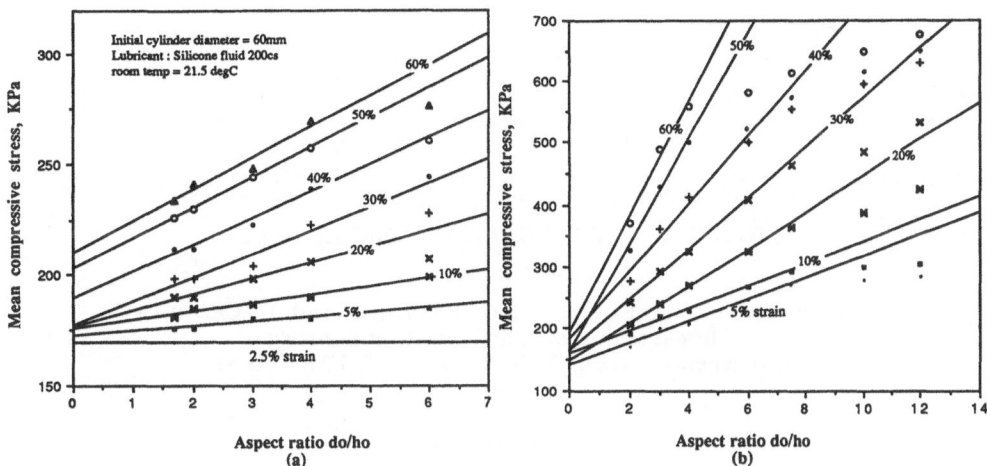

Figure 1. The mean pressure as a function of aspect ratio at different strains for (a) lubricated and (b) unlubricated Plasticine discs at a strain rate of $0.1s^{-1}$.

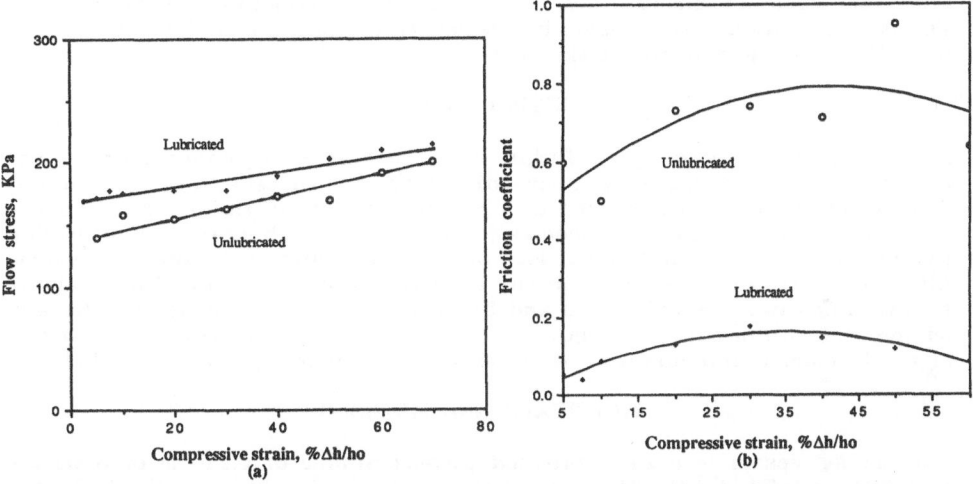

Figure 2. (a) The stress-strain curves at a zero aspect ratio obtained from figure 1. (b) The friction coefficients calculated from figure 1 as a function of strain.

REFERENCES

1. Sherman, P. , Rheological evaluation of the textural properties of foods. Prog. Trends Rheol. , 1988, II, pp. 44-53.

2. Van Rooyen, G.T. and Backofen, W.A., A study of interface friction in plastic compression. Int. J. Mech. Sci. , 1960, Vol. I, pp. 1-27.

MIXING IN RHEOLOGICALLY COMPLEX FLUIDS

E. BRITO-DE LA FUENTE, J.C. LEULIET* and L. CHOPLIN
Department of Chemical Engineering
Université Laval, Québec, QC, G1K 7P4, CANADA

ABSTRACT

In this study, the effect of elasticity on both power consumption and mixing and circulation times is investigated for the case of an helical ribbon impeller. In order to separate the effect of shear-dependent viscosity from the effect of elasticity, we used an elastic-constant viscosity fluid. The main results are that elasticity increases power consumption but it does not affect mixing and circulation times.

INTRODUCTION

Mixing rheologically complex fluids is increasingly important since those fluids are encountered in a great diversity of processes and industries. Several rheological complexities can be found in those fluids like a shear-dependent viscosity and a viscoelastic behavior, and their effects on mixing performance have been the subject of only few studies in the last years. Although it has been reported in the literature that elasticity affects mixing by altering power requirements and by modifying flow patterns and thereby mixing and circulation times, however, more experimental data are needed in order to clarify and quantify these effects on mixing performance [1].

MATERIALS AND METHODS

The mixing vessel is a flat-bottomed glass cylinder of 0.210 m of diameter and 0.435 m of height. The ratio between the height of the liquid in the vessel and the height of the helical ribbon was equal to 1.14. The impeller diameter was equal to the pitch (0.185 m). The torque, T, is measured by a non-contact strain gauge torquemeter (Mod. MCRT-24002T, S. Himmelstein Co.). Mixing time (defined as the time needed to reach a specified degree of homogeneity or the time required for a tracer to disperse) was measured by the thermal technique [2], using a thermistor thermometer and two thermistor probes placed at fixed positions inside the vessel. The mixing time, t_m, is considered to be reached when:

$$\forall t \geq t_m : |\theta(t) - \theta(f)| \leq \Delta\theta\infty/2 \qquad (1)$$

* Permanent address: INRA/LGIA, 369 Jules Guesde, 59650 VILLENEUVE D'ASCQ, FRANCE.

where $\theta(t)$ is the temperature at time t, $\theta(f)$ the final temperature and $\Delta\theta\infty$ the difference between $\theta(f)$ and the initial temperature. Circulation time, t_c, (related to the pumping capacity of the impeller) was estimated directly from the periodicity of the thermistor response curve.

Two newtonian solutions consisting of mixtures of polybutene (MW: 1300; Parapol 1300 - Exxon Chemical Co.) and kerosene were used together with an elastic-constant viscosity fluid obtained by adding 0.05% (w/w) of polyisobutylene (MW: 1.2×10^6; Vistanex L-120, Exxon Chemical Co.) to one of the Newtonian solutions. The amount of PIB was chosen in order to ensure that the viscosity, η, remains essentially unchanged. All the rheological properties were determined with a Rheometrics Inc. System IV rheometer in the fluid mode with cone and plate geometry at 20°C. For all fluids, the viscosity was found to be independent of shear-rate (the values are given in Table 1). For the PIB-containing solution, the first normal stress difference, N_1, was found to be proportional to the square of the shear-rate in the covered range.

TABLE 1 - Fluids used and viscosity at 20°C (% in w/w)

Fluid No. 1	Polybutene 97% + Kerosene 3%	η = 52.4 Pa.s
Fluid No. 2	Polybutene 91% + Kerosene 9%	η = 9.92 Pa.s
Fluid No. 3	Fluid No. 2 + PIB 0.05%	η = 10.04 Pa.s

In order to characterize the fluid's elasticity, we used the Weissenberg number, W_i, defined by:

$$W_i = \Psi_1 N/\eta \tag{2}$$

where Ψ_1 is the first normal stress coefficient and N is the rotational speed.

RESULTS

Power consumption

Figure 1 shows that the Chavan and Ulbrecht's prediction [3] for power consumption is in good agreement with our Newtonian experimental data. The elasticity of fluid No. 3, increases the power consumption. This augmentation is also observed for the torque number, Γ, when this is expressed as a function of the Reynolds number, Re (see Figure 2). Collias and Prud'homme [4] reported a similar effect using a Rushton turbine.

Mixing and Circulation times

Experimental mixing times for Newtonian and elastic fluids, in the laminar region, were fitted by:

$$N t_m = 55 \pm 22 \tag{3}$$

Although the scatter of the data may be considered important, however, it may be explained on the basis of the imprecisions found in measuring the temperature of equilibrium and therefore in calculating $\Delta\theta\infty$. Considering the low values of Re covered (Re < 6), the fit is reasonable. Using the criteria proposed by Hoogendoorn and den-Hartog [2], the scatter is greater.

No significant difference in the behavior of Newtonian and elastic fluids regarding the ratio: mixing time/circulation time was observed. This ratio was:

$$n_c = N t_m/N t_c = 2.4 \tag{4}$$

For Newtonian fluids, the ratio n_c found in the literature fluctuates from 3 to 4. For most of the authors [1,4,5], elasticity increases both mixing and circulation times. This increase may depend on the ratio W_i/Re. In our case, elasticity did not affect mixing time and this may be explained on the basis of the lower W_i/Re ratios used in this work (El = W_i/Re = 0.0046).

REFERENCES

1. Ulbrecht, J.J. and Carreau, P., in "Mixing of Liquids by Mechanical Agitation", Vol. 1, Chapter 4, J.J. Ulbrecht and G.K. Patterson ed., Gordon and Breach Publishers, N.Y., 1985.
2. Hoogendoorn, C.J. and den-Hartog, A.P., Chem. Eng. Sci., 1967, Vol. 22, pp. 1689-1699.
3. Chavan, V.V. and Ulbrecht, J.J., Ind. Eng. Chem. Process Des. Dev., 1973, Vol. 12, pp. 472-476.
4. Collias, D.J. and Prud'homme, R.K., Chem. Eng. Sci., 1985, Vol. 40, No. 8, pp. 1495-1505.
5. Skelland, A.H.P., in "Handbook of Fluids in Motion", Chapter 7, N.P. Cheremisinoff and R. Gupta ed., Ann Arbor Science, N.Y., 1983.

FIG. 1

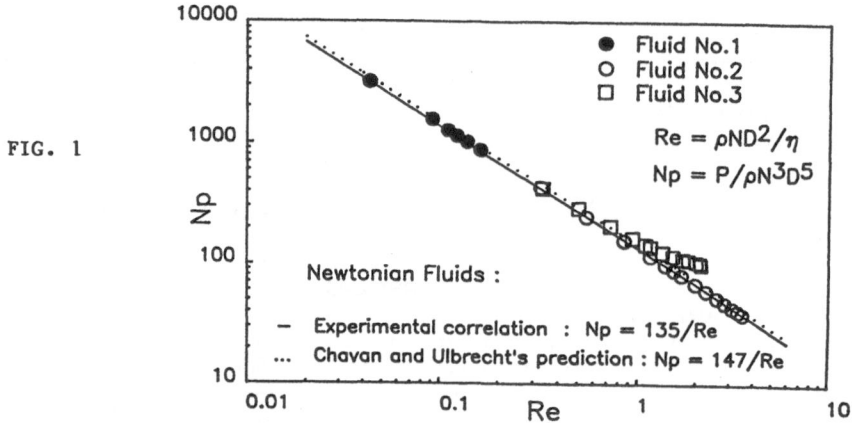

FIG. 2

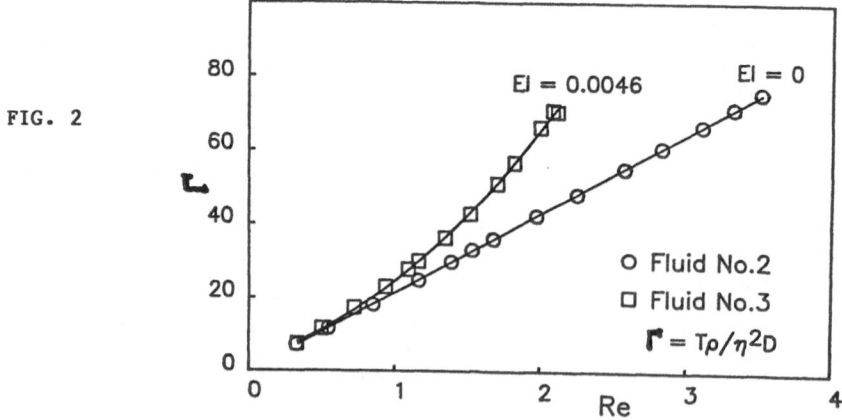

ANISOTROPIC CONDUCTION OF HEAT IN A POLYMERIC MATERIAL

B. H. A. A. VAN DEN BRULE and S. B. G. O'BRIEN
Philips Research Laboratories
P. O. Box 80000, 5600JA-Eindhoven (NL)

ABSTRACT

A model to relate the thermal conductivity tensor to the deformation of an amorphous material is presented. The basis of the theory is formed by the network theory for polymer melts and rubbers. With this model the average orientation of the molecular segments can be calculated. Combined with an expression to relate the molecular orientation to the heat flux this enables us to calculate the thermal conductivity tensor. The practical importance of these results is illustrated by a calculation of the temperature rise caused by viscous dissipation in a flowing polymer melt.

INTRODUCTION

In the past decade a number of computer programs have been developed to simulate processes like injection moulding, extrusion etc. The reliability of the numerical calculations, of course, is largely dependent on the accuracy of the model used to describe the thermal and mechanical properties of the flowing polymeric material. It is well known that the the flow of a polymer melt is very sensitive to small changes in the thermal conductivity. Most of the existing thermal conductivity data have been obtained on non-oriented samples. There is, however, experimental evidence [1-3] that molecular orientation may change the thermal conductivity in different directions by an amount which is large enough to change the flow field significantly.

THE THERMAL CONDUCTIVITY TENSOR

Recently, a model to calculate the thermal conductivity tensor as a function of the kinematics of the flow field has been developed [4]. The basis of this theory is formed by the network theory for rubbers and polymer melts [5]. Since the network structure contains all major inter-molecular interactions it is assumed that it is sufficient to consider the network structure solely in order to calculate the thermal conductivity. To calculate the energy transport through the network it is necessary to make an assumption to relate the energy flow through a segment to the macroscopic temperature gradient. In this paper we assume that the net amount of energy ϕ flowing between two junctions is proportional to the temperature difference between the junctions and inversely proportional to the length Q of the segment connecting the junctions. Analogous to the derivation of the stress tensor, the thermal conductivity tensor can be obtained. The result is

$$\underline{\underline{\lambda}} = \alpha \sum_i n_i < \underline{Q}\, \underline{Q} >_i, \tag{1}$$

where n_i is the number density of i-segments and the brackets denote an averaging wih respect to the distribution of the i-segments. If, for simplicity, we assume the network to consist of one type of segment only and with constant creation and loss rates g and h, the expression for the thermal conductivity tensor becomes

$$\underline{\underline{\lambda}} = \lambda \int_{-\infty}^{t} h\, e^{-h(t-t')} \underline{\underline{R}}(t,t') . \underline{\underline{U}}(t,t') . \underline{\underline{R}}^T(t,t')\, dt', \tag{2}$$

where $\underline{\underline{R}}$ and $\underline{\underline{U}}$ are obtained by a decomposition of the deformation gradient. The isotropic value of the thermal conductivity is given by $\lambda = g\alpha<Q>/3h$.

VISCOUS HEATING IN SHEAR FLOW

To demonstrate the influence of anisotropic heat conduction we consider a polymeric liquid flowing between two infinitely large parallel plates a distance $2a$ apart. The temperature of the walls is kept constant at T_0. For the type of network assumed above the equation for the extra stress becomes identical to the upper convected Maxwell model with a relaxation time $1/h$ and a viscosity η. Both the viscosity and the relaxation time are functions of the temperature. For convenience we introduce the following dimensionless parameters:

$$y' = y/a \;\; ; \;\; u' = u/U \;\; ; \;\; T' = (T-T_0)/T_0$$
$$\eta' = \eta(T)/\eta_0 = e^{-\beta T'} \; ; \;\; h' = h(T)/h_0 = e^{\beta T'}$$
$$\nabla p' = -a^2 \nabla p/(3\eta_0 U) \;\; ; \;\; \kappa = \lambda_{yy}/\lambda \tag{3}$$

where the subscript o indicates the reference temperature T_0 and U is the average velocity of the fluid. The equations describing the velocity and temperature distribution are then

$$du'/dy' = -3y' \nabla p'/\eta' \tag{4a}$$

$$\nabla p' = \left(3 \int_0^1 y'^2/\eta' dy' \right)^{-1} \tag{4b}$$

$$\kappa = \int_0^{\infty} e^{-\xi} \frac{1}{\sqrt{1 + \xi^2 \Gamma^2/4}}\, d\xi \;\; , \;\; \text{with} \;\; \Gamma = \frac{1}{h'}\frac{du'}{dy'}\, \text{De} \tag{4c}$$

$$\frac{d}{dy'}\left(\kappa\,\frac{dT'}{dy'}\right) = -Br\;\eta'\left(\frac{du'}{dy'}\right)^2 \qquad\qquad (4d)$$

where the Deborah and Brinkman number are defined as De = $U/(ah_0)$ and Br = $\eta_0 U^2/\lambda T_0$. For a given temperature profile the velocity distribution can be calculated from (4a-4b). Once, the velocity profile is known a new temperature distribution can be calculated with (4c-4d). In this way the system can be solved by numerical iteration. The results are presented in the figure.

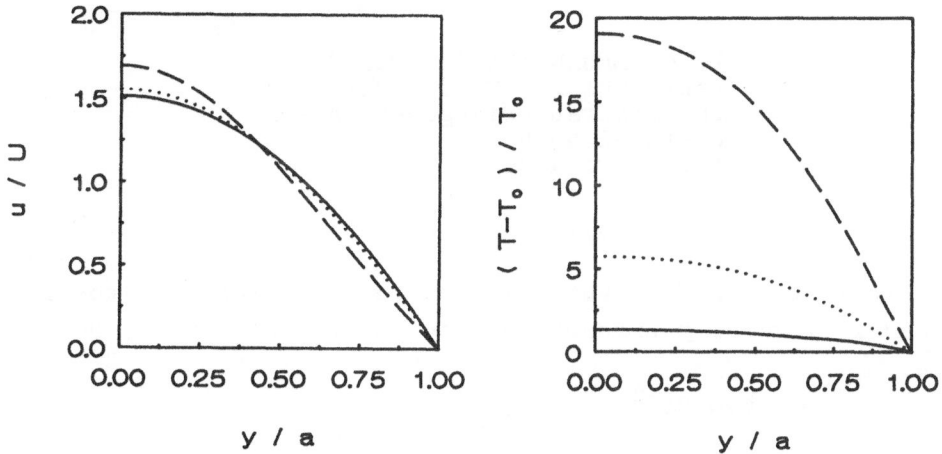

In these calculations the flow rate, and thus the Brinkman number, is kept constant; Br=1. The relaxation time of the fluid and thereby the Deborah number has been varied.(full line De=1, dotted line De=10, dashed line De=100). The temperature coefficient is chosen to be β=0.05. It can be seen that molecular orientation has a significant effect the temperature distribution. This can be understood if one realizes that orientation occurs primarily in the region near the wall and thus more or less insulates the flow from the environment. It is also illustrative to look at the pressure gradient necessary to maintain the flow rate constant at different Deborah numbers: De=1: $\nabla p'$=0.96; De=10: $\nabla p'$=0.87; De=100: $\nabla p'$=0.61. This, of course, is explained by the fact that at high Deborah numbers the temperature of the fluid is higher thus lowering the viscosity.

REFERENCES

1. Knappe, W., <u>Adv. Polym. Sci.</u>, 1971, **7**, 477.
2. Tautz, H., <u>Exper. Tech. der Phys.</u>, 1959, **7**, 1.
3. Hands, D., <u>Rubber Chem. Technol.</u>, 1980, **53**, 80.
4. van den Brule, B.H.A.A., <u>Rheol. Acta</u>, 1989, **28**, 257.
5. Bird, R.B., Curtiss, C.F., Armstrong, R.C., Hassager, O., <u>Dynamics of Polymeric Liquids (vol.2).</u> John Wiley & Sons, New York, 1987.

THE FLOW OF SURFACTANT SOLUTIONS THROUGH POROUS MEDIA: UNIVERSAL LAWS

P. O. BRUNN AND J. HOLWEG
Lehrstuhl für Strömungsmechanik
Universität Erlangen-Nürnberg
Egerlandstraße 13
D-8520 Erlangen

INTRODUCTION

For laminar tube flow the friction factor f for a Newtonian fluid is inversely proportional to the Reynoldsnumber Re, i.e.

$$f \propto Re^{-1}, \text{ laminar} \tag{1}$$

On the other hand, if the flow is turbulent, then the rough ness of the tube is important. For highly turbulent tube flows f becomes a function of rough ness only, such that f approaches a constant for tubes with "appreciable" rough ness,

$$f = \text{const, highly turbulent.} \tag{2}$$

Applying the tube flow model to the flow through packed beds implies the validity of equation (1) for $Re \ll 1$, while equation (2) should result for $Re \to \infty$. Since experiments are performed only for certain (finite) ranges of Re a correlation of the form

$$f = C_1/Re + C_2 Re^{C_3}, \tag{3}$$

with $C_3 \to 0$ for $Re \to \infty$ seems appropriate. For the 8 decades, $10^{-5} \leq Re \leq Re^3$, our own experimental results are best fitted by the choice C_1=181, C_2=2.01 and C_3=-0.04. Within the Reynoldsnumber range indicated, this is the mastercurve for the flow in packed columns of an incompressible newtonian fluid

A typical flow curve for the flow of a surfactant solution is shown in figure 1.

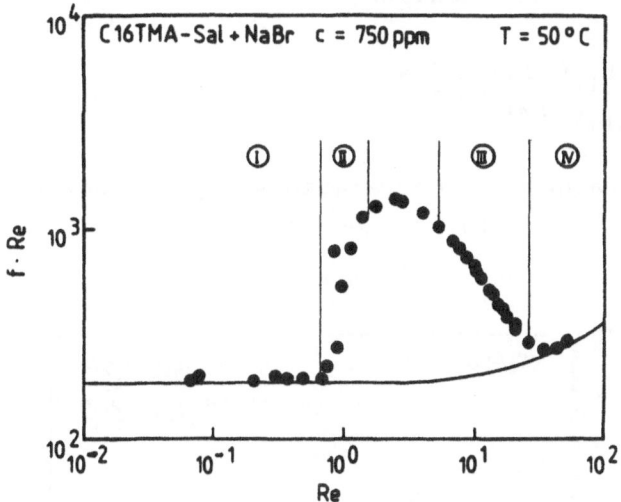

Figure 1: The flow of the equimolar surfactant solution C16TMABr + NaSal, through a porous medium

It becomes quite clear from that graph that the correlation eq. 3, holds in regions labeled I and IV, respectively. In region II as well as in region III non-newtonian behavior quite clearly manifests itself. For what is to follow we shall assume that time constants, termed λ_{II} and λ_{III}, somehow characterize non-newtonian effects. Scaling λ_i, i=I or II, with the time scale characteristic of fluid inertia /1/, leads to the (first) elasticity number El_i as dimensionless parameter for the non-newtonian effects observed in regions II and III, respectively.

Using dimensional reasoming shows that the following should
be true:

a) Re $= F_{II}(El_{II})$ in region II, and   (4)

b) $fRe^2 = F_{III}(El_{III})$ in region III,   (5)

where the F_i, i=I or II, are unknown functions. Determining
the time constants from results valid for simple shear
leads to power law behavior,

$$F_{II}(x) \propto x^{-1} \text{ and}$$
$$F_{III}(x) \propto x^{0.1}.$$

These are the desired correlations. Figure two shows the
results   for region II.

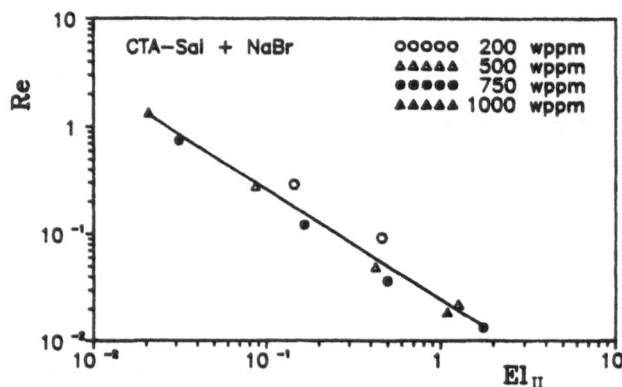

Fig. 2: The mastercurve Re=const.El_{II} for region II.
Increasing temperature (at one concentration)
leads to decreasing El_{II}

REFERENCES

/1/ Brunn, P.O., Classification scheme for suspensions
with particular emphasis on tube flow. In Recent
developments in structured continua, eds.
D. De Kee & P.N. Kaloni, Pitman Research Notes in
Mathematics, Series 143, Longman House, Harlow 1986,
pp. 71-104.

THE RHEOLOGY OF WEAKLY-ATTRACTING COLLOIDAL PARTICLES AT HIGH CONCENTRATIONS

RICHARD BUSCALL and IAN J. McGOWAN
ICI Corporate Colloid Science Group,
The Heath, Runcorn, Cheshire WA7 4QE

INTRODUCTION

A central aim in dispersion rheology is to understand and quantify the effects of particle interactions on the bulk properties of concentrated colloidal dispersions. Progress towards this aim depends in part upon obtaining rheological data for model dispersions that are well-characterised enough for the particle interaction potentials to be estimated theoretically. Much of the recent work in this vein has been concerned with systems with purely repulsive interactions. Thus a fairly substantial body of data now exists concerning the shear viscosity and linear viscoelasticity of model systems for which the interactions can be regarded as hard sphere-like. There are also available data illustrating the effect of soft electrostatic repulsion on the elasticity and flow of ordered latex.

In the case of attractive forces it is helpful at the outset to distinguish between weak attraction of the type associated with a secondary-minimum in the DLVO potential, or with depletion flocculation, and the strong van der Waals attraction experienced by fully-destabilised particles in the primary minimum. This distinction is helpful if not essential, in part because there are problems with arriving at even an order-of-magnitude estimate of the van der Waals attraction between particles in "contact", and also because there are expected to be qualitative as well as quantitative differences in rheology between systems where the attraction is weak enough to cause reversible aggregation, and those where it is strong enough for diffusion to be eliminated entirely. This paper, which draws on data from three recent studies, is concerned with the effect of weak attractive forces on the rheology of concentrated latex. In one case (SM hereafter) the attraction is van der Waals attraction in the secondary-minimum [1,2,5], in the second (DF) the attraction derives from the presence of added, non-adsorbing polymer (depletion flocculation) [3], in the third (B) it is due to polymer-bridging [4]. A more detailed discussion can be found elsewhere [5].

STORAGE MODULUS

Here we re-examine the scaling of the storage modulus data of Partridge et al. **[1,2]** who measured the limiting storage modulus G_∞ for three latices having particle diameters of 0.97, 1.4 and 1.92 μm and for volume-fractions between 0.3 and 0.6. The calculated attractive well-depths were proportional to the particle diameter and between –7 and –14 kT. The experimental data for each latex were compared with predictions based on an equation of the type derived by Zwanzig and Mountain **[6]** for simple liquids, which reads,

$$ G_\infty = \frac{3\phi}{4\pi a^3}\left(kT + \frac{\phi}{10}\int_0^\infty g(z)\frac{\partial}{\partial z}\left[z^4\frac{\partial U}{\partial z}\right]dz \right) $$

[1]

$$ \approx \frac{3\phi^2}{40\pi a^3}\int_0^\infty g(z)z^4\frac{\partial^2 U}{\partial z^2}dz \qquad ; \quad z = \frac{2r}{d} $$

This equation relates the modulus to the pair-potential U(r) but it requires the pair distribution function g(z) which also depends upon U(r) but which is less accessible. One expedient way to proceed is to approximate g(z) by that for the hard-sphere fluid on the basis that this should be reasonable for very weak forces and/or rather high concentrations. On this basis eqtn 1 reduces approximately to,

$$ G_\infty \propto \frac{4\phi^2 + 2\phi^3 - \phi^4}{(1-\phi)^3} $$

[2]

fig.1 semi—log plot of G versus φ (inset log—log)

Figure 1 shows that this agrees well with the experimental data implying on the face of it that g_m is approximately proportional (but not necessarily equal) to g_m^{HS}. The apparent agreement should not however be taken to mean that the structure is not significantly perturbed. Over the range of volume-fraction for which G is available the data can be fitted near equally as well by a power-law using an exponent of about 3.7. The significance of this type of dependence is that it shown by particulate gels formed by the coagulation of fully-destabilised particles in the primary-minimum [7]. In systems of this latter type the attractive forces are strong enough to cause the aggregation to be fully-irreversible and there is no doubt that the short-range structure is signicantly perturbed in the sense that the particles are in strict contact with their immediate neighbours at all volume-fractions. Power-law behaviour is predicted at low to moderate volume-fractions (of order 0.1) as a consequence of the fractal microstructure, but its apparent persistence right up to close-packing has been a puzzle [7]. It is now clear that the problem is that a cross-over from a fractal microstructure to a compact microstructure will never be noticed given less than perfectly precise data. Given this the only sensible conclusion to draw is that the concentration-dependence of the modulus is not sensitive to short-range structure, nor to a notional cross-over from reversible to irreversible aggregation.

BINGHAM STRESS and VISCOSITY

The effect of attractive forces is to enhance considerably the propensity for shear-thinning. The Bingham stress (or extrapolated yield stress) might be taken as a measure of the extra stress required to maintain shear flow. The data obtained for the SM and DP systems support a very similar dependence of the Bingham stress on the strength of attraction $S \propto U_{min}$ (fig.2), viz., $\sigma_B \propto S^{1.9}$.

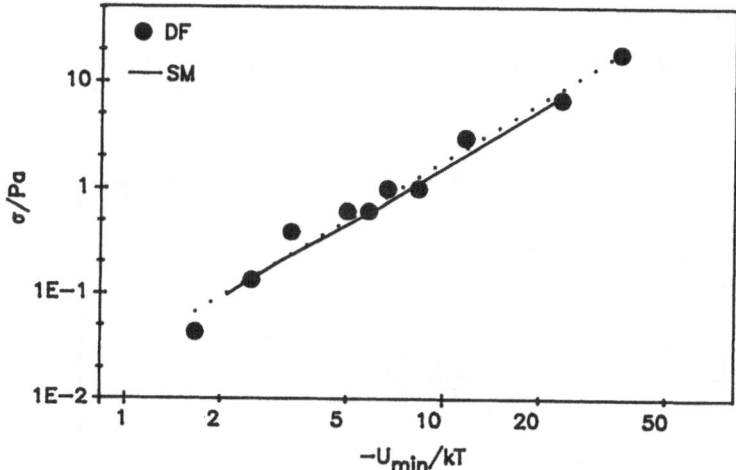

fig.2 Bingham stress versus U_{min}

Other comparisons can be made. For example Partridge determined the "zero-shear" relative viscosity for one particle-size (d = $0.97 \mu m$ and a series of volume-fractions. Similar data are also available for a similar-sized latex flocculated by the addition of sodium carboxymethyl cellulose [4], however in this case little definite is known regarding the particle-attraction except that is very weak. These results are replotted in

84

fig. 3. Also shown are data for the DF systems, these representing the effect of varying S at constant volume-fraction, whereas the other two sets illustrate the effect of varying the latter at constant S. The data are very suggestive of a single curve, if true this is a rather remarkable result. It suggests that the low-shear viscosity depends upon volume-fraction but not upon particle-size or strength of attraction (within limits). Further work is required in order to establish how general this is.

REFERENCES

1. S.J. Partridge, Ph.D. Thesis, 1985, School of Chemistry,
University of Bristol.

2. J.W. Goodwin, R.W. Hughes, S.J. Partridge and C.F. Zukoski,
J . Chem. Phys., 1986, 85, 559.

3. D. Belbin, R. Buscall, C. A. Mumme-Young and J. M. Shankey,
Plastics. Rubber. Inst. preprint,(1985), 7, 1.

4. R. Buscall and I.J. McGowan, Faraday Discuss. Chem. Soc.,1983, 76, 277.

5. R. Buscall and I.J. McGowan, Faraday Discuss. Chem. Soc.,1990, 90,
to be published.

6. R.W. Zwanzig and R.D. Mountain, J. Chem. Phys., 1965, 43, 4464.

7. R. Buscall, P.D.A. Mills, J.W. Goodwin and D.W. Lawson,
J. Chem. Soc. Faraday Trans. 1, 1988, 84, 4249.

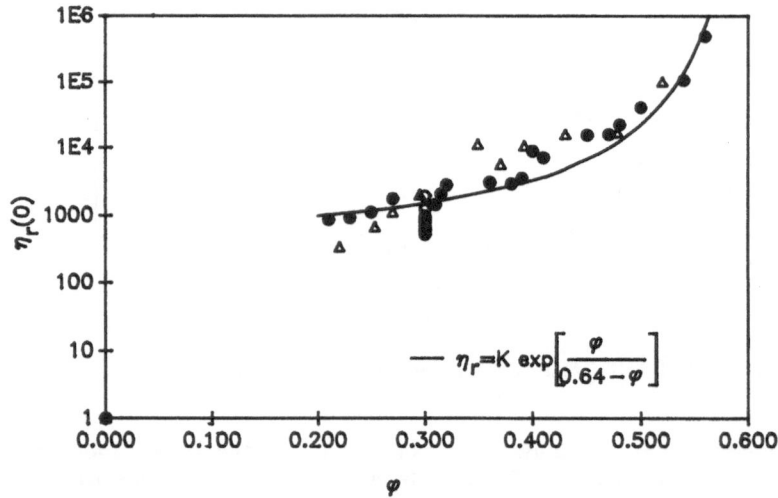

fig.3 low shear viscosity vs vol. fraction

The Rheological Properties of Extrusion-cooked Wheat Starch in the Moisture Range 27-36%

J Castle, R C E Guy, P A P Hastilow *, D A Janes.
F M B R A Chorleywood.Herts
* South Bank Polytechnic London SE1

Introduction

The rheological properties of extrusion-cooked wheat starch have been measured in parallel experiments using an instrumented slit die on a twin-screw extruder and a Bohlin VOR rheometer,with parallel plate geometry in oscillatory mode, over the frequency range 0.2-20 Hz. Shear rates approaching those used in the production of snack foods and crispbreads have been achieved,for melt samples of wheat starch over a range of moisture contents, with the aim of correlating on-line viscometric measurements with laboratory measurements on a research rheometer.

Previous work (1), (2), (3), has shown that starch melts behave as a pseudoplastic material with a power law behaviour, and indices between 0.2 and 0.6. These were all in-line measurements with capillary or slit dies. Han (4) has shown that the data from slit dies can also enable the normal stress difference to be calculated from the extrapolated exit pressure.

Experimental

Preliminary experiments using a short slit die 30mm width 3mm depth and 10mm length showed that samples in the form of flat strips could be obtained. Discs punched from these were suitable for use on the Bohlin rheometer, provided that the temperature had not exceeded 90°C and no expansion had occurred. Initially an instrumented capillary die 12.mm diameter and 300 mm length was used to measure the rheological characteristics of the starch for moisture contents 25% and 30%.and shear rates from 26 to 52 sec^{-1}. Based on these results a new set of slit dies was designed to cover the shear rate range from 50 to 200 sec^{-1} using a single feed rate of 400 g/min. This was to enable a consistent pressure profile to be maintained along the screw so that material entering the die would have been subjected to the same shear history for all the shear rates used.

A factorial design was selected to cover the shear rates from 50 to 200 sec^{-1}, moistures 27%, 30%, 33%, and 36% wb and screw speeds 175, 200, 225, and 250 rpm. All samples were produced from ABRASTARCH (supplied by ABR Chemicals)using a Baker Perkins MPF50D Twin screw Extruder of L:D ratio 15:1. The shear rate range was achieved by using a 30 mm wide slit die with interchangable inserts giving die depths of 2mm, 3mm, and 4mm repectively all had a parallel length of 150mm a 45^0 conical entry and were threaded to take pressure/temperature probes at 1, 5, and 11 mm from the exit plane of the die. Additional probes were also placed along the extruder barrel and at the tip of the screw. The pressure and temperature at these points were monitored together with data on screw speed, water and starch feed rates, motor torque and temperature at the die exit. When stable extrusion conditions had been achieved signals from all these were logged at 2 Hz for a period of 30 seconds onto an 80286-based PC-compatible microcomputer, through a 24 channel analogue to digital converter.

To enable parallel measurements to be made on both the extruder and the Bohlin rheometer the latter was set up within 1m of the extruder and a belt haul-off was used to collect the extrudate. A short sample was cut from the extrudate and a 25mm diameter disc was punched out and transferred rapidly to the pre-heated rheometer. The dynamic viscosity η' was determined by sweeping over the frequency range 0.2-20Hz whilst the pilot plant data was logged. The sample was inserted into the rheometer and tested within 2 minutes of removal from the extruder. This timing was a compromise between allowing sufficient time for sample relaxation and minimising specimen moisture loss.

During the experimental runs it was apparent that considerable viscous shear heating was taking place in the dies and that the full range of experiments could not be achieved without producing expanded products which would be unsuitable for rheometeric measurements. Therefore the data for some of the conditions selected were collected using auxiliary cooling of the dies by iced water/cloths. This was not satisfactory and some sample temperatures reached 120°C. To achieve better temperature control would probably need a die with coring for chilled cooling water . Even then it would possibly only work for the narrower dies since poor internal heat transfer can produce large temperature gradients in the emerging extrudates.

The data collected was analysed to produce the apparent viscosity as a function of shear rate from the slit dies and the in-phase dynamic viscosity from the Bohlin data, additionally the normal stress in the die was estimated from the exit pressure. Some specimen data for the samples of moisture content 36% wb formed by extrusion at a screw speed of 200 rpm show that the wheat starch is pseudoplastic and may be described by the equation

$$\eta = \eta_0 \gamma^{n-1}$$

with $\eta_0 = 100$ kPas for extrusion data and
$\eta_0 = 190$ kPas for the Bohlin Rheometer data
Both sets of data suggested a power index of 0.1 .

References.

1. Guy,R.C.E.and Horne,A.W.,Extrusion and co-extrusion of cereals.In Food Structure, its Creation and Evaluation, eds J.M.V.Blanshard, and J.R.Mitchell,Butterworth Press 1988
2. McMaster,T,J.,Senouchi,A.,   and   Smith,A.C.,Measurement   of rheological and ultrasonic properties of food and synthetic polymer melts. Rheol Acta 1987 26 308-315.
3. Padmanabhan,M., and Bhattacharya,M., Analysis of Pressure Drop in Extruder Dies.Journal of Food Science 1989 54 709-713.
4. Han,C.D.,Slit rheometry.In Rheological Measurement eds A.A.Collyer and D.W.Clegg,Elsevier Applied Science Publishers, London 1988.

FURTHER INVESTIGATIONS ON THE MODELLING OF GRAVITY-DRAWN BOGER FLUID JETS

M.S. CHAI and Y.L. YEOW

Department of Chemical Engineering, University of Melbourne, Parkville, Victoria 3052, Australia

ABSTRACT

The flow field generated by a viscoelastic jet stretching under its own weight has been used to study the elongational behaviour of a number of constant viscosity elastic fluids (Boger fluids). Previous studies have shown that the Oldroyd-B model failed to describe the jets adequately. An attempt is made here to evaluate the performance of a number of more complicated constitutive models in predicting the jet profiles. The models reported in this study include the FENE-P, PTT, Giesekus and multiple-relaxation-time (MRT) KBKZ equations. As in the earlier studies, the fluid parameters are estimated from the measured shear properties and are used to predict the jet profiles. The jet profile calculation is simplified by making the standard one-dimensional approximation. The use of the MRT KBKZ-type constitutive equation for modelling Boger fluids is recommended in preference to the more traditional Oldroyd-Maxwell and other single-time-constant equations.

INTRODUCTION

Because of their unique rheological characteristics, Boger fluids provide useful insights into the elongational behaviour of viscoelastic fluids in complex flow fields, whereby observation of elastic effects can be made without the added complication of significant shear-thinning and inertial effects (see [1] for an up-dated review of works carried out on this class of fluids). The Oldroyd-B constitutive equation is capable of providing quite a satisfactory description of the rheological properties of Boger fluids in shear flow. However, it is less successful in predicting their elongational behaviour. Earlier, it has been shown that the Oldroyd-B equation performs better than the Newtonian and Maxwell constitutive equations in modelling the gravity-drawn jets of Boger fluids, but a substantial discrepancy still exists between numerical simulations and experimental observations [2]. This has prompted a search for constitutive equations that will provide a more satisfactory description of the shear and elongational properties of Boger fluids. There are a number of suggestions in the literature that the use of constitutive equations with a spectrum of relaxation times may be a good first step towards this goal [3].

This paper explores the use of a well-tested MRT KBKZ-type equation [4] in modelling the elongational behaviour of Boger fluids. For comparison, the results obtained from the Oldroyd-B, FENE-P, PTT and Giesekus equations are also included. Full details of this investigation and results on other constitutive equations not included in this paper can be found in [1] and [2].

TEST FLUID AND GRAVITY-DRAWN JET EXPERIMENT

The preparation and characterization of Boger fluids are well documented in the literature [1,2] and will not be repeated here. The relevant fluid parameters of one such fluid (0.31%PIB/4.83%Tetradecane/94.86%PB) are given in Table 1. It is noted that ε, the additional elongational parameter of the PTT, and β, that of the KBKZ equation, require elongational viscosity data. The values of these two parameters listed in Table 1 are reasonable estimates based on experimental data for a similar Boger fluid [1]. The main measurement of the gravity-drawn jet experiment is the shape of the jet at different flow rates, obtained by a computerized digitization technique. The measured shape is compared against those computed from various assumed constitutive equations. For experimental details, see [1] and [2].

TABLE 1. Parameters of various rheological equations for Boger fluid*

Equations	Fluid Parameters
Oldroyd-B	$\eta_o = 16.57\,Pas,\ \lambda_1 = 0.721\,s,\ \lambda_2 = 0.652\,s$
FENE-P	$b = 400,\quad \eta_o, \lambda_1, \lambda_2$ as above.
PTT	$\varepsilon = 0.0002,\ \zeta = 0.002,\quad \eta_o, \lambda_1, \lambda_2$ as above.
Giesekus	$\alpha = 0.003,\quad \eta_o, \lambda_1, \lambda_2$ as above.
KBKZ	$a_1 = 0.84\,Pa,\quad \lambda_1 = 2.3\,s$ $a_2 = 11.81Pa,\quad \lambda_2 = 0.21\,s$ $a_3 = 3.73E4\,Pa,\quad \lambda_3 = 3.0E-4\,s$ $\alpha = 2000,\quad \beta = 0.001$
* $\rho = 890\,kg/m^3$	

FULLY-DEVELOPED JETS

The form of the KBKZ-type equation adopted in this study is similar to that proposed by Papanastasiou *et al.* [3]. The extra stress $T(t)$ is given by:

$$T(t) = \int_{-\infty}^{t} \left[\sum \frac{a_k}{\lambda_k} \exp\left(-\frac{t-t'}{\lambda_k}\right) \right] \frac{\alpha}{(\alpha-3) + \beta I + (1-\beta)II} B_t(t')dt', \qquad k = 1,2\ldots..n. \tag{1}$$

In this study, the use of $n = 3$ was found to be adequate. The jet problem is solved using the standard one-dimensional approximation, in which the origin of the fully-developed jet is taken to be at two tube diameters downstream of the exit plane of the capillary. For steady-state jets, the integral through past time of eqn. (1) can be transformed into an equivalent spatial form [3]. To solve the equation of motion, it is necessary to specify the upstream kinematics of the jet. In this study an elongational history of the form $U(Z) = U_0 \exp(CZ)$ has been assumed. A finite difference scheme is used to evaluate the velocity and stress at regular intervals along the jet [1].

Two sets of the measured jet profiles and predictions of the KBKZ and other constitutive equations are shown in Figs. 1 and 2. The experimental conditions are indicated in the figures. These results are typical of the large number of comparisons carried out for four different Boger fluids each over a range of flow rates [1]. Clearly, the jet profiles predicted by the KBKZ equation are in better agreement with the experimental measurements, indicating that this equation is a more promising candidate for modelling Boger fluids in any future numerical simulation of viscoelastic flows.

90

References

1. Chai, M.S., *Ph.D. Thesis*, University of Melbourne, 1990.
2. Chai, M.S. and YEOW, Y.L., J. Non-Newtonian Fluid Mech., 1988, **29**, 433-442.
3. Petrie, C.J.S., *"Elongational Flows"*, London, Pitman, 1979.
4. Papanastasiou, A.C., Scriven, L.E. and Macosko, C.W., J. Rheol. 1983, **27**, 387-410.

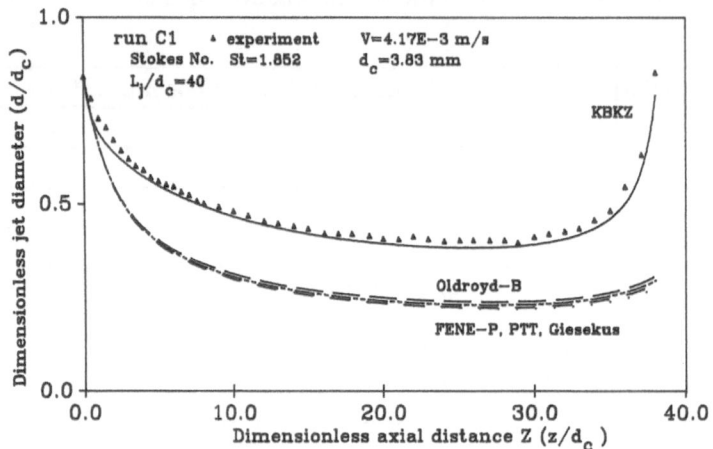

Figure 1 : Fully–developed jet profiles: experiment and predictions.

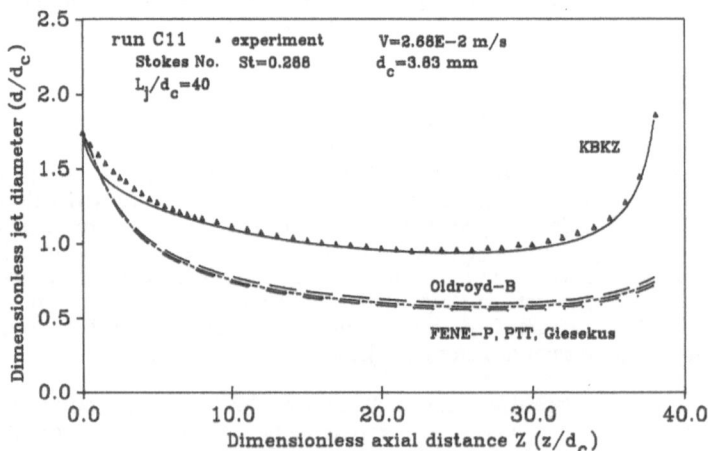

Figure 2 : Fully–developed jet profiles: experiment and predictions.

VISCOMETRIC PROPERTIES OF A DENSE AQUEOUS SILICA SLURRY CONTAINING IRON(III) AND TIN(IV) OXIDE

H.D. CHANDLER and R.L. JONES
School of Mechanical Engineering & Dept. of Biochemistry,
University of the Witwatersrand, Johannesburg,
P O Wits, 2050, South Africa

ABSTRACT

The present study involves the determination of the rheological properties of an aqueous silica base slurry containing small amounts of iron(III) oxide and tin(IV) oxide. The results are analysed in terms of an activation energy approach. In the range of oxide concentrations used, the two metal oxides affect the flow properties independently of each other.

INTRODUCTION

In previous work on describing rheograms of non Newtonian, shear thinning mineral slurries [1,2], the flow was assumed to depend on two processes, a fluid flow term involving the probability of there being suitable spaces into which particles or groups of particles could migrate and a "solid" term depending on charge interactions between particles. Both processes occur together in a flow event so the strain rate can be expressed as :-

$$1/\dot{\gamma} = 1/\dot{\gamma}_f + 1/\dot{\gamma}_s \tag{1}$$

where $\dot{\gamma}_f$ and $\dot{\gamma}_s$ are the strain rates associated with the fluid and solid components respectively. The fluid flow component appears to be a function only of the solids volume fraction and maximum packing density. For the solid flow component, the strain rate can be expressed in terms of an attempt frequency, $\dot{\gamma}_o$, and the probability of overcoming charge interactions as :-

$$\dot{\gamma}_s = \dot{\gamma}_o exp - \frac{\Delta F}{RT} \left[1 - \left(\frac{\tau}{\hat{\tau}} \right)^P \right]^q \tag{2}$$

where τ is the shear stress, ΔF is the activation energy, $\hat{\tau}$, is a flow stress barrier and p and q represent the shape of obstacles to flow.

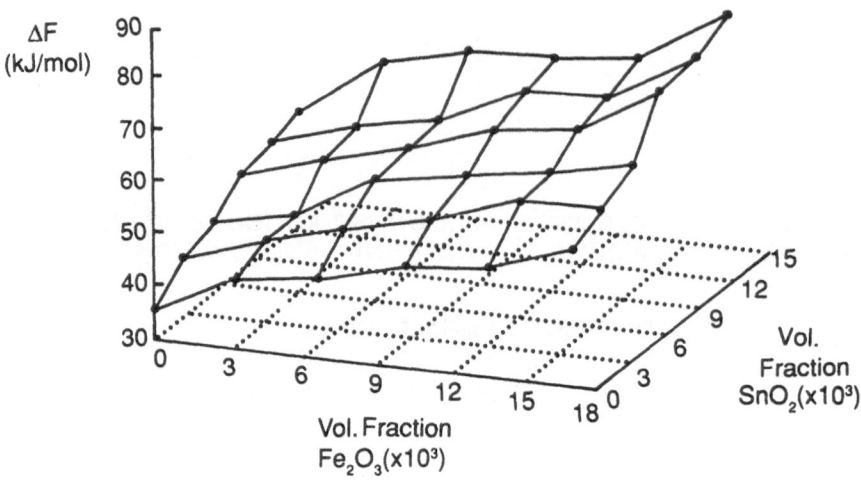

Figure 1: Plot of free energy for flow against volume fractions of iron(III) oxide and tin(IV) oxide. Values of ΔF calculated from rheograms shown as filled circles

Previous work [2] was concerned with the behaviour of slurries containing water, silica and iron(III) oxide. The present work concerns slurries containing water, 0.42 vol. fraction silica to which small quantities of iron(III) oxide and tin(IV) oxide were added. Rheograms were obtained from a Haake rotary viscometer modified so as to prevent settling of the solids during testing [3].

RESULTS

Values of the material parameters and ΔF and $\hat{r}$ were found from the rheograms using eqns. (1) and (2). The value of the frequency factor $\dot{\gamma}_o$ was taken to be 10^6. Material parameters are dependent on the frequency factor and are thus relative. Values of p and q are, respectively, 1/2 and 3/2 which give linear plots of the argument of the exponential in eq. (2) to the power 2/3 vs square root shear stress.

Figure 1 shows relative values of ΔF plotted against volume fractions of iron(III) oxide and tin(IV) oxide. Experimental points show some scatter but indicate that the free energy is linearly dependent on the additive volume fractions and can be represented as :-

$$\Delta F = \Delta F_o + K_1 \phi_1 + K_2 \phi_2 \tag{3}$$

where ΔF_o represents the free energy of silica alone, ϕ_1 and ϕ_2 are the volume fractions of iron(III) and tin(IV) oxide respectively and the constants K_1 and K_2 represent the chemical potentials of the two oxides.

The flow stress values do not show any clear correlation with volume fractions of each oxide but $\ln \hat{\tau}$ appears to be a linear function of the total solids volume fraction.

The effect of adding small quantities of iron(III) oxide and tin(IV) oxide to a silica water slurry is to cause thickening. In terms of the model described, the thickening effect appears to be due to the effect of the additives on the free energy for flow. Each oxide appears to make an independent contribution to the free energy which indicates that the dominant interaction effects are due to silica - iron(III) oxide interaction and silica - tin(IV) oxide interaction with interaction between the two oxide species having a negligible effect at these low concentrations. Many slurries of technological interest consist of a major solid component with small quantities of other species and the present analysis suggests a method for predicting the performance of a multi-component system from tests on two solid component systems.

REFERENCES

1. Chandler, H.D. and Jones, R.L., An entropy approach to determining the viscosities of concentrated slurries.
   Philos. Mag. Letters, 1988, 57, 81-84

2. Chandler, H.D. and Jones, R.L., Flow kinetics in concentrated suspensions of silica, iron(III) oxide, water.
   Philos. Mag. A, 1989, 59, 645-60

3. Overend, I.J., Horsley, R.R., Jones, R.L. and Vinycomb, R.K., A new method for the measurement of rheological properties of settling slurries.
   Proc IX Int. Conf. on Rheology, 1984, 583-616

THE USE OF VISCOMETERS AS MIMICS OF FLOW PROCESSES

D C-H CHENG
Warren Spring Laboratory, Stevenage, Herts, UK
(On behalf of Fisons PLC, Scientific Equipment Division)

ABSTRACT

This paper concerns the development of guidelines for routine industrial viscosity and thixotropic measurement. It describes some practical problems encountered. Put simply, it does not seem possible to specify simple thixotropic models and reduce measurement to the determination of material parameters. For dense suspensions, the viscosity is so bound up with the concentration profile developed in viscometers that it is not currently possible to separate them. These problems are of a fundamental nature and require research still. In routine measurement, it is suggested that, for the results to be relevant and useful, viscometers should be used to simulate flow processes and the conditions of testing should closely match those of the application situation.

INTRODUCTION

The objective of this study was to determine routine testing procedures that can be implemented on commercial viscometers and rheometers. This paper discusses some practical problems in the measurement of viscosity of suspensions and of thixotropy. This involves the review of some fundamental issues such as the aims of the measurement and the assumptions that are made. Experimental and theoretical research results from the literature are referred to and the results of two new experimental studies are introduced. Details can be found in two reports quoted below [1,2].

THIXOTROPIC MEASUREMENT

An assumption in thixotropic measurement is that the sample is inelastic and that the behaviour is governed by a time-dependent structure (λ). The constitutive relationship is given by an equation of state relating shear stress (τ), to shear rate (D): $\tau = \tau(D,\lambda)$; and a rate equation, $d\lambda/dt = g(D,\lambda)$, where t is time and g() a function of D and λ. It can be represented by a contour map on a τ-D plot with constant-λ and constant-g locii. A variety of ways of measurement has been developed.

One method is to use the step-shear ráte experiment. This gives the contour map directly. The data may be presented to a computer in tabular form and used directly in flow modelling and engineering calculations. But for other applications, in product formulation and quality and process control, it is important to model fit the data to give a small number of material parameters that can be used in correlation with product composition and processing conditions, and to define acceptance limits in quality and process control.

Now, the step-shear rate experiment is too elaborate for material parameters determination. It is much simpler to use the model testing alternative. Here, the viscometric behaviour is predicted using a simple mathematical model. Then, by matching measurement results with model prediction, the model ie material parameters can be determined. The simplest model of thixotropy is the 4-constant Moore model. However, real materials are more complex than this. Alternative models, of increasing complexity can be used, such as the Cross model (5 constants), Worrall-Tuliani (5), Cheng (6), and combinations of these giving 7-constant models.

Recently I have carried out a study specifically to attempt this approach [1]. The measurements were performed on a 10% aqueous bentonite slurry. A computer controlled Haake RV20 instrument with cone and plate and coaxial cylinder sensors was used. It was found that the recovery rate was very slow and, during the normal working day, the bentonite was effectively irreversibly thixotropic and samples cannot be used indefinitely. Then results from different samples were found to show much scatter, indicating that the bulk sample was not as homogeneous as it appeared to the eye. It made the pooling of results for parameter determination unsatisfactory. Another important finding was that the constant-λ curves were not straight lines. This meant that none of the model equations named earlier were applicable. These all assume that the constant-λ curves are either Newtonian or Bingham plastic. We have to start building new models with non-Newtonian constant-λ curves.

In industry, it is desired that the model to use should be well established beforehand and the problem in routine measurement is simply to determine the model parameters using the minimal testing. It is clearly not satisfactory if the industrialist (QC chemist or design engineer) has first to determine the functional form of the model. Even if there is a list of models for him to try out, or if a systematic approach may be used to generate models of increasing complexity, the question remains about what he should do if these fail. This requires the industrialist to research into the material rheology, which is clearly not his function. The problem of thixotropic measurement is therefore still at the research stage. Until a solution is found, the industrialist requires guidance on how to proceed. It is suggested that he should use the viscometer to mimic flow processes. This is elaborated below.

VISCOSITY OF SUSPENSIONS

The viscosity of suspensions depends on the solids concentration. However, at high concentrations, the dense suspensions of industrial interest, it is well-known that the suspensions show a wide variety of complicated behaviour. In particular, the viscosity appears to depend on the geometry (shape and size) of the sensor used in measurement.

Recently, I have carried out some measurements specifically to illustrate this behaviour [2]. The sample was an industrial PVC plastisol. It was tested in a Haake RV20 instrument using a range of six coaxial cylinder and cone and plate sensors. The results show, firstly, below a critical shear rate, results for different test conditions coincided onto a master curve that showed a yield stress. But the yield stress varied systematically with the sensor used, increasing from 0.5 Pa for wide gap to 5000 Pa for narrow gap. As D was increased, results for all the sensors appeared to merge giving a Newtonian section of about 10 Pa s. At very high D above about 6000 s^{-1}, the flow curve became dilatant. Secondly, as the critical D was approached, hysteresis loops were obtained. The area increased as D was increased. On exceeding the critical D, the flow curve showed a catastrophic breakdown. The critical D varied with ramp rate: the slower the rate, the lower the value. The critical D varied with sensor gap, increasing with decreasing radius and angle.

These results dramatically showed that suspension viscosity appears to depend on sensor geometry. The reason for this behaviour lies in the assumption made in the equation for D, that the sample composition is uniform in the gap. This is clearly not valid in general for suspensions. There is much experimental and theoretical évidence for particle migration and phase separation. The catastrophic changes have the same basis as the mechanism of flow in granular solids. The measurement result is an integrated effect of the viscosity-concentration relationship, (η,c), and the concentration profile in the gap, $c(r,\theta)$. The latter varies with geometry and so the apparent viscosity changes accordingly. (We do not even need to invoke structural changes due to shear-induced agglomeration or deagglomeration to explain this).

The correct interpretation of viscometric measurement is then a question of separating (η,c) from $c(r,\theta)$. At present, we do not know how to do this in general terms. And, even if we can achieve the separation, we need to know how to predict $c(r,\theta)$ for non-viscometric flow fields before we can apply the (η,c) results. One approach would be through some kind of modelling, but much work needs to be done into the kinetics of particle migration. Again, the problem is still at the research stage.

As regards industrial measurement of suspension viscosity, it is as if, by choosing the appropriate sensor, one can obtain any result that one wishes. This is clearly not satisfactory, but what guideline can one give the industrialist? It is again suggested that the viscometer should be used to mimic flow processes. This is discussed in the next section.

USE OF VISCOMETERS AS MIMICS OF PROCESSES

Fundamental problems exist still in the measurement of thixotropy and viscosity of suspensions and also other multiphased materials including solutions of macromolecules, emulsions and other dispersions. Research is in progress, but in the meantime the industrialist (QC chemist, engineering designer, applied rheologist) needs to know how to carry out measurements that are relevant in application. It is suggested that the viscometer is to be seen, not so much as an instrument that will measure the inherent viscosity property of a sample, but rather is to be used to investigate microstructure behaviour under shear. To understand the rheological behaviour of a fluid overall, it should be tested using a variety of shearing conditions and in different viscometric geometries. For application to specific flow processes, the test conditions should match the shearing in the flow processes - in other words, the viscometer should be used to simulate or mimic flow processes. It is not possible to be more specific on exactly what to do as each application has to be considered individually. The best that can be done is to give examples. Take the roll coating of the PVC plastisol for floor covering manufacture, the viscometric gap has to match the narrow gap under the doctor blade and the critical shear rate has to be found so that it is not exceeded on the plant. Both the initial and final viscosity from the hysteresis loop are relevant, the former controls the fluid feed and passage under the doctor blade, and the latter governs the flow out and release of air from the coated layer. The subsequent levelling takes place at low shear stresses and thus low shear rate testing is also needed. Fuller discussion of the PVC processing is given in the report quoted earlier on.

REFERENCES

1. Cheng, D. C-H., Thixotropic measurement on a bentonite slurry. Res. Rept No. LR 738 (MP/BM), Warren Spring Laboratory, Stevenage, 1989.

2. Cheng, D. C-H., Viscosity measurement on a PVC plastisol as an example of dense suspensions. Res. Rept No. LR 721 (MP/BM), Warren Spring Laboratory, Stevenage, 1989.

Acknowledgement: This work was carried out under a consultancy contract for Fisons PLC, Scientific Equipment Division. The company's permission for publication is gratefully acknowledged.

FLOW OF POWER LAW FLUIDS THROUGH ASSEMBLAGES OF SPHERICAL PARTICLES

R.P. CHHABRA
Department of Chemical Engineering
Indian Institute of Technology
Kanpur, India 208016

ABSTRACT

Drag coefficients of particle assemblages have been obtained for the creeping flow of power law liquids through uniform beds of spherical particles. The theoretical results reported herein encompass wide ranges of fluid behaviour ($1 > n > 0.2$) and bed voidage ($0.99 > \epsilon > 0.3$), and these have been validated by performing detailed comparisons with the experimental results available in the literature.

INTRODUCTION

Slow flow of non-Newtonian fluids through multi-particle assemblages represents an idealisation of many industrially important processes such as filtration of polymer melts and slurries, enhanced oil recovery processes, etc. Despite its overwhelming pragmatic significance, the progress in this area has been rather slow. In modelling the fluid flow and other transfer processes in multi-particle assemblages, the main obstacle has been a realistic description of the inter-particle interactions. One approach which seeks to surmount this difficulty, and has had considerable success [1], is the so-called free surface cell model due to Happel [2]. In this model, the manybody difficult problem is replaced by a much simpler onebody equivalent. Hence this approach will be used here to take into account the inter-particle interactions. The complete governing equations for the creeping motion (Reynolds number = 0.001) of power law liquids through particle assemblages have been solved numerically. Theoretical results are expressed using the dimensionless groups such as Reynolds number, drag coefficient, bed voidage, etc.

PROBLEM STATEMENT AND FORMULATION

Consider the steady flow of an incompressible power law liquid through an assemblage of rigid spherical particles. The free surface cell model stipulates each particle (of radius a) to be surrounded by a hypothetical spherical fluid envelope of radius r_∞ such that the voidage of each cell is same as the average porosity (ϵ) of the assemblage. Happel [2] proposed the cell boundary to be frictionless thereby emphasizing the non-interacting nature of cells. The continuity and momentum equations for this flow configuration can be written as :

$$\nabla . V = 0 \tag{1}$$

$$\rho V . \nabla V = - \nabla p + \nabla . \{ \mu (\nabla V + (\nabla V)^T) \} \tag{2}$$

where for a power law liquid, $\quad \mu = K (2 I_2)^{(n-1)/2}$ (3)

The boundary conditions for this flow configuration are:

at $r = a$, $V_r = U_0 \cos \theta$ and $V_\theta = - U_0 \sin \theta$ (4a)

At $r = r_\infty$, $V_r = o$ and $\tau_{r\theta} = o$ (4b)

The cell radius r_∞ is related to a via the following equation :

$$r_\infty = a (1 - \epsilon)^{-1/3} \tag{5}$$

Equations 1 to 5 have been solved numerically using the finite element method to obtain the values of drag coefficient as a function of power law index (1 to 0.2) bed voidage (0.3 to 0.99) and for Re = 0.001.

RESULTS AND DISCUSSION

Figure 1 shows the variation of drag coefficient with n for a range of values of bed voidage. As expected for a given value of ϵ , the value of drag coefficient decreases with the decreasing values of the power law index. The present results are in reasonable agreement with both the upper and lower bound analysis of Mohan and Raghuraman [3] as well as the approximate treatment of Kawase and Ulbrecht [4] only for a small degree of deviation from Newtonian behaviour (1 > n > 0.8). For lower values of n (0.8 > n > 0.2), the present results are believed to be more accurate than those available in the literature [3,4]. Finally, the present theoretical predictions have been validated by using the literature data in the range 0.4 < n < 0.8 but only for ϵ = 0.4. On the whole, the discrepancy between this theory and experiments is of the order of 20%.

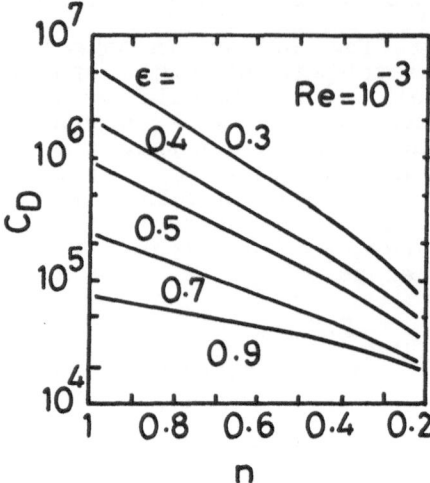

Figure 1.   Variation of drag coefficient with n and ε.

CONCLUSIONS

The free surface cell model has been used to simulate the slow flow of power law fluids through multi-particle assemblages. Extensive theoretical predictions of drag coefficient embracing wide ranges of conditions of bed voidage and fluid behaviour have been obtained and validated using the appropriate results available in the literature.

REFERENCES

1.   Prasad, D., Narayan, K.A. and Chhabra, R.P., Creeping fluid flow relative to an assemblage of composite spheres, Int. J. Eng. Sci., 1990, in press.

2.   Happel, J., Viscous flow in multi-particle systems AlChE J., 1958, **4**, 197-201.

3.   Mohan, V. and Raghuraman, J., A theoretical study of pressure drop for non-Newtonian creeping flow past an assemblage of spheres, AlChE J., 1976, **22**, 259-64.

4.   Kawase, Y. and Ulbrecht, J., Drag and mass transfer in non-Newtonian flows through multi-particle systems at low Reynolds numbers, Chem. Eng. Sci., 1981, **36**, 1193-1202.

FURTHER OBSERVATIONS OF ANOMALOUS ENTRY FLOW PATTERNS THROUGH A PLANAR CONTRACTION

KUNJI CHIBA, TORU SAKATANI, SATORU TANAKA and KIYOJI NAKAMURA
Department of Mechanical Engineering, Osaka University,
Suita, Osaka 565, JAPAN

ABSTRACT

The present paper adds further experimental evidence to show that the planar contraction flows of shear-thinning polyacrylamide aqueous solutions are far more complex than initially imagined, especially the presence of non-trivial 3-dimensional component has been confirmed for unstable flow regime, such as several bundles of stream near the side walls.

INTRODUCTION

One of the most widespread and commercially important non-Newtonian flow problems is the abruptly converging flow.

Sufficient data on entry flows has been published to demonstrate the viscoelastic effects[1,2] and it seems to be commonly accepted that the extensional behavior predominates the entry flows. The entry flow problem, however, is not well understood and recently becomes more complex.

In our recent experiments, both steady flows and unstable flows of polymer solutions through a planar contraction have been observed not only in the central plane but also in other planes to reveal the 3-dimensionality of the flow. The present paper adds further experimental evidence to show that the flow characteristics are far more complex than initially imagined.

EXPERIMENTAL

Two kinds of channels were used; w_u=40 mm, h=300 mm and w_u=20 mm, h=150 mm as shown in Fig. 1. The contraction ratios were 10:1, 10:2 and 10:3. The test fluids were 0.1 and 0.5 wt% aqueous solutions of polyacrylamide (PAA, SEPARAN AP-30). The test fluids were seeded with polystyrene powder and a planar sheet of light illuminated the flow. The observation planes were not only the central plane (xy-plane), but also the xz-plane and the parallel plane to the xz-plane in the vicinity of the side wall, y≈19 mm or 9 mm, furthermore the parallel plane to the yz-plane near the entrance face, x≈-2 mm. The wall shear rate and the Reynolds number for fully developed Power-law fluid flow through a contraction are chosen as the parameters of experimental conditions.

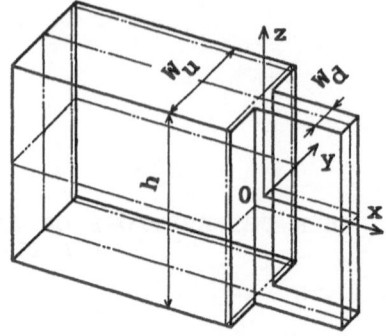

Figure 1. Planar contraction and observation planes.

FLOW PATTERNS THROUGH A CONTRACTION

The flow visualizations near the side wall enable us to investigate the change of flow patterns along the channel height. Fig. 2(a) clearly shows that the salient corner vortex almost never changes in length along the channel height except the region near the upper wall and lower wall. In contrast, Fig. 2(b) also reveals that further increasing the flow rate results in a slow increase in corner vortex length along the channel height, i.e. the tendency of 3-dimensional flow can be observed even within the steady flow regime.

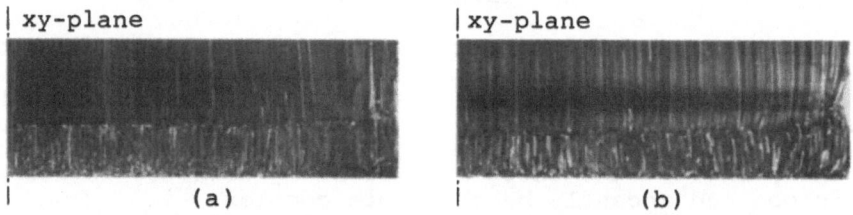

| xy-plane | xy-plane |

(a)   (b)

Figure 2. Side-viewed flow patterns near the side wall for
0.5 wt% PAA solution;
(a) Q=21.1 cc/s, Re=0.16, through a 10:1 contraction,
(b) Q=400 cc/s, Re=4.55, through a 10:3 contraction.

Unstable flow is characterized by the flow that salient corner vortices grow and decay alternately. Furthermore, several bundles of stream in the vicinity of the side wall can be observed in Fig. 3(b). In Fig. 3 the structure of the unstable flow and the alternate flow mode can be supposed as follows: the recirculating flow region Ⓐ corresponds to the region ⓐ and the higher velocity region Ⓑ corresponds to the bundle of stream ⓑ. Also bundles of stream move up and down along z-direction. This movement corresponds to the alternate growth and decay mode of the salient corner vortices.

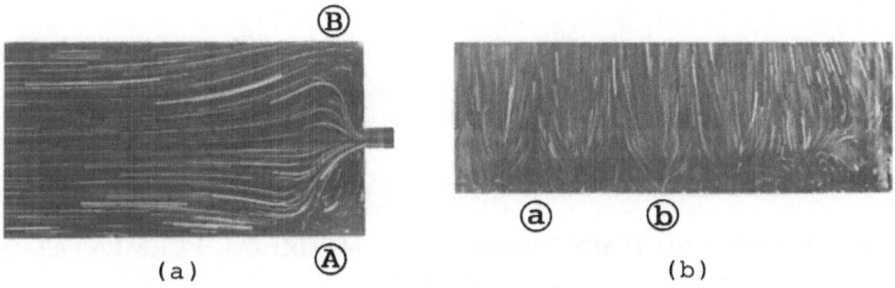

(a)   (b)

Figure 3. Typical unstable flow patterns;
(a) top-viewed streaklines in the central plane,
(b) side-viewed streaklines near the side wall.

In Fig. 4 the flow patterns near the side wall and along the entrance face (yz-plane) also show that the bundle-like streams converge as the salient corner is approached, then they diverge quickly while flowing along the entrance face. The flow pattern on the opposite side of the contraction is out of focus in Fig. 4(b), however, it can be supposed that the bundle-like streams lie alternately on either side of the contraction.

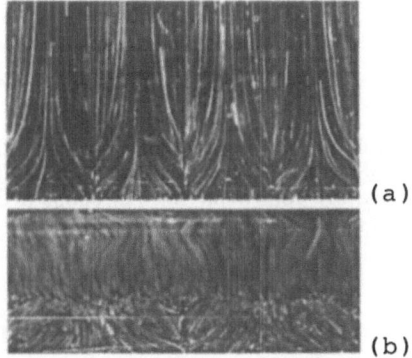

(a)

(b)

Figure 4. Typical streaklines;
(a) near the side wall,
(b) near the entrance face.

CONCLUSIONS

At the sufficiently low flow rates the planar entry flows are 2-dimensional, however, the tendency of 3-dimensional flow begins to appear as unstable flow regime is approached.

On the other hand, for unstable flow regime the side-viewed photographs clearly indicate the presence of non-trivial 3-dimensional component, such as several bundles of stream in the vicinity of the side walls. We suppose that the bundle-like streams lie alternately on either side of the contraction and they move up and down along the channel height.

REFERENCES

1. D.V. Boger, D.U. Hur and R.J. Binnington, J. Non-Newt. Fluid Mech., **20**(1986) 31.
2. R.E. Evans and K. Walters, J. Non-Newt. Fluid Mech., **20** (1986) 11.

RHEOLOGY AND MASS TRANSFER IN MYCELIAL FERMENTATION BROTH

H. Chmiel, W. Krischke, U. Schmid, M. Schüttoff, E. Walitza, V. Wessling
Fraunhofer-IGB, Nobelstr. 12, 7000 Stuttgart, FRG

ABSTRACT

Rheological properties of *streptomyces tendae* (*s.t.*) broth determined from macro- and micro-rheological techniques are presented. The influence of viscoelasticity on mass transfer is shown by 'follow-up' fermentations in 3 reactor types; stirred vessel, jet loop and propeller loop reactors.

INTRODUCTION

The significance of viscoelasticity on transport phenomena in homogeneous fluids is well known (1,2). Since mycelial fermentation broths exhibit strong viscoelasticity, this can be regarded as an important factor in the fermentation process. Published literature on mycelial broth viscoelasticity is sparse and the correlation between viscoelasticity and transfer coefficients has not yet been established. The mechanisms which lead to the observed flow behaviour of broths are also not well understood. As the dimensions of the elastic structures in homogeneous fluids and heterogeneous broths range on very different scales, different viscoelastic effects in the bioreactor may be the result.

Application of conventional rheometry to mycelial broths is accompanied by problems resulting from heterogenity. For this reason micro-rheological methods are being developed in our institute. One purpose of this approach is to attain a better insight into the macro-rheological measurements. Another is to provide a further means to assess the stress relaxation time, it being the appropriate quantity to establish the dimensionless elastic parameter, the Deborah number. As we earlier emphasized (3), the Deborah number is equivalent to the tangent of the viscoelastic phase angle. Therefore our attention is focused on this quantity and its influence on mass transfer.

MATERIAL, METHODS AND RESULTS

The most striking features of macro-rheological measurements of mycelial broth are shear thinning and nonlinear viscoelasticity. Relevant influencing parameters are biomass content, morphology of mycelium, physical properties of mycelium and viscosity of the liquid phase. Shear thinning and viscoelasticity are shown in figures 1 and 2 for a typical mycelial micro-organism $s.t.$ The measurements were done using a concentric cylinder rheometer. Interactions between the rheometer wall and mycelium are so weak that wall slip occurs at low shear rates (<1 s^{-1}) and high biomass contents. Cylinder surfaces with special coatings give higher apparent viscosities (up to the 2-fold value), the viscoelastic phase angle is also increased.

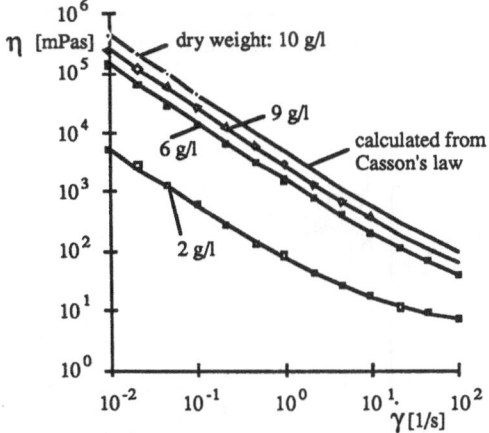

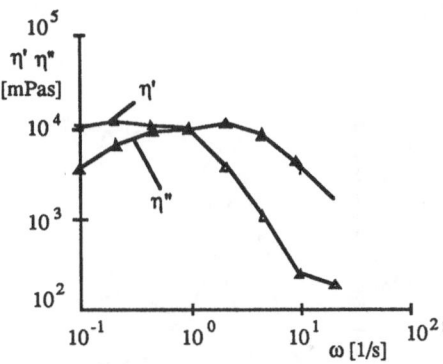

Fig.1. Apparent viscosity of filamentous suspensions of $s.t.$ with different dry weights. The measured points fit to the Casson's law (solid curves).

Fig. 2. Viscous and elastic component of complex viscosity of a filamentous suspension of $s.t$, magnitude of oscillating shear rate 0.21 s^{-1}, dry weight 9.0 g/l mean, particle diameter 1.0 mm.

Stress relaxation times derived from figure 2 range between 10 and 1s, but this depends strongly on magnitude of shear rate. Typical system specific times in a stirred tank bioreactor are about 1 -10 s^{-1}.

For a correct interpretation of macro-rheological findings, each of the above mentioned parameters has to be known. However determination of morphology and physical properties of the mycelium is difficult. Morphology is described by several characteristics. In addition to the terms 'filamentous', 'pelleted', 'single hyphae' and a mean floc size, porosity and permeability of mycelium are used. These are all highly significant for mass transfer within a mycelial floc. Different methods resulted in a mycelial porosity of about 77-82% for $s.t.$, this was nearly independent of age and morphology. Thus for a suspension of *s.t.* with 10 g/l biomass content, from a mycelial porositiy of 80% a total void fraction of 97% can be calculated. This apparently small void fraction however is related to high viscosity due to the high porosity of mycelium. Determination of elastic moduli of flocs or even of single hyphae, are another subject of our micro-rheological investigations.

In stirred tank, jet loop and propeller loop reactors, viscoelasticity and mass transfer was measured during fermentation. Apart from the morphological changes which are usually observed in an individual fermentation during growth, quite different morphologies occured due to the different stresses acting on the microorganisms from the respective reactor conditions. First results show that the influence of viscoelasticity differs from that found in homogeneous aqueous polymeric liquids. Figure 3 demonstrates the time course of the elastic component of viscoelasticity and oxygen transfer rate in a fermentation of *chaetomium cellulolyticum*. The strong increase of the elastic component is accompanied by a strong decrease of the oxygen transfer. This is due to dampening of turbulence and reduction in gas dispersion by elasticity.

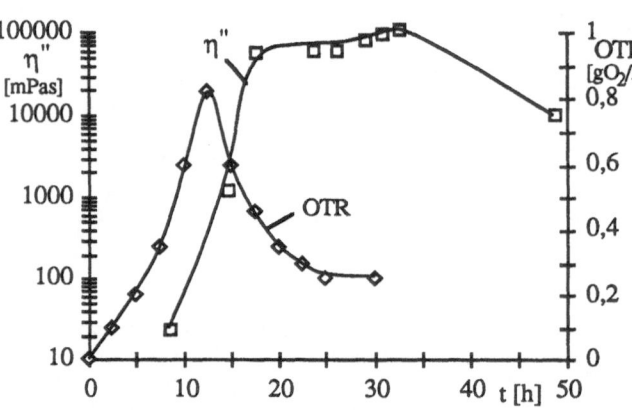

Fig. 3 Oxygen transfer rate and the elastic component of viscoelasticity vs time. Fermentation of *chaetomium cellulolyticum* in stirred tank reactor.

REFERENCES

1. Ranade, V.R. and Ulbrecht J.J., Influence of Polymer Additives on the Gas-Liquid Mass Transfer in Stirred Tanks, AIChE Journal, 1978, **24**, 5, 796-803.

2. Nakanoh, M., Yoshida, F., Gas Absorbtion by Newtonian and Non-Newtonian Liquids in a Bubble Column, Ind. Eng. Chem. Des. Dev., 1980, **19**, 190-195.

3. Oolman, T., Walitza, E., Hanssmann, E. and Chmiel, H., Viscoelasticity and Mass Transfer in Bioreactors, In Biochemical Engineering, H. Chmiel, W.P. Hammes and J.E. Bailey, editors, 1987, 416-420.

RHEOLOGY OF UNHEATED GELLAN GELS

L. CHOPLIN[1], A. TECANTE[1], S. BISSON[1] and M. MOAN[2]

[1] Department of Chemical Engineering, Université Laval, Ste-Foy, QC, G1K 7P4, CANADA
[2] Laboratoire de Mécanique Physique, Université de Brest, 29287-Brest, FRANCE

ABSTRACT

Gellan is a linear anionic extracellular polysaccharide, which forms gels with different strength depending on the gum and salt concentrations and the cations present. By using sequestrants such as EDTA, it is possible to prepare unheated gellan gels. These gels are physical gels and are thermoreversible. The rheology of unheated gellan gels has been examined as well as the rheological behavior of concentrated gellan aqueous solutions in presence of EDTA around the gel point.

EXTENDED ABSTRACT

Gellan is a linear anionic gel-forming extracellular polysaccharide secreted by the bacterium *Pseudomonas Elodea*. The primary structure has been shown to be composed of the following tetrasaccharide repeating-unit [1,2]:

$$\rightarrow 3)\text{-}\beta\text{-Glc}p\text{-}(1{\rightarrow}4)\text{-}\beta\text{-Glc}pA\text{-}(1{\rightarrow}4)\text{-}\beta\text{-D-Glc}p\text{-}(1{\rightarrow}4)\text{-}\alpha\text{-L-Rha}p\text{-}(1 \rightarrow$$

Of the repeating units, almost 25% contain an O-acetyl group linked to C6 of one of the β-D-glucopyranosyl residues.

Both the acetylated (native) and the deacetylated (commercially available under the trademark GELRITE) polysaccharides form thermoreversible gels. The strength of the gel vary according to the O-acetyl content: soft and elastic gels are obtained from the acetylated form, whereas hard and brittle gels are obtained from the fully deacetylated form [3]. The strength of the gellan gel depends also on the gum and salt concentrations and on the presence of cations.

The fully deacetylated gellan (GELRITE) is considered as a particularly interesting model substance for detailed studies of the gelation mechanism because of its simple well-defined primary structure, contrarily to other gel-forming polysaccharides [4].

108

The general procedure for gelation involves a preheating at a temperature of the order of 90°C followed by a cooling. The temperature at which the gel is formed is called the setting temperature and is usually in the range of 30 to 45°C for gellan. Deacetylated gellan gels show large melting/setting hysteresis: the melting temperature being either below or above 100°C depending on the gelation conditions. Without added cations in deionized water, gellan does not gel upon cooling. In tap water, ionic concentration may be high enough to induce gelation, but the inclusion of a sequestrant is desirable to control the reactivity of the ions.

In the case of gellan, it is quite difficult to obtain true aqueous solutions, except in the case of addition of tetramethylammonium (TMA) salt and/or a previous time-consuming purification procedure [5]. The presence of a sequestrant such as ethylene-diaminetetracetic acid (EDTA), by facilitating the dissolution of the deacetylated (or partially deacetylated) form of the polysaccharide even in salt-containing aqueous solutions, and without purification allows the preparation of true solutions. Moreover, subsequent cooling induces gelation without the necessity of preheating. This is particularly interesting if we consider that molecular degradation might be a consequence of preheating [6]. Evidence of this molecular degradation has been obtained by comparing the viscous behavior of a gellan aqueous solution in non-gelling conditions with and without preheating.

We have studied the rheological properties of concentrated gellan aqueous solutions in presence of EDTA (5 g/L) around the gel point. The physical thermoreversible gelation has been followed rheologically through the evolution of non-divergent rheological functions, like the dynamic moduli in the linear viscoelasticity domain. Above the gel point, the loss modulus (G") is higher than the storage modulus (G'), and the classical frequency dependence in the terminal zone (low frequency) is observed. Well below the gel point, G' is significantly higher than G", both moduli showing the following power-law behavior over 4 decades of frequency:

$$G' \sim G'' \sim \omega^{0.2} \qquad (1)$$

Moreover, G' and G" appear to become insensitive to temperature. Between these two extreme rheological behaviors, all intermediate behaviors can be obtained depending on the temperature. The gel point has been localized when the loss tangent becomes independent of the frequency. This corresponds to a loss modulus higher than the storage modulus, that is: gel point occurs at a temperature higher than the temperature corresponding to the crossover of the loss and storage moduli, whatever the frequency.

REFERENCES

[1] O'Neill, M.A., Selvendran, R.R. and Morris, V.J., Carbohydrate Research, 124, 123 (1983).
[2] Jansson, P.E., Lindberg, B. and Sandford, P.A., Carbohydrate Research, 124, 135 (1983).
[3] Kang, K.S. and Veeder, G.T., US Patent 4326053 (1982).
[4] Grasdalen, H. and Smidsrod, O., Carbohydrate Polymer, 7, 371 (1987).
[5] Milas, M. and Rinaudo, M., Carbohydrate Research, 158, 191 (1986).
[6] Brownsey, G.J., Chilvers, G.R., I'Anson, K. and Morris, V.J., Int. J. Biol. Macromol., 6, 211 (1984).

RHEOLOGICAL BEHAVIOUR OF A THERMOTROPIC LIQUID CRYSTAL COPOLYESTER: HYDROXYBENZOIC ACID / 2,6-HYDROXYNAPHTHOIC ACID (HBA/HNA)

F.Cocchini
EniChem R&S, via Fabiani 2, 20097 San Donato Milanese, Italy

M.R.Nobile and D.Acierno
Istituto di Ingegneria Chimico-Alimentare, Facoltà di Ingegneria,
Università di Salerno, 84100 Salerno, Italy

ABSTRACT

The rheology of the all-aromatic copolyester thermotropic liquid crystal 73/27 HBA-HNA has been studied over a large interval of shear rates.
At small shear rates the transient behavior has been analyzed with a cone and plate rheometer. The shear stress and the first normal stress difference have been found to scale with the strain. Steady values for both viscosity and first normal stress difference have been obtained. The steady shear viscosities have been correlated to capillary data at higher shear rates. Three zones can be found in the steady viscosity vs shear rate curve: a shear thinning region at small shear rates, an almost newtonian region, and a second power-law region at higher shear rates. In the shear rate interval studied with the cone and plate rheometer the steady first normal stress difference remains positive.

INTRODUCTION

The rheology of liquid crystal polymers is known to be very unusual with respect to flexible polymers. Phenomena such as shear thinning viscosity at low shear rates, negative first normal stress difference, little extrudate swelling despite high elasticity in the linear viscoelastic limit, are infact observed both in lyotropic and thermotropic LCP's [1,2,3].
The rod-like shape of LCP molecules gives rise to a mesophase, with a preferred direction (the "director") along which the molecules tend to be aligned. This event induces a strong coupling between flow and orientation of the director. So far only few models taking into account this feature are available in litterature [4,5]. Recently Marrucci and Maffettone [5] have been able to explain the occurrence of negative first normal stress difference with a rod model within a single domain. On the other hand, the LCP overall feature usually shows a polydomain morphology which is ill defined [6] and strongly affects the flow behaviour, being responsible for long relaxation times.
Most of the papers that have been appeared in literature deal with lyotropic LCP's. Among the thermotropic materials a more detailed study has been performed only on PET-PHB

systems which have the flexible ethylene linkage [3]. Studies on all-aromatic liquid crystal polymers are mainly limited to capillary rheometer and oscillatory measuraments [7]. In this work transient and steady measuraments of shear and normal stresses on the 73/27 copolymer of hydroxybenzoic acid (HBA) and 2,6-hydroxynaphthoic acid (HNA) are presented.

EXPERIMENTAL

The sample chips of 73/27 HBA-HNA, commercialized as VECTRA A950 by Hoechst-Celanese, have been dried in vacuum at 120 °C overnight. The material has been studied at small shear rates using the rotational rheometer Rheometrics RMS 800 equipped with a cone and plate of 25 mm radius and 0.1 cone angle. The experiments have been performed in inert gas atmosphere. At higher shear rates the capillary rheometer Gottfert Rheograph 2002 has been used with a 30/1 mm round hole die and a 100 x 10 x 1 mm slit die.

Reliable normal stress difference measurements need a good confidence on the zero level for the normal thrust. The compression and the squeezing flow during the loading of the sample between the cone and plate of the rotational rheometer cause large positive values of the normal force which slowly relax to zero. The loading stretches the polydomains of the LCP which takes hours to recover a random equilibrium structure. However, it is not practicable to leave the material at high temperatures for hours. Moreover, the quiescent state obtained after a long relaxation is still expected to depend on the previous thermal and mechanical history [6].

Shearing the sample during the long relaxation causes a fast decrease of the normal thrust. If the total strain applied is small (< 100 units of strain) the normal force increases again at the cessation of the shear, partially recovering the previous magnitude. Otherwise the level of the normal force is mantained even after the cessation of flow, and eventually it relax further to zero.

As a good compromise the transient tests have been performed after a preshearing of 1200 seconds at a shear rate of 0.02 sec^{-1}, 400 seconds at a shear rate of 0.5 sec^{-1}, and a relaxation of other 400 seconds. This procedure has given a complete relaxation and a good reproducibility of the normal forces. It reasonably drives the sample to a defined structure independent of all previous stress history.

RESULTS AND DISCUSSION

Stress transient measurements have been performed at various shear rates, at 310 °C, on samples prepared as described in the previous section.

In figure 1.a and 1.b the shear stress and the first normal stress difference N_1 are plotted vs the strain for shear rates of 0.2 , 0.5 and 1 sec^{-1}. Both shear and normal stresses scale approximatively with the strain at different shear rates. In figure 1.a the shear stresses have an overshoot and a slow increase to the steady values, reached at about 200 units of strain. In figure 1.b the normal stress differences tend to a steady positive level within 25 unity of strain, without appreciable overshoot. Nevertheless oscillations of the signal persist at higher strains.

The viscosity and the normal stress differences N_1 obtained at a strain 200 have been reported in figures 2.a-b. In figure 2.a capillary data are also reported. They have been obtained by leaving the pressure signals to level off at the different values of piston speed. Due to the long strain needed to achieve steady level of stress in the material, it can be inferred that the inner material, which is subject to small shear rates, does not equilibrate passing through the capillary. Thus a strict comparison between capillary and rotational rheometer data must be taken judiciously. Moreover, due to this event, a geometry dependence of the experimental data might exist, as reported in ref. [7], and as can be observed by comparison of slit die and round hole die data in figure 2.a. Taking this in mind, three zones can be found in the viscosity data reported in fig. 2.a. A power-law region at small shear rates, an almost newtonian region, and a second shear thinning region at higher shear rates. This behaviour is similar to that of other lyotropic and thermotropic LCP's (region I, II and III of Onogi and Asada [8]). Region I is usually interpreted as due to the polydomain morphology. The existence of region I even in our data on presheared samples should suggest that the preshearing is not sufficient to give a nearly monodominial morphology, or that an oriented structure is not stable at small shear rates. This latter feature should be supported by some theoretical evidence of the tumbling of the director at small shear rates [9,10].

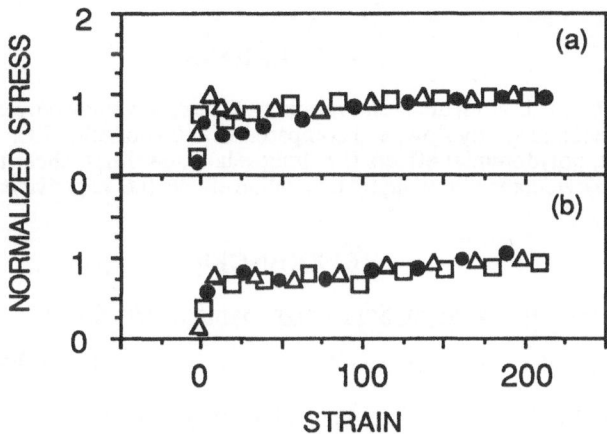

Figure 1. Normalized transient stress of HBA/HNA 73/27 at 310 °C. a) Shear stress and b) first normal stress difference vs strain at different shear rates:

$\triangle \; \dot{\gamma} = 0.2;$ $\square \; \dot{\gamma} = 0.5;$ $\bullet \; \dot{\gamma} = 1.$

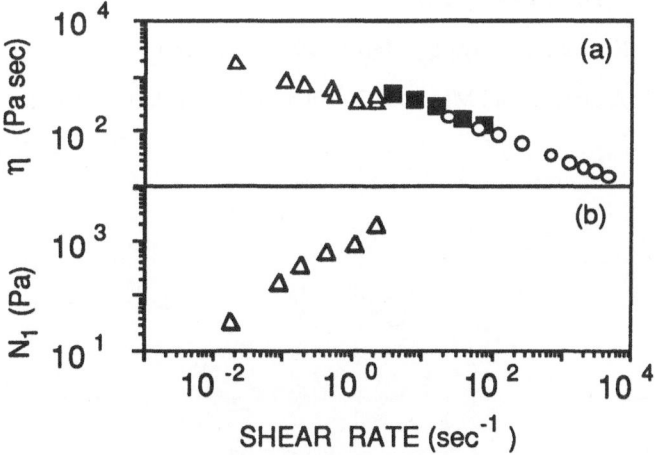

Figure 2. a) Viscosity and b) first normal stress difference of HBA/HNA 73/27 at 310 °C: $\triangle$ cone and plate; $\blacksquare$ slit die; $\circ$ round hole die.

In figure 2.b the normal stress differences have been reported. They are positive in the range of shear rates considered. It can be noticed a small inflection of the values at shear rates corresponding to the plateau of the viscosity in fig. 2.a. The same behaviour has been observed in other LCP's with all-positive values of N_1 [1], and predicted theoretically [5]. The first normal stress differences tend to decrease but not enough to become negative. At small shear rates a linear dependence on the shear rate can be observed. Also this feature is often observed in other LCP's.

CONCLUSIONS

The rheology of an all-aromatic thermotropic LCP has been studied over a large range of shear rates in transient and steady flow, and compared with that of other LCP systems. For the material considered, polydominial effects (i.e. long relaxation times, shear thinning at small shear rates) seems to overcome purely liquid crystalline effects (i.e. negative normal stresses).

REFERENCES

1.   Kiss G. and Porter R.S. J.Polym.Sci., Polym.Symp., 1978, **65**, 193-211.

2.   Mewis J. and Moldenaers P. Mol.Cryst.Liq.Cryst., 1987, **153**, 291-300.

3.   Gotsis A.D. and Baird D.G. J.Rheology, 1985, **29**, 539-556.

4.   Doi M. J.Polym.Sci., Polym.Phys., 1981, **19**, 229.

5.   Marrucci G. and Maffettone P.L. Macromolecules, 1989.

6.   Chapoy L.L., Marcher B. and Rasmussen K.H. Liq.Cryst., 1988, **3**, 1611-1636.

7.   Wissbrun K.F., Kiss G. and Cogswell F.N. Chem.Eng.Comm., 1987, **53**, 149-173.

8.   Onogi S. and Asada T. in Rheology, ed. G.Astarita, G.Marrucci and L.Nicolais, Plenum, New York, 1980, p.127.

9.   Doi M. and Kuzuu N. J.Phys.Soc.Japan, 1983, **52**, 3486.

10.   Cocchini F., Aratari C. and Marrucci G., submitted to Macromolecules.

INJECTION MOULDING OF CONFECTIONERY MATERIALS

A A Collyer*, M J Bennett*, R H Mayhew*, D W Clegg**,
          C R Elson*** and S G Jones***
  *   Department of Applied Physics
 **   Department of Metals and Materials Engineering
       Sheffield City Polytechnic, Pond Street,
       Sheffield S1 1WB, England, United Kingdom.
***   Leatherhead Food Research Association, Randalls Road,
       Leatherhead, Surrey, England, United Kingdom.

ABSTRACT

Gelatin fruit gums based on a Maynard's standard recipe were
successfully injection moulded at 85% solids by use of a 16% gelatin
(100 bloom) mixture at a 1.42:1 gelatin water ratio, with a cylinder
temperature of 60°C, a 15 s plunger dwell time and a mould cooling time
of five minutes. All samples were assessed at the demoulding stage for
their mouldability, and after a week had elapsed the samples were
assessed by organoleptic and instrumental analysis for their textural
qualities.

After initial experiments had been carried out to find the optimum
gelatin:water ratio, a Haake Rotovisco ball and cup viscometer was used
to observe the structural build up with time in the chosen recipe.
Some experiments were carried out to accelerate the gelling process by
replacing 33% of the gelatin with a fast-gelling starch. This proved
feasible at 75% solids content when the starch was pre-gelled.

INTRODUCTION

The traditional production of wine gums involves the energy-intensive
and time-consuming process of stoving, in which a molten mixture of
gelatin/sugar syrup is deposited into starch moulds at 76% solids and
dried using air heated to 55°C, until the final solids content of 85%
is achieved.

Previous work [1] has shown that injection moulding at the final solids
content is possible, thereby removing the need for stoving. The
problems associated with this are that the cycle times may be large as
a result of the slow build up of structure, and in the present work the
use of fast-gelling starches with gelatin is discussed.

EXPERIMENTAL PROCEDURE

(a)   Process Variables

Recipes containing 85% solids were produced with gelatin contents varying from 10.4 to 16.9%. These syrups were injection moulded on a Florin Ltd plunger-type injection moulding machine, with temperature control over the range 30-80°C [1-3]. The mould cavity had the dimensions 58.0 x 12.5 x 6.25mm, with liquid paraffin as a mould release agent.

Experiments were carried out on the best of the recipes (58% gelatin 100 bloom/42% water) in which only one of the process variables was changed at a time. The variables investigated were cylinder temperature, moulding time and plunger dwell time for a ram pressure of 6.8 x 10^5 Nm^{-2}.

All products were assessed for processability (shape retention, demouldability, and firmness), organoleptic characteristics (hardness, chewiness and gumminess as adjudged by a panel) and performance in a two-cycle compression test [5] measuring hardness, recovery and cohesiveness. This was carried out on an Instron Universal Testing Machine, Model No. 1140.

(b)   Rheological Studies

A Haake Rotovisco RV3 viscometer was used in the ball and cup mode to measure the yield stress of the gelling mixtures as a function of time. After various rest periods the viscometer was switched on and the shear rate increased until the shear stress became too high. The down curve was then obtained.

(c)   Starch/gelatin mixtures

Some experiments were carried out on syrups in which 33% of the gelatin was replaced by a fast-gelling starch. The starches used were Emgum 948 (JEW Starches Ltd, Heywood, Lancs.) and Snowflake 083210 (CPC Ltd, Manchester).

RESULTS AND DISCUSSION

The recipe contained 16% gelatin (100 bloom) mixture at 1.42:1 gelatin to water ratio [1-3]. With this recipe the best balance of properties was for the following process conditions : cylinder temperature 60°C; plunger dwell time of 15s; and a mould cooling time of 5 min. at 15°C. There is, therefore, a need for a faster gelling system.

Figure 1 shows a typical Rotovisco trace for an 80% solids mixture at 60°C. This shows a yield stress, which disappears as the structure collapses under the increasing stress. The downcurve shows no yield value and once the structure has been destroyed the material shows Newtonian behaviour.

115

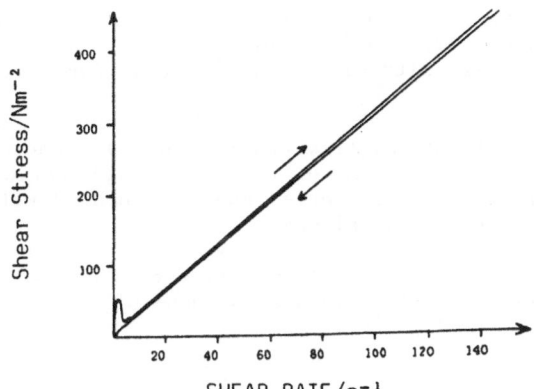

SHEAR RATE/s⁻¹

Figure 1 : A typical Rotovisco trace (80% solids at 60°C)

(a) . Gelatin
(b) ▫ Gelatin/1/3 Pregelatinised Emgum
(c) x 2/3 Gelatin/1/3 Emgum
(d) o 2/3 Gelatin/1/3 Snowflake

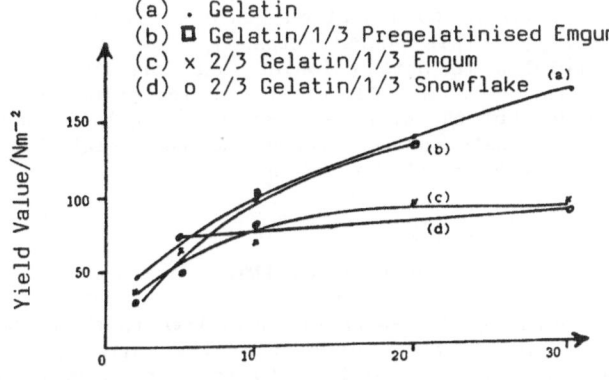

Standing Time/min

Figure 2 : The variation of yield stress with time at 60°C

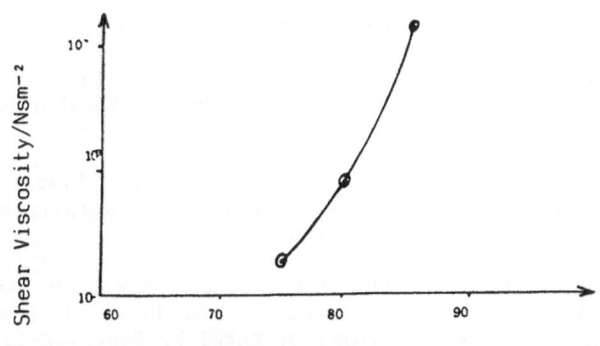

SOLIDS CONTENT/%

Figure 3 : The variation of viscosity with solids content
after a five minute rest at 60°C

The yield stress at 60°C increases with time as the structure builds up (Figure 2) for syrups containing 75% solids. Figure 3 shows the variation of viscosity with solids content for a resting period of 2 min. at 60°C.

When Emgum 948 or Snowflake 083210 were introduced, it was necessary to pre-gelatinise them in a 1:1 ratio with water, after which they were dried and then cooled with the other ingredients in the usual way.

Both these starches gave lower stresses when not pre-gelatinised, but the Emgum showed similar values to those of the gelatin above, when it was pre-gelatinised.

CONCLUSION

It is possible to injection mould fruit gum materials but cycle times are long. This necessitates the use of fast gelling starches with the gelatin to obtain longer and more rapidly developed yield stresses. The work carried out so far indicates this may be possible at 75% solids, but the starches must be pre-gelatinised. The situation of competition between the starch and the gelatin for the available water will be greater at 85% solids.

ACKNOWLEDGEMENTS

One of the authors (M J Bennett) would like to thank SERC for the award of a CASE grant to carry out part of the work, and we should like to thank CPC Ltd and JEW Starches Ltd for provision of the fast-gelling starches.

REFERENCES

1. Wright, S.J.C. and Dodgson, A.G., 'Novel Methods for the Manufacturing of Confectionery Products,' Leatherhead Food Research Association Ref. No.484, 1984.

2. Jones, S.A., Elson, C.R. and Bennett, M.J. 'Novel Methods for the Manufacture of Confectionery Products,' Leatherhead Food Research Association Rep. No. 577, 1987.

3. Collyer, A.A., Bennett, M.J., Clegg, D.W., Mayhew, R.H., Elson, C.R. and Jones S.A., Rheology of Food, Pharmaceutical and Biological Materials, Univ. of Warwick, Sept. 12-15, 1989.

4. Szezesniak, A.S., 'Classification of Texture Characteristics,' J.Food Sci., 28, 385, 1963.

5. Bourne, M.C., 'Texture Profile Analysis,' Fd.Technol., 32, 7, 63, 1978.

INERTIAL EFFECTS IN VISCOELASTIC FLOWS

M.J. CROCHET, V. DELVAUX
Mécanique Appliquée, Université Catholique de Louvain
2 Place du Levant, 1348 Louvain-la-Neuve, Belgium

ABSTRACT

By means of the numerical simulation of viscoelastic flows, we show that some experimentally observed effects, such as delayed die swell and anomalous transport properties in the flow around circular cylinders, are related to the change of type of the vorticity equation, when the viscoelastic Mach number exceeds one.

INTRODUCTION

Most numerical simulations of viscoelastic flow published over the last few years have concentrated on low Reynolds number situations. By analogy with generalized Newtonian flow analysis, inertia effects are usually neglected in the momentum equations. However, a number of recent papers by Joseph and his collaborators (a review may be found in [1]) have shown the importance of the *viscoelastic Mach number M*, which compares the velocity of the fluid to the velocity of shear waves. When M is larger than one, the vorticity equation changes from the elliptic to the hyperbolic type, and one may expect some peculiar phenomena associated with hyperbolicity, as in the propagation of sound waves in a compressible fluid when the velocity exceeds the speed of sound. The viscoelastic Mach number is equal to the product of the Weissenberg number $We = \lambda V / L$ times the Reynolds number $Re = \rho V L / \eta$, where λ is a relaxation time, ρ is the density, η is a viscosity, while L and V are characteristic length and velocity, respectively. Since $M = We\,Re$, we may therefore expect some interesting phenomena to occur even at low values of the Reynolds number, provided We is large enough.

Two such phenomena are the *delayed die swell* [2,3] and anomalous transport properties associated with the *flow of a viscoelastic fluid past a circular cylinder* [4]; they are described below. Such peculiar flows have been evidenced experimentally with polymeric solutions for which an appropriate constitutive equation is not readily available. In the present paper, we concentrate on the classical models of Maxwell-B and Oldroyd-B fluids which are clearly not appropriate for reproducing the properties of the experimental fluids used in [2-4]. Our goal is to show that, even with our simple constitutive equations, a viscoelastic Mach number larger than one suffices for generating the phenomena observed with the experimental fluids. In the second section, we concentrate on the analysis of delayed die swell [5], while in the third we analyze the flow round a circular cylinder [6].

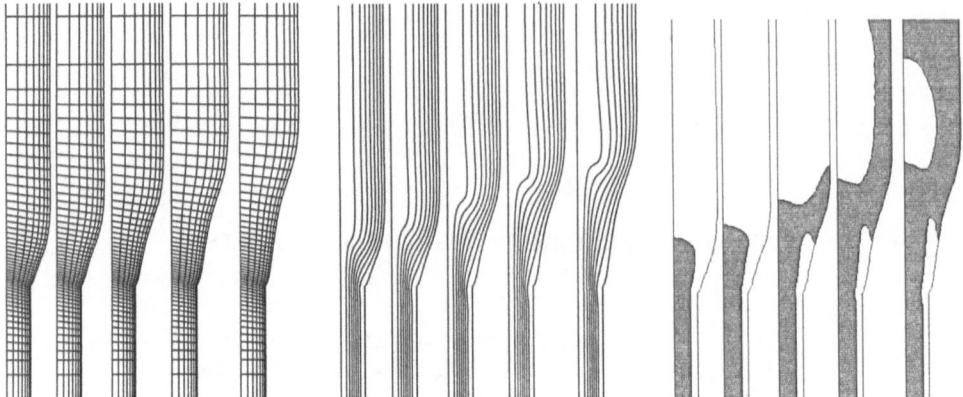

Fig. 1. Delayed die swell for an Oldroyd-B fluid with M ranging from 2.3 to 5.9.
Finite element meshes; streamlines; supercritical flow regions in dark area.

DELAYED DIE SWELL

When a viscoelastic fluid is extruded from a pipe, it is known that the diameter of the extruded jet usually increases; the phenomenon is well known experimentally, while many theoretical papers have been able to predict the phenomenon (a review may be found in [7]). At low rates of extrusion, the swell starts at the pipe exit. At high rates of extrusion, or for large values of We, the tendency of the jet to swell at the exit is suppressed and there is a delay in the swell [1]. The delayed die swell occured in all the experiments described in [3] beyond a critical value of the flow rate. For a given fluid and a given geometry, it is thus clear that the elasticity number

$E=We/Re=\lambda\eta/\rho D2$ remains a constant throughout the experiment. We wish to add inertia to our earlier calculations, and to verify whether a high enough Mach number induces delayed die swell. Since it is known that a high level of accuracy is difficult to attain at high values of We, we have selected a low value of E, i.e.0.03, in order to generate the phenomenon at a relatively low value of We. For some of the experimental points given in [3], the Weissenberg number is of the order of 10 to 100.

The numerical method used for simulating delayed die swell is explained in [5]. Briefly, we use the mixed method of Marchal and Crochet [8] for calculating the stress, velocity and pressure fields. For the free surface calculation, we perform an implicit integration based on Newton's method which takes into account the hyperbolic character of the kinematic condition. We have obtained the same results with three different finite element meshes shown in [5]; the number of degrees of freedom varies from 14000 to 29000.

The main results are shown in Fig. 1. For a given initial finite element mesh, we show, for an Oldroyd-B fluid, the evolution of the die swell when M grows from 2.3 to 5.9. On the left, Fig. 1 shows the successive finite element meshes while, on the right hand-side, the dark area represents the growing hyperbolic region. The streamlines show that, near the axis of the tube, the material particles which travel faster than shear waves do as if they were ignoring the end of the cylindrical wall.

FLOW ROUND A CIRCULAR CYLINDER

Let us now consider the flow of a viscoelastic fluid around a circular cylinder, for which experimental results using a dilute polymer solution were made available by James and Acosta

119

[4]. They showed that, beyond a critical Reynolds number, the drag coefficient does not decrease as a function of the Reynolds number, as it does with a Newtonian fluid. Moreover, by measuring heat transfer between the cylinder and the surrounding fluid, they found that the Nusselt number reaches an asymptotic value beyond the critical Reynolds number. The reason for such an anomalous behavior has been attributed in [9] and [1] to critical conditions where the velocity of the flow exceeds that of shear waves. Joseph [1] suggests that the domain surrounding the cylinder can be decomposed into a region of silence for the vorticity and a region of disturbed vorticity.

We have calculated the flow of a Maxwell-B fluid around a circular cylinder of unit radius between parallel walls fifty radii apart [6]. In order to garantee the accuracy of our results, we have performed the calculation on four finite element meshes, with a number of degrees of freedom varying from 15000 up to 46000. All our results converge on the drag coefficient and the Nusselt number as a function of the Reynolds number. Fig. 2a shows the evolution of the drag coefficient as a function of the Reynolds number for three values of the elasticity number, i.e. 0 (Newtonian), .75 and 2.08. The vertical lines correspond to the critical values of the Reynolds number at which M=1. We find that, for viscoelastic flow with inertia, the curves separate precisely at the critical value of the Reynolds number. Fig. 2a shows the behavior of the Nusselt number as a function of Re for three values of the Prandtl number, and for E=2.08. We find that the Nusselt number exhibits an asymptotic behavior precisely beyond the critical Reynolds number indicated by a vertical dotted line.

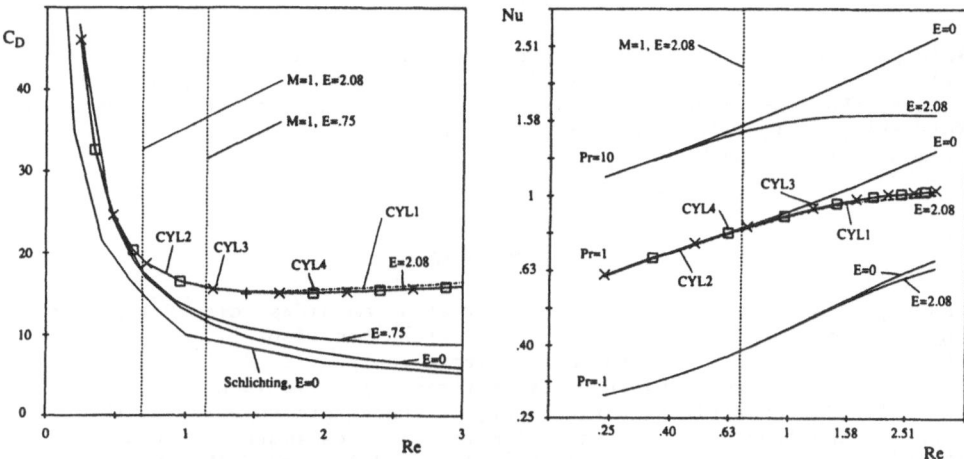

Fig.2 a. Drag coefficient CD versus Re for various values of the elasticity number E; b. Nusselt number versus Re for three values of the Prandtl number in the Newtonian case (E=0) and the viscoelastic case (E=2.08).

REFERENCES

1. Joseph D.D., *Fluid dynamics of viscoelastic liquids*, Springer Verlag, 1990.
2. Giesekus H., *Rheol. Acta*, 1968, **8**, 411-421.
3. Joseph D.D., Matta J.E., Chen K., *J. non-Newtonian Fluid Mech.*, 1987, **24**, 31-65.
4. James D.F., Acosta A.J., *J. Fluid Mech.*, 1970,.**42**, 269-288.
5. Delvaux V., Crochet M.J., *Rheol. Acta*, 1990, **29**, 1-10.
6. Delvaux V., Crochet M.J., Numerical properties of anomalous transport properties in viscoelastic flow, *J. non-Newtonian Fluid Mech.*, 1990, to appear.
7. Keunings R., in *Fundamentals of computer modeling for polymer processing*, ed. Tucker C.L., Hanser Verlag, 1990.
8. Marchal J.M., Crochet M.J., *J. non-Newtonian Fluid Mech.*, 1987, **26**, 77-114.
9. Ultman J.S., Denn M.M., *Trans. Soc. Rheol.*, 1970, **14**, 307-317.

NON-LINEAR EFFECTS IN CONTROLLED STRESS OSCILLATION

J.M. DAVIES*, K.GOLDEN** and T.E.R. JONES*
* Department of Mathematics and Statistics, Polytechnic South West,
Drake Circus, Plymouth, Devon PL4 8AA.
** Department of Theoretical Mechanics, University of Nottingham,
University Park, Nottingham, NG7 2RD.

ABSTRACT

This paper will be concerned with a theoretical study of non-linear
effects in controlled stress oscillation. A theory is developed which
extends the generalised Maxwell model to include a yield stress component.
This theory will be used to determine the non-linear effect of yield
stress on the complex viscosity data.

INTRODUCTION

Many types of materials such as concentrated emulsions, dispersions, gels
and slurries 'appear' to exhibit yield stress behaviour [1]. It has been
shown that dynamic mechanical measurements can provide useful information
about the structure of yield stress materials [2]. It is known that a
generalised linear viscoelastic theory is inadequate to describe the
dynamic behaviour of these types of materials. Yoshimura and Prud'homme
[2] have proposed an elastic Bingham fluid model to describe the
non-linear behaviour that arises in oscillatory shear experiments.

In this paper we propose a modified form of the generalised linear
viscoelastic model which includes a yield stress component. This model is
used to describe non-linear oscillatory effects. We shall also examine the
response of this model to an applied oscillatory stress. In order to
illustrate the effect of yield stress on dynamic data, we present the
results for our proposed model with a single Maxwell element.

THEORY

The model consists of a yield stress component in parallel with a finite
number of Maxwell elements. Therefore a critical stress needs to be
exceeded before deformation can take place. Following the analogy of the
Bingham model it is convenient to define a complex plastic viscosity η_p^*
to characterise the oscillatory shear behaviour. The shear stress/strain

rate relationship is given by

$$\sigma = \sigma_0 \frac{\dot{\gamma}(t)}{|\dot{\gamma}(t)|} + \int_{-\infty}^{t} G(t-t')\, \dot{\gamma}(t')\, dt', \qquad \dot{\gamma}(t) \neq 0 \qquad (1)$$

where σ is the shear stress, σ_0 is the yield stress, $\dot{\gamma}$ is the strain rate and G is now defined as the plastic relaxation modulus.

The response of each Maxwell element is described by a viscous component η_i and an elastic component with an associated relaxation time λ_i $(i=1,2,..,n)$. The plastic relaxation modulus is therefore given by

$$G(t-t') = \sum_{i=1}^{n} \frac{\eta_i}{\lambda_i}\, e^{-(t-t')/\lambda_i} \qquad (2)$$

For an applied stress of $\sigma_a \sin \omega t$ the strain rate can be determined from the linear differential equation

$$\sum_{k=1}^{n} \eta_k \prod_{i=1}^{n} \left[1 + \lambda_i \frac{d}{dt} \right] \dot{\gamma} = \prod_{i=1}^{n} \left[1 + \lambda_i \frac{d}{dt} \right] \sigma_a \sin(\omega t) - \sigma_0 \frac{\dot{\gamma}}{|\dot{\gamma}|} \;,\; \dot{\gamma} \neq 0 \quad (3)$$

It can be shown that the general solution to this equation, over one period of oscillation (after steady state has been achieved), is given by

$$\dot{\gamma} = \begin{cases} \displaystyle\sum_{i=1}^{n-1} a_i e^{m_i t} + \mathrm{Im}\left[\frac{\sigma_a\, e^{j\omega t}}{\eta_p^*} \right] - \frac{\sigma_0}{\eta_0} & t_1 < t < t_2 \\[4mm] \displaystyle\sum_{i=1}^{n-1} a_i e^{m_i t} + \mathrm{Im}\left[\frac{\sigma_a\, e^{j\omega t}}{\eta_p^*} \right] + \frac{\sigma_0}{\eta_0} & t_3 < t < t_4 \end{cases} \qquad (4)$$

where $\dot{\gamma}$ is zero for all other time t, Im denotes the imaginary part of a complex number, m_i are constants which are determined from the values of m which satisfy the equation

$$\sum_{i=1}^{n} \frac{\eta_i}{(1+m\lambda_i)} = 0 \;, \quad \text{and} \quad \eta_p^* = \sum_{i=1}^{n} \frac{\eta_i}{(1+j\omega\lambda_i)} \qquad (5)$$

The constants of integration a_i in equation (4) are determined from satisfying the conditions that the strain γ and the stress in each Maxwell element is continuous for all time t.

Fig. 1 shows strain and strain rate waveforms, obtained from equation (4), for a single Maxwell element. The times t_1, t_2, t_3, and t_4 which are referred to in this figure and in equation (4) are obtained from the theory by constantly monitoring the difference between the applied shear stress and the total stresses arising from the Maxwell elements. This difference is then compared with the yield stress value and the times are noted. The strain waveform in Fig. 1 show a similar trend to the elastic strain predictions of Yoshimura and Prud'homme [2]. The flat topping of the strain waveform is characteristic of the behaviour that is observed when testing yield stress materials under controlled stress oscillation.

RESULTS AND DISCUSSION

We shall now consider the effect of yield stress on the dynamic flow
behaviour as predicted from the theory discussed in the previous section
(for a single Maxwell element). We shall use the definition that the
complex viscosity for this model is defined by the ratio of the
fundamental stress amplitude to the fundamental complex strain rate
amplitude. This amplitude is determined from equation (4) by a fourier
series analysis. In Fig. 2 we present the variation of normalised complex
viscosity with a normalised stress amplitude (σ_a/σ_0). The complex
viscosity has been normalised with respect to the plastic complex
viscosity defined in equation (5). The results, in this figure, clearly
demonstrate that the effect of yield stress is to produce a non-linear
complex viscosity behaviour. Therefore we propose that the plastic complex
viscosity function should be used to characterise the dynamic behaviour of
yield stress materials.

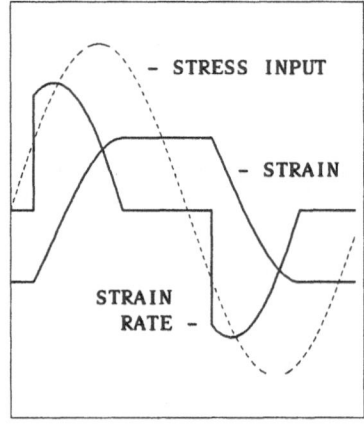

Figure 1.

Theoretical prediction of
strain and strain rate
for an applied sinusoidal
stress waveform.

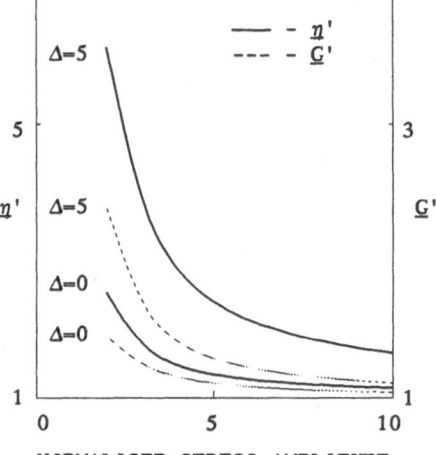

NORMALISED STRESS AMPLITUDE

Figure 2.

Theoretical variation of
normalised dynamic viscosity $\underline{\eta}'$
and normalised dynamic rigidity
$\underline{G}'$ with normalised stress
amplitude (σ_a/σ_0) where $\Delta=\lambda\omega$.

REFERENCES

1. Yoshimura, A.S., Prud'homme, R.K., Princen, H.M. and Kiss, A.D.,
   A comparison of techniques for measuring yield stresses. J. Rheol.,
   1987, **31**, 699-710.

2. Yoshimura, A.S. and Prud'homme, R.K., Response of an elastic Bingham
   fluid to oscillatory shear. Rheol. Acta., 1987, **26**, 428-436.

ON THE CORNER AND LIP VORTICES IN ABRUPT CONTRACTIONS

Benoît DEBBAUT
Unité de Mécanique Appliquée - Université Catholique de Louvain
Place du Levant 2
B - 1348 Louvain-la-Neuve - Belgique

ABSTRACT

We examine the effects of inertia upon the circular abrupt 4:1 contraction flow of a generalized Newtonian fluid, the extensional viscosity of which has been increased by a dependence upon the third as well as the second invariants of the rate of deformation tensor. It is shown that the combination of inertial and elongational effects leads to the development of a small but strong lip vortex which may envelop the stagnant corner vortex.

INTRODUCTION

Several papers have already been devoted to the vortex growth mechanism in an abrupt plane or circular contraction. In particular, it was shown in [1,2] that increasing inertia resorbs the stagnant corner vortex for a Newtonian liquid. In [3,4] Debbaut et al. have discussed the effects of the extensional viscosity upon the stagnant corner vortex growth by means of a numerical investigation. Such effects were also discussed in [5] from a theoretical point of view.

Lip vortices have been observed experimentally. For large contraction ratios, Walters [6] has been able to generate a small lip vortex with a Boger fluid [7]. Boger et al. [8,9] have commented on the interaction between both lip and stagnant corner vortices. For planar contraction flows, Evans and Walters [10] have suggested that the strong lip vortex envelops the stagnant corner vortex. This phenomenon is then followed by a classical vortex growth mechanism.

In the present paper we wish to examine the effect of combining the opposite extensional and inertial contributions upon the vortex growth mechanism for the circular abrupt 4:1 contraction flow. We use a generalized Newtonian model, the extensional viscosity of which is increased through its dependence upon the third as well as the second invariant of the rate of deformation tensor. As already mentionned in [4], the model used here is not advocated for general use since the third invariant vanishes in plane flow.

BASIC EQUATIONS

Let us denote the rate of deformation tensor d as follows

$$d = \frac{1}{2}(\nabla v + \nabla v^T) \tag{1}$$

where v is the velocity field. For an incompressible fluid, d is traceless. We can define invariants of $\boldsymbol{d}$ as follows [3,4] :

$$\dot{\varepsilon} = 3\,\mathbf{III}_{\boldsymbol{d}} / \mathbf{II}_{\boldsymbol{d}}\ ,\quad \dot{\gamma} = 2\sqrt{\mathbf{II}_{\boldsymbol{d}}}\ ,\quad \text{with}\quad \mathbf{II}_{\boldsymbol{d}} = \frac{1}{2}\operatorname{tr}\boldsymbol{d}^2\ \text{and}\ \ \mathbf{III}_{\boldsymbol{d}} = \det(\boldsymbol{d})\,. \tag{2}$$

One can easily show that in a uniaxial extensional flow, $\dot{\varepsilon}$ is the stretch rate, while $\dot{\gamma}$ is the velocity gradient in a simple shear flow. We define a generalized Newtonian fluid for which the viscosity η depends upon $\dot{\varepsilon}$ instead of $\dot{\gamma}$. The following constitutive equation is selected for the stress tensor $\boldsymbol{T}$:

$$\boldsymbol{T} = 2\,\eta_0 \cosh(5\,\lambda\dot{\varepsilon})\,\boldsymbol{d} \tag{3}$$

where λ is a characteristic time constant. It was shown in [3,4] that the extensional behaviour of this model is similar to that of the Maxwell-B model up to $\lambda\dot{\varepsilon} = 0.45$. It has also been shown that its increased extensional viscosity enhances the vortex growth in the circular 4:1 contraction flow.

The constitutive equation (3) is coupled with the momentum equations where inertia terms are taken into account. We define the Reynolds number Re and the Elasticity number E (although we do not consider memory effects) as follows

$$Re = \frac{\rho\,U\,D}{\eta_0},\qquad E = \frac{\lambda\,\eta_0}{\rho\,D^2} \tag{4}$$

where ρ is the fluid density, U denotes the mean velocity in the downstream tube of the contraction of diameter D. When fthe flow rate increases, Re increases while E, which characterizes the fluid and the geometry, remains constant.

The numerical simulation is performed by means of the velocity-pressure formulation over the finite element meshes of Fig. 1. In particular, we observe that MESH3 is characterized by fairly small quadrilaterals near the reentrant corner.

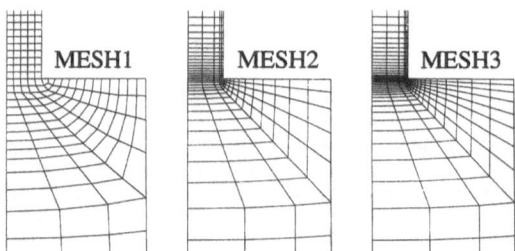

Figure 1. Finite element meshes in the vicinity of the reentrant corner.

RESULTS

We investigate with model (3) the contraction flow for two values of the Elasticity number E : $E = 1/80$ and $E = 1/60$. For a given fluid, this second situation suggests a flow through a contraction of smaller dimension. For both values, Fig. 2 displays the vortex development as a function of Re, or the flow rate.

When $E = 1/80$, the stagnant corner vortex decreases with increasing Re, and a small lip vortex appears at $Re = 19.2$. At $Re = 25.9$, this lip vortex has grown in size and intensity.

For $E = 1/60$, the stagnant corner vortex also decreases for moderate values of Re. At $Re = 12$, we find a small lip vortex which becomes more intense and grows with increasing Re. At $Re = 18$, this strong lip vortex has enveloped the remaining stagnant corner vortex.

The vortex growth mechanism simulated here is similar to that sketched by Boger in [8]. Let us point out at this stage that this lip vortex growth has been quantitatively observed on all three meshes. In particular, for $E = 1/80$ and $Re = 25.9$, the lip vortex is covered by more than 40 finite elements on MESH3. This emphasizes that it is not a numerical artefact.

It is well known that inertia is responsible for decreasing the stagnant corner vortex [1,2]. It appears that the observed lip vortex results from the growing extensional effects which progressively dominate the inertial ones, as the flow rate increases with Re.

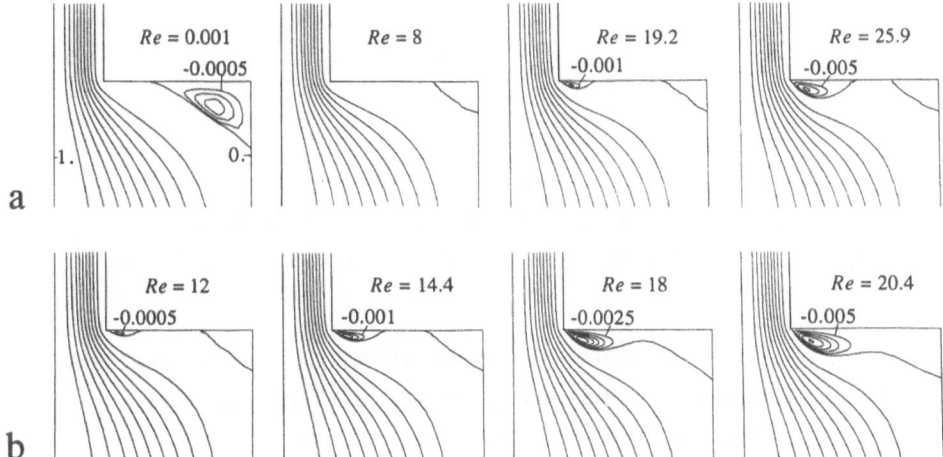

Figure 2. Streamlines obtaines at $E = 1/80$ (a) and $E = 1/60$ (b) for various values of Re.

CONCLUSIONS

The inertial contraction flow of a generalized Newtonian fluid with increased extensional viscosity has been investigated. For particular values of the Elasticity number, we find that the combination of the opposite inertial and elongational effects generates a small and strong lip vortex which may envelop the remaining stagnant corner vortex. Viscoelastic effects have not been considered here ; they should be taken into account in a later work.

REFERENCES

1. Christiansen E.B., Kelsey S.J. and Carter T.R., Laminar tube flow through an abrupt contraction, A.I.Ch.E.J., 1972, 18, 372-380.
2. Vrentas J.S. and Duda J.L., Flow of a Newtonian fluid through a sudden contraction, Appl. Sci. Res., 1973, 20, 241-260.
3. Debbaut B., Crochet M.J., Barnes H.A. and Walters K., Extensional flows of inelastic liquids, in Proc. of the Xth Int. Cong. on Rheology, vol. 1, ed. P.H.T. Ulherr, Published by the Australian Society of Rheology, Sydney 1988, pp. 291-293.
4. Debbaut B. and Crochet M.J., Extensional effects in complex flows, J. non-Newtonian Fluid Mech., 1988, 30, 169-184.
5. Binding D.M., An approximate analysis for contraction and converging flows, J. non-Newtonian Fluid Mech., 1988, 27, 173-189.
6. Walters K., Some modern developments in non-Newtonian fluid mechanics, in Advance in rheology, I : theory, Proc. IXth Int. Congress on Rheology, Mexico, eds. B. Mena, A. Garcia-Rejon and C. Rangel-Nafaile, Universidad Nacional Autonoma de Mexico, 1984, pp. 31-38.
7. Boger D.V., A highly elasitc constant-viscosity fluid, J. non-Newtonian Fluid Mech., 1977, 3, 87-91.
8. Boger D.V., Hur D.U. and Binnington R.J., Further observations of elastic effects in tubular entry flow, J. non-Newtonian Fluid Mech., 1986, 20, 31-49.
9. Boger D.V., Viscoelastic flows through contractions, Ann. Rev. Fluid Mech., 1987, 19, 157-182.
10. Evans R.E. and Walters K., Further results on the lip-vortex mechanism of vortex enhancement in planar contraction flows, J. non-Newtonian Fluid Mech., 1989, 32, 95-105.

DEFORMATION OF PLASTIC DISPERSE
SYSTEMS (PDS) AT LOW SHEAR RATES

YU.F.DEINEGA, G.B.FROISHTETER, YA.A.VERBITSKY
Ukr.SSR Acad. Sci. Institute of Colloid
Chemistry and Chemistry of Water

ABSTRACT
Integrated study results involving the analysis of rheological
and electrical properties of greases having various nature are
reported. Thixotropic properties of greases are estimated with
reference to their performance in low-speed rolling bearings.

The region of low shear rates is very critical for many technological applications of PDS. This paper discusses this problem with reference to typical PDS, i.e. greases of various
nature obtained by thickening mineral oils with lithium, sodium, barium and aluminium soaps of individual saturated carboxylic acids.

The rheological analysis was performed in rotary plastic
flow viscosimeter with rifled measuring surface having the
uniformity of stressing within 94%. Dynamometers of various
rigidity were used. Fig. 1 shows flow curves for the two PDS
(industrial grease samples) when the shear rate decreased below critical values for tested grease samples it was impossible to reach stable flow conditions. In the shear rate region
below critical values (the latter being close to $\dot{\gamma}$ = 0.252
sec^{-1} and $\dot{\gamma}$ = 0.035 sec^{-1} for sodium and lithium greases, respectively) shear stress alterations are observed which become
higher with further decrease of shear rate. At $\dot{\gamma}$ > 0.252 sec^{-1}

and $\dot{\gamma} > 0.035$ sec^{-1} the conditions of stable flow can be reached, which is described by the smooth curves of $\tau = f (\dot{\gamma})$. The critical rate value for the stable flow conditions was found to be determined by the grease nature. According to experimental findings obtained for reference PDS the critical shear rate was found to change by more than two orders of magnitude depending on the soap cation (from 75.6 sec^{-1} for lithium and o.12 sec^{-1} for aluminium soaps). At shear rates below critical values the shear stress was found to change with time, the time — stress curve having serrated pattern. Apart from grease nature the critical shear rate depend up on the dynamometer rigidity as well. However, even using very rigid dynamometers stable flow conditions cannot be reached at shear rate values below critical.

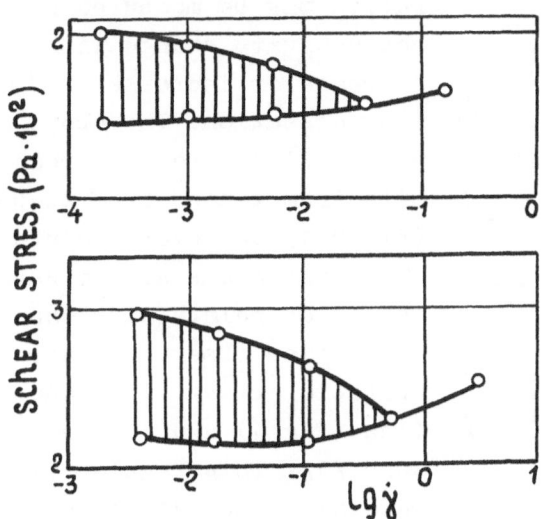

Fig. 1. Shear stress vs deformation rate; 1) Na-grease, 2) Li-grease.

The fact that at low and very low shear rates stable flow conditions cannot be achieved both with PDS in general, and greases,in particular is determined by the rheological properties of these materials, i.e. the frittleness of their structural sceleton and its ability to thixotropic restoration. Electrical conductivity of various nature greases were made to study the above regularities under dynamic conditions. The grease nature was understood as being determined by grease interphase conductivity connected both with the integrity of its structural sceleton and the electrical parameters at the interface. The measurements were performed in rotary plastic flow condensor-viscosimeter; the results obtained are shown in Fig.2. The curves show the conductivity change under various stressing stages; untrained rest condition (a𝖻); onset of structural sceleton breakdown (𝖻c); stable flow conditions (cd); thixotropic restoration of structural sceleton after abrupt flow cessation at d . The rate of thixotropic structural resto-

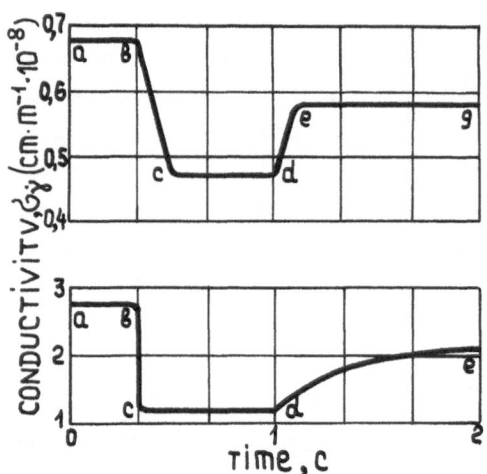

ration may be measured using conductivity change with sharp decrease or increase of δy and flow cessation. The above rate varies within rather broad limits from tenths of a second (Fig.2,1) to several hours for different greases. The data obtained correlate with the regularities observed under unstable grease flow conditions (Fig.1) and in the bearings /1/. This regularities were also observed when analysing electrical conductivity of greases at various shearing intensity.

Fig.2. Kinetics of conductivity changes ($\dot{\gamma}$ =7.55 sec^{-1}); 1) Na-grease; 2) Li-grease.

In order to confirm the correlation between electrical conductivity and structural transformations in greases the analysis of the transformations was performed using viscous flow activation parameters /2/. The structural component of activation energy were compared with conductivity change values. Fair correlation was found within wide alteration range.

Integrated study of rheological and electrical properties of greases allows to obtain their thixotropic properties and may be used as a basis for target-oriented selection of the most efficient grease compositions for the low-speed rolling bearings.

REFERENCES

1. Vinogradov, G.V., Deinega, Yu.F., Verbitski, Ya.A. Rheol. Acta, 1967, 6, N 3, pp. 252-289.
2. Froishteter,G.B., Triliski, K.K., Ishchuk, Yu.L., Stupak,P.M. Rheological and Thermophysical Properties of Greases.Gordon and Breach Science Publishers,New-York-London,1988,p.288.

THEORETICAL CONSIDERATIONS ON THE STEADY ROD CLIMBING AT VANISHING INFLUENCE OF GRAVITY

A. Delgado, S. Berg, R. Kröger, H.J. Rath

Center of Applied Space Technology and Microgravity (ZARM)
University of Bremen, FB 4, P.O. Box 33 04 40
Im Fallturm/Hochschulring, D-2800 Bremen 33

ABSTRACT

Delgado [4] and Delgado & Rath [5] have shown that the surface elevation connected with the rotating rod climbing (Weissenberg effect) conducts to a steady ellipsoid-like body of revolution when the gravity effect is higly reduced. In the present paper a theory has been developed which describes this phenomenon correctly for the steady-state case.

INTRODUCTION

The ability of viscoelastic fluids to climb a partially submerged rotating rod (angular velocity w_0) in the opposite direction of gravity is well known. There exists a large body of literature in connection with this phenomenon. Giesekus [1] and Serrin [2] gave first theoretical explanations of the Weissenberg effect. An excellent overview of the state of art in this field is given by Joseph et al. [3].

Delgado [4] and Delgado & Rath [5] have investigated for the first time the behaviour of a viscoelastic fluid in an environment of highly reduced effects of gravity g (microgravity). Carrying out rheological experiments under microgravity conditions as available not only aboard orbital space crafts but also in earth related facilities as drop towers and special aircrafts offers considerable advantages in many cases. For example, problems connected with the management of test fluids and with unpredictable memory effects resulting from this can be avoided. Futhermore, a lot of physical quantities can be studied free of gravity induced disturbances like sedimentation, thermal and solutal bouyancy. In addition, the investigation of asymptotic fluid behaviour can be carried out with decisive advantages.

[4, 5] studied the Weissenberg effect for the limiting case of very high Froude numbers Fr → ∞, which can be realized when the acceleration g vanishes.

In [4, 5] was observed that the elevation of the surface conducts to the formation of a steady ellipsoid-like body of revolution (see fig. 1). These authors postulate that the fluid motion within the ellipsoid is neither affected substantially by the rest of the viscoelastic fluid nor by the surrounding air rotating which is not able to maintain high momentum transfer. Therefore, the ellipsoid-like shape of surface should express the balance between surface and centrifugal forces, chiefly. The theoretical description given below demonstrates these predictions to be correct.

THEORETICAL DESCRIPTION

Within the scope of a second order theory for the steady state rotating rod problem the pressure p must equilibrate the radial forces (direction of $\underline{e}_r$, see fig. 1) due to centrifugal force ($\rho r\omega$) and the first and second normal stress components N_1 and N_2. Additionally there is an axial pressure component ($\underline{e}_h$-direction) when gravity is present (ρ: density)

$$\text{grad } p = [\rho r\omega^2 - \frac{N_1}{r} + \frac{dN_2}{dr}]\ \underline{e}_r - \rho g\ \underline{e}_h \qquad (1)$$

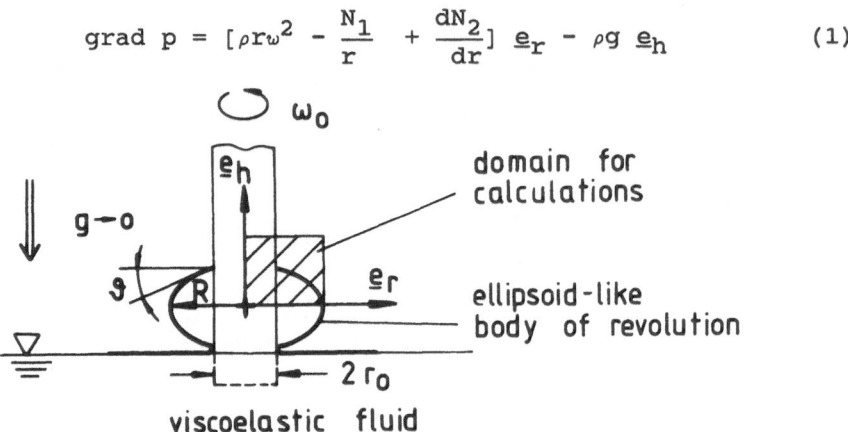

Figure 1. Schematic representation of the Weissenberg effect under microgravity

For the asymptotic case g → 0, the axial pressure component vanishes. N_1 and N_2 can be expressed in quadratic terms of the shear rate γ. The shearing effects due to the surrounding gas phase are expected to be very small and, therefore, N_1 and N_2 are postulated to be negligible in good approximation. Assuming a rigid solid motion in the gas phase and employing the Laplace formula for calculating the pressure discontinuity due to surface tension σ, following expression for the dimensionless surface distortion h* (see fig. 1 for the origin of the coordinate system) is obtained:

$$h*(r*) = \int \frac{H(r*)}{\sqrt{1-H^2(r*)}}\ dr* \qquad (2)$$

with

$$H(r*) = [(1 - r_0* \cos\vartheta - We(1-r_0*^4))(r*^2-r_0*^2)/(1-r_0*^2)$$
$$+ We(r*^4-r_0*^4) + r_0* \cos\vartheta]/r*$$

The dimensionless coordinates h* and r* have been scaled with the radius of the body of revolution R. This solution fulfils the boundary conditions at the radius r_0* and r* = 1.

$$r* = r_0*: h*' = \cot \vartheta; \qquad r* = 1: \lim h*' \to \infty \qquad (3)$$

The prime (') denotes differentiation with respect to r*.

DISCUSSION OF RESULTS

As can be deduced from equation (2) the shape of the ellipsoid-like body of revolution depends on the Weber number We = $\Delta\rho\ \omega_0 R^3/(8\sigma)$, the radius of the rotating rod r_0* and the wetting angle ϑ. $\Delta\rho$

denotes the density difference between the viscoelastic fluid and the surrounding gas phase. The integral expression (2) is of hyperelliptic type for which no general analytical solution has been found because terms of the degree up to r^{*8} are present in the denominator. In contrast to this full analytical solutions can be given for the case We = 0. Additionaly, equation (2) can be easily integrated numerically for arbitrary parameters by a suitable method. More details about that will be published in [6].

For demonstrating the suitability of the theory developed here for describing the behaviour observed by Delgado [4] and Delgado and Rath [5] for the asymptotic case g → 0 some exemplary results are shown in fig. 2. The values of We, r_0^* and ϑ given in fig. 2 are considered to correspond closely to the experimental range in question.

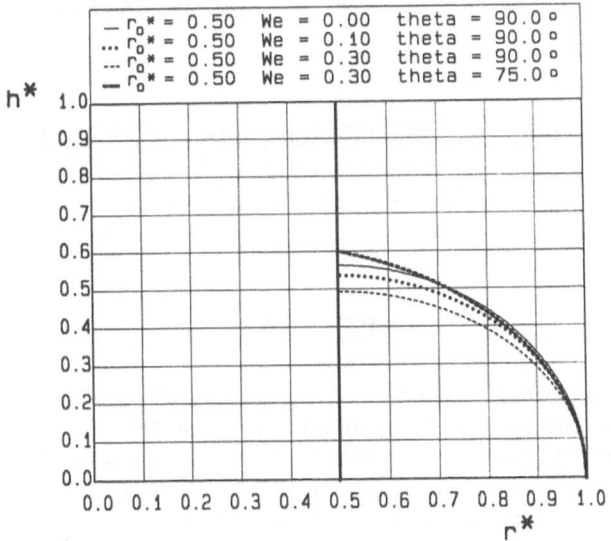

Fig. 2 Exemplary results of calculations

As demonstrated in fig. 2, the theory developed here represents an adequate method for investigating the steady rotating rod climbing at the asymptotic case Fr → ∞. The surface shape calculated is ellipsoid-like as observed in [4].

REFERENCES

1. Giesekus, H., Einige Bemerkungen zum Fließverhalten elasto-viskoser Flüssigkeiten in stationären Schichtenströmungen, Rheologica Acta, 1961, 1, 404.

2. Serrin, J., Poiseuille and Couette Flow of Non-Newtonian Fluids, Z. angew. Math. Mech., 1959, 39, 295.

3. Joseph, D.D. and Beavers, G.S., Free Surfaces Induced by the Motion of Viscoelatic Fluids. In The Mechanics of Viscoelatic Fluids, ed. R.S. Rivlin, AMD 22, ASME, New York, 1977, 59.

4. Delgado, A., First Study of Non-Newtonian Flow Behaviour in Microgravity. Appl. Microgravity Techn., 1988, I, 4, 209.

5. Delgado, A., Rath, H.J., Erste Untersuchungen viskoelastischer Substanzen unter Schwerelosigkeit. Deutsche Rheologentagung 1989, Darmstadt, 1989.

6. Delgado, A., Rath, H.J., Steady rotating rod climbing for the asymptotic case Fr → ∞. To be published.

A COMPUTER CODE FOR SIMULATING TRANSIENT NON-ISOTHERMAL FLOWS OF NON-NEWTONIAN FLUIDS

D.DING, P.TOWNSEND and M.F.WEBSTER
Department of Mathematics and Computer Science,
University College, Swansea.

ABSTRACT

Over the past two years or so, work has been under way at Swansea to develop a computer code for simulating transient flows of non-Newtonian materials. The code is based on an algorithm presented in Townsend and Webster [1] and makes use of a Taylor-Galerkin formulation together with a pressure correction scheme to account for the equation of continuity. The code has already been used for predicting a variety of two-dimensional flows of Newtonian, power-law and visco-elastic materials and work is in progress to extend consideration to three-dimensional geometries.

In this paper we shall concentrate on non-isothermal flows and will describe how the code has been validated using a number of well-documented problems whose solution is known. We will then go on to describe results for more complex situations where no analytical solution is available. Where possible we shall compare our predictions with either experimental results or with other numerical predictions.

INTRODUCTION

Computer codes are now used widely in industry to predict flows of rheologically complex materials. In many cases the simulation procedure may adopt simplifying assumptions - axial symmetry, isothermal conditions, simple fluid models, and so on. With the development of ever more powerful computers the ability of codes to accurately predict complex flows of industrial significance is growing all the time. In particular, a computer code developed at Swansea is capable of simulating time evolving processes with non-Newtonian materials, complex geometries and non-isothermal conditions. The mathematical details are contained in [1-3].

The development of the code has progressed through a number of stages, from an initial simple two-dimensional Newtonian isothermal version. At each stage it has been necessary to validate the code against known results, either in closed form or computed via some other numerical technique. It is often difficult to find results to compare with, particularly for the transient evolution of flows. Only very limited closed form solutions are available and comparatively little numerical work has appeared in the literature for complex non-Newtonian flows.

MODEL PROBLEMS FOR NON-ISOTHERMAL FLOW

Of interest to ourselves and our industrial collaborators is the interplay between thermal effects and a non-Newtonian shear dependent viscosity. In the conference presentation we will give a number of examples of relatively simple flows which incorporate such effects. Even for such simple flows, validation of numerical results is still a major problem which may only be addressed by performing suitable flow experiments. For the purposes of this extended abstract we describe one typical problem that incorporates the features discussed earlier.

One of the major difficulties is to specify a problem which is physically realisable, particularly where, as in our case, the transient development of the flow is of interest. Figure 1 shows a schematic diagram of such a problem.

134

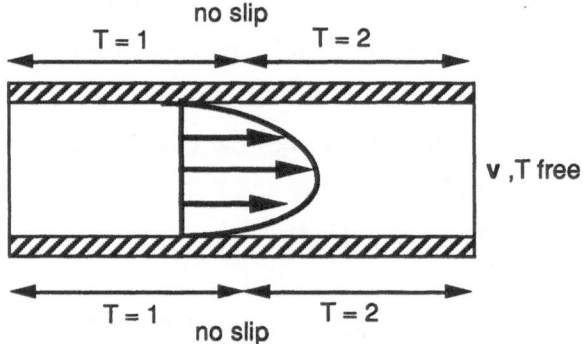

Figure 1. Non-isothermal, transient Poiseille flow

Essentially it represents a simple Poiseuille flow in a channel whose walls are initially held at a constant temperature. (Note: non-dimensional values used throughout). At some instant in time the temperature of the downstream half of the channel is increased and one observes a development of the flow through intermediate stages to a new steady state. Results will be presented showing the various effects of thermal properties and shear thinning. A particularly interesting result is obtained when the Péclet number is sufficiently small for diffusion effects to be comparable with those of convection.

REFERENCES

1. Townsend, P. and Webster, M.F., An algorithm for the three-dimensional transient simulation of non-Newtonian fluid flow, Proc. NUMETA 87, Nijhoff, Dordrecht, Holland, vol. II, T12/1-11, 1987.

2. Webster, M.F. and Townsend, P., Development of a transient approach to simulate Newtonian and non-Newtonian flow, Proc. NUMETA 90, Elsevier Applied Science, vol. II, 1990, pp. 1003-1012.

3. Hawken, D.M., Tamaddon-Jahromi, H.R., Townsend, P. and Webster, M.F., A Taylor-Galerkin based algorithm for viscous incompressible flow, Int. J. Num. Meth. Fluids, vol. 10, no. 3, 1990, pp. 327-351.

ANISOTROPIC TURBULENCE IN AQUEOUS SURFACTANT SOLUTIONS

J. DOHMANN
Lehrstuhl Energieprozeßtechnik, Fachbereich Chemietechnik
Universität Dortmund
P. O. Box 500 500, D-4600 Dortmund 50, F.R.G.

ABSTRACT

Experiments on the influence of a cationic surfactant additive (C_{14}TASal+NaBr) on grid generated turbulence and on the wakes of circular cylinders indicate a change in the structure of the turbulence. The results described here show that the fluctuations perpendicular to the main flow direction are reduced and that the eduction of coherent vortices is shifted to higher critical Reynolds numbers.

FLUID STRUCTURE AND RHEOLOGICAL BEHAVIOUR

Aqueous solutions of some cationic surfactants contain micelles when a critical concentration is exceeded. This critical micelle concentration can be determined using various methods, such as measurements of electrical conductivity [1], ultrasonic velocity [2] and refractive index [3]. The anisotropy of the electrical conductivity of a 2.4 mM solution of C_{14}TASal (Tetradecyltrimethylammoniumsalicylate) at 20 $^{\circ}$C in shear flow shows that the rod-like micelles are orientated in the flow direction; the conductivity in the flow direction is increased whereas it is reduced in directions perpendicular to this. Small-angle neutron scattering experiments also show strong anisotropy, even in turbulent shear flows [4].

At rest and at low shear rates the solution behaves like a Newtonian fluid. Above a critical shear rate, changes occur in the micellar structure with a steep increase in viscosity being observed. The solution shows rheopectic and viscoelastic behaviour. From the temperature and concentration dependence of the relaxation time, it can be concluded that the shear induced structure is in dynamic equilibrium with the forces of the flow [3].

GRID GENERATED TURBULENCE

In Newtonian fluids the turbulence field produced by a grid of parallel rods is a good approximation to homogeneous isotropic turbulence. The symmetry condition reduces the Reynolds stress tensor to its main components. The turbulence is nearly isotropic, so the tensor is determined by its only component $\overline{u_1' u_1'}$, called the turbulence intensity. It is also a measure of the energy of the turbulent motion. The intensity depends on the distance x/M from the grid according to the empirical decay law $\overline{u_1'^2}$ = A $(x/M)^{-n}$, where A is a constant depending on the initial conditions. The decay order n is a constant with a weak dependence on initial conditions and varying in the range between 1<n<2.

In aqueous surfactant solutions the turbulence intensities $\overline{u_2'^2}$ and $\overline{u_3'^2}$ are attenuated, as demonstrated in Fig 1. The decay of the turbulent energy cannot be described by a simple power law [5]. The decay order is not constant but rather depends on the distance from the grid.

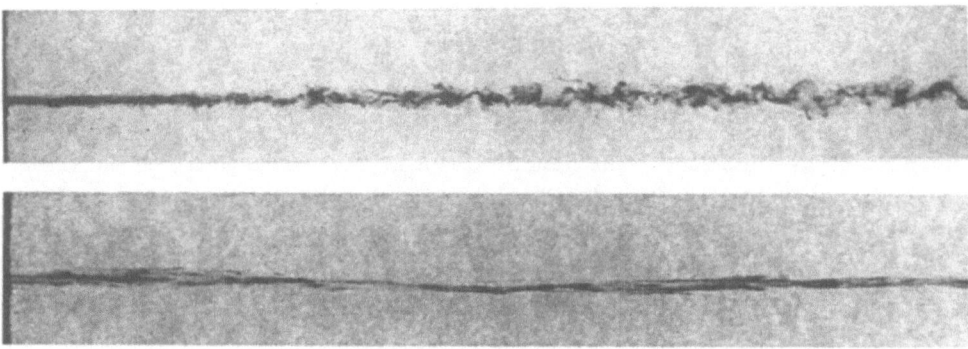

Fig.1 Dye-tracer flow visualization behind a ceramic honeycomb (M=1.25mm, d=0.25mm L=150mm, blockage ratio σ=36) in water and a 2.4mM C_{14}TASal surfactant solution. The flow is from left to right with a mean speed of 0.75m/s in a 40x40mm channel. Note the strong one-dimensional nature of the flow.

KARMAN VORTEX STREET

The inhibited turbulent fluctuation perpendicular to the main flow direction was studied in the wakes of several circular cylinders. The shedding frequency f of the Kármán vortices was determined using fast Fourier analysis of the velocity signals. The Strouhal number was calculated as St=f·d/U (d-rod diameter, U -mean velocity) and is shown in Fig. 2 as a function of the Reynolds number. The data for water show qualitatively good agreement with other published results [6]. The surfactant additive inhibits the eduction of vortices, which corresponds with the suppressed turbulent fluctuations. The critical Reynolds number is shifted from Re_c=45 to Re_c=2500.

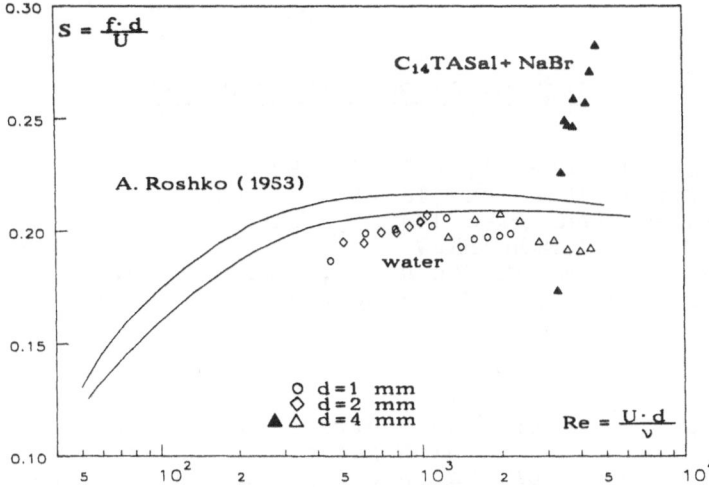

Fig.2 Strouhal number as a function of Reynolds number

CONCLUSIONS

Both grid generated turbulence and the turbulent wakes of cylinders indicate a dramatic change in the mechanism of turbulent motion when the critical micelle concentration of certain cationic surfactants in water is exceeded. The suppressed fluctuations in the cross-flow directions suggest the existence of a "one-dimensional" turbulence.

REFERENCES

[1]   Rehage, H., Dissertation 1982, Universität Bayreuth. F.R.G.

[2]   Zielinski, R. et al.; J. Colloid Interface Sci., **129**, (1989), pp. 175–184

[3]   Wunderlich, I.; Dissertation 1986, Universität Bayreuth, F.R.G.

[4]   Bewersdorff, H.-W. et al., Physica B, **156** & **157**, (1989), pp. 508–511

[5]   Dohmann, J.; Strauß, K.; ZAMM, **70**, (1990), T364–T367

[6]   Roshko,A.; NACA, technical note 2913 (1953)

SHOCK DISINTEGRATION OF POLYMER SOLUTIONS

I.A.DUKHOVSKII, P.I.KOVALEV
A.F.Ioffe Physical Technical Institute, USSR Academy
of Sciences, Leningrad 194021, USSR
A.N.ROZHKOV
Institute for Problems in Mechanics, USSR Academy of
Sciences, pr.Vernadskogo 101, Moscow 117526,   USSR

ABSTRACT

Different stages of the interaction between steel sphere moving at supersonic velocity and slow jet of polymer solution
were investigated by means of high speed photography. It was
observed that polymer solution breaks up into sizable pieces
whereas Newtonian liquid forms finely dispersed phase. There
is the pulling of filament at the first stage of interaction
and then disintegration occurs in the case of capillary thinning liquid filament of polymer solution. It was theoretically evaluated that the internal stresses of order of 1 GPa
arose in the filament before disintegration. Observed effects
could be explained by composite structure of polymer solution
and orientational processes in thinning filament.

INTRODUCTION

Quantitative experimental observations of the impact of rapidly moving solid body with liquid object were solely concerned
with Newtonian liquids: water, glycerine etc. The present
work was undertaken to study the disintegration of liquids
with elastic properties - polymer solutions. It is known that
these liquids demonstrate unusual behaviour in strong flow
and shock collision is an example of such strong flow.

MATERIALS, METHODS AND RESULTS

The experimental technique is described in details in (1).
Tested liquids were water solutions of poly(ethylenoxide)
(PEO) with molecular mass about 4 000 000 and weight concentration 1-4 per cent. Glycerine was used as Newtonian liquid.
Briefly the experiment may be described as follows. A steel
sphere of 30 mm diameter was accelerated by a launching system up to a velocity of 670 m/s and then collided with a thin

jet of tested liquid. Shadow photographing of the collision
process was performed using the equipment for two-frame recording (1).
    Typical results of experiments with polymer solutions
are shown in Fig.1.

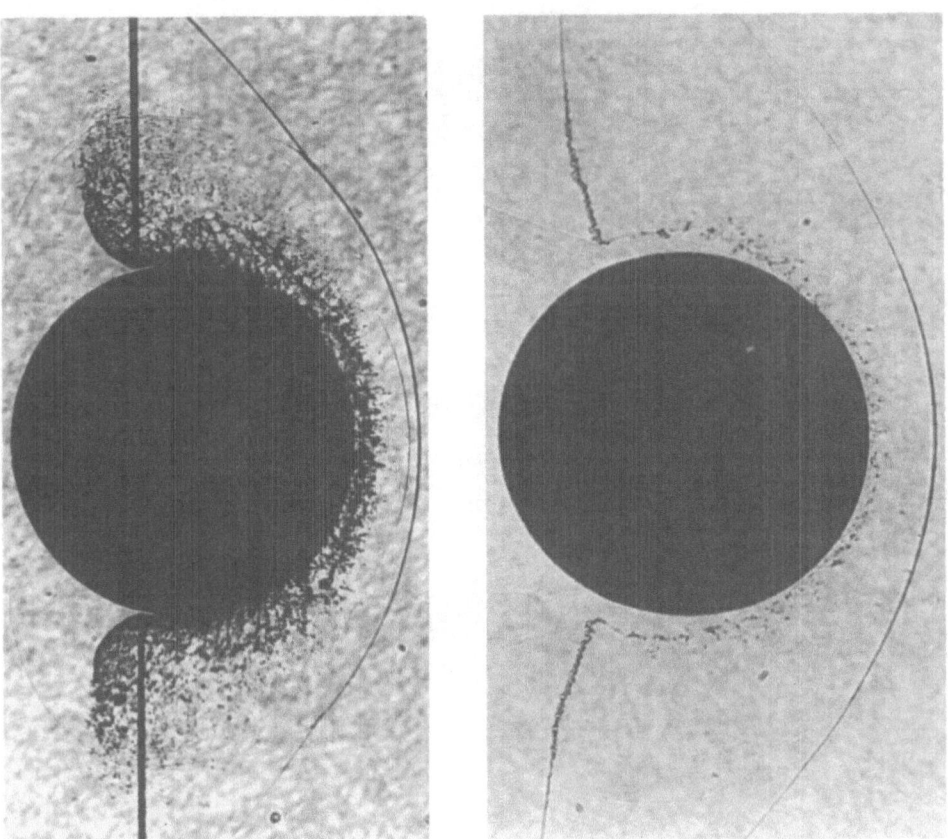

Figure 1. Disintegration of PEO-4% solution (left).
    Interaction of thinning filament of PEO-2%
    solution with a rapidly moving sphere (right).
    Sphere motion: left to right.

DISCUSSION

The disintegration processes of glycerine and polymer solution have quite different features while the shear viscosities
of these liquids differ not much.   Large particles are
formed in the process of disintegration of polymeric liquids.
On the other hand Newtonian liquids form finely dispersed
particles (2). Obviously this difference in fracture properties is connected with composite nature of polymer solutions.
     There is the deflection of part of filament by the
sphere as shown in Fig.1 (right). This deflection is not
connected with action of air-drag forces. Results of (2) and
experiments with small spheres of 10 mm diameter confirm it.
In this case inertial forces in filament are balanced by internal forces in it. Theoretical evaluations show (2) that
internal stresses are of order of ρc^2 , where ρ -density,
c - deflection wave velocity. One can conclude that internal
stresses in filament reach the value of order of 1 GPa. Probably the "hyperstrength" of polymer solution filament is caused by orientation of flexible macromolecules along the
axis of filament during the process of filament thinning.

REFERENCES

1. Dukhovskii, I.A., Komissaruk, V.A., Kovalev, P.I., Mende,
   N.P., High speed photography of the interaction of a water
   drop with a supersonic sphere. Opt. Laser Technol., 1985,
   17, No 3, 148-150.

2. Dukhovskii, I.A., Kovalev, P.I., Rozhkov, A.N., Shock
   disintegration of polymer solution jets. Doklady Akad.
   Nauk SSSR, 1989, 307, No 4, 865-868 (in Russian).

MEASUREMENT OF THE NORMAL STRESS COEFFICIENTS BY MEANS OF AN OPTICAL TECHNIQUE

DOMINIQUE DUPUIS
Laboratoire de Physique et Mécanique Textiles.URA CNRS N° 1303
E. N. S. I. T. M.
11, rue Alfred Werner F-68093 MULHOUSE Cedex

ABSTRACT

The double exposure speckle photograph is applied to the Weissenberg effect in order to determine the two normal stress coefficients. Two different geometries are employed: coaxial cylinders and half-filled concentric spheres. These first results prove the feasibility of the method and display the limits of the theory which is used to fit the experimental data.

INTRODUCTION

The measurement of the two normal stress differences is difficult, especially for low viscous fluids: cone-plate geometries are not very suitable for them. Then, it is possible to work under other flow situations such as the Weissenberg effect which is a consequence of the normal stresses [1]. We apply to it an optical technique: the double exposure speckle photograph. It gives the angle θ between a plane tangent to the free surface of the fluid and an horizontal plane. The details of the experiment have been already published [2] but, we recall briefly its principle and give some new results.

EXPERIMENTAL

A speckle pattern given by the ground glass G (fig.1) illuminates a transparent flow cell. On a photographic plate P' are recorded two images of the plane P situated at the distance ε from the free surface, the fluid being at rest and flowing. Then, an optical Fourier analysis at each point of the photography allows the detection of the splitting of the speckles caused by the changes of the free surface. Young's fringes are obtained. Their interfringe i is proportionnal to $1 / \theta \varepsilon$. From Bohme's theory [3], de-

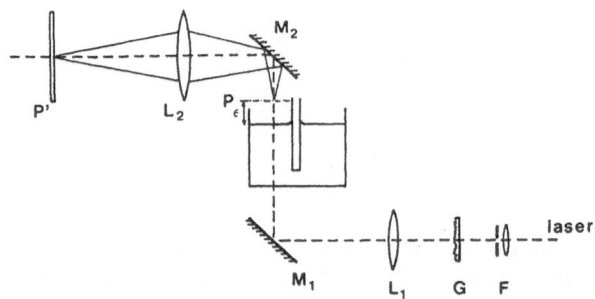

Figure 1: F: spatial filter; G: ground glass; M_1, M_2: mirrors;
L_1, L_2: lens; P': photographic plate.

velopped for slow flows, small changes of the free surface, second order fluids and neglecting surface tension, it can be
shown that $1/i\varepsilon\omega_1^2$ (ω_1 = rotation speed of the inner surface) is
independent from ω_1 and relied to $\psi_{10}+4\psi_{20}$ for coaxial cylinders
and, separately to ψ_{10} and ψ_{20} for concentric spheres (ψ_{10} and
ψ_{20} are the two normal stress coefficients).

RESULTS

In this paper, we give only some results obtained with concentric spheres geometry. They have to be compared to those concerning coaxial cylinders which have been already published [2].
$1/i\varepsilon\omega_1^2$ is plotted versus $R = r/r_1$ (r=radial position; r_1=radius
of the inner surface) for different values of the maximum shear
rate $\dot{\gamma}_m$. Firstly (fig.2), we consider a non elastic fluid (Emkarox FC 31-140, oil provided by Pechiney). The values of $1/i\varepsilon\omega_1^2$
are negative. The continuous line is given by Bohme's theory
with $\psi_{10}=0$ and $\psi_{20}=0$. This is in good agreement with the previous results [2].

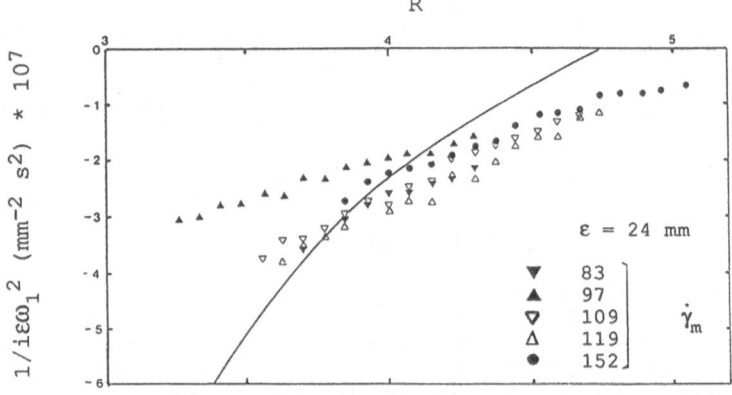

Figure 2: Emkarox FC 31-140; half-filled concentric spheres; inner, outer radius = 0.3 cm, 2.8 cm.

Secondly (fig.3), we consider the fluid M1 which a second order one [3,4]. The values of $1/i\varepsilon\omega_1^2$ are positive. The continuous line is given by Bohme's theory with ψ_{10} = 1.25 Pa.s^2 and ψ_{20} / ψ_{10} = -0.1.But, the value obtained by rheogoniometer's measurements is ψ_{10}= 0.45 Pa.s^2 [4]. The difference can be explained, as in the case of coaxial cylinders, by the influence of surface tension which ignored in Bohme's theory.

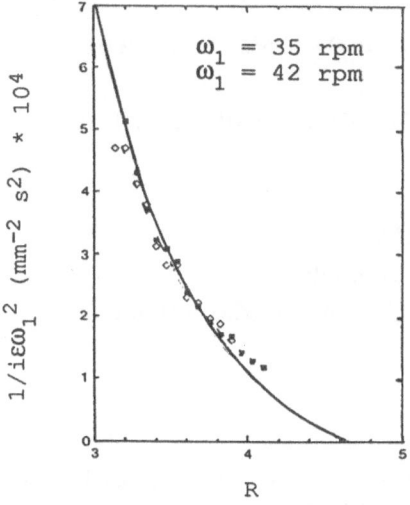

ω_1 = 35 rpm
ω_1 = 42 rpm

$1/i\varepsilon\omega_1^2$ (mm^{-2} s^2) * 10^4

R

Figure 3: Fluid M1; half-filled concentric spheres; inner, outer radius = 0.3 cm, 2.8 cm.

CONCLUSION

These first results are quite promising. Further improvements will be provided, as in the case of coaxial cylinders, by calculation of the profile of the free surface taking into account the effect of surface tension[5].

REFERENCES

1.Le Roy P., Pierrard J.M., L'étude de lois de comportement viscoélastique liquide au moyen d'écoulements de révolution. Revue de l'Industrie Minérale,1970,II,pp.287-293

2.Dupuis D.,Normal stress measurement by speckle photography on the Weissenberg effect,J.n.n.f.m.,1989,33,pp.95-104; Weissenberg effect of the fluid M1; study by an optical method, J.n.n.f.m., to appear.

3.Bohme G.,A theory of secondary flow phenomena in non-newtonian fluids,E.S.A. Tech. Transl.,1975(ESA Org. Report n° TT 157;1974)

4. Hudson N.E., Ferguson J., The steady and oscillatory shear properties of M1, J.n.n.f.m., to appear.

5.Joseph D.D.et al., The free surface on a liquid between cylinders rotating at different speeds, Arch. Ration. Mech. Anal.,1973,49,pp.321-401.

FLOW OF ENTANGLED POLYDIMETHYLSILOXANES THROUGH CAPILLARY DIES:
CARACTERISATION AND MODELISATION OF WALL SLIP PHENOMENA

N. El Kissi & J. M. Piau
Institut de Mécanique de Grenoble
B. P. 53X Domaine Universitaire - 38041 Grenoble Cedex - France.

ABSTRACT

The objective of our study is to clearly identify the succession of the different flow regimes, to analyse their stability and to reveal the conditions of the appearance of the slip phenomenon, for a serie of silicon fluids. We also propose a simple but complete model of flow with slip at the wall specifying its application conditions.

EXPERIMENTAL APPARATUS

We used three highly entangled polydimethylsiloxanes (see Table 1). For each of the fluids, the flow under controlled pressure through a perfectly transparent acrylic die (2mm in diameter and 20mm long), will be considered. A reservoir was machined perpendicular to the capillary, near the entrance region, containing fluid loaded with silicon carbide particles. This fluid can be injected into the capillary under steady flow conditions. In this way, stream lines close to the capillary wall could be visualized.

EXPERIMENTAL RESULTS

For each of our fluids we have plotted the flow curves. There is a characteristic slope discontinuity in all curves at a pressure of about 26 bar. In addition, for LG2 and LG3, which are more entangled than BG, the discontinuity of flow curves is accompanied by a hysteresis and a sudden jump in flow rate, reaching a factor of as much as 100.

Before the discontinuity, flow visualization upstream and into the capillary reveal no instability. Downstream, we first observe a smooth and transparent extrudate. When pressure increases, scratches appear on the jet surface. They grow as the flow regime is increased, giving rise to cracks, localised on the free surface, at the outlet from the capillary. We have shown that the formation of the cracks is associated with the occurence of wrap-around [1]. For BG, the relaxation time is sufficiently small and the size of cracks decreases as they are carried along the extrudate and then degenerate into a roughness surface named sharkskin [2].

Simultaneously with the flow curve discontinuity, the upstream flow becomes unstable. For BG, the rod is excited by these instabilities which trigger the well known melt-fracture phenomenon [2]. For LG2 and LG3, more entangled than BG, visualization of flow inside the capillary clearly shows that the fluid marker injected into the flow, very distinctly moves near the wall, indicating the occurence of slip. In addition, at the capillary outlet, the jet suddenly accelerates and the cracks immediately disappear. The extrudate becomes opaque and there is far less swelling.

MODELISATION OF WALL SLIP

We propose a theoretical model for wall slip taking into account the polymer wall static friction stress [3]. Note τ_S this stress, and τ_R the shear stress at the capillary wall. At small flow rates, τ_R is lower than τ_S and wall velocity, U_R, is zero. If τ_R reaches a critical value equal to τ_S, slip then occurs and U_R becomes positive. Then, flow rate increases while τ_R decreases at first. Finally at higher velocity, τ_R varies as a function of U_R in accordance with a bell-shaped curve [4]. The slip law described above can be represented by the relation :

$$\tau_R = \tau_S - A\left[\, 1 - \exp(-\lambda U_R)\right] - B\left\{\exp\left[-\alpha u_1^2\right] - \exp\left[-\alpha\,(U_R - u_1)^2\right]\right\}$$

where A, B, λ, α and u_1 are constants.

Applied to the particular case of our products, this model reproduces the experimental results as can be seen on figure 1.

REFERENCES

1 - N. El Kissi, J. M. Piau, J. Non-Newtonian Fluid Mech., accepted for publication
2 - J. M. Piau, N. El Kissi, B. Tremblay, J. Non-Newtonian Fluid Mech., 34 (1990), 145.
3 - N. El Kissi, J. M. Piau, C. R. Acad. Sci. Paris, 309, série II, 1989, 7.
4 - D. F. Moore. The friction and lubrication of elastomers. Pergamon Press. 1972.

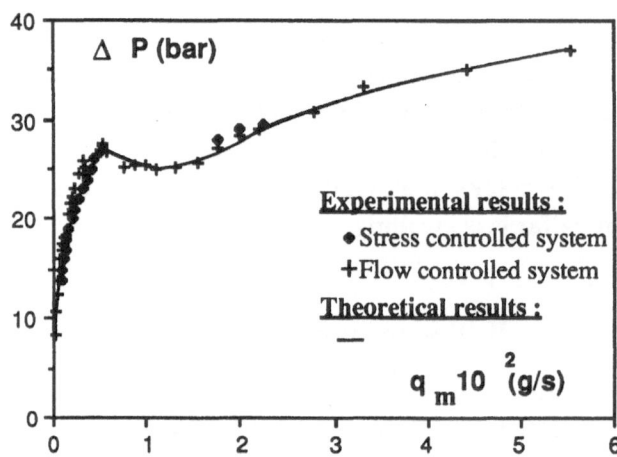

Figure 1 : Total pressure drop versus mass flow rate for the flow of LG2 through a capillary (D=2mm ; L/D=10).

TABLE 1

Rheological data of Polydimethylsiloxanes

Fluid	Mw (g / mol)	Mw / Mn	Mw / Me
BG	$4,28 \ 10^5$	2,9	32
LG2	$8,35 \ 10^5$	4,0	62
LG3	$1,84 \ 10^6$	7,9	136

Inferring a retardation spectrum from creep data by a modified maximum entropy method

C. Elster, J. Honerkamp
Fakultät für Physik der Universität
Hermann-Herder-Str. 3 D-7800 Freiburg / FRG
und Freiburger Materialforschungszentrum F · M · F
Stefan-Meier-Str.31 D-7800 Freiburg / FRG

ABSTRACT

In addition to the regularization method the maximum entropy method is also a tool to solve ill-posed problems such as the determination of relaxation or retardation spectra from rheological measurements. We modify the maximum-entropy method to make it more reliable. The use of the method is demonstrated by testing it with synthetic data and by applying it to realistic creep data.

The determination of the relaxation spectrum from the measurements of the dynamic moduli $G'(\omega)$, $G''(\omega)$ is a particular example where an integral equation of the first kind need to be solved. This task belongs to the class of inverse problems that are known to be ill-posed. In general there is a linear relationship between a measurable quantity $G(\omega)$ and a distribution function $h(\tau)$ of the form

$$G(\omega) = \int\limits_0^\infty d\tau K(\omega, \tau) h(\tau),\tag{1}$$

where the kernel function $K(\omega, \tau)$ is known from the theory. There will be noisy observations g_i^σ of the quantity $G(\omega)$ at $\omega_i, i = 1, \ldots, N$, so that we may assume

$$g_i^\sigma = G(\omega_i)(1 + \sigma_0 \eta_i) = G(\omega_i) + \sigma_i\tag{2}$$

where η_i is a standard Gaussian random number and σ_0 may be called the relative error, σ_i the absolute error.

The determination of the function $h(\tau)$ from these observations $\{g_i^\sigma, i = 1, \ldots, N\}$ cannot be accomplished with the help of the naive linear regression method by choosing $h(\tau)$ so that the discrepancy

$$\chi^2 = \sum_{i=1}^N \frac{1}{\sigma_i^2} \left(g_i^\sigma - \int\limits_0^\infty d\tau K(\omega_i, \tau) h(\tau) \right)^2\tag{3}$$

is minimized. Such a solution would not depend smoothly on the data $g_i^\sigma, i = 1, \ldots, N$. Practically, if one approximates the integral by

$$\int\limits_0^\infty d\tau K(\omega_i, \tau) h(\tau) = \sum_{\alpha=1}^M K_{i\alpha} h_\alpha \tag{4}$$

where $h_\alpha = h(\tau_\alpha)$ and $\tau_\alpha, \alpha = 1, \ldots, M$ may be some set of well-chosen points, it turns out that small changes in the data (well within the experimental errors) will largely change the solution h_α. This indicates that a solution obtained by linear regression is not reliable and more sophisticated tools must be applied.

The breakdown of the naive linear regression method has been elaborated, e.g. in [1], where one of these more sophisticated tools, the regularization method, has been applied to the determination of the relaxation spectrum from the data of the dynamic moduli $G'(\omega)$, $G''(\omega)$ in polymer rheology. This regularization method is also used in a wide spread programm package CONTIN elaborated by Provencher [2]. The difficult point in this method is always the choice of the regularization parameter. In [3] a new method for this choice was introduced which proved to be much more robust and reliable than the earlier ones.

Another method for treating such ill-posed problems is the maximum-entropy - method. This approach has already been widely used in image processing [4], in quasielastic light scattering [5] or in the determination of particle size distribution from SANS-data [6]. The maximum-entropy-method makes use of the fact that for any distribution $h(\tau)$ (which is an intrinsically positive quantity) one may define the entropy

$$S[h] = - \int\limits_0^\infty d\tau h(\tau) \ln[h(\tau)/h_0] \tag{5}$$

where h_0 is some constant. If there is no information about the function $h(\tau)$ we would assume $h(\tau)$ to be a constant. This is exactly what we get if we maximize $S[h]$. Hence, finding the distribution function with maximum entropy leads to a solution which most truly resembles the information that one has and does not introduce more structure into the solution.

In our case we know a little bit more about the distribution function, namely that it should be compatible with the observations, so that for the overall discrepancy between the observations g_i^σ and the expected values of these (on the basis of the distribution function)

$$\chi^2 = \sum_{i=1}^N \frac{1}{\sigma_i^2} \left(g_i^\sigma - \sum_0^M K_{i\alpha} h_\alpha \right) \tag{6}$$

one obtains values less then or equal to N. This ensures that in the mean each individual discrepancy is within the experimental error.

Hence, the maximum entropy method selects the function $h(\tau)$ or the set $\{h_\alpha\}$ for which the entropy S is a maximum, subject to the constraint

$$\chi^2[h] = N. \tag{7}$$

For the case that M, the number of points where $h(\tau)$ shall be estimated, is smaller than N, the maximum entropy method fails, if the data are unrepresentative so that

$$z = \min_h \chi^2[h] \tag{8}$$

is larger than N. Therefore a condition different from (7) leads to a more robust version of the maximum entropy method, namely

$$\chi^2[h] = \frac{N}{N-M}\, z \qquad (9)$$

where z ist calculated by (8). This version can also be applied if the relative error σ_0 is unknown.

Fig.1 shows a reconstruction of a spectrum (dashed line) by the traditional maximum entropy method (Fig.1a) and by the modified version (Fig.1b) with the condition (9). Also the retardation spectra from creep data can be estimated [7].

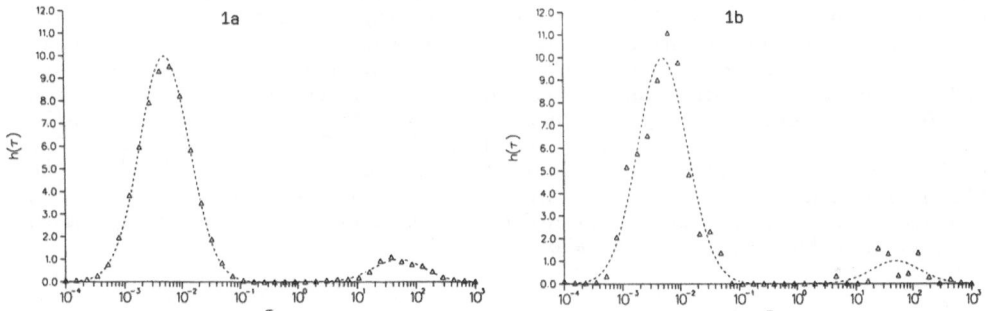

Fig.1 Reconstruction of a spectrum (solid line) from simulated data: (a) by the traditional maximum entropy method, (b) by the modified version

REFERENCES

[1] Honerkamp, J., Weese, J.: Macromolecules 22,4372–4377(1989)

[2] Provencher, S.W.: Computer Physics Communications 27,213–227 (1982)

[3] Honerkamp, J., Weese, J.: Cont. Mech. and Thermod.2.17-30 (1990)

[4] Gull, S.F., Skilling, J.: IEE Proc. 131, 646–659 (1984)

[5] Livesey, A.K., Licinio, P., Delaye, M.: J. Chem. Phys.84, 5102– 5107 (1986)

[6] Potton, J.A., Daniell, G.J., Rainford, B.D.: J. Appl. Cryst. 21, 663–668(1988)

[7] Elster,C.,Honerkamp,J.: Freiburg preprint (1990)

TORSION PENDULUM FOR MEASURING THE DYNAMIC MODULI IN A STEADY SHEAR FLOW

D. van den Ende, F. Ganzevles, J. Mellema, C. Blom
Rheology Group, Faculty of Applied Physics,
University of Twente, The Netherlands

ABSTRACT

To investigate the viscoelastic behaviour of fluid dispersions under steady shear flow conditions, an apparatus for parallel superimposed oscillations has been constructed which consist of a rotating cup containing the liquid under investigation in which a torsional pendulum is immersed. By measuring the resonance frequency and bandwidth of the resonator both in the liquid and in air, the frequency and steady shear rate dependent dynamic moduli G' and G'' can be obtained. By exchange of the resonator lumps it is possible to use the instrument at 4 different frequencies: 85, 284, 740 and 2440 Hz while the steady shear rate can be varied from 1 to 55 s^{-1}. After discussing the design and calibration procedure, measurements on a 0.4% aqueous solution of polyacrylamide are presented. These measurements clearly show the steady shear rate of G'.

INTRODUCTION

Fluid dispersions exhibit under steady flow viscous and/or elastic effects, which can be quite different from the linear rheological behaviour at rest, due to changes in the microstructure. To investigate such behaviour an apparatus has been constructed which measures the complex dynamic shear modulus of superimposed oscillations under steady shear flow. The apparatus consists of a rotating cylindrical cup, containing the liquid under investigation in which the cylindrical lump of a torsional pendulum is immersed coaxially. The measuring method is based upon the determination of the resonance frequency and the bandwidth of the resonator immersed in the liquid and in air respectively. From the difference between the measurements in the two media the liquid impedance can be calculated and from this the frequency and shear rate dependent shear moduli G' and G'' are obtained.

According to the simple fluid theory of Noll [1] the stress in an incompressible liquid is a tensor-valued functional of the history of the deformation tensor. Considering a parallel plate geometry with the lower plate at rest and the upper one moving in its plane with constant speed, the flow between the plates is purely viscometric. In this case the functional reduces to a (tensor valued) function of the steady shear rate $\dot{\gamma}$ and the undetermined isotropic pressure p. An additional small amplitude oscillatory motion of the lower plate, parallel to the velocity of the upper one, results in a superposition of a small amplitude harmonic oscillating shear flow on a viscometric flow, a so-called "nearly viscometric flow". For such a flow stress and deformation can be separated in a viscometric and a non-viscometric part [2]. The non-viscometric parts of the stress and the deformation tensor can be handled as in the linear viscoelastic theory, now resulting in a complex shear modulus $G^*(\dot{\gamma},\omega)$ depending on both the angular frequency, ω, and the steady shear rate. Then for surface loading the liquid impedance on the oscillating plate can be calculated. In the apparatus considered however, the loading on the lump of the surrounding liquid is composed of two parts, one due to the cylindrical shear waves from the side of the lump, the second due to the planar shear waves from top and bottom. Using a first order approximation for the cylindrical part as given by [3] one obtains: $G^*(\dot{\gamma},\omega)=(Z/B)^2\cdot(1+2i(E/\omega\rho B)Z)/\rho$ with $i=\sqrt{(-1)}$ while B and E are geometrical constants of the pendulum [4], ρ the density of the liquid and Z the liquid impedance.

CALIBRATION AND PERFORMANCE OF THE APPARATUS

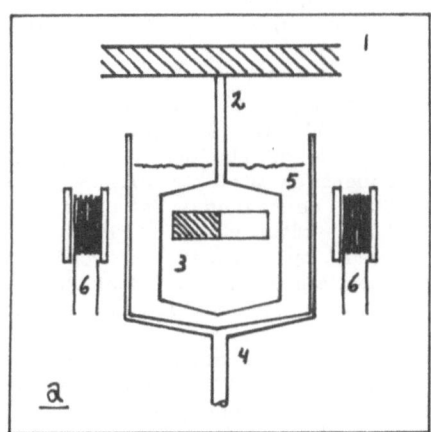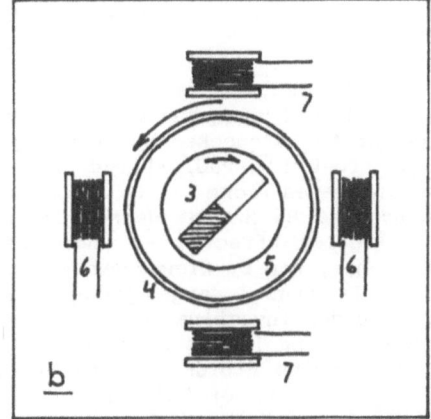

Figure 1. Schematic front (a) and top (b) view of the apparatus. 1: base plate, 2: torsion bar, 3: lump with magnet, 4: rotating sample holder, 5: the liquid, 6: detection coils, 7: excitation coils.

Figure 1 gives a schematic picture of the instrument. In the past much attention has been paid on the mechanical and thermal stability of the torsional pendulum, so resonator details are copied from existing pendula [4]. New is the construction of a rotating sample holder and the improvement of the detection in order to suppress disturbing signals originating from the rotating parts of the instrument. With a permanent magnet in the lump the torsional vibrations can be excited and detected by coils which are positioned just outside the sample holder, in such a configuration that the crosstalk between excitation and detection coils is minimized. Using a lock-inn amplifier the part in phase with the input current of the output voltage is measured as a function of the frequency.

From the shift in the peak frequency $(\nu_o^o-\nu_o)$ and the change of the bandwidth $(\Delta\nu-\Delta\nu^o)$ with respect to measurements in air the liquid impedance Z can be deduced by solving the equation of motion of the pendulum: $Z=2\pi I\cdot((\Delta\nu-\Delta\nu^o) + i(\nu_o^o-\nu_o))$, with I the moment of inertia.

The geometrical constants $A=2\pi I/B$ and E are obtained from measurements on Newtonian liquids with known viscosity and density and compared with values calculated from the nominal dimensions of the pendula. The results are collected in tabel 1 together with the resonance frequencies and the Q-factors of the different pendula. The errors in the measured E values are quite large but the influence upon the accuracy of G^* is small because the value of the correction term containing E is at most 5%. These measurements show an accuracy better than 2% over the whole range of $\dot\gamma$ at the 4 frequencies for liquids with a viscosity between 1.5 and 60 mPas.

TABLE 1
The apparatus constants

pendulum number	resonance freq. [Hz]	Q	A-calc [kg/m^2]	A-meas [kg/m^2]	E-calc [m^{-1}]	E-meas [m^{-1}]
1	84.966	7000	40±1	39.5±0.2	102±2	116±10
2	283.211	15000	40±1	39.6±0.1	102±2	110±20
3	739.432	9000	19.3±0.5	19.5±0.2	235±5	240±20
4	2440.44	15000	19.3±0.5	19.2±0.1	235±5	200±40

A 0.4% aqueous solution of polyacrylamide, PAA, is used as a demonstration of a measurement on a viscoelastic liquid. In this experiment the linearity is checked by changing the amplitude of the oscillating lump from 10^{-5} to $5\cdot10^{-4}$ rad. At the operating frequencies $\eta'=G''/\omega$ does not vary if $\dot\gamma$ increases form 1 to 55 s^{-1} but $\eta''=G'/\omega$ decreases. This is most pronounced at 85 Hz: η'' decreases from 16 to 11 mPas while $\eta'=14.5$ mPas. This elastic effect can not be observed by linear viscoelastic measurements, nor by flow curve measurements, so superposition measurements in this frequency range produce new additional information about the material under investigation.

Summarizing it has been demonstrated that the apparatus operates well with low viscous fluids, due to its high mechanical and thermal stability and high Q-factors of the pendula. The instrument is appropriate to study rheological phenomena under steady shear flow. For PAA the measurements show clearly the influence of this shear flow upon G'.

REFERENCES

1. Truesdell C., Noll W., Non-linear field theories in mechanics, in Springer Encyclopedia op Physics III/3, Springer, Berlin, 1965
2. Schowalter W.R., Mechanics of non Newtonian fluids, Pergamon Press, Oxford, 1978
3. Schrag J.L., Johnson M.J., Rev. Sci. Instr. 1971, **42**, 224
4. Blom C., Mellema J., Rheol. Acta, 1984, **23**, 88

ON THE RHEOLOGICAL PROPERTIES OF RUBBER COMPOUNDS CONTAINING DIFFERENT CARBON BLACK CONCENTRATIONS

SERDAR ERTONG   PAUL SCHÜMMER
Institut für Verfahrenstechnik, RWTH Aachen
Turmstraße 46, 5100 Aachen, FRG

ABSTRACT

SBR 1712 rubber compounds containing different volume fractions of carbon black N330 are measured by means of a Sandwich–Type Creep Rheometer and a modified Weissenberg Rheometer. In steady shear flow the filled melts exhibit yield values superposed with a highly non–Newtonian behaviour which is related to the viscosity of the unfilled elastomer by a hydrodynamic factor. The complex dynamic moduli of the compounds display a strong strain dependence. At high frequencies there is hardly an influence of the carbon black concentration on the loss tangent. For the filled rubber systems with explicit yield stress a modification of the classical temperature–time–superposition principle is recommended.

INTRODUCTION

Unvulcanized rubber–carbon black compounds play an important role in rubber industry especially in the tyre manufacture. A highly filled compound can be classified as a plastic–viscoelastic fluid consisting of a reinforced polymer network and a secondary particle network. The hydrodynamic rubber–carbon black interactions give rise to the discrete shear moduli G_i of the unfilled elastomer which can be described by a hydrodynamic factor dependent on the effective volume fraction $f(\phi_{eff})$. Assuming the validity of Guth's analogy it can be maintained that the relaxation times τ_i are not affected by the presence of the filler [1]. The viscosity function of the filled system $\tilde{\eta}$ is related to the matrix viscosity η by $f(\phi_{eff})$ if no particle–particle interactions occur. Otherwise the secondary network must be considered which is responsible for the yield stress S_Y and the thixotropic behaviour. Proposing a linear superposition of two networks [2] one can describe the viscosity function of the filled elastomer by :

$$\tilde{\eta}(\dot{\gamma}) = f(\phi_{eff}) \cdot \eta(\dot{\gamma}) + S_Y/\dot{\gamma} \qquad (1)$$

For the dynamic moduli at small amplitudes $\dot{\gamma}$ it can be shown :

$$\tilde{G}'(\omega,\dot{\gamma}) = f(\phi_{eff}) \cdot G'(\omega) + S_Y/\dot{\gamma} \;\; ; \;\; \tilde{G}''(\omega) = f(\phi_{eff}) \cdot G''(\omega) \qquad (2)$$

The parameters S_Y and $f(\phi_{eff})$ may also depend on the shear rate or stress. For the determination of the temperature invariant master curves the two networks must be shifted seperately. The temperature dependence of the yield stress is assumed to be Arrhenius–type, and it can be considered by an additional factor s_T.

$$\tilde{\eta}(\dot{\gamma},T) = f(\phi_{eff}) \cdot \eta(\dot{\gamma} \cdot a_T, T_0)/a_T + S_Y(T_0)/(\dot{\gamma} \cdot s_T) \qquad (3)$$

MATERIALS AND METHODS

Four SBR 1712 compounds with theoretical volume fractions of 0, 10, 18, and 27 % N330 are investigated (M_w = 5.9·10[5]; M_w/M_n = 2.7). In order to prevent the wall slippage a modified Weissenberg Rheometer is used [3]. The dynamic measurements are carried out on cone–and–plate geometry. The viscosities at very low shear rates and the yield stresses are measured by means of a Sandwich–Type Creep Rheometer [4].

RESULTS

Figure 1 shows the master curves which are obtained by considering the two networks seperately. The data can be fitted by the pure elastomer viscosity and the yield stresses according to equation (1). The concentration dependence of the hydrodynamic factors follows the Dougherty–Krieger relation [5] quite well.

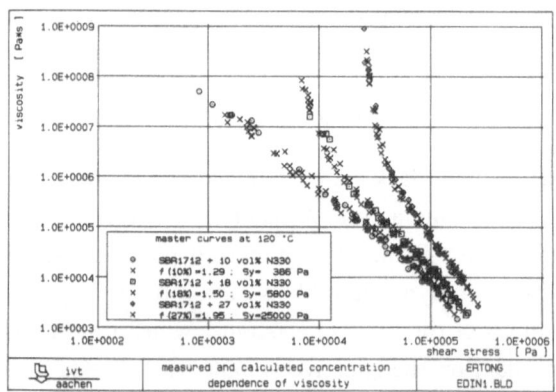

Figure 1. steady shear viscosity of the compounds

In the case of complex moduli a strong strain dependence is observable. It is caused by the reduction of ϕ_{eff} and of the bound rubber at high amplitudes. Therefore, the master curves are determined for constant strains of 10 and 20%. With increasing volume fraction the storage modulus of the pure rubber undergo a nearly parallel shift to higher values, as shown in figure 2. The concentration dependence of loss tangent corresponds qualitatively to the equation (2). At high frequencies and stresses respectively the influence of the yield stress is negligible in comparison to polymer network response. As shown in figure 3, a difference between the compounds occurs only in the low frequency region. Here, increasing carbon black concentration gives rise to the solid response of the filled elastomers.

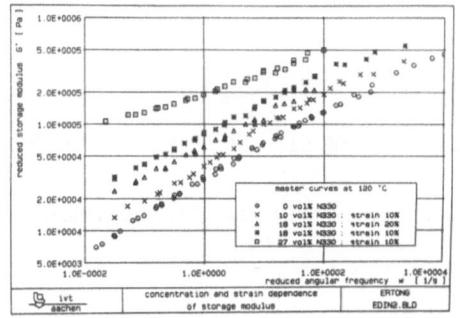 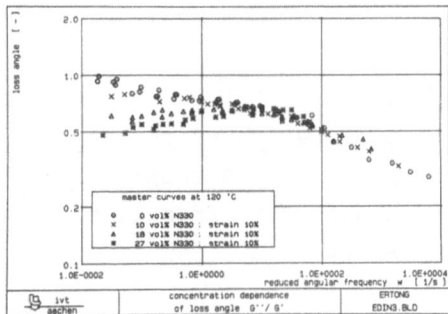

Figure 2 and 3. storage modulus and loss tangent of the compounds

DISCUSSION AND CONCLUSIONS

A qualitative explanation of the measured data is possible by a simple assumption about the linear superposition of two different networks. The steady shear viscosities of the compounds can be described by introducing two further parameters. For the verification of Guth's analogy concerning the relaxation times of the filled rubbers relaxation experiments are intended. Another important question is about the possible variation of the hydrodynamic factor by the structural properties and the activity of the carbon black.

ACKNOWLEDGEMENT

The authors acknowledge the Deutsche Forschungsgemeinschaft for the financial support of this research.

REFERENCES

1. White, J.L., Tanaka, H., Comparison of a plastic–viscoelastic constitutive equation with rheological measurements on a polystyrene melt reinforced with small particles. J. Non–Newtonian Fluid Mech., 1981, 8, 1–10.

2. Stephens, T.S., Winter, H.H., Gottlieb, M., The steady shear viscosity of filled polymeric liquids described by a linear superposition of two relaxation mechanisms. Rheol. Acta, 1988, 27, 263–72.

3. Ertong, S., Schümmer, P., Rheological measurements of unvulcanized rubber compounds using a modified Weissenberg Rheometer. Proc. X Int. Congr. Rheology, Sydney, 1988, pp. 315–7.

4. Laun, H.M., Meissner, J., A Sandwich–Type Creep Rheometer for the measurement of rheological properties of polymer melts at low shear stresses. Rheol. Acta, 1980, 19, 60–7.

5. Krieger, I.M., Rheology of monodisperse latices. Advan. Colloid Interface Sci., 1972, 3, 111–36

The application of rheometry in the testing and characterization of speciality thermoplastics

Dr. Aloyse J.P. Franck, M. McGoldrick; Rheometrics Europe GmbH, Frankfurt

The engineering thermoplastics are high performance materials, finding use in applications were traditional engineering materials are being replaced. Typically, these are materials for use in high temperature applications, where good mechanical strength and good chemical resistance are needed.

The traditional engineering plastic materials were thermosetting materials, as against the presently available speciality thermoplastics. These materials include, PEEK, PES, Polycarbonate, etc., with further blends, copolymers and filled systems. The application of these materials requires information on the service temperature range, mechanical performance, the effect of additives, chemical compatibility of alloys and blends, and processing effects on the end product. These are important material parameters needed in design and production control.

One of the most convenient methods in determining necessary data is by the use of dynamic mechanical analysis. This involves applying a defined deformation to a sample at a set frequency and measuring the material's response, this is carried out from below the material's glass transition temperature T_g to above it's melt temperature T_m, from this one obtains information as to how the loadbearing properties of the material changes with respect to temperature. The evaluated material properties being elastic modulus, which represents the loadbearing ability of the material and the loss factor, which yields important information as to the onset and temperature range of transitions within the material. This is important design information, in determining the allowed loading of a material, temperature limit and in studying the effect of added fillers or process cycle changes.

A typical example of the sensitivity of DMA is illustrated in the case of PEEK. PEEK is an exception in comparison to the majority of the amorphous engineering thermoplastics, in that it can crystallize. Semi-crystalline PEEK has the advantage over amorphous PEEK in that it's elastic modulus retains a significant loadbearing value above it's T_g, which can be further increased through the addition of fillers. However, for the material to crystallize the cooling cycle must be adequately slow, using DMA one can determine amorphous from semi-crystalline PEEK. Semi-crystalline PEEK as against amorphous, has a higher elastic modulus below it's T_g temperature and a smaller loss factor peak at it's T_g temperature.

DMA has for a long time proved it's reliability in testing the compatibility of blended materials, whether the materials were blended to improve the the service temperature or modulus characteristics. Depending on the application, miscibility can be a desired or undesired property. In the case of high impact polystyrene, poly-butadiene rubber and polystyrene exist as a 2-phase IPN. Each material when "fingerprinted" using DMA over a broad temperature range, shows a peak in the loss factor corresponding to it's T_g. A blended material of two materials incompatible with one another, shows two loss factor peaks in it's DMA curve, at the T_g temperatures corresponding to it's original components. In the case of HIP, the two components exist as two phases, a β-transition at −80°C corresponding to the butadiene rubber and at 110°C for Polystyrene. A completely compatible system shows one loss factor peak, between the individual component peaks.

The effect of fillers is an important area to consider. Traditionally fillers were employed on an economic basis, so as to reduce the amount of resin required. However fillers also play an important role in the material properties. The addition of fillers can increase the elastic modulus substantially and reduce creep effects in the design. The addition of filler can increase the elastic modulus, so that above the T_g, the material still has substantial strength. The use of high strength carbon fibres and other materials, has seen the arrival of high strength composite material. This is achieved through the high strength adhesion between matrix and resin. Again DMA can provide a tool in fibre orientation optimization.

The testing of engineering thermoplastics and composites, can be achieved in a variety geometries. However to eliminate the effects of thermal expansion and softening of the material, which can influence one's result, a DMA instrument requires some refinements to accurately characterize materials in torsional rectangular and tension-compression tests. In the case of torsional rectangular testing , axial force on the sample due to thermal expansion must be corrected for. As the sample expands at different rates depending on; filler content, temperature below or above the T_g, this cannot be achieved by a simple preprogrammed expansion coefficient.

Dynamic mechanical analysis can give information as to the processibility of homopolymers. Further, one can determine the performance, morphology and characteristics of blends, copolymers and filled materials. Data important for design and process control.

VISCOSITY MEASUREMENT DURING COOLING OF A CRYSTALLIZING GLASS

Chr. Friedrich* and T. Seidel
Academy of Sciences of the GDR, Institute of Mechanics, Karl-Marx-Stadt 9010,
PF 408,GDR, *University of Freiburg i. Br., Institute of Makromol. Chemistry, FRG

ABSTRACT

The influence of shear rate and temperature on viscosity and microstructure of a crystallizing glass during cooling is explained. A self-developed concentric cylinder vicscometer which works up to temperatures of 1300°C has been used. Different kinds of experiments are performed: shearing with constant rate above and below the liquidus temperature T_L at constant cooling rate and shearing with different rates at constant temperature below T_L. The experiments show that it is possible to influence on viscosity during cooling at temperatures near and below T_L and thereby on microstructure (sice and shape of crystals).

INTRODUCTION

The knowledge of viscosity-temperature-relationship for glasses or glass-ceramics is of fundamental importance for the choice of processing operations . Because of that it is of interrest to investigate the influence of shear on the form of $\eta-T$ curves.This idea was realized by some authors [1-3] at the example of binary metal alloys. The crystallization behaviour under shear was also investigated for food stuff (chocolade, cacaobutter) [4,5] and for polymers [6-8]. For silicatic melts only one paper is known which deals with this problem [9]. It has been found a slight dependence of shear rates on viscosity during cooling of a hawaiian theoleiit in temperature range from 1120°C to 1300°C. For that reason we want to study the influence of shear rate on η-T relationship for a crystallizing glass experimentally and see how the microstructure (sice and shape of crystals) is altered.

MATERIAL and METHOD

The investigated crystallizing glass (glass ceramic "Ilmavit") is a multicomponent silicatic melt with a liquidus temperature of 1170° C. At room temperature the material can be tooled. Detailed information about chemical structure and behaviour at room temperature can be found in [10]. For viscosity measurement we used a self-developed concentric cylinder viscometer with working parameters described in [11]. To show that the viscometer works well in a large temperature range we determined the viscosity of a standard glass with a known viscisity-temperature relationship. These results are displayed in fig.1. The experiments had been performed in following manner: the "Ilmavit"-powder was melted at a constant rate and then tempered at a temperature of 1350°C for 15 minutes. After that different kindes of experiments were performed.

A) cooling with constant rate (T = -2.44 K/min) and after reaching T_L shearing with constant rate γ.

B) cooling with that constant rate to a temperature below T_L and at this constant temperature shearing with different shear rates.

C) cooling and shearing with constant rates to T_L.

When the experiments were finished the melt was chilled to freeze the actual microstructure. This microstructure was detected from polished cuts with the help of a microscope.

RESULTS and DISCUSSION

Some results of experiments A) and B) are presented in fig. 1. In case of the experiment A) the individual $\eta-T$ points are seen. These points were fitted for constant γ by following equation.

$$\lg \eta = \frac{a}{(bT)^c + 1} + d \tag{1}$$

The results of the fit are drawn as thin full lines. It is seen that the viscosity drops down while the shear rate increases. Furthermore it seems to be that at "low" temperatures (T<1050°C) the viscosity reaches a plateau. This fact is regarded by choosing eq.1. The

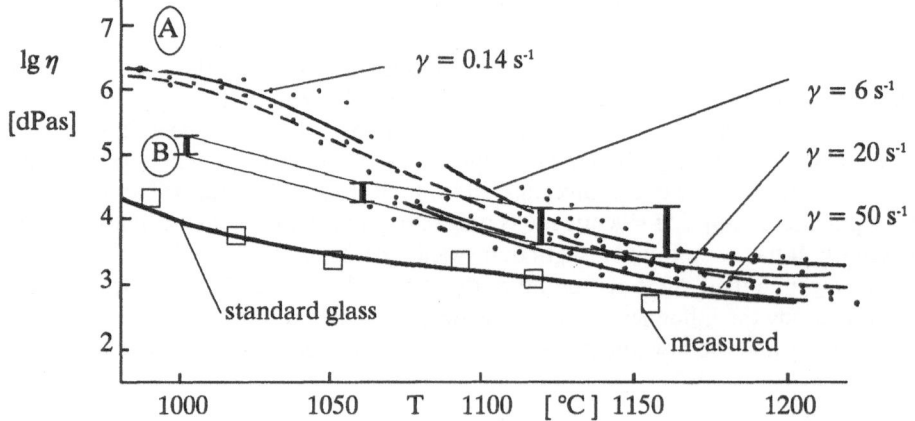

Figure 1. Viscosity-temperature dependence for a crystallizing glass, experiment A) and B) and for a standard glass.

shear rate independent fit of all values is presented by the dashed line. In experiment B) the viscosities are measured in dependence of the shear rate for four different temperatures (1160°C, 1120°C, 1060°C and 1000°C) .These results are given by vertical bars at these temperatures. The upper point of the bar corresponds to a shear rate of about 1s⁻¹ and the lower of about 10s⁻¹. In the "high" temperature area (T>1100°C) experiments A) and B) show the same viscosities. At low temperatures (or left from the inflection point) the "thermostatic" viscosities are lower than the "thermodynamic" ones. No viscosities are given here for experiment C) because the accuracy of the measurements was not good. For this reason only the microstructure was considered.

A comparison of the produced microstructure among one another is given in fig.2.

 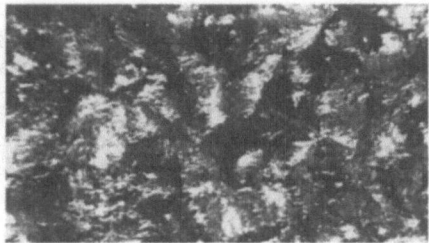

Figure 2. Microstructure of a unsheared (right) and sheared ($\gamma = 50$ s⁻¹) probe . The magnification factor is about 9.

For the experiment A) two cuts are seen which differ only in shear rate. On the left side the fine grained microstructure (magnification factor of about 9) is seen which corresponds to $\gamma = 50$s⁻¹. The right side shows the same material without shearing. Surprisingly was not only the great influence of shear rate on microstructure, but also occur crystals unknown for this material. In experiment C) the influence of shear rate on microstructure is not so strong. Details and photographs are contained in [11]. From that the conclusion can be drawn that subliquidus shearing and superliquidus shearing near T_L is a interessting method to influence on η-T relation and microstructure of the melt.

REFERENCES

1. Spencer, D.B., Mehrabian, R., Flemings, M.C., Metall. Trans. 1972, **3**, 1925-1932.
2. Flemings, M.C., Rieck, R.G., Young, K.P., AFS Intern. Cast. Metals J., 1976, 11-22.
3. Joly, P.A., Mehrabian, R., J. Mater. Sci., 1976, **11**, 1393-1418.
4. Ziegleder, G., IZL, 1985, **6**, 412-416.
5. Tscheuschner, H.D., Markov, E., IZL, 1986, **5**, 315-320.
6. McHugh, A.J., J. Appl. Polym. Sci., 1975, **19**, 125-140.
7. Mackley, M.R., Keller, A., Phil. Trans. Royl. Soc., 1975, **278**, 29-66.
8. Janeschitz-Kriegl, H., et al., Kautschuk und Gummi, 1987, **40**, 301-307.
9. Shaw, H.R., J. Petrology, 1969, **10**, 510-535
10. Reif, D., Stiebert, M., Hirst, H., 28. Int. Wiss. Koll. TH Ilmenau, 24.-28. 10. 1983.
11. Friedrich, Chr., Seidel, T., Report Nr. 25, Institut for Mechanik, AdW der DDR, KMStadt ,1990.

INFLUENCE OF OIL PHASE CONCENTRATION AND TEMPERATURE ON LINEAR VISCOELASTIC PROPERTIES OF MAYONNAISE

C. GALLEGOS
Departamento de Ingeniería Química. Facultad de Química.
Universidad de Sevilla. 41012 Sevilla (SPAIN).
L. CHOPLIN
Département de Génie Chimique. Université Laval.
Québec (CANADA) G1K 7P4.

ABSTRACT

Mayonnaise is a concentrated oil-in-water emulsion, that shows viscoelastic behaviour due to the presence of a three-dimensional network structure. This network is very sensitive to shear. Dynamic shear in the linear region prevents structure breakdown. Thus, linear dynamic viscoelastic properties, as a function of oil phase concentration and temperature, were studied in samples containing egg yolk as the only emulsion stabilizer. Results are discussed in relation to the previous thermorheological history of the sample. Particular attention has been given to the aspects of structure breakdown and recovery after subjecting the sample to strains outside the linear viscoelastic region.

INTRODUCTION

Mayonnaise is an oil-in-water food emulsion, that shows viscoelastic behaviour, related to the formation of a three-dimensional network structure. This is due to, both, the existence of a high concentration of dispersed phase and presence of stabilizers (lipoproteins, gums) that influence the flocculation of oil droplets (1).

Transient stress tests in steady shear show the non-linear viscoelastic behaviour of mayonnaise, with the appearance of a stress overshoot, which depends on shear rate (2-4). A quantitative evaluation of this behaviour through constitutive rheological equations needs a previous linear viscoelastic characterization of these emulsions (5-6).

This characterization can be carried out using stress relaxation or dynamic tests. The latter are of a high significance level and permit the study of the material response in conditions close to the undisturbed state.

In this paper dynamic viscoelastic properties are studied, as a function of oil concentration and temperature, in samples that only contain egg yolk as stabilizer.

Four samples of mayonnaise with different oil contents were
prepared. The oil concentration was only varied between 80 and 75% by
weight, because lower contents yielded unstable emulsions. Two of them
contained 80% in oil, in order to study the influence of the batches.
Compositions of samples are shown in Table I.

TABLE I

Compositions of mayonnaise samples, expressed in percentages by weight.

	MAY-80	MAY-77.5	MAY-75
Oil	80.0	77.5	75.0
Egg Yolk	9.0	9.0	9.0
Vinager	3.3	3.3	3.3
Lemon Juice	0.3	0.3	0.3
Salt	0.6	0.6	0.6
Water	6.8	9.3	11.8

Emulsions were manufactured in a colloidal mill, model Delmix
MZM/VK-7 from FRYMA. The vertical section of the mill is an exact replica
of a machine of industrial manufacture.

Rheological measurements were carried out in a "Rheometrics-System
Four" from Rheometrics, using a plate and plate sensor system, with a
radius of 25 mm. Temperature was varied between 15 and 40ºC.

RESULTS AND DISCUSSION

From the experimental results obtained in strain sweep tests, it can
be deduced that, both, low contents in oil and high temperature produce a
significant decrease in the linear viscoelastic region limits;
nevertheless, the effect of temperature is less significant at low oil
concentration.

Linear viscoelastic behaviour of mayonnaise is gel-like, but Cole-
Cole plots of $\eta"$ versus η' show a slight tendency to a decrease
in the first parameter, within the region of low frequencies. Moreover,
different attempts carried out with the aim of valuating a yield stress,
plotting $|\eta^*|\omega$ versus ω, show a dramatic decrease in the
ordinate axis values within the low frequencies range.

Dynamic viscosity values, η', decrease with oil concentration
and temperature, but the latter has much more influence on this
parameter. In summary, temperature preferably changes viscous character
of the emulsion, but oil content modifies predominantly the elasticity
($\eta"$).

Strains beyond linear limits produce, above a certain yield value,
an irreversible structural breakdown of the emulsion, which depends on
oil content and temperature. This structural breakdown yields lower
values of the viscoelastic functions, but does not modify, in a

TABLE II
Values of the power low parameters $\eta^* = \eta_1^* (\omega/\omega_1)^b$ for the different mayonnaise compositions studied.

W %	T(ºC) 15 η^*_1 (Pa. s)	b	20 η^*_1 (Pa. s)	b	30 η^*_1 (Pa. s)	b	40 η^*_1 (Pa. s)	b
80	447	0.95	407	0.95	316	0.95	324	0.96
77.5	380	0.96	355	0.96	363	0.96	339	0.96
75	158	0.95	200	0.94	214	0.93	148	0.95

$\omega_1 = 1$ rad/s

significant way, their relationship with frequency. The structural recovery of the samples after a change in strain amplitude, from the non-linear to the linear region, can be described by a series of two first order kinetic functions, the former showing a kinetic constant much higher than that of the latter.

ACKNOWLEDGMENT

These results form part of a research project sponsored by the "Junta de Andalucía" (Spain). Authors acknowledge its finantial support.

REFERENCES

1. Kiosseglou, V.D. and Sherman, P., Influence of yolk lipoproteins on the rheology and stability of o/w emulsions and mayonnaise. 1.- Viscoelasticity of groundnut oil-in-water emulsions and mayonnaise. J. Texture Stu., 1983, 14, 397-417.

2. Campanella, O.H. and Peleg, M., Analysis of the transient flow of mayonnaise in a coaxial viscometer. J. Rheol., 1987, 31, 439-452.

3. Gallegos, C.; Berjano, M.; García, F.P.; Muñoz, J. and Flores, V., Destrucción estructural con la cizalla en mayonesas. EFCE Publication Series, 1987, 68, 221-228.

4. Gallegos, C.; Berjano, M.; García, F.P.; Muñoz, J. and Flores, V., Aplicación de un modelo cinético al estudio del flujo transitorio en mayonesas. Grasas y Aceites, 1988, 39, 254-263.

5. Campanella, O.H. and Peleg, M., On the relationship between the dynamic viscosity and the relaxation modulus of viscoelastic liquids. J. Rheol., 1987, 31, 511-513.

6. Berjano, M., Comportamiento Viscoelástico no lineal de mayonesas comerciales. Ph. Thesis, 1989, Universidad de Sevilla.

THE ONSET OF VISCOELASTIC RAYLEIGH-BENARD CONVECTION

B. Gampert, J. Domjahn
Universität Essen, Strömungsmechanik
4300 Essen 1, Schützenbahn 70, West Germany

ABSTRACT

Rayleigh-Bénard convection in aqueous solutions of polyacrylamide is visualized using holographic interferometry. The results confirm and extend earlier findings of the present authors which showed that the critical Rayleigh-number Ra_C for the onset of free convection in viscoelastic fluids can be significantly increased as compared to Newtonian fluids depending on the properties of the polymer and the polymer solution. The present results were obtained in an improved test chamber with well defined electrical conductivity of the solvent.

INTRODUCTION

The onset of steady Rayleigh-Bénard convection in a layer of Newtonian fluid placed between two horizontal plates and heated from below is determined by the critical value Ra_C of the Rayleigh number $Ra = (g\beta h^3 \Delta T)/(a\nu)$, see [1]. At the onset of this kind of free convection the layer of fluid resolves itself into a number of parallel rolls the axes of which are orientated in the direction of the shorter wall. For infinitely extended layers in Newtonian fluids Ra_C is 1708.

With increasing temperature difference a sequence of instabilities appears. First an oscillatory motion of the roll cells can be observed. With enlarged Ra the flow becomes more and more three dimensional, later chaotic and finally fully turbulent.

The subject of the present paper is the onset of Rayleigh-Bénard convection in aqueous solutions of polyacrylamide. Such solutions show viscoelastic flow properties. The question whether the numerical value of Ra_C for these fluids is the same as for Newtonian fluids has been already investigated in earlier papers of the present authors, [2],[3]. It was shown that Ra_C can be significantly increased depending on the physico-chemical properties of the polymer and the polymer solution which determine the rheological characteristics of materials.

As Ra_C is changed non-Newtonian flow properties are obviously effective and the onset of Rayleigh-Bénard convection does not only depend on Ra_c but on additional rheological properties.

Liang & Acrivos [4] and Tien et al. [5] have published experimental results concerning the Rayleigh-Bénard convection in some viscoelastic fluids. While in the first paper no influence of polymer additives on the stability could be observed for aqueous solutions of polyacrylamide, a destabilizing effect was reported in the second paper for CMC. The concentration of the solutions investigated was much higher than in our experiments (0.7 - 7.5 %).

As rheological properties strongly depend on various physico-chemical parameters further experiments were necessary in order to obtain a general picture of the Rayleigh-Bénard convection of viscoelastic fluids.

Theoretical investigations concerning viscoelastic fluids up to now were confined to linear stability analysis, see [6]. Under these conditions no influence of viscoelastic properties on Ra_c had been found.

MATERIALS AND METHODS

Holographic interferometry has been applied in order to study the temperature field.

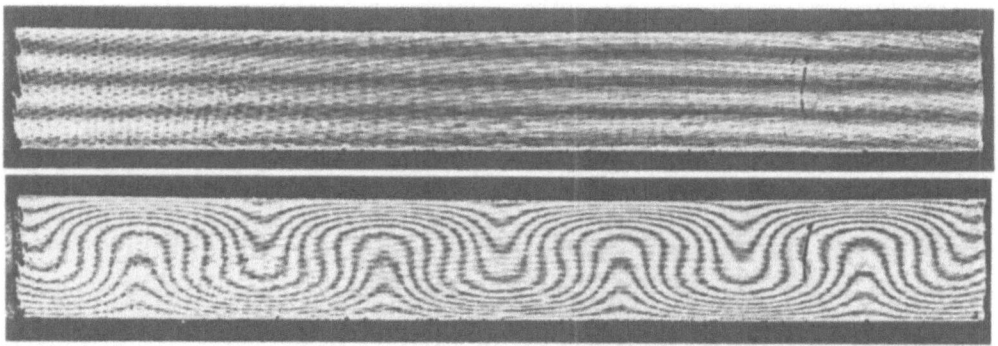

Fig. 1. Interference patterns for a) pure heat conduction, b) steady two-dimensional convection with eight rolls. The bright and dark bands correspond to isotherms.

In the present paper aqueous solutions of technical ionic PAAm P2540 have been investigated. The viscosity function was measured applying a viscosimeter from Fa. Contraves Type LS-30 in the shear rate region between 10^{-2} and 10^2 1/sec. The same zero-shear-rate viscosity was obtained when a Zimm-Crothers viscosimeter was used.

RESULTS AND DISCUSSION

Our results published in [3] (see Fig. 2) were obtained applying solution of PAAm P2450 solved in Essen tap water. The data which showed a Ra_c number increase of up to 50 % scattered broadly due to differences in the electrical conductivity of Essen tap water. The present data were obtained using a mixture of Essen tap water and non-ionic water in such a way that the electrical conductivity was constant $\kappa = 200\mu S/cm$. In addition the

copper surfaces of the test chamber were coated with nickel in order to
passivate the surface. Under these conditions the data were reproduced much
better. The Rayleigh-number now shows an increase of about 20 % (for the
solutions with reduced electrical conductivity), which has a tendency to
diminish slightly towards higher concentration values.

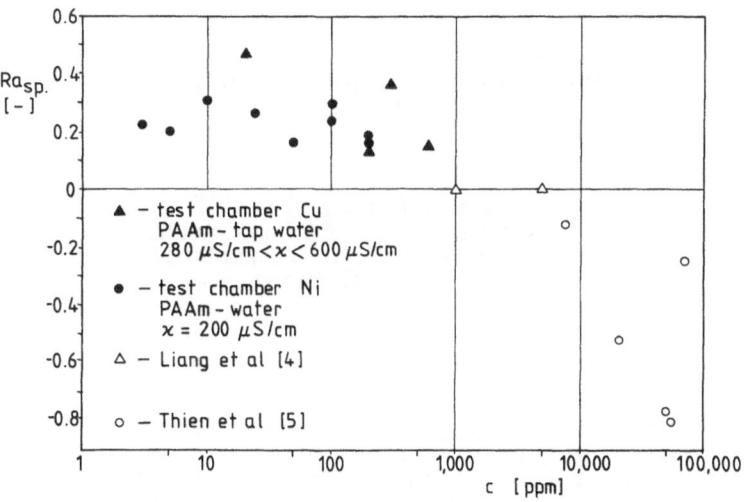

Fig. 2. Ra_{sp} as function of the polymer concentration for different electrical conductivities of the solvent, $Ra_{sp} = (Ra_{c,pol.sol.}/Ra_{c,H_2O}) - 1$.

The results by Tien et al. which show a drastic destabilisation effect are
connected with a Ra number which had been modified by introducing the power
law index of the viscosity function into the Ra number. An interpretation
of these results cannot be given at the moment. In our experiments it was
observed that changes of Ra_c even appear when the shear viscosity shows
Newtonian behaviour in the shear rate region present during Rayleigh-Bénard
convection. Thus the phenomenon cannot be explained by shear thinning effects but should be connected with the presence of elongation rate areas in
the flow field.

REFERENCES

1. Chandrasekhar, S., Hydrodynamic and Hydromagnetic Stability, Oxford
   University Press, Oxford, 1961.
2. Gampert, B. and Wagner, P., In Flow Visualization III, ed. Yang, W.-I.,
   Hemisphere, Washington, 1985, pp. 733 - 737.
3. Gampert, B. and Domjahn, J., In Flow Visualization V, ed. Reznicek, R.,
   Hemisphere, Washington, in progress.
4. Liang, S. F. and Acrivos, A., Experiments on buoyancy driven convection
   in non-Newtonian fluid. Rheol. Acta, 1970, Bd. 9, Heft 3, pp. 447 - 455.
5. Tien, C., Tsuei, H. S. and Sun, Z. S., Thermal instability of a horizontal layer of non-Newtonian fluid heated from below. Int. J. Heat Mass
   Transfer, 1969, Vol. 12, pp. 1173 - 1178.
6. Kolkka, R. W. and Ierley, G. R., On the convected linear stability of a
   viscoelastic Oldroyd B fluid heated from below. J. Non-Newt. Fluid Mech.,
   1987, Vol. 21, pp. 201 - 223.

Relaxation times of aqueous polyacrylamide solutions

B.GAMPERT and M.LIST
Universität Essen, Strömungsmechanik
4300 Essen 1, Schützenbahn 70
West Germany

ABSTRACT

Relaxation times and drag reduction results for aqueous solutions of highly clean laboratory polymerized polyacrylamide are presented. Relaxation times are determined in the region of moderate and high polymer concentrations from steady shear flow measurements following the concepts of Carreau-Yasuda and of Mochimaru. Drag reduction results are expressed by the Deborah number concept applying the theoretical relaxation times from Bird's theory. It is shown that in this presentation results for different pipe diameters follow a single curve.

INTRODUCTION

Carreau and Yasuda [1] and Mochimaru [2] have presented concepts for determining relaxation times in polymer solutions of moderate and high concentration by shear flow measurements. In order to test and to compare the two concepts well characterized experiments in a large region of parameters are needed. Further it is of interest to see how theoretical relaxation times for highly dilute solutions fit into the picture. These theoretical relaxation times moreover can be analyzed by their application. Therefore Bird's relaxation time [3] is used to describe drag reduction results by a Deborah number concept [4],[5].

MATERIALS AND METHODS

Non-ionic polyacrylamide samples with linear molecules and comparatively narrow distributions of molecular mass were prepared under laboratory conditions by a radical solvent-polymerization [6]. Weight averaged molecular masses $\bar{M}_W$ have been obtained from viscosity measurements in a Zimm-Crothers-viscosimeter. By applying a relationship for the special polymerization process given in [6] $\bar{M}_W$ could be determined from the intrinsic viscosity $[\eta]$. Steady shear flow measurements are performed in two coaxial cylinder viscosimeters: Contraves-Low Shear 30 and Contraves-Rheomat 115 with double shearing system in a shear rate range of D~0.01-1000 1/s. The drag reduction experiments were performed in a pressure driven single-pass-system with different pipe diameters. The influence of polymer additives on the pressure loss

Δp was investigated as a function of the Reynolds number for different PAAm samples of varying molecular and solution parameters.

RESULTS AND DISCUSSION

Typical results from steady shear flow measurements are presented in Figure 1. As an example data are shown for $\bar{M}_W=7.4 \cdot 10^6$ g/mol and polymer concentrations between c=100 ppm and c=15000 ppm. It can be seen that the shear viscosity of drag reducing highly dilute aqueous solutions of PAAm (c<100 ppm) is only slightly increased (< 10 %) as compared to the water value and shows a quasi - Newtonian behaviour.

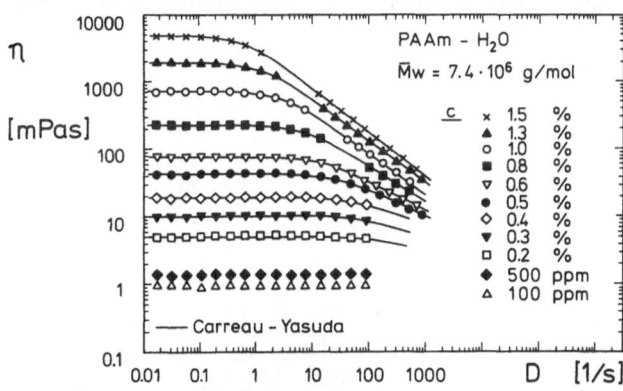

Figure 1.
Shear viscosity dependence of a non-ionic polyacrylamide sample ($\bar{M}_W$= 7.4 · 10^6 g/mol) as a function of shear rate D and concentration c (—— Carreau-Yasuda viscosity function [1], drawn every second measuring point).

Over a wide range of concentration, molecular weight and shear rate the viscosity data of laboratory-polymerized PAAm can be closely described by the Carreau-Yasuda-model [1], see Fig. 1. From the viscosity data of Fig. 1 relaxation times for semi-dilute and concentrated solutions were determined applying the Carreau-Yasuda model [1] and the model from Mochimaru [2] based on the FENE-dumbell concept.

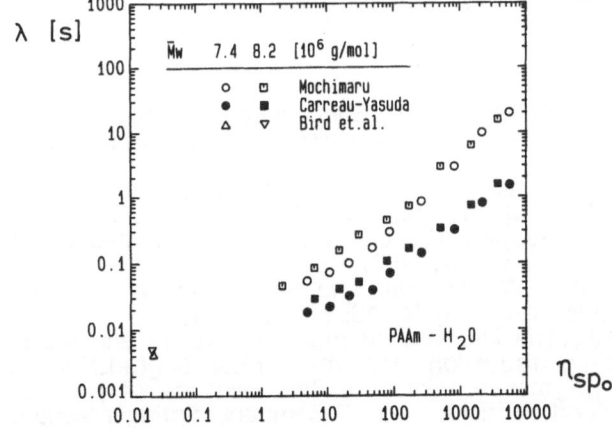

Figure 2.
Relaxation times λ for semi-dilute and concentrated PAAm solutions as a function of zero-shear-rate-specific viscosity η_{sp_0} for different molecular weights.

In the log-log presentation of Fig. 2 the relaxation times λ depend almost linearly on the zero-shear-rate-specific viscosity η_{sp_0} with a slope of approximately one. Data obtained by the Mochimaru model are about one order of magnitude larger than those obtained by the method of Carreau-Yasuda while the kind of the λ - η_{sp_o} - dependence is in good accordance. There are definite changes in the slope of the λ - η_{sp_o}- relationships at $\eta_{sp_o} \approx 100$. For highly dilute PAAm solutions the relaxation times are not directly accessible by steady shear flow measurements. Thus relaxation times for these polymer solutions (c~5 ppm) have been determined theoretically from the FENE-dumbell model by Bird et.al. [3]. The Bird values are roughly in accordance with the trend expressed by the experimental data as can be seen by extrapolating the sequence of experimental data towards low η_{sp_o} values , see Fig. 2. A Deborah number De = λ / λ_T was defined with Bird's theoretical single molecule relaxation time λ and the characteristic turbulence time ($\lambda_T \sim \nu/u_T^2$) [4],[5]. The coefficent in this Deborah number concept was assumed to be one. The turbulent pipe flow drag reduction results were expressed by a reduced drag reduction DR red = DR / (c[η]) [5] and plotted versus the Deborah number.

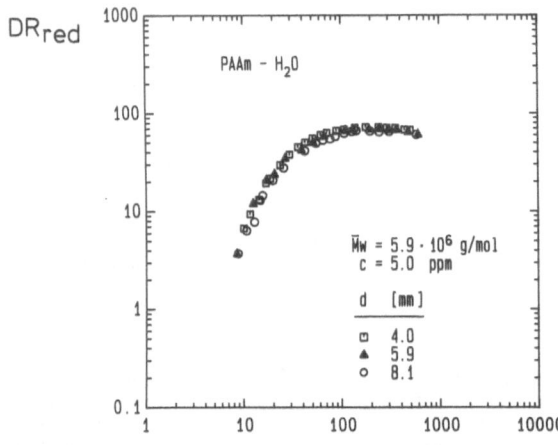

Figure 3.
Reduced drag reduction DRred for a PAAm sample with $\bar{M}_W$=5.9 · 10^6 g/mol and c = 5 ppm as a function of Deborah number De and pipe diameter d.

In Fig. 3 it can be seen that the drag reduction results for three different pipe diameters in this presentation follow a single curve with a uniform onset-Deborah number De≈5. This could mean that the constant in the Deborah number concept should have a value of about 0.1.

REFERENCES

1. Yasuda, K., Armstrong, R.C. and Cohen, R.E., Shear flow properties of concentrated solutions of linear and star branched polystyrenes. Rheol. Acta, 1981, 20, 163-178.
2. Mochimaru, Y., Possibility of a master plot for material functions of high-polymer solutions. J. Non-Newt. Fluid Mech., 1983, 13, 365-84.
3. Bird, R.B., Armstrong, R.C., Hassager, O., and Curtiss, C.F. Dynamics of polymeric liquids. John Wiley&Sons, New York, 1977.
4. Durst, F., Haas, R. and Interthal, W., Laminar and turbulent flows of dilute polymer solutions: A physical model. Rheol. Acta, 1982, 21, 572-77.
5. Gampert, B. and Wagner, P.J., The influence of molecular weight and molecular weight distribution on drag reduction and mechanical degradation in turbulent flow of highly dilute polymer solutions. In The influence of polymer additives on velocity and temperature fields, ed. B.Gampert, Springer-Verlag, Berlin, 1985, pp. 71-85.
6. Kulicke, W.-M., Kniewske, R. and Klein, J., Preperation, characterization, solution properties and rheological behaviour of polyacrylamide. Prog. Polymer Sci., 1982, 8, 373-468.

CONSTITUTIVE EQUATIONS FOR DYNAMICALLY LOADED POLYMER SOLIDS AND THEIR APPLICATION IN A FINITE ELEMENT PROGRAM

Egon Geißler
Fraunhofer Institut für Chemische
Technologie (ICT)
Joseph-von-Fraunhofer-Str. 7
P. O. Box 1240
D-7507 Pfinztal-Berghausen, FRG.

Herbert Weber
Universität Karlsruhe (TH)
Institut für Mechanische Verfahrens-
technik und Mechanik (MVM)
P. O. Box 6980
D-7500 Karlsruhe, FRG.

ABSTRACT

Polymeric materials with good damping properties are used in many technology fields. For example, vibration sensitive products are often cushioned by soft polymers. The polymer damps the mechanical loadings acting on the product during transportation. To analyze the dynamic behavior of such polymeric structures numerically, the finite element method (FEM) is used. For this purpose suitable constitutive equations must be implemented into the program. In this paper constitutive equations for polymeric cushioning materials are introduced with respect to their special loading conditions.

INTRODUCTION

The cushioning structures under consideration are loaded by their own weight and by externally applied dynamic loads. According to these loading conditions, the displacement field consists of a large static component, coupled with superimposed vibrations which are assumed to be harmonic and of small amplitudes. The displacement field and the deformation history are described mathematically by the following equations:

$$\vec{u}(t) = \begin{cases} \vec{u}^m & : \text{ for } -\infty < t \leq 0 \\ \vec{u}^m + \vec{u}^0 e^{i\omega t} & : \text{ for } 0 < t \end{cases} \tag{1}$$

$$\underset{\sim}{G}(t) = \begin{cases} \underset{\sim}{G}^m & : \text{ for } -\infty < t \leq 0 \\ \underset{\sim}{G}^m + \underset{\sim}{\bar{\epsilon}}^0 e^{i\omega t} & : \text{ for } 0 < t \end{cases} \tag{2}$$

In eqn (1) $\vec{u}^m$ is the displacement field of the equilibrium state and $\vec{u}^0$ is the complex amplitude vector of the superimposed vibration. This displacement field results in the strain history in eqn(2). G is Green's strain tensor [1], G^m represents Green's static equilibrium strain tensor and $\bar{\epsilon}^0$ is the complex amplitude tensor of the linearised amplitude tensor of the dynamic strains [2].

SPECIAL CONSTITUTIVE EQUATIONS

The cushioning materials may be regarded as 'simple materials '[3]. This means that Cauchy's stress tensor $\sigma(t)$ is a functional of the strain history G. Using a simple integral representation for this functional, the stress response of the deformation history described in eqn (2) has the following form:

$$\sigma(t) = \sigma^m(G^m) + \bar{\sigma}^0(G^m, \bar{\epsilon}^0)e^{i\omega t} \tag{3}$$

Here σ^m describes the equilibrium state of stress resulting from the prestrain G^m. The complex stress amplitude tensor $\bar{\sigma}^0$ is induced by the dynamic loading.

$$\bar{\sigma}^0 = \bar{\lambda} tr\,\bar{\epsilon}^0 1 + 2\bar{\mu}\bar{\epsilon}^0 + +\bar{\Psi}_2 tr(G^m\bar{\epsilon}^0)1 + \bar{\Psi}_3 G^m tr\,\bar{\epsilon}^0 + \bar{\Psi}_5(G^m\bar{\epsilon}^0 + \bar{\epsilon}^0 G^m) \tag{4}$$

Eqn (4) has the form of an extended Hooke's law. It contains only the Lamé's functions $\bar{\lambda}$ and $\bar{\mu}$.

$$\bar{\lambda} = \bar{\Psi}_0 + \bar{\Psi}_1 tr\,G^m \tag{5}$$

$$2\,\bar{\mu} = \bar{\Psi}_4 + \bar{\Psi}_3 tr\,G^m \tag{6}$$

This form (4) is more general than Hooke's simple formulation. In eqn (4), (5) and (6) the coefficients $\bar{\Psi}_k$, $k = 0,\ldots,5$, are the complex dynamic moduli dependent on the circular frequency of excitation ω.

For a numerical calculation, the material parameters must be determined. Appropriate experiments to identify these parameters are described by H. Weber and A. Geißler [4].

FINITE ELEMENT FORMULATION

A finite element analysis is used to calculate the frequency response which characterizes the damping behavior of polymeric structures [5]. For the FEM formulation, the weighted residual method is used [6].

This formulation results in the following equation for the components of the complex displacement amplitude field:

$$\left(-\omega^2 \int \rho[H]^T[H]\,dV + \int[B]^T[\overline{C}][B]\,dV\right)\{\bar{u}^0\} = \{\overline{F}^0\} \tag{7}$$

Where $[H]$ is the shape function matrix, $[B]$ is the stress displacement matrix and $[\overline{C}]$ is the complex elasticity matrix. $\{\bar{u}^0\}$ is a column matrix which contains all components of the nodal point displacement amplitude vectors. The column matrix $\{\overline{F}^0\}$ is the matrix of the nodal point force amplitudes.

The components of the complex elasticity matrix $[\overline{C}]$ must be calculated for the special constitutive law (4). In this case they are complex, depending on the static prestrain G^m and the excitation frequency ω.

The elasticity matrix is nonsymmetric. It consists of a symmetric and a nonsymmetric part. The symmetric part depends on all complex moduli $\overline{\Psi}_k$, $k = 0, \ldots 5$, the nonsymmetric part depends on the difference of the complex moduli $\overline{\Psi}_3$ and $\overline{\Psi}_2$. If the moduli $\overline{\Psi}_3$ and $\overline{\Psi}_2$ are equal, a complex potential function exists for the complex stress amplitude tensor.

The coefficient matrix of the equations for the assemblage is nonsymmetric when the elasticity matrix is nonsymmetric. In this case the coefficient matrix is divided into a symmetric and a nonsymmetric part. The total system of equations is solved iteratively.

CONCLUSIONS

This paper presents an FEM-formulation for vibration loading of polymeric structures. Special constitutive equations are implemented in a FEM-program.

In this case, the nonsymmetric coefficient matrix possesses a symmetric band-profile structure. The total system of equations is solved iteratively.

Acknowledgement

The authors are greatly thankful to Deutsche Forschungsgemeinschaft (DFG) for supporting this work.

REFERENCES

1. Becker, E., Bürger, W., Kontinuumsmechanik, Teubner, Stuttgart, 1975.

2. Weber, H., Geißler, A., Kugler, H.-P., Einsatz der experimentellen Mechanik in der statischen und dynamischen Prüfung unverstärkter Kunststoffe, VDI-Berichte Nr. 679, 1988, pp. 289-300.

3. Lockett, F. J, Nonlinear Viscoelastic Solids, Academic Press, London, 1972.

4. Weber, H., Geißler, A., Characterisation of a Certain Class of Polymers, Proc. of the 4th SAS-World Conf., Vol. 2, Paris,1988, pp. 431-439.

5. Zienkiewicz, O. C., The Finite Element Method, Mc Graw-Hill, London, 1977.

6. Finlayson, B. A., Weighted Residual Methods and their Relation to the Finite Element Method in Flow Problems, In: Finite Elements in Fluids, Vol. 2, Gallagher, R. H., Oden, J. T., Taylor, C., Zienkiewicz, O. C. (eds), Int. Conf. on the Finite Element Method in Flow Analysis, University of Wales, 1974, pp. 1-29.

SINGULAR FINITE ELEMENTS FOR FLUID FLOW PROBLEMS WITH STRESS SINGULARITIES

GEORGIOS C. GEORGIOU
Unité de Mécanique Appliquée
Université Catholique de Louvain
2 Place du Levant, 1348 Louvain-la-Neuve, BELGIUM
AND
WILLIAM W. SCHULTZ AND LORRAINE G. OLSON
Department of Mechanical Engineering and Applied Mechanics
The University of Michigan, Ann Arbor MI 48109, USA

ABSTRACT

A singular finite element method is presented for fluid flow problems with stress singularities caused by boundary discontinuities. Special elements are employed in the neighbourhood of the singularity, with the basis functions constructed to approximately embody the radial form of the local solution. The stick-slip problem and the die-swell flows of a Newtonian liquid have been solved with the proposed method, and improved accuracy has been achieved with relatively coarse meshes compared to standard finite elements. In the die-swell problem, the convergence of the free surface is dramatically accelerated.

INTRODUCTION

This paper concerns the use of singular finite elements for flow problems with stress singularities caused by sudden changes in the boundary conditions and/or in the boundary geometry. The particular problems examined here are the stick-slip and the die-swell problems.

When standard finite elements are used, the accuracy and the convergence rate in the neighbourhood of a singularity are not in general satisfactory. In fluid mechanics, the numerical inaccuracies due to a singularity are often severe and may propagate to the global solution. The stresses cannot comply with the local solution and thus are tainted by spurious oscillations [1]. In the die-swell problem, the position of the free surface and the resulting die-swell ratio depend on the mesh refinement in the region around the singularity; a coarser mesh gives more swelling [2].

The contamination of the global solution is far more serious in non-Newtonian flows than in their Newtonian counterparts. Numerical inaccuracies caused by singularities can lead to numerically stiff iteration schemes, to the formation of fictitious limit points, or to artificial changes of type of the governing equations [3, 4]. The high Weissenberg number problem is partially due to the excessive approximation error caused by the singularity [5].

The objective of this work is to develop singular finite elements in order to improve the accuracy and the convergence rate of the numerical solution. The basic idea is to incorporate the form of the singularity (obtained by a local analysis) into the numerical scheme. Special elements are used around the singular point with the basis functions constructed in such a manner to approximately embody the radial form of the local solution. Ordinary elements are used in the rest of the domain.

METHOD

The singular elements for the Newtonian stick-slip and the die-swell problem are described in detail in [1] and [2]. As shown by Moffat [6], the flow around the exit of the die is described by two solution sets:

$$\psi = r^{\lambda+1} \left[\cos(\lambda+1)\theta - \cos(\lambda-1)\theta \right], \qquad \text{for} \qquad \lambda = \frac{1}{2}, \frac{3}{2}, \frac{5}{2}, ..., \qquad (1)$$

$$\psi = r^{\lambda+1} \left[(\lambda-1) \sin(\lambda+1)\theta - (\lambda+1) \sin(\lambda-1)\theta \right], \qquad \text{for} \qquad \lambda = 2, 3, 4, ..., \qquad (2)$$

where ψ is the streamfunction and (r,θ) are the polar coordinates centered at the lip of the die. The first term of (1) results in an inverse square root singularity for the pressure and the velocity derivatives.

The velocity basis functions are constructed to embody the four leading terms of the local solution in the radial direction and to be compatible with the biquadratic ordinary elements in the circumferential direction. The pressure basis functions are of the same form as the derivatives of the velocity basis functions with respect to both directions. Thus the singular elements are of triangular shape and consist of 13 velocity and 8 pressure nodes, with no pressure node at the singular point.

RESULTS AND DISCUSSION

The Newtonian stick-slip and the die-swell problems are solved with both ordinary and singular element meshes to make comparisons.

When ordinary elements are used for the stick-slip problem, the normal stress along the wall and the slip surface is characterized by spurious oscillations which move towards the singular point with increasing amplitude as the mesh is refined. These oscillations are eliminated when introducing the singular elements [1].

In the die-swell problem, the position of the free surface is the obvious choice for comparisons. The predicted free surface profiles with various ordinary and singular meshes are plotted in Figure 1. All the singular meshes give practically the same results and predict the same die-swell ratio (1.186). The ordinary element solution appears to converge slowly to the singular element solution. Thus, the singular elements speed up the convergence of the free surface considerably.

The extension of the singular finite elements to other flows is straightforward provided that the radial form of the local solution is known. Flows of an Oldroyd-B fluid are currently under investigation.

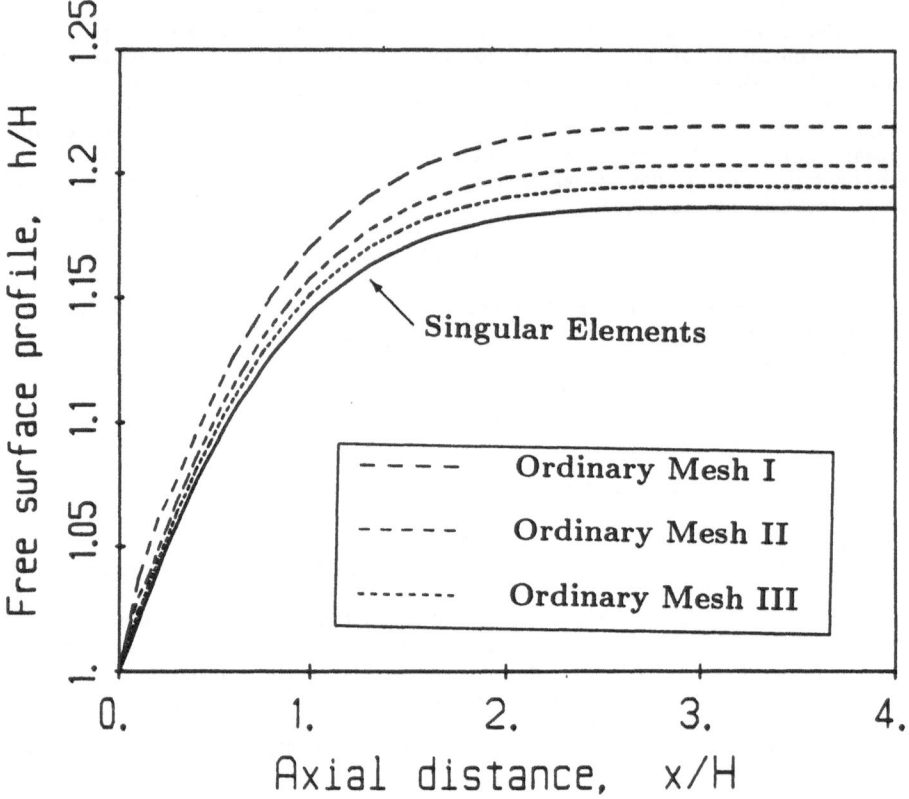

Figure 1. Free surface profiles at zero Re and zero surface tension.

REFERENCES

1. Georgiou, G.C., Olson, L.G., Schultz, W.W. and Sagan, S., A singular finite element for Stokes flow: the stick-slip problem. Int. j. numer. methods fluids, 1989, 9, 1353-67.

2. Georgiou, G.C., Schultz, W.W. and Olson, L.G., Singular finite elements for the sudden-expansion and the die-swell problems. Int. j. numer. methods fluids, 1989, 19, 357-72.

3. Marchal, J.M. and Crochet, M.J., A new mixed finite element for calculating viscoelastic flow. J. Non-Newt. Fluid Mech., 1987, 26, 77-114.

4. Lipscomb, G.G., Keunings, R. and Denn, M.M., Implications of boundary singularities in complex geometries., J. Non-Newt. Fluid Mech., 1987, 24, 85-96.

5. Keunings, R., On the high Weissenberg number problem. J. Non-Newt. Fluid Mech., 1986, 20, 209-26.

6. Moffat, H.K., Viscous and resistive eddies near a sharp corner. J. Fluid Mech., 1964, 20, 1-18.

RHEOMETER STRAIN AND STRAIN-RATE ERRORS

DAVID W. GILES and MORTON M. DENN

Center for Advanced Materials, Lawrence Berkeley Laboratory

and

Department of Chemical Engineering, University of California at Berkeley

Berkeley, CA 94720 USA

ABSTRACT

We have found systematic errors which can affect measured rheological properties in the instrumentation of two commercial rheometers. Users of commercial instrumentation should establish the extent to which these and similar errors may exist in their systems.

INTRODUCTION

We have found systematic errors in the operation of two commercial rheometers, the Rheometrics Mechanical Spectrometer Model 705M (*RMS*) and the Rheometrics Stress Rheometer Model 8600 (*RSR*). The errors appear to be inherent in the designs, though we do not know the extent to which such errors exist in other examples of these models or in similar rheometers produced by the same or other manufacturers. Users of commercial instrumentation should establish the extent to which these and similar errors may exist in their systems.

MECHANICAL SPECTROMETER

The error in the RMS is in the steady strain-rate drive system, which has an oscillatory component superimposed on the steady component. The oscillations are illustrated in Figs. 1a and 1b at strain rates of 0.001 and 0.01 s^{-1}, respectively, for a polydimethylsiloxane test fluid (code 040890) at 23°C. The measurements were carried out in the cone-and-plate geometry with 25mm platens and a 0.1 rad cone angle, using the 100 gm-cm transducer. The oscillation amplitude is about twelve percent of the mean

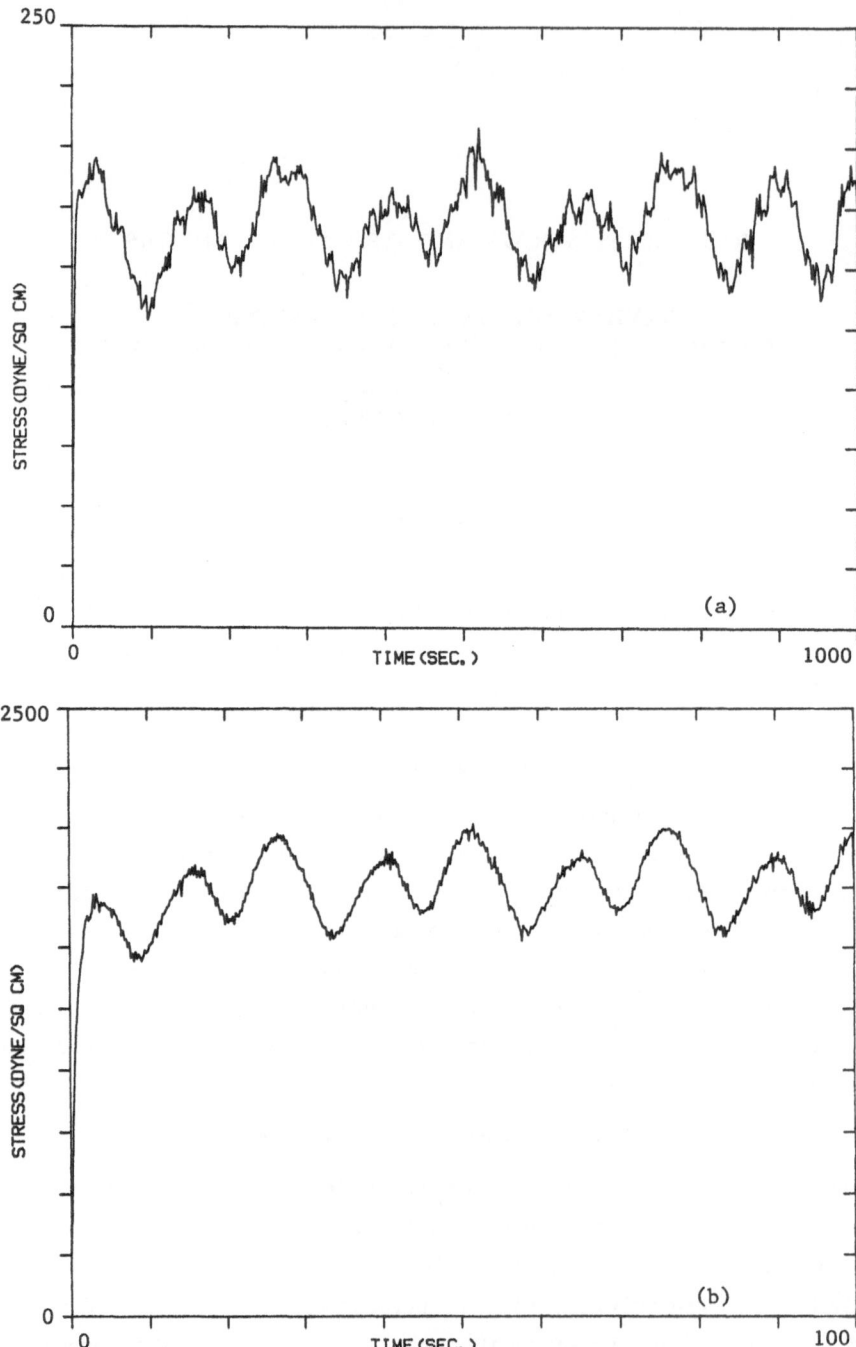

Figure 1. RMS measurement of stress vs time for PDMS at (a) 0.001 s^{-1} and (b) 0.01 s^{-1}. Note the changes in the time and stress axes.

stress in each case, and the stress is proportional to strain rate. The time axes are scaled by strain rate; the full scale is 1,000 s in Fig. 1a and 100 s in Fig. 1b, so the total strain is the same. The two experimental traces nearly superimpose on these axes, demonstrating that the error is in the rotational displacement, with a frequency proportional to the strain rate. It should be noted that the period of the oscillation is very small relative to the period of rotation of the platen.

An error of this magnitude will probably have little effect on the measured steady-state rheological properties of flexible-chain polymers, where the linear region can be quite large and oscillations of this type will average to zero. The effect on the measured properties of highly nonlinear materials, such as thermotropic liquid crystalline polymers, cannot be established without independent shear measurements that are truly steady. We do note that we have obtained overlap of RMS and RSR steady shear data for a fully-nematic liquid crystalline copolymer of 80% p-hydroxybenzoic acid and 20% poly(ethylene terephthalate) to within the apparent precisions of the two instruments as discussed here [1].

STRESS RHEOMETER

We have discussed the error in the strain measurement of the RSR in detail elsewhere [2]. A single, linear conversion is used at low rates to interpret the output of two independent strain sensors, each of which operates over a different angular range. This causes a position-dependent strain error which can be of order five percent. A position-dependent calibration can be developed.

ACKNOWLEDGMENT

This work was supported by the Director, Office of Energy Research, Office of Basic Energy Sciences, Materials Science Division of the U. S. Department of Energy under contract number DE-AC03-76SF00098.

REFERENCES

1. Kalika, D. S., Giles, D. W., and Denn, M. M. Shear and time-dependent rheology of a fully-nematic thermotropic liquid crystalline polymer, *J. Rheology*, 1990, **34**, 139-154.

2. Giles, D. W. and Denn, M. M., Strain-measurement error in a constant-stress rheometer, *J. Rheology*, 1990, **34**, in press.

HERBERT FREUNDLICH AND THIXOTROPY
THE LONG HORIZON OF HIS PIONEERING INSIGHT

DAVID GILLINGS
22 Courtenay Drive, Emmer Green,
Reading RG4 8XH, UK

In the few years before the British Society of Rheology was founded, as the British Rheologists' Club, a terminology had been evolving, in fact, as a language in both general and formal terms, familiar only to those in the specialisation. Two such terms adopted at the time were destined for wide application - Rheology itself, recognised as bringing coherence to rather disparate sectors of science; and Thixotropy, the term proposed by Herbert Freundlich to describe phenomena investigated by his colloid research school at the Kaiser Wilhelm Institut (KWI), Berlin. It was first applied particularly to sol-gel transformations, but extended generally to relate to departures from the 'normal' Newtonian viscous flow. Within ten years, the term was well established, so that the first of a series of colloid science monographs was 'Freundlich on Thixotropy' [1].

The development of the subject area was by then quite diverse, with applications ranging from industrial processes to biological systems. To complement Freundlich's descriptive survey in 'Thixotropy', there was emphasis on questions of the mechanism of the phenomena, the origin of effective molecular forces, and resulting responses of the colloid systems. Freundlich had written the monograph while at University College London (UCL) as an émigré, at Frederick Donnan's instance, from the Germany of the time. The work became a nucleus for continuation on rheology and its applications, activated by Charles Goodeve, who was eminent among the co-founders of the Rheologists' Club. His school achieved much in only a few years before World War II, and he gathered a few of us who had been with him at UCL over the transition from Freundlich's time as early members of the Club. One may judge that such growth of Rheology would hardly have happened without Freundlich's presence as an acknowledged leader in the field.

Research on thixotropy had been a predominant part of the programme at the KWI and continued at UCL. Donnan made this evident in the review of the Freundlich work in the Royal Society memoir [2], after his untimely death in 1941. The memoir went on to describe a new development also originating at KWI on ultrasonic techniques applied to colloid research, made possible by a high-intensity oscillator, which fortunately for us continuing the work, had been brought by Freundlich to UCL. Demonstrations on this equipment of effects of ultrasonics on colloids were quite striking. It worked under an adequate range of intensity, frequency and some other conditions for the formation of dispersions and emulsions, investigations on thixotropy, and other observations. Much was mainly qualitative, but some quantitative estimations were attempted, with the aim of deriving relations between ultrasonic intensity and changes in colloid systems, for example in sol viscosities. Although direct ultrasonic intensity measurement was hardly possible with equipment of that time, operation was stable enough for useful inference from input power indications. The most consistent factor which could be established was the role of collapse of cavities, as a dominant mechanism of most dispersion effects. This was not, however, universal and relations between ultrasonic intensity and frequency, and observed effects could by no means be formulated in any simple terms.

In 'Thixotropy', Freundlich referred to the transformation stages from stable sols to the gel state in terms of changing potential energy/interparticle distance relations, but observed that '... (only) fairly general insight is gained thus ... (one can hardly regard descriptions of thixotropy on these lines)... as a method for direct test of any theory of van der Waals' forces...'. While always reserved as to the speculative elements of insight, Freundlich seemed to retain a 'strong impression' that some serviceable relation might be discovered between oscillatory parameters of ultrasonic fields, and measurable effects originating not from cavitation, but from oscillations as such. The last series of observations at UCL did suggest 'direct' effects in gelation sols, and this was supported in a review of other work on long chain polymers in 1940 [3].

The range of interest in ultrasonics developed over the past 50 years has been very wide, in fields as disparate as non-destructive tests of engineering materials, and 'active' therapeutic use in medicine. It is, however, only recently that a term comprehending investigations of well-defined physico-chemical phenomena - Sonochemistry - has come into widely accepted usage. At the same time, a contrast is rather apparent between the marked advances in many technologies applicable to research on molecular force relations, and the still seemingly empirical basis of specification and design of ultrasonic plant and equipment for effecting dynamic changes in any form of colloid or other disperse system, whether on laboratory or engineering scale.

Might not a case be made for taking stock on the question - can the insights into colloid structures and dynamic properties already derived from rheology be taken further on lines suggested by ultrasonic research, possibly by specification of 'designer' molecules with structures which can minimise problems to be expected in ultrasonic absorption methods applied to typically complex colloid systems.

If the case can be sustained to regard 'Thixotropy' in every sense as a point of departure in a field in itself innovative in bringing a range of basic sciences together, what - if any - might be corresponding generalities that will have matured over the next 50 years? Whatever the answer, BSR can still look forward to the Centenary, still there to keep alive the memory of Freundlich's and his contemporaries' age.

REFERENCES

1. Freundlich, H., 'Thixotropy': Hermann & Cie, Paris, 1935. Monograph Series, 'Actualities ...'
2. Donnan, F.G., Obituary Notices, The Royal Society, London, 1942, p. 27.
3. Mark, H., Ultrasonics in Polymer Research, Journal of the Acoustic Society of America, 1945, 16, p. 183 ff.

A POWER-LAW MODEL OF THIXOTROPY

J.C. GODFREY
Department of Chemical Engineering
University of Bradford
Bradford BD7 1DP

M. F. EDWARDS
Unilver Research
Port Sunlight
Merseyside L63 3JW

ABSTRACT

A simple model of thixotropy is described which follows the transition in flow properties of a fluid which is highly viscous and highly non-Newtonian at rest and becomes less viscous and less non-Newtonian after shearing. The instantaneous flow properties at any time during shearing are approximated by the power law. For a complete history of the fluid a family of power law flow curves, with a common intersection point at high shear rate, can be approximated. This allows time dependent properties to be represented by changes in the flow behaviour index and to be modelled as a simple, shear rate dependent, rate process. The description can be applied to both breakdown and recovery processes.

INTRODUCTION

The model is based on experimental data collected in the two types of experiment most commonly used in the study of thixotropic behaviour. In the stress-time experiment conducted at constant shear rate, an initial stress $[\tau_i]$ corresponding to the fluid condition of startup, and an equilibrium stress $[\tau_e]$ corresponding to the fluid condition at steady state after a long period of shearing, can be observed. The time dependent stress decay can be approximated by a simple rate process:

$$d\tau/dt = -k \ [\tau - \tau_e]^a$$

where the 'order' parameter [a] usually seems to be greater than unity; here ≈ 6. The initial stress is a function of the test shear rate and the property of the fluid in its rest condition can be approximated in some cases by a power law model:

$$\tau_i = K_i \, [\dot{\gamma}]^{n_i}$$

The equilibrium stress is also a function of shear rate but the relationship is not that of a flow curve as higher shear rates produce more reduction of thixotropic structure. Thus although the equilibrium relationship can frequently be approximated by:

$$\tau_e = K \, [\dot{\gamma}]^{n}$$

this equilibrium locus is not a true flow curve.

The equilibrium locus makes a convenient practical reference test condition for the measurement of flow curves. Each equilibrium stress/shear rate combination corresponds to one structural condition of the sheared fluid. Increases or decreases in shear rate allow stress measurements which approximate to the fluid response at constanct structure. By returning to the reference shear rate the known equilibrium stress and structural condition can be re-established as a basis for further measurement. A series of measurements based on different reference conditions yields a family of flow curves corresponding to a range of structural conditions. For some fluids this behaviour may be approximated by a family of power law curves extrapolating to a single point $[\tau^1, \dot{\gamma}^1]$

POWER LAW MODEL

Given a range of constant structure flow curves and the establishment of an intersection point, the stress time experiment at constant shear rate can be reinterpreted as a flow behaviour index – time relationship running from n_i [corresponding to the rest condition of the fluid] to $n_e[\dot{\gamma}]$ [corresponding to the flow behaviour index of the sheared fluid]. The value of n_i may be regarded as constant for a range of experimental shear rates starting with the test fluid at a standardised initial condition. The value of $n_e[\dot{\gamma}]$ will be a function of shear rate, higher shear rates producing more structural change and thus higher values of $n_e[\dot{\gamma}]$. These values of $n_e[\dot{\gamma}]$ correspond to true flow curves, not an equilibrium locus.

Thus it is possible to regard the breakdown of a thixotropic fluid at constant shear rate as a rate process in terms of n:

$$dn/dt = k_n \, [n_e - n]^b$$

Both the rate constant k_n and the equilibrium flow behaviour index, n_e, are functions of the test shear rate but the 'order' term b could be regarded as a constant [b = 4.5] for the data examined here.

RESULTS AND DISCUSSION

Measurements were made for a range of fluids and test geometries with the least complex data being obtained for a fuel oil sample and 1° cone and plate. At rest this fluid could be approximated by $\tau_i = 44.3 \, [\gamma]^{0.41}$, in S.I. units. At different structural levels, after shearing, a family of power law curves could be described by the common intersection point: $\tau^1 = 750 \, Nm^{-2}$, $\dot{\gamma}^1 = 660 \, s^{-1}$. Stress-time measurements were made in the shear rate range 0.9 –90 s^{-1} and values of dn/dt in the range 10^{-10} to 10^{-5} were observed. The degree of structural change, as reflected by the flow behaviour index, was very shear rate dependent: $n_e - n_i = 0.17$ at 0.9 s^{-1}, $n_e - n_i = 0.58$ at 90 s^{-1}. The shear rate dependent rate constants varied from 14 at 0.9 s^{-1} to 1.1 at 90 s^{-1}. This does not indicate that higher rates of breakdown occur at lower shear rates but is because the model is a high order process with dn/dt proportional to $[n_e - n]^{4.5}$. The equilibrium locus could be described by $\tau_e = 16 + 1.7 \, \dot{\gamma}^{0.9}$.

The parameters obtained gave a good description of both the stress time and the instantaneous flow curve data. Even in this simple form the model requires a large number of parameters to describe the shear rate dependency: fluid properties at rest, the equilibrium locus and the rate constant. However the model could be used in computer simulation of some flow conditions of a thixotropic fluid e.g. tube flow. The model could also be extended fairly simply to add a yield stress characteristic:

$$\tau = \tau_y + K[\dot{\gamma}]^n$$

with the possibility of allowing τ_y to decrease as n increases in a realistic manner. When used to evaluate recovery data it is interesting to note that values of dn/dt are similar in magnitude [opposite in sign] to those recorded at the same shear rate in breakdown measurements.

RHEOLOGICAL PROPERTIES OF LIQUID CRYSTALLINE HYDROXYPROPYLCELLULOSE AQUEOUS SOLUTIONS

NINO GRIZZUTI
Dipartimento di Ingegneria Chimica
Università di Napoli Federico II
Piazzale Tecchio, 80125 Napoli, Italy

ABSTRACT

Rheological measurements of aqueous solutions of Hydroxypropylcellulose (HPC) have been performed by means of a cone and plate rheometer, in a concentration range where a liquid crystalline phase is formed. The steady-state experiments confirm some peculiar features of liquid crystalline polymeric systems, among them, the absence of a Newtonian plateau at low shear rates, and the non-monotonic shape of the first normal stress difference vs. shear rate curve. The dependence of such properties from HPC concentration is also studied.

INTRODUCTION

A full understanding of the flow behaviour of liquid crystalline polymers (LCPs) is still far from being reached. Compared to "normal", isotropic polymeric systems, LCPs exhibit a number of peculiar rheological properties. Among them, the so-called "three region flow" shape of the viscosity vs. shear rate curve [1], and the non-monotonic behaviour of the first normal stress difference with shear rate [2-4,6].

The origin of this anomalous behaviour can be reconduced to two main properties of such materials:
i) there is a strict interaction between flow and molecular orientation; in fact, the dynamics of the molecular orientation distribution is strongly altered by both the type and the magnitude of the flow;
ii) LCPs are characterized by the so-called "polydomain" structure. The uniform orientation of the director field, which is typical of all liquid crystalline systems, is observed only over relatively small regions, called domains, whose size is in the micron scale. Different domains possess different director orientations, so that LCPs appear to be macroscopically isotropic at rest.

The aim of this work is to present a set of rheological measurements on HPC aqueous solutions in the liquid crystalline phase. An attempt to

relate the rheological results to the predictions of a very recent
theoretical model will also be made.

EXPERIMENTAL RESULTS AND DISCUSSION.

Aqueous solutions of HPC at 60% wt concentration were prepared by
dissolution and long time mixing. 50% wt and 45% wt solutions were then
obtained by further dilution. The limiting concentration for mesophase
formation is reported to be about 41% wt [5]. Above such a threshold, HPC
solutions are cholesteric at rest. Rheological measurements were performed
on a Weissemberg Rheogoniometer, equipped with a cone and plate geometry.

In Fig.1, the steady-state viscosity, η, is plotted as a function of
the shear rate, $\dot{\gamma}$. No Newtonian plateau is observed, even at very low shear
rates, for all the concentrations investigated. It can be noticed, however,
that the low shear rate slope of the 60% concentration curve is much
steeper than that of the two lower concentration ones. In fact, a slope of
-0.80 is found for the most concentrated sample, compared to values of
-0.46 and -0.33 for the 50% and 45% solutions, respectively.

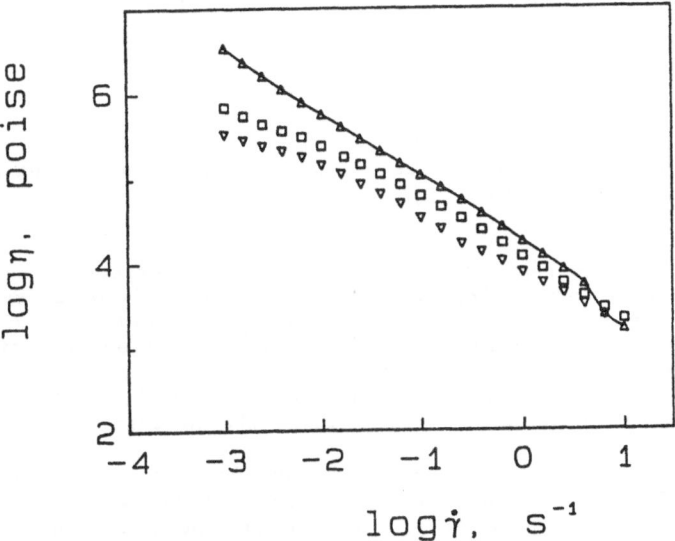

Figure 1. Steady shear viscosity vs. shear rate for different HPC weight
concentrations. ($\triangle$) 60%, ($\square$) 50%, ($\triangledown$) 45%.

Another, apparently minor, peculiarity of the 60% concentration sample is
represented by the inflection point in the viscosity curve, at a shear rate
of about 6 s^{-1}, as shown in Fig.1. This small, but perfectly reproducible,
"hesitation", corresponds to a qualitative change also in the behaviour of
the first normal stress difference, N_1, as reported in Fig.2.

Fig. 2 shows that, for the 60% solution, N_1 initially increases with
increasing shear rate, then reaches a maximum and finally decreases at high
shear rates. Conversely, the normal stress of the lower concentration

samples increases linearly (in a log-log scale) with increasing shear rate (the straight line in Fig.2 has a slope of 0.66). Comparison of Figs 1 and 2 shows that the dip in the viscosity curve takes place at a shear rate where the N_1 curve starts decreasing.

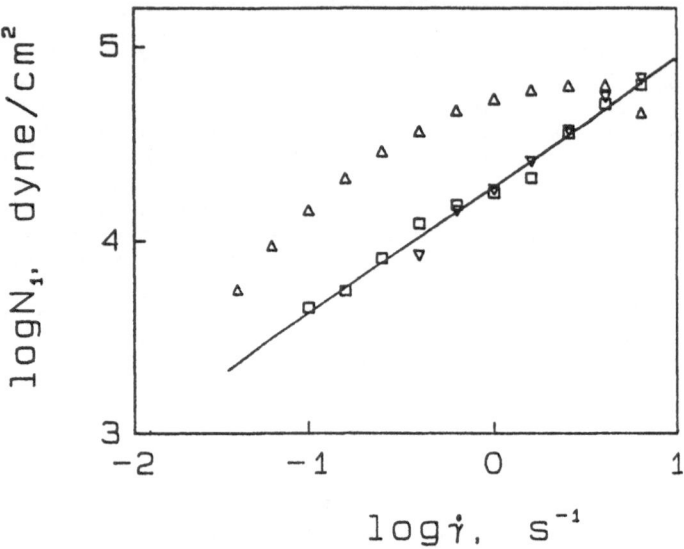

Figure 2. First normal stress difference vs. shear rate for different HPC weight concentrations. Symbols as in Figure 1.

The simultaneous occurrence of the above mentioned effects seems to be peculiar to many LCP systems [2,3,6]. Such a behaviour is predicted by recent theoretical developments of the Doi model for rigid rods [7-9]. According to the theory, the material switches from a situation, prevailing at low and intermediate shear rates, where the director dynamics never reaches a time-independent regime (tumbling), to a completely stationary regime (non-tumbling) at high shear rates, where the director alignes to the flow direction.

The theory correctly predicts the anomalous behaviour of η and N_1 (see, for example, Fig.1 of Ref.8). In particular, the transition between the two flow regimes takes place approximately in the same shear rate region where the maximun in N_1 and the inflection point of the η curve are observed.

The absence of such a behaviour, and the observed moderate slope of the viscosity curves at low shear rates for the less concentrated solutions might be due to the presence of an isotropic phase, i.e., the concentration is not sufficiently high to produce a fully liquid crystalline phase.

REFERENCES

1. Onogi, S. and Asada, T., Rheology and Rheo-optics of polymer liquid crystals, in Rheology, eds G. Astarita, G. Marrucci and L. Nicolais, Plenum, New York, 1980, vol.I, p.127.

2. Kiss, G. and Porter, R.S., Rheology of concentrated solutions of Poly(γ-benzyl-glutamate), _J. Polym. Sci.: Polym. Simp._, 1978, **65**, 193.

3. Moldenaers, P. and Mewis J., Transient behavior of liquid crystalline solutions of Poly(benzylglutamate), _J. Rheol._, 1986, **30**, 567.

4. Navard, P., Formation of band textures in Hydroxypropylcellulose Liquid Crystals, _J. Polym. Sci.: Polym. Phys. Ed._, 1986, **24**, 435.

5. Werbowyi, R.S. and Gray, D.G., _Macromolecules_, 1984, **17**, 1512.

6. Grizzuti, N., Cavella, S. and Cicarelli, P., Transient and steady-state rheology of liquid crystalline Hydroxypropylcellulose solutions, submitted to _J. of Rheology_.

7. Marrucci, G. and Maffettone, P.L., Description of the liquid-crystalline phase of rodlike polymers at high shear rates, _Macromolecules_, 1989, **22**, 4076.

8. Marrucci, G. and Maffettone, P.L., Nematic phase of rodlike polymers. I. Predictions of transient behavior at high shear rates. II. Polydomain predictions in the tumbling regime, submitted to _J. of Rheology_.

9. Larson, R., Arrested tumbling in shearing flow of liquid crystal polymers, _Macromolecules_, 1990, in press.

ISOTROPIC AND ANISOTROPIC SHEAR FLOW IN CONTINUOUS FIBRE THERMOPLASTIC COMPOSITES

D J GROVES, D M STOCKS, A M BELLAMY
Wilton Materials Research Centre
PO Box No 90 Wilton Middlesbrough
Cleveland TS6 8JE England

ABSTRACT

The melt viscoelasticity of continuous fibre thermoplastic composites is characterised isotropically, when using apparent Maxwell viscosities yield stresses are found in the range 10Pa to 1000Pa, and as separate parallel and transverse viscoelasticities relative to the fibres, when the anisotropy of viscous and elastic responses can differ.

INTRODUCTION

A method of characterising the shear flow in continuous fibre thermoplastic composites using a commercial rotational rheometer has previously been described (1). However, the shear strain was imposed in all directions relative to fibre orientation, so the data are essentially isotropic. This was confirmed by the similarity of data for laminates with all parallel fibres and with alternate layers at 0° and 90°. With such obvious anisotropy of fibres it is important to characterise the rheology in both the parallel to fibre and transverse directions. Theory to predict these parallel and transverse components in shear using off centre specimens in the same rotating parallel disc geometry, has been provided by Rogers (2). A preliminary study of two fibre and polymer systems, 60% carbon fibre in a polyetheretherketone matrix (A), and 35% glass fibre in polypropylene (B), is used to compare the isotropic and anisotropic interpretations.

EXPERIMENTAL, RESULTS AND DISCUSSION

All measurements were made with samples between parallel disc platens, rotating or oscillating about the cylindrical axis, using Rheometrics Dynamic or Mechanical Spectrometers. All isotropic work with centred specimens used 25mm diameter platens, while the off centre specimens for determination of anisotropy were rectangular and placed symmetrically in pairs between 50mm diameter platens. Specimens were cut from

precompacted sheets of material, and held at temperature between the
platens in a nitrogen atmosphere for between 3 and 5 minutes prior to
measurement. The pairs of off centre specimens were positioned using
removable templates to determine position and angle of fibre orientation.
Single off centre specimens tended to move radially from the centre,
implying slight platen distortion, so all reported data are from balanced
pairs.

Apparent Isotropic Parameters
The usual in phase and quadrature moduli, G' and G" were measured
dynamically as a function of angular frequency at a range of shear
strains. The dynamic viscosity, η', is strain dependent, but expressing
data as an apparent Maxwell viscosity the strain dependence is
eliminated, furthermore, the Maxwell viscosity approximates to the steady
shear viscosity as shown in Fig 1 for composite B. This is similar to
composite A which has been described elsewhere (1). If expressed as a
function of stress, the viscosity of B falls rapidly from a stress around
10Pa suggesting a yield, whereas composite A had a yield at 1KPa. The
difference in yield stress might qualitatively be associated with the
difference in fibre volume fraction.

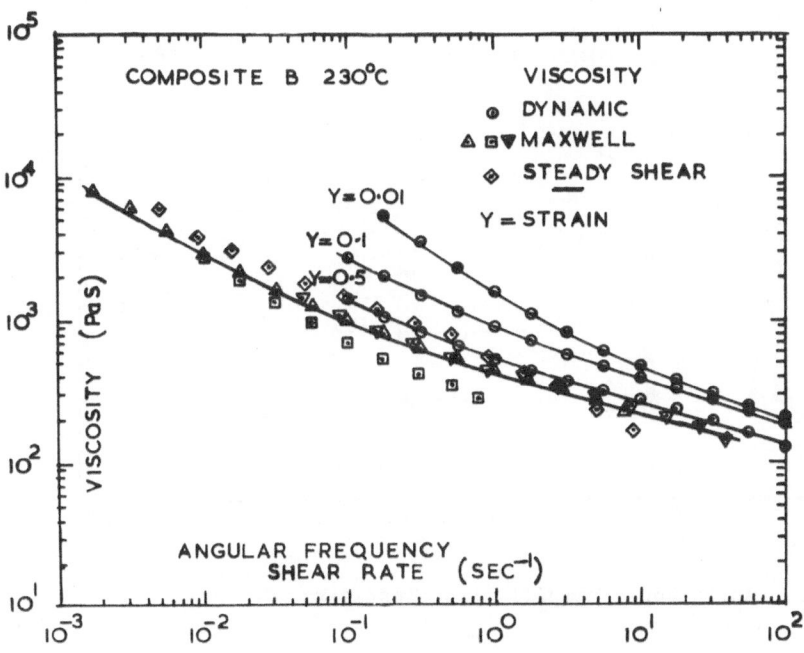

Figure 1. Isotropic viscosities for composite B

Anisotropic Parameters

With offset specimens, the torque was measured dynamically as a function of shear strain amplitude. The fibre orientation in pairs of specimens was matched to avoid mixed phase angle contributions. Rogers (2) has provided analysis for several experimental procedures, though we have used only the dynamic viscoelastic method for this preliminary investigation. Using suffix L and T to indicate along fibre and transverse directions data at an angular frequency of 1 rad/sec show that dynamic moduli G'_L, G'_T, and dynamic viscosities η'_L, η'_T are all strain dependent. It is seen from Fig 2 that the two composites A and B differ in both the magnitude and anisotropy of their viscoelastic responses. Expressed again as Maxwell parameters composite B has both a larger viscosity and modulus parallel to the fibre direction, while composite A has a larger viscosity in the transverse direction and a larger modulus along the fibre direction. These differences might be attributed to the polymer-fibre system or simply the volume fraction. A precise correlation between the anisotropic and isotropic data is difficult with different fibre deformations, and though we note that corresponding data have the same order of magnitude, both may differ from large area values.

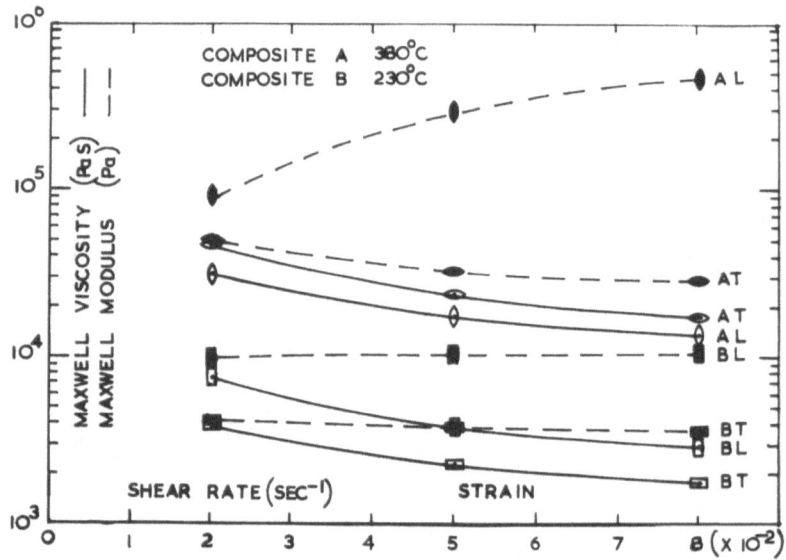

Figure 2. Anisotropic Maxwell viscoelasticity

REFERENCES

1. Groves, D.J. A characterisation of shear flow in continuous fibre thermoplastic laminates. Composites, 1989, 20, 28-32.
2   Rogers, T.G. Rheological characterisation of anisotropic materials. Composites, 1989, 20, 21-27.

DYNAMIC VISCOELASTICITY OF MAYONNAISE CONTAINING DIFFERENT EGG PRODUCTS

ANTONIO F. GUERRERO
Departamento de Ingeniería Química. Facultad de Química.
Universidad de Sevilla. 41012 Sevilla (SPAIN)

HERSHELL R. BALL
Department of Food Science. North Carolina State University.
Box 7624, Raleigh, NC 27695-7624 (USA)

ABSTRACT

The linear viscoelastic properties of five mayonnaise products were evaluated in oscillatory shear. Mayonnaise was prepared with native egg yolk and with rehydrated commercially dried yolk in two concentrations of solids, as well as with an experimentally produced reduced-cholesterol egg yolk. The dynamic viscoelastic properties obtained for the low cholesterol mayonnaise product are comparable to those presented by the other products studied, as well as by commercially accepted mayonnaise.

INTRODUCTION

Egg proteins and phospholipid act as emulsifiers which allow the formation of stable oil/water emulsions, giving the desired appearence, mouth-feel and texture to mayonnaise. Other components of egg yolk contribute to the flavour and colour, which may not be intensified by any other ingredient, of mayonnaise. Egg yolk also contributes cholesterol, in a proportion of about 5% of yolk.

There is considerable consumer interest in having traditional foods, such as mayonnaise, with low cholesterol content. The consumer would prefer that the low or cholesterol-free food have all of the sensory qualities of the traditional food.

Although several authors have conducted research on the emulsion stability and rheological properties of mayonnaise at steady or oscillatory shear, few studies refer to either the use of dried egg yolk or reduced-cholesterol egg yolk [1].

This communication focuses on the dynamic viscoelastic properties of mayonnaise made from different egg products which use native, dried yolk or an experimentally produced reduced-cholesterol yolk.

EXPERIMENTAL

Five different mayonnaise products were evaluated. The composition was held constant (73% soybean oil, 10% vinegar, 1% salt and 16% egg) while the type of egg product used as the emulsifier was varied. Three kinds of yolks containing approximately 50% by weight of solids were used in combination with native egg white, which contained about 12% of solids, to form three commercial type whole egg products (NW, DW and LW) containing 24% of solids and two commercial type egg yolks (NY and DY) with solid concentration of 43%. The yolks used were native egg yolk (N), a rehydrated, commercially available, dried yolk (D) and a rehydrated, experimentally produced, low-in-cholesterol yolk (L). Cholesterol in the dried egg yolk was reduced by up to 90% by supercritical fluid extraction with CO_2. The emulsifying power of L yolk at 43% of solids was found to be too poor to form mayonnaise in the experimental conditions.

All the products were stored at 4ºC; and 12 days after preparation, rheological measurements were carried out using three replicates for each batch. Dynamic measurements were performed in a Bholin Rheometer, VOR Model, using a Mooney-Ewart type sensor system, C14. A strain sweep test was previously conducted at 20ºC and 1 Hz for every mayonnaise in order to find the viscoelastic linear range. A small value for maximum strain amplitude, (0.006), which assures permanence in the linear range, was selected for oscillatory tests at frequencies ranging from 0,005 to 5 Hz.

RESULTS AND DISCUSSION

The mayonnaise products studied present a decrease in dynamic viscoelastic properties with ageing time. This decrease is believed to depend on the preparation procedure and has been attributed to the flocculation and the coalescence of the oil drops. Two phases of coalescence were described by Sherman (2) for freshly prepared emulsions. The first phase is rapid and was assumed to be due to the initially large total drop surface area and to insufficient adsorption of the emulsifier, while the second phase, in which the measurements were taken, is slow.

A statistical study was carried out using three batches of NW at 20ºC. Analysis reveals that there is a significant difference, although moderate, among the storage modulus values for the three batches; while the difference is almost unsignificant for the loss modulus. Therefore, the variability among these batches must be considered in order to compare the different products studied.

The five products studied, including LW which contained 0.026% of cholesterol or less, present dynamic properties comparable to values for commercially accepted mayonnaise (3). Within the range of frequencies studied, the elastic component is much more relevant than the viscous one for each mayonnaise, since the values obtained for the phase angle are very low, ranging from 2 to 8 degrees.

The storage modulus, the dynamic viscosity and the complex viscosity present a power-law variation with frequency for each mayonnaise, although some slight deviations for G' may be found at the higher frequency. The parameters obtained for this equation at 20ºC, which are the values for the storage modulus, the dynamic viscosity or the complex viscosity at a frequency equal to 1 rad/s, G'[1], η'[1] or η*[1], and the power-law exponents "a", "b" or "c", are shown in Table 1.

TABLE 1
Parameters for the power law equations

	G′ [1] (Pa)	100. a	R²	η′ [1] (Pa. s)	-b	R²	η* [1] (Pa. s)	-c	R²
NW	310	4. 75	0. 999	28. 2	0. 926	0. 997	312	0. 950	0. 99997
DW	289	4. 41	0. 997	23. 1	0. 868	0. 998	291	0. 952	0. 99997
LW	309	4. 08	0. 996	27. 1	0. 916	0. 998	312	0. 955	0. 99995
NY	386	7. 06	0. 999	43. 9	0. 884	0. 999	389	0. 927	0. 99998
DY	462	3. 57	0. 994	29. 2	0. 858	0. 993	465	0. 961	0. 99996

For the accuracy of the experimental procedure, mayonnaise products NW, DW and LW do not present significant differences in the viscoelastic parameters G′[1], η′[1], or η*[1]. On the other hand, parameter "a", which is an index of the dependence which frequency exerts on the elastic component, decreases when the egg is replaced by either of the non-native products. This effect becomes quite clear when a commercial type yolk is used in the preparation of mayonnaise.

Thus, parameter "a" presents half the value for DY than for NY. Moreover, G′[1] and η*[1] values are significantly higher for mayonnaise DY than NY. Therefore, mayonnaise DY has a higher elastic component, although with an inferior frequency dependence, than mayonnaise NY. On the contrary, the viscous component is much lower for DY than NY, as may be deduced from the η′[1] values in Table 1.

An increase in the solid concentration for the egg from 24% (for NW, DW and LW) to 43% (for NY and DY) produced a rise in the storage modulus and the dynamic viscosity of mayonnaise, although this effect may be unsignificant at low frequencies.

In conclusion, the increase in elasticity produced by the replacement of the native egg product for a non-native egg may be associated with the higher consistency which rehydrated, dried or reduced-cholesterol, egg yolk possess over native yolk. This superior consistency is a consequence of the thermal treatment to which the yolk was subjected during the drying process, since a certain degree of denaturation of proteins takes place. When yolk is replaced by whole egg this effect diminishes because of the dilution of yolk proteins. It must be pointed out, that the low in cholesterol mayonnaise prepared for the current study does not differ significantly in the values for elastic or viscous components from a mayonnaise made from native whole egg.

REFERENCES

1. Harrison, L. J. and Cunningham, F. E., Factors influencing the quality of mayonnaise: A Review. J. Food Quality, 1985, 8, 1-20.

2. Sherman, P., Rheological properties of emulsions. In Enciclopaedia of Emulsion Technology. Vol. II. Applications, ed. P. Becher, Marcel Dekker, Inc., New York, 1985, pp. 405-437.

3. Berjano, M., Comportamiento Viscoelástico no lineal de mayonesas comerciales. Ph Thesis, 1989, Universidad de Sevilla.

TIME-RELATED FLOW BEHAVIOUR OF SHEAR THICKENING SYSTEMS

Dimiter Hadjistamov, Ciba-Geigy AG, CH-4133 Schweizerhalle

ABSTRACT

The time-related flow behaviour of shear thickening systems was examined in the stressing experiment (stress growth experiment) with subsequent stress relaxation. The observed systems show on relaxation no residual shear stress also in the shear thickening region. The viscosity increase in this region is shear-induced, because of an impeded rearrangement of the structure.

INTRODUCTION

Shear thickening is defined as the increase of viscosity with increasing shear rate (1). A summary of examples for systems with shear thickening flow behaviour can be found in the publications of Chaffey (2), Efremov (3) and Barnes (4).

Shear thickening flow behaviour is observed in most instances only over a specific narrow shear rate region. Below and above this particular (shear thickening) shear rate region, shear thinning flow behaviour is observed. The systems with shear thickening flow behaviour exhibit also mixed flow behaviour - with shear thinning, shear thickening and again shear thinning regions following one after another (5). We classified the systems with mixed flow behaviour, just as the systems with shear thinning flow behaviour, according to the shape of the flow curve in three different flow types: typ B_m - mixed structural-viscous, type C_m - mixed pseudoplastic and type D_m-mixed plastic flow behaviour (see Fig.1)(5).

The time-related behaviour was examined in the stressing experiment ($\dot{\gamma}$ = const) with subsequent stress relaxation ($\dot{\gamma}$ = 0). In principle, two stress growth curves can be distinguished in the stressing experiment - viscoelastic stress growth curve without maximum (VOM) and with maximum (VMM)(6,7).

TIME-RELATED FLOW BEHAVIOUR

The aqueous solution of 1.31% polyvinylalcohol (M = approx. 72000 g/mol) and 4.48% di-sodiumtetraborate-10-hydrate has mixed structural-viscous flow behaviour (see Fig.2)(5). It is interesting that no residual shear stress is observed after cessation of the steady shear flow (see Fig.3).

The sytems with 7.5%, 10% and 12.5% Aerosil 380 in the epoxy resin Araldite GY 260 (Diglycidylether of Bisphenol A) exhibit mixed pseudoplastic flow behaviour (see Fig.4). The critical shear rate for the onset of the shear thickening region decreases with increasing Aerosil 380 concentration (see.Fig.4). The shear stress of the 10% Aerosil 380 suspension declines towards the zero point on relaxation - no residual shear stress is observed (see Fig.5).

The aqueous solution of 2.16% polyvinylalcohol and 0.97% di-sodiumtetraborate-10-hydrate exhibits mixed plastic flow behaviour with more than one shear thickening region (see Fig.6). This system is really an exeption. This system shows on relaxation also no residual shear stress (see Fig.7).

N.B. The relative residual shear stress (the ratio of absolute residual shear stress value to steady state shear stress) of shear thinning systems with pseudoplastic and plastic flow behaviour declines gradually with increasing shear rate, i.e. there is a gradual destruction of the structure (6,7).

Polyvinylalcohol and di-sodiumtetraborate-10-hydrate form a chemical complex in aqueous solution. The thixotropic agent Aerosil 380 builds a hydrogen-bonded three dimensional structure in the epoxy resin Araldite GY 260. These systems show in a narrow shear rate region shear thickening behaviour with increasing viscosity and normal stress differences. But there is no residual shear stress observed in this region after cessation of the steady shear flow of the stressing experiment. The residual shear stress represents the structural strengthening, capable of balancing the twisted torsional bar. Referring to these facts, we can conclude that the viscosity increase in the shear thickening region does not result from the formation of a better three-dimensional structure. The structure in the shear thickening region is shear-induced, because of an impeded rearrangement of the structure, i.e. the "normal" destruction of the structure is disturbed. This "overstructure" is reversible - does not exist after cessation of the steady shear flow.

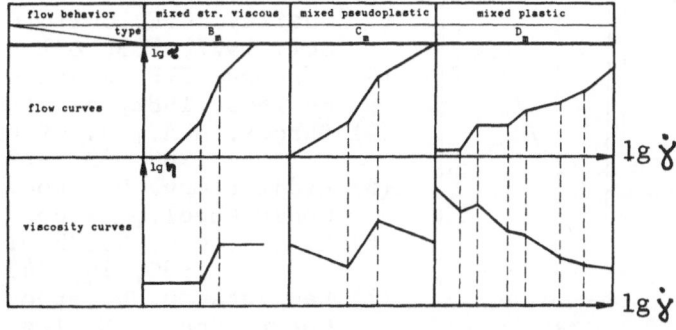

Fig.1 Rheol. profiles of systems with mixed flow behaviour

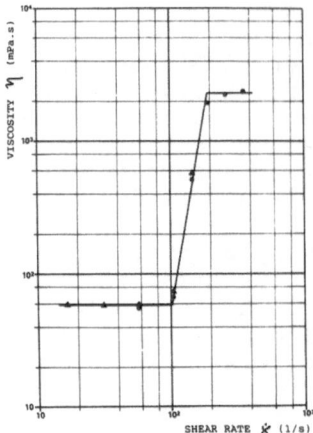

FIG.2 VISCISITY CURVE OF POLYVINYLALCOHOL /
BORATE (1.31% / 4.48%) IN WATER
(Rheomat-30, ▲DIN 45, ● DIN 25, 25°±0.1°C)

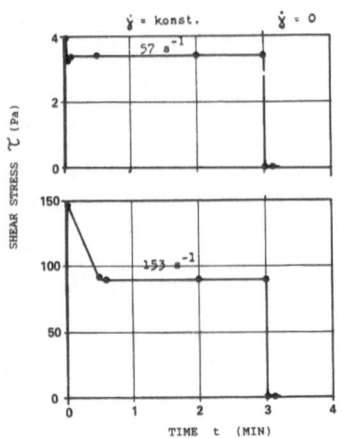

FIG.3 STRESSING EXPERIMENT AND STRESS RELAXATION
OF POLYVINYLALCOHOL/BORATE (1.31%/4.48%) IN WATER
(Rheomat-30, DIN 45, 25°±0.1°C)

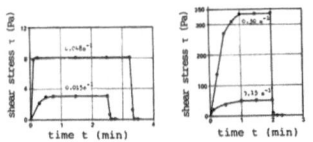

FIG.5 STRESSING EXPERIMENT AND STRESS RELAXATION
OF SYSTEM WITH 10% AEROSIL 380 IN ARALDITE GY 260
(Weissenberg-Rheogoniometer, cone-plate, 25°±0.3°C)

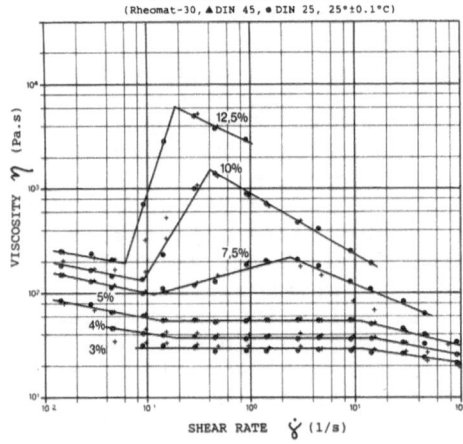

Fig.4 VISCOSITY CURVES OF SYSTEMS WITH DIFFERENT AEROSIL 380-
CONCENTRATIONS IN ARALDITE GY 260   (Weissenberg-
Rheogoniometer, + cone-plate, ● plate-plate, 25°±0.3°C)

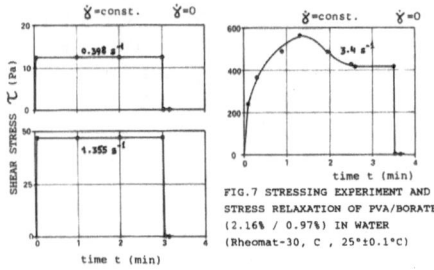

FIG.7 STRESSING EXPERIMENT AND
STRESS RELAXATION OF PVA/BORATE
(2.16% / 0.97%) IN WATER
(Rheomat-30, C , 25°±0.1°C)

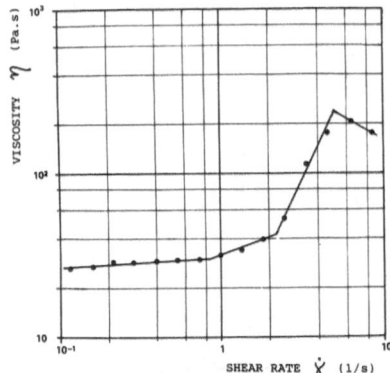

FIG.6 VISCISITY CURVE OF POLYVINYLALCOHOL/
BORATE (2.16% / 0.97%) IN WATER
(Rheomat-30, C , 25°±0.1°C)

REFERENCES
(1) DIN-Norm 13 342, Part 10
(2) Chaffey, C.E.,Colloid&Polymer
    Sci. 1977, 255, 691
(3) Efremov, I.F.,Russian Chemical
    Reviews, 1982, 51, 160
(4) Barnes, H.A., J. of Rheology,
    1989, 33, 329
(5) Hadjistamov, D.,Proc. 9th Int.
    Congr.Rheol.,Mexico, 1984, 277
(6) Hadjistamov, D.,Rheol. Acta,
    1980, 19, 345
(7) Hadjistamov, D.,Proc. 8th Int.
    Congr. Rheol. Naples, 1980,
    P. 609

A CONSTITUTIVE EQUATION THEORY OF VISCOELASTIC FLUIDS

HAN SHIFANG

Research Lab.of Non—Newtonian Fluid Mech., Chengdu Branch,Academia Sinica,610015,Chengdu,P.R.China

ABSTRACT

A new point of view for constitutive equation theory is proposed.The group of Rivlin—Ericksen tensors A_m is used to describe the deformation hsitory of the material with relaxation effects. Application of the theory is given for the viscometric and extensional flows.

THEORY OF CONSTITUTIVE LAW

Another point of view for constitutive theory is described as following: the deformation history of the material with relaxation effects is decribed by the group of mth Rivlin—Ericksen tensors A_m. The general form of the constitutive equation for the viscoelastic fluids should be written as

$$H[S,\overset{\diamond}{S},A_1,\overset{\diamond}{A}_1,A_2,\overset{\diamond}{A}_2,\cdots,A_m,\overset{\diamond}{A}_m,\lambda_1,\lambda_2,\cdots,\lambda_\eta]=0 \qquad (1)$$

where the H is a tensor—valued function.
where $\diamond$ denotes a generalized convective derivative which is defined as

$$\overset{\diamond}{A}=\alpha_1\overset{\triangledown}{A}+\alpha_2\overset{\triangle}{A}$$

where the$\triangledown$ —————upper—convected derivative;$\triangle$ —————lower—convected derivative.

Using the theory and generalize the model of convected Maxwell—Oldroyd fluid one can propose a constitutive equation for the viscoelastic fluids

$$S + \lambda_1 \overset{\diamond}{S} + \mu_\circ \; trSA_1 + \beta_\circ \; trSA_2 = R_1 + R_2 \tag{2}$$

where

$$R_1 = \alpha_\circ \; I + \alpha_1 A_1 + \alpha_2 A_2 + \alpha_3 \overset{\diamond}{A}_1 + \alpha_4 \overset{\diamond}{A}_2 \tag{3}$$

$$R_2 = \alpha_5 A_1^2 + \alpha_6 A_2^2 + \alpha_7 \overset{\diamond}{A}_1^2 + \alpha_8 \overset{\diamond}{A}_2^2 + \alpha_9 (A_1 A_2 + A_2 A_1) \tag{4}$$
$$+ \alpha_{10} (\dot{A}_1 \overset{\diamond}{A}_2 + \overset{\diamond}{A}_2 A_1) + \alpha_{11} (A_1 \overset{\diamond}{A}_1 + \overset{\diamond}{A}_1 A_1)$$
$$+ \alpha_{12} (\overset{\diamond}{A}_1 A_2 + \overset{\diamond}{A}_2 A_1) \alpha_{13} (A_2 \overset{\diamond}{A}_2 + \overset{\diamond}{A}_2 A_2)$$

where the α_i, i = 1 to 13, are polynoms of invariants of tensors A_1, A_2, A_1, A_2 and its products.

A special case is considered when constitutive equation is given as

$$S^{ik} + \lambda_1 \overset{\triangledown}{S}{}^{ik} = \eta_\circ \; (A_1^{ik} + \lambda_2 \overset{\triangledown}{A}_1^{ik} + \beta_1 A_2^{ik} + \beta_2 \overset{\triangledown}{A}_2^{ik}) \tag{5}$$

where the $\eta_\circ$, $\lambda_1, \lambda_2, \beta_1, \beta_2$ are material functions for the present fluid model. The equation(2) is called constitutive equation of relaxation–differential type.

VISCOMETRIC FLOW

The velocity field of a simple viscometric flow is given by

$$v^1 = x^2, \qquad v^2 = v^3 = 0 \tag{6}$$

Using the constitutive equation(5) the first normal stress difference σ_1 and second normal stress σ_2 are given as

$$\sigma_1 = 2\eta_\circ \; [(\lambda_1 - \lambda_2 - \beta_1) + 2\lambda_1(\lambda_1\beta_1 - \beta_2)\gamma^2] \; \gamma^2 \tag{7}$$

$$\sigma_2 = 2\eta_\circ \; \beta_1 \gamma^2 \tag{8}$$

The shear stress s^{12} has a form of

$$s^{12} = \eta\gamma \tag{9}$$

where the apparent viscosity η is given as

$$\eta = \eta_o \ [1 + 2(\lambda_1\beta_1 - \beta_2)\gamma^2] \tag{10}$$

EXTENSIONAL FLOW

The multiaxial extensional flow investigated by author the velocity field of which is gived by

$$v^1 = k_1x_1, \qquad v^2 = k_2x_2, \qquad v^3 = k_3x_3 \tag{11}$$

The elongational viscosity η_e can be calculated from (26) which is defined as $\eta_e = \sigma_1 / k_1$.

$$\eta_e = \eta_o \left[\frac{[1 - 2(\lambda_2 - \beta_1)k_1 - 4\beta_2 k_1^2]}{1 - 2\lambda_1 k_1} \right.$$
$$\left. - \frac{[1 - 2(\lambda_2 - \beta_1)k_2 - 4\beta_2 k_2^2]k_2 / k_1}{1 - 2\lambda_1 k_2} \right] \tag{12}$$

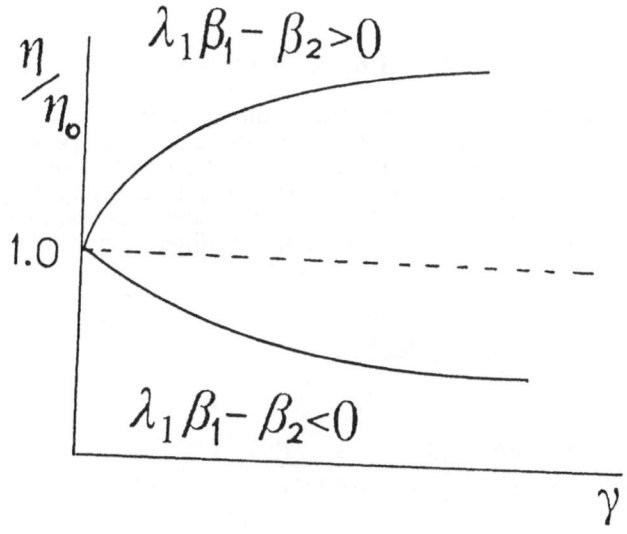

UNSTEADY FIBRE SPINNING OF VISCOELASTIC FLUID

HAN SHIFANG

(Research Lab.of non—Newtonian Fluid Mech.,Chengdu Branch,Academia
Sinica, Chengdu,610015,Sichuan,P.R.China)

ABSTARCT

Unsteady fibre spinning of upper—convected Maxwell fluid is described by a system of four quasilinear hyperbolic equations. Numerical results have been obtained by LW difference scheme. The unsteady behaviour and instability of fibre spinning are investigated.

GOVERNING EQUATIONS AND APPROACH TO THEORY

Fibre spinning is a principal operation in the textile industry.

The process of fibre spinning is mathematically described by a system of mementuum equations and and constitutive equations of upper—convected Maxwell fluid in the cylindrical coordinate system (r,θ,z). Integrating the governing equations over the cross section of the fibre and intrducing area average of the velocity components and stress components a set of quasilinear hyperbolic equations are obtained for the unsteady fibre spinning flow which is given as

$$\frac{\partial A}{\partial \tau} + \frac{\partial(AF)}{\partial y} = 0 \tag{1}$$

$$\frac{\partial F}{\partial \tau} + F\frac{\partial F}{\partial y} = \frac{1}{ReA}\frac{\partial T}{\partial y} \tag{2}$$

$$T + We[\frac{\partial T}{\partial \tau} + \frac{\partial(TF)}{\partial y} - 2\frac{\partial F}{\partial y}T - 3\frac{\partial F}{\partial y}S] = 3A\frac{\partial F}{\partial y}, \tag{3}$$

$$S + We[\frac{\partial S}{\partial \tau} + \frac{\partial(SF)}{\partial y} + \frac{\partial F}{\partial y}S] = -A\frac{\partial F}{\partial y} \qquad (4)$$

where F,A,T,S are dimensionlees velocity,area of section of the fibre ,extra ,stress and stress components. Re is the Reynolds number and We is the Weissenberg number. The y and τ are dimensionlees variables of z and time.It was assumed that $V_r = V_r(r,z,t), V_z = V_z(z,t)$ in order to derive equations (1)—(4).

The system of the hyperbolic equations (1)—(4) are solved numerically by LW scheme as:

1st step

$$\frac{\partial F}{\partial \tau} = F^{n+1/2}_{i+1/2} - 0.5(F^n_{i+1} + F^n_i)/0.5\triangle\tau \qquad (5)$$

$$\frac{\partial F}{\partial Y} = (F^n_{i+1} - F^n_i)/\triangle y \qquad (6)$$

2nd step

$$\frac{\partial F}{\partial \tau} = (F^{n+1}_i - F^n_i)/0.5\triangle\tau \qquad (7)$$

$$\frac{\partial F}{\partial y} = (F^{n+1/2}_{i+1/2} - F^{n+1/2}_{i-1/2})/\triangle y \qquad (8)$$

DISTURBANCE OF FINITE AMPLITUDE

Two type of disturbances are proposed which are given as :
disturbance extrusion velocity

$$V' = 1 + V_m Sin(\omega\tau), \qquad 0 < \tau < \tau_m \qquad (9)$$

where τ_m is disturbance duration.
disturbance velocity along the axis of the fibre

$$V'(Y) = V^e(y) \, (1 + V_m Sin(\omega y)) , \qquad (10)$$

where $V^e(y)$ is the velocity profile along the axis for steady flow.

It should be noted that amplitude of the disturbance is not of a small magnitude. It is of a finite amplitude. In Fig. it is shown that instability in extrusion velocity is caused by the disturbance(9).

CONCLUSIONS

In the present paper the unsteady fibre spinning of viscoelastic fluids is described by a system of one–dimensional aproximate equations. LW difference scheme is used to get the numerical solution. The major conclusion is that the numerical method, i.e.LW difference scheme is available for the stability analysis of fibre spinning of the fluid. It is possible to investigate the instability of viscoelastic fluid by numerical methods for the case of finite amplitude.

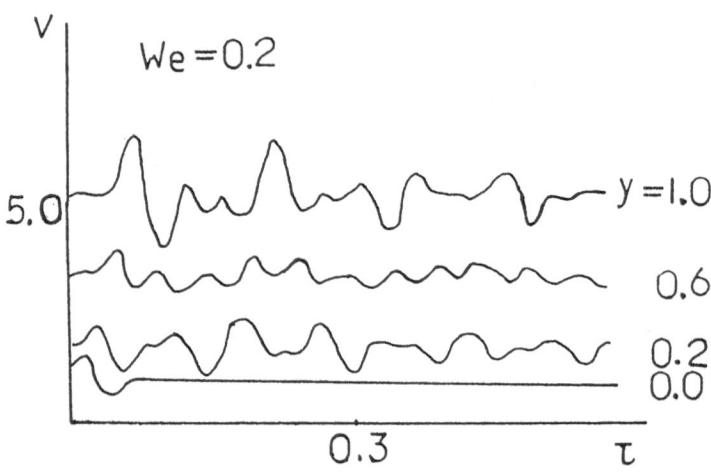

REFERENCE

1. Han shifang Contunuum Mechanics of Non–Newtonian Fluids, Sichuan Academic Press (1988) Chengdu,China

DRAG REDUCTION BY ADDITION OF KONJAKU FLOUR FOR TURBULENT FLOW IN A CIRCULAR TUBE

HAN SHIFANG AND WANG YUBIN

Research Lab. of Non—Newtonian Fluid Mech., Chengdu Branch Academia Sinica, Chengdu,610015,P.R.China

ABSTRACT

An experimental investigation on drag reduction in turbulent tube flow is presented. The equeous konjaku flour solution is used as an additive of polymer. The rheological behaviour of konjaku solution is investigated.

RHEOLOGY OF KONJAKU FLOUR

The konjaku flour consists of Glucose and Mannose which connect a polymer of linear structure. The molecular weight of the konjaku is 7.10^5. In the following table is shown the molecular weight of different additives.

polymer	Molecular weight .10^6
Guar Gum	0.2
Locust Bean Gum	0.31
Sodium Carboxyme thylcellulose	0.2—0.7
Polyethylene Oxide Polyox WSR—35	0.2
Polyacrylanide	1—2
Konjaku flour	0.7

It can be seen from the Table that the molecular weight of the konjalu is of the order of the typical polymer additives. The molecular conformation of the konjaku is given as following:

$$
\begin{array}{cc}
\begin{array}{c}
\text{CHO} \\
\text{H} - | - \text{OH} \\
\text{HO} - | - \text{H} \\
\text{H} - | - \text{OH} \\
\text{H} - | - \text{OH} \\
\text{CH}_2\text{OH}
\end{array}
&
\begin{array}{c}
\text{CHO} \\
\text{HO} - | - \text{H} \\
\text{HO} - | - \text{H} \\
\text{H} - | - \text{OH} \\
\text{H} - | - \text{OH} \\
\text{CH}_2\text{OH}
\end{array} \\[4pt]
\text{Glucose} & \text{Mannose}
\end{array}
$$

The kinematic viscosity of the konjaku flour in aqueous solution is shown in Fig 1. It can be seen from the Fig. that the kinematic viscosity increases with the concentration and decreases with the temperature. Non—linear bahaviour is found at low temperature.

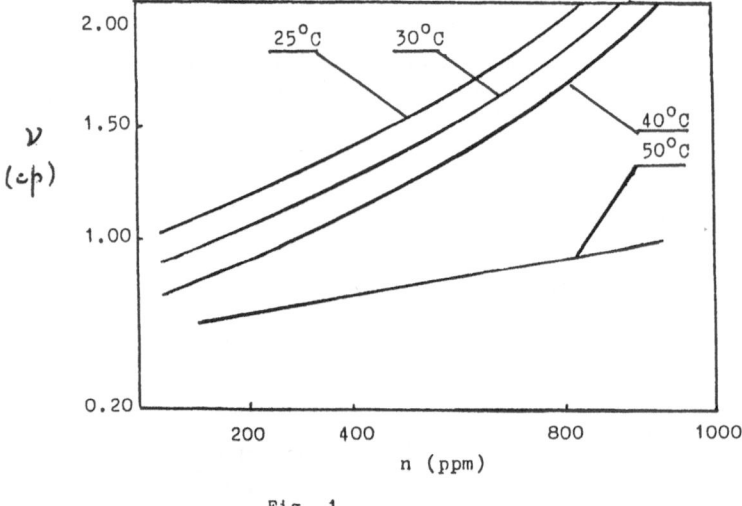

Fig. 1

EXPERIMENTAL DEVICE

A special experimental system is designed to study the effects of drag reduction. The system consists of fluid tank,pressure tank,konjaku solution tank, manometers,test pipe, damping vessel. The test section consists of two coaxial cylinders. The dilute aqueous solution of konjaku flour as an additive was injected under pressure through the holes at the surface of the inner pipe into the the turbulent boundary layer. Four concentrations(0,250,500,1000 ppm) of the aqueous konjaku solution were used for the friction factor measurements. The experimental results are showr in Fig.2.

CONCLUSIONS

1).The natural product, konjaku flour,can be used as an additive in turbulent flow for drag reduction;
2).A new engineering application of konjaku is possible which is used as an additive;
3). It is experimentally substantiated that the additive can be injected through the hole at the wall into the boundary layer to observe the drag reduction.

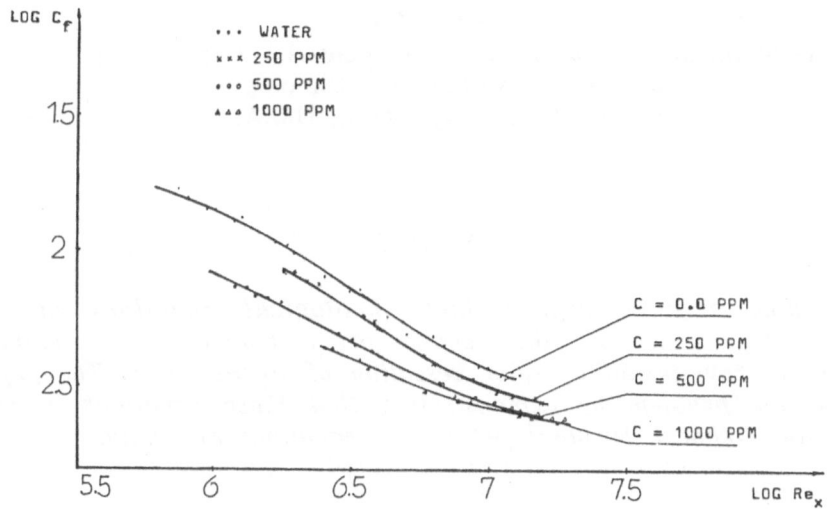

INVESTIGATION OF INSTABILITY IN RHEOLOGICAL EQUATIONS OF THE DIFFERENTIAL TYPE

A. HARNOY

Department of Mechanical and Industrial Engineering
New Jersy Institute of Technology
Newark, New Jersy 07102, U.S.A.

ABSTRACT

Previous studies have argued that rheological equations of the differential type, such as the second order model are inadequate because they show instability after cessation of steady shear. The paper elucidates this problem by demonstrating that these equations become stable at low Deborah Numbers, where the equations are valid.

INTRODUCTION

The merit of differential type rheological equations is their ability to offer relatively simple solutions of flow fields for complex geometries and unsteady flow. These equations are useful in many important engineering applications and have resulted in qualitative explanations to various viscoelastic effects which have been observed experimentally, including delayed fluid separation at laminar boundary layers flow past submerged bodies and viscoelastic response for squeeze film flow[1].

However, it has been shown [2,3,4] that a major problem exists in the differential type equations. If the signs of the fluid parameters are selected, so that the storage modulus is positive, the equation shows instability at initial value problem after cessation of steady shear. Coleman Duffin and Mizel[3] showed that this type of instability (CDM instability) results by substituting the second order fluid equation into the equation of motion. For many rheologists, the CDM instability casts doubt on the validity of differential type fluid models and hence all analytical results based on them become questionable.

Metzner et al [5] have argued that the second order fluid model is valid only at low values of Deborah Number De, whereas the cessation of steady shear involves a sudden acceleration resulting in high De. Recalling that De=$\lambda/\Delta t$, where λ is the relaxation time of the fluid and Δt is a characteristic time of the flow. However, this explanation is not acceptable to many scientists. They consider the CDM instability to be independent of De, because a functional relationship is not known between the onset of instability and De. This work attempts to resolve this controversy. In fact, it shows that CDM instability does not exist at De<<1, where differential type equations are valid.

ANALYSIS

Let us consider the Maxwell viscoelastic model, which may be demonstrated by a spring and dashpot in series. The Maxwell equation can be expanded to a differential type series[1], in which the deviatoric stress τ' is an explicit function of increasing order time derivatives of the strain rate $\mathbf{e}$,

$$\tau' = 2\eta[\, \mathbf{e} - \lambda\, D/Dt\, (\mathbf{e}) + \lambda^2\, D^2/Dt^2\, (\mathbf{e}) + \ldots\ldots + (-\lambda)^{n-1}\, D^{n-1}/Dt^{n-1}(\mathbf{e})\,] \qquad (1)$$

where η is the fluid viscosity (both η and λ being greater than zero). D/Dt is defined by Harnoy [6], but other definitions of D/Dt lead to the same conclusion, since only $\partial/\partial t\, (\mathbf{e})$ plays a role in this problem. Eq. (1) can be truncated for De<<1, where Δt is evaluated by :

$$O(\Delta t) = \frac{D^{n-1}/Dt^{n-1}\, (\mathbf{e})}{D^n/Dt^n\, (\mathbf{e})} \qquad . \qquad (2)$$

The second order approximation, where only the first two right hand terms in eq. (1) are retained was tested by Huilgol [4] who showed that the equation can be unstable. His work is extended here to find the limits of the problem. Let us test a shear flow in the x direction between two parallel plates having a clearance of π/b, where the velocity is given by,

$$u = A \sin by \quad . \qquad (3)$$

Appropriate pressure gradients must be provided to maintain the flow and after its sudden cessation, the motion is governed by inertia and friction stresses, $\rho\, \partial u/\partial t = \partial/\partial y(\tau_{xy})$. Substituting τ_{xy} according to the second order approximation of eq. (1), yields,

$$\rho \; \partial u/\partial t \; = \; \eta \; \partial^2 u/\partial y^2 \; - \; \lambda \eta \; \partial^3 u/\partial t \partial y^2 \; . \tag{4}$$

A solution can be found by seperation of variables in the form,

$$u = A \; e^{at} \; \sin by \tag{5}$$

which satisfies the initial value, eq. (3), at t=0. Substituting (5) into (4) results in

$$a = \eta \; b^2/(\lambda \eta b^2 - \rho) \quad . \tag{6}$$

The condition for stability is a<0 where the flow decays by viscous dissipation. A viscous fluid, (λ=0) is stable, having $a = -\eta b^2/\rho$. For our flow, according to eq.(5), we get, $O(\Delta t) = |1/a|$, resulting from eq.(2). For $\lambda << \Delta t$ the first approximation of Δt is that of viscous fluid, resulting in

$$De = O(\lambda \eta b^2/\rho) \quad , \qquad at \; De<<1, \tag{7}$$

or $\lambda \eta b^2 << \rho$, at De<<1, which results in a<0 in eq.(6) and hence the flow according to eq.(5) will decay.

CONCLUSIONS

The CDM instability does not exist in truncated differetntial type rheological equations at De<<1 where these equations are valid. At higher De, this instability problem results from the omission of significant terms in the series (1). This analysis considers steady flow according to eq. (3). In the case of an arbitrary initial flow, the solution contains a sum of harmonic waves involving a mix of De values. Previously, this fact obscured the role of De in the CDM instability.

A similar analysis shows that the odd order equations (namely the third order, the fifth order etc., approximations of the series (1)) do not show CDM instability at higher De. However, truncation of eq. (1) is valid only at De<<1. The same conclusions apply to other differential type equations, including the second order fluid equation of Coleman and Noll.

References
1. Harnoy A, *J. of Rheology*, 1989, **33** pp. 93-117
2. Ting, T.W., *Arch. Ratl. Mech. Anal.* 1963, **14** p.1
3. Coleman, B.D., Duffin R.J., and Mizel V.J., *Arch. Ratl. Mech. Anal.* 1965, **19** p.100
4. Huilgol, R.R., *Rheol. Acta*, 1979, **18** pp. 451-455
5. Metzner, A.B., White J.L. and Denn M.M., *A.I. Ch.E.J.* 1966, **12** p. 863
6. Harnoy, A, *J. Fluid Mech* 1976, **76**, pp. 501-517

THE CALIBRATION AND USE OF THE STRATHCLYDE RHEOMETER

FOR THE STUDY OF PLASTISOLS AND RELATED SYSTEMS

D. HAYWARD[1], A. McKINNON[1], R.A. PETHRICK[1], F.S. BAKER[2], J. FERGUSON[1], R.E. CARTER[2] and J.H. DALY[3]

1. Department of Pure & Applied Chemistry, Thomas Graham Building, University of Strathclyde, Cathedral Street, Glasgow G1 1XL.
2. Carter Baker Enterprises Ltd., 2 Marshbrook Business Park, Marshbrook, Church Stretton, Shropshire, SY6 6QE.
3. Dow Benelux N.V., P.O. Box 48, 4530 AA Terneuzen, The Netherlands.

ABSTRACT

A versatile method for the determination of both the increase in viscosity and the time to gellation for a plastisols systems together with the calibration of the equipment is described in this paper.

INTRODUCTION

An increase in viscosity leading to the formation of a stable three dimensional gelled network is a characteristic property of many paints, powder coatings, composites and thermosetting filled systems.[1,2] This process, or in the case of a drying polymer the related solvent loss, leads to an increase in viscosity with a value of approximately 10^6 Pa.s being reached at the gel point.[3,4] A further important application of the gelation phenomena is the transformation of a dispersion of polymer particles in a poor solvent (plastisol) to a solid. This process is used for the fabrication of many flexible elastomeric articles, the polymer particles being polyvinylchloride and the solvent − a phthalate plasticizer. Ideally, the increase in the viscosity over the fluid could be measured using a conventional cone and plate or rotating cylinder viscometer. However, the expensive nature of such equipment makes it inadvisable to cure to a rigid solid the sample which is being investigated. It is therefore desirable to have an alternative inexpensive technique which allows determination of the gellation time of materials without hazarding the apparatus used for the measurement. In addition, information obtained concerning the rate of energy dissipation and also the temperature of the sample during the cure can add significantly to our understanding of the total cure process.

EXPERIMENT AND THEORY

The apparatus used involves an extension of the conventional probe method and uses observation of the damping of the motion of the probe as a measurement of the viscosity.

The method has been described elsewhere [5,6] and in this paper we consider tests of the validity of the theory and also the use in a study of a plastisol system. The motion of the probe, p_2, can be represented in terms of the real, p_2', and imaginary, p_2'', parts of its motion;

$$p_2' = p_1/(1 + (\eta\omega C/k)^2) \qquad \text{and} \qquad p_2'' = p_1(\eta\omega C/k)/\{1 + (\eta\omega C/k)^2\}$$

where η is the complex viscosity, ω is the frequency, C is a geometric factor related to the probe/material contact area and k is the spring constant.

Calibration of the Apparatus

In order to test the validity of this method the damping coefficients for a liquid whose viscosity has been accurately determined as a function of temperature were obtained. The liquid used was capable of being supercooled and is an oligomer of polyphenyylenoxide– SANTOVAC 5. The sample used had been previously measured by the viscoelastic group at Glasgow University [7,8] and the data for the damping coefficients at various values of the viscosity are shown in figure 1. The viscosities of the material were also measured using a Rheometrics RMS800 viscometer and it can be seen that very good correspondence is obtained between the two sets of measurements. The constants in the above equation had been previously obtained from measurement of the viscosities of a series of silicone fluids, whose viscosities had been previously measured using the Rheometrics viscometer. The agreement with the simple theory is good over the range of viscosities examined, allowing for the arbitrary geometry factor A. A more rigorous analysis would allow calculation of the latter factor (Ferry [9]) but for our purposes the calibration with materials of known viscosity is sufficient.

Investigation of a plastisol system

As an example of the application of the method a plastisol system was studied. The plastisol consists of particles of PVC dispersed in a poor solvent, an alkyl phthalate. On heating, the particles of PVC swell and expand to form an entangled matrix structure. At low temperature, figure 2, the rate of swell is slow and this is indicated by the slow monotonic increase in the viscosity as the size of the particles dispersed in the plasticizer increase. At higher temperature the rate of swelling is dramatically increased and this is once more reflected in the observed rheology. At high temperature the curves indicate that a limiting viscosity is attained, whereas at lower temperatures the viscosity appears to continually increase to some high value. At high temperature the resultant matrix will have a glass transition temperature which is below the processing temperature and a flexible product is obtained. At low temperature the glass transition temperature of the matrix is above that of the processing temperature and a rigid matrix is obtained. The curves therefore not only give information on the rate of gelation of the plastisol but also indicate the nature of the final product.

REFERENCES

1. Macosko, C.W., Brit. Polym. J., 1985, 17, 239–245.

2. Mussatti, F.G. and Macosko, C.W., Poly. Eng. Sci., 1973, 13, 236–240.

3. Gillham, J.K., Developments in Polymer Characterization, Ed. Dawkins J.V., Vol 3, Applied Science Publishing Ltd., London 1982, Chap 5, pp. 159–227.

4. Gillham, J.K., <u>Crosslinking and Scission in Polymers</u>, Ed. O. Guven, Nato ASI Series C, 1990, Vol 292, pp. 171–198.

5. Affrossman, S., Hayward, D., McKee, A., MacKinnon, A., Pethrick, R.A., Lairez, D., Vitalis, A., Baker, F.S. and Carter, R.E., <u>Flow and Cure of Polymers – Measurements and Control</u>, RAPRA, 1990.

6. Affrossman, S., Collins, A., Hayward, D., Trottier, E. and Pethrick, R.A., <u>JOCCA</u>, 1989, 452–453.

7. Barlow, A.J., Erginsav, A. and Lamb, J., <u>Proc Roy Soc. A</u>, 1969, **309**, 473–469.

8. Mason, P.W., <u>Trans ASME</u>, 1947, **69**, 359–370.

9. Ferry, J.D., <u>Viscoelastic Properties of Polymers</u>, Wiley, New York, 1980, pp. 116.

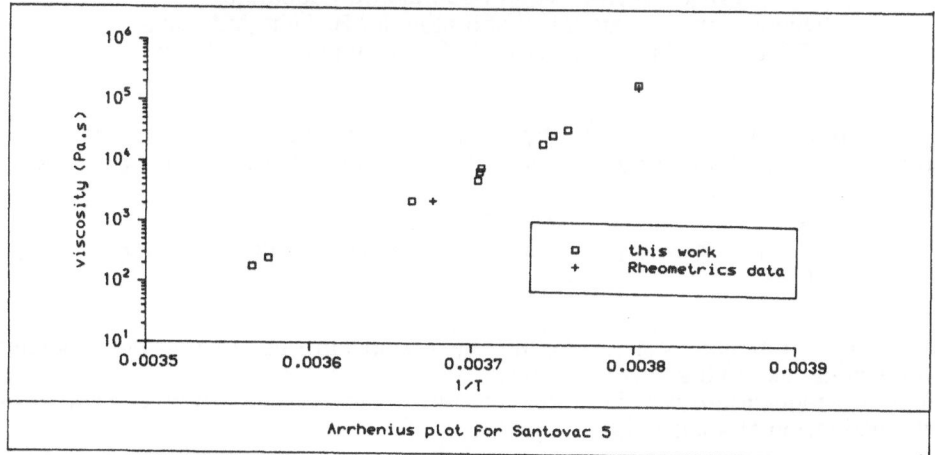

Figure (1)

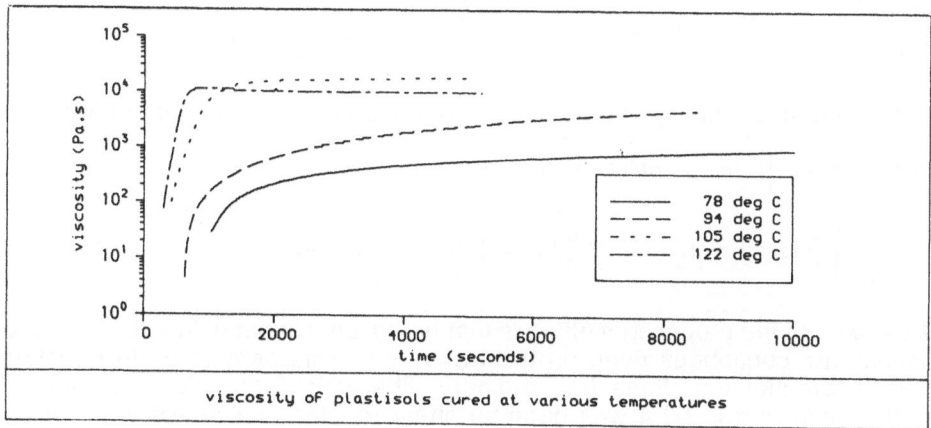

Figure (2)

ON PROPAGATION OF DISCONTINUITIES IN NONLINEAR HEREDITARY MEDIA

SERGEJ HAZANOV
Senior Scientist,
Ecole Polytechnique Fédérale de Lausanne
Department of Materials, Laboratory for Building Materials
32, ch. de Bellerive, CH - 1007 Lausanne (Switzerland)

We are studying the problem of shock wave propagation in a semi-infinite bar, made of nonlinear hereditary material with the constitutive equation of Volterra-Freché type :

$$\varepsilon = \int_{-\infty}^{t} k_1 (t - \tau) \, d\sigma (\tau_1) + \int_{-\infty}^{t} k_2 (t - \tau_1, t - \tau_2) \, d\sigma (\tau_1) \, d\sigma (\tau_2) + \dots \quad (1)$$

The equation of motion is standard, the initial condition - absence of perturbations in the material before the wave arrival.
At first only regular kernels of heredity are taken into account. Then for fixed x the Taylor expansion of solution is :

$$\sigma = \sum_{n=0}^{\infty} \frac{1}{n!} \left(\tau - \frac{x}{c} \right)^n \frac{\partial^n b}{\partial t^n} = \sum_{n=u}^{\infty} \frac{1}{n!} \left(t - \frac{x}{c} \right)^n \left[\frac{\partial^n b}{\partial t^n} \right] \quad (2)$$

Here [f] is the discontinuity of function on the wave front. Hence, the problem is to calculate all the $\left[\partial^n \sigma / \partial t^n \right]$, utilizing well-known relations [1] :

$$\frac{d}{dt} [f (x, t)] = [f_t] + G [f_x], \quad \frac{d}{dt} [u] = 0, \quad [\sigma] = -\rho \, G [u_t] \quad (3)$$

The essence of the proposed method is that hereditary elements in our constitutive equations are continuous even on the wave front. This permits us to transform integro-differential equations into ordinary differential, with discontinuities as unknown variables. In [2] it was done for linear regular constitutive equations of nonageing type.
After some algebraic transformations (1) takes the form :

$$\varepsilon = \sigma\, k_1\,(0) \;+\; \int_{-\infty}^{t} k_1'\,(t-\tau)\,\sigma\,(\tau)\,d\tau \;+$$

$$\sigma \left\{ \int_{-\infty}^{t} k_{2,1}\,(t-\tau_1,\,0)\,\sigma\,(\tau_1)\,d\tau_1 + \int_{-\infty}^{t} k_{2,2}\,(0,\,t-\tau_2)\,\sigma\,(\tau_2)\,d\tau_2 \right\} + \sigma^2\,(t)\,\{\ldots\} + \ldots \tag{4}$$

from where we can deduce :

$$[U_{xt}] = k_1\,(0)\,\Big[\sigma_t + \psi\,([\sigma])\Big] \tag{5}$$

$$\frac{d}{dt}\,[\sigma] + \frac{\rho c^2}{4}\,\psi\,([\sigma]) = 0 \tag{6}$$

For the discontinuities of higher orders the same procedure gives (and that's important !) linear differential equations, though nonhomogeneous ones. For example, for typical ageing material (concrete) :

$$\varepsilon = J_0\sigma + \int_{-\infty}^{t} K\,(t,\tau)\,\sigma\,(x,\tau)\,d\tau \tag{7}$$

$$\frac{d}{dt}\Big[\sigma^{(n)}\Big] + \frac{\rho c^2}{2}\,k\,(t)\Big[\sigma^{(n)}\Big] = F_n(t),\quad k\,(t) = K\,(t,t), \tag{8}$$

$$F_n(t) = \frac{1}{2}\frac{d^2}{dt^2}\Big[\sigma^{(n-1)}\Big] - \frac{\rho c^2}{2}\sum_{j=0}^{n-1} C_{n+1}^{j+1}\,k^{(n-j)}(t)\Big[\sigma^{(j)}\Big]$$

What concerns practically important kernels with weak, integrable singularity (parabolic models), the apparatus of fractional derivatives can be used :

$$D^{\gamma}f\,(t) = \frac{d^p}{dt^p}\left(\frac{1}{\Gamma\,(p-\gamma)}\int_{0}^{t}(t-\tau)^{p-\gamma-1}\,f\,(\tau)\,d\tau\right),\; \gamma \in [\,p-1,\,p\,]\;\; p = 0,\,1,\,2\ldots \tag{9}$$

Equation (1) is rather complicated for applications and we'll utilize more comfortable nonlinear (but non-ageing) models of Rosowsky and Rabotnov [3] :

$$\varepsilon = J_0\sigma + \int_{-\infty}^{t} k\,(t-\tau)\,\psi\,(\sigma)\,d\tau \tag{10a}$$

$$\varphi\,(\varepsilon) = J_0\sigma + \int_{-\infty}^{t} k\,(t-\tau)\,\sigma\,(\tau)\,d\tau \tag{10b}$$

Here the simplest singular kernels (of Abel type) were chosen, because we are interested only in the vicinity of the wave front. Elementary analysis, using (9) shows that shocks of any order can propagate here only with the "elastic" speed. From (10) and (9) :

$$\frac{d}{dt}\left[D^{(1-\alpha)}U\right] = \frac{1}{\rho c}\frac{d}{dt}\left[D^{-\alpha}\sigma\right] + k\,c\,\psi\left(\lfloor\sigma\rfloor\right) \tag{11}$$

$$\left[D^{(1-\alpha)}U\right] = \frac{d}{dt}\left[D^{-\alpha}U\right] - c\left[D^{-\alpha}Ux\right],\ \psi\left(\lfloor\sigma\rfloor\right) = 0, \lfloor\sigma\rfloor = 0 \tag{12}$$

and hence the impossibility of shocks for constitutive equation (10a). Similarly for discontinuity of the first order :

$$\frac{2}{\rho c}\frac{d}{dt}\left[D^{1-\alpha}\sigma\right] + k\,c\,\psi'(0)\left[\sigma_t\right] = 0 \tag{13}$$

that can be satisfied only for $\left[\sigma_t\right] = 0$. The same procedure gives that $\left[\partial^n\sigma/\partial t^n\right] = 0$ and the behaviour of solution in the vicinity of the front is similar to the behaviour of the function exp $(-1/t^\alpha)$ in the vicinity of 0.

For the constitutive equation (10b), the scheme is absolutely the same. For instance, for acceleration waves.

$$\phi''(0)\left[\sigma_t\right]\left[D^{-\alpha}\sigma\right] = 2\,\rho c^2\frac{d}{dt}\left[D^{1-\alpha}\sigma\right] + k\rho^2 c^4\left[\sigma_t\right] \tag{14}$$

and the solution here is also infinitely smooth in the vicinity of the front. From the results mentioned it follows that near the wave front we have

$$\psi(\sigma) = \sum_{n=1}^{\infty}\frac{\psi^n(0)}{n!}\sigma^n \cong \psi'(0)\,\sigma\ ;\ \phi(\varepsilon) \cong \phi'(0)\,\varepsilon \tag{15}$$

and for the first approximation of nonlinear solution one can take the solution of linearized problem, pretty good investigated. The results, concerning the smoothness of wave shapes, coinside well with the linear ones.

REFERENCES

1. Tomas, T. Plasticheskoye techenie i razrushenie tverdych tel M., MIR, 1964.
2. Achendach, J., Reddy, D., A note on wave propagation in linearly viscoelastic media. Z. angew. Math. und Phys., 1967, Bd 18, H.I.
3. Rabotnov, Y. Elementy nasledstvenoy mechaniki tverdych tel. M., Nauka, 1977.

3D FLOW SIMULATION IN NON-NEWTONIAN FLUIDS

PETER HENRIKSEN AND OLE HASSAGER
Institut for Kemiteknik
Danmarks tekniske Højskole
DK 2800 Lyngby, Denmark

ABSTRACT

A finite element program for the simulation of 3D steady non–Newtonian flow is developed. Velocities are approximated by trilinear continuous basis functions. A discontinuous representation is used for the pressure. The swirling flow in a channel of square cross section with a bend of $\pi/2$ is used as a test problem. The geometry is proposed as a simple benchmark problem for the testing of 3D codes. The flow near a semipermeable membrane is also investigated. Different suction boundary conditions are implemented. The interaction between main flow, fluid rheology and suction boundary conditions is discussed.

INTRODUCTION

The implementation of the finite element program was tested on a 3D problem in conjunction with results obtained with an independent finite element solver developed by the Professors P.Townsend and M.F.Webster from University College of Swansea. This test problem, the 'Quarterbend', is proposed as a standard 3D benchmark problem by the authors. Also investigated is the velocity, pressure and concentration fields for a rectangular channel with a semipermeable membrane. This kind of crossflow filtration often used when concentrating macromolecules, has large industrial applications. The purpose of the simulations is to improve the understanding of the transport process of macromolecules back from the membrane surface to the mainstream.

A 3D Benchmark. The Quarterbend

For a problem to be well suited as a 3D benchmark it has to be truly three dimensional with numerical obtainable variations in all three directions. It must also have boundary conditions that may easily be implemented by different finite element formulations just as it is advantageous if it is experimentally realisable.

The Quarterbend in Figure 1 has a velocity profile imposed at inlet given by the analytical solution to the flow in a straight rectangular channel(see Happel and Brenner [1])

$$\nu_z^* = \frac{1}{2}(1-x^{*2}) - \sum_{n=0}^{\infty} \frac{2(-1)^n}{c_n^3 \cosh(c_n)} \cos(c_n x^*)\cosh(c_n y^*) \ , \ c_n = \pi(\tfrac{1}{2}+n) \qquad (1)$$

x^*, y^* and ν_z^* are the non–dimensional coordinates and z–velocity

$$x = (a/2)x^*, \ etc \qquad ;\nu_x = \frac{\Delta p(a/2)^2}{L\mu}\nu_x^*, \ etc \qquad ;p = \frac{\mu V}{(a/2)}p^* \qquad (2)$$

p is the pressure, μ the viscosity and a the width and height of the channel, see Figure 1. $V = \Delta p(a/2)^2/(L\mu)$ is the reference velocity and $dp/dz = -\Delta p/L$ is the pressure gradient in the fully developed entry flow. The analytical volume rate of flow becomes

$$q = \frac{4\Delta p(a/2)^4}{\mu L}\left(\frac{1}{3} - \sum_{n=0}^{\infty}\frac{2}{c_n^5}\tanh c_n\right) = \frac{4\Delta p(a/2)^4}{\mu L}\beta \qquad (3)$$

and the Reynolds number is defined by $Re = q\rho/(a\mu)$, ρ being the the density, whereby the steady Navier–Stokes equation becomes

$$\frac{Re}{2\beta} \, v^* \nabla^* v^* = -\nabla^* p^* + \nabla^{*2} v^* \tag{4}$$

At outlet all three velocity components are free, thereby imposing a zero force on this surface. Symmetry conditions have been assumed on the centerplane allowing us to consider only the upper half of the cross section as indicated in Figure 1.

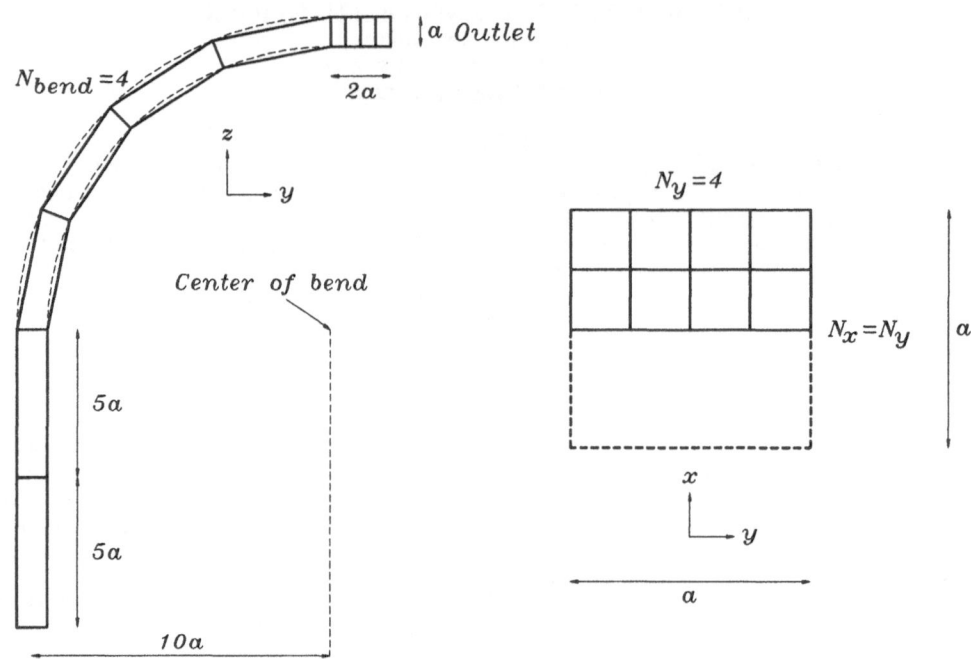

Figure 1. The Quarterbend geometry. The left part shows the yz–plane and the right part the xy–plane. This example has $N_{bend}=4$ and $N_y=4$.

Flow Near a Semipermeable Membrane

We consider the separation of a 2–component fluid with concentration c of the non–permeable species. The boundary condition at the membrane is implemented as a force F on the surface

$$F = a^{-1} \mathbf{n} \cdot v + (\Pi(c)+P_0) \tag{5}$$

where a gives the flow properties of the membrane,(a function of the membrane resistance and its thickness) $\mathbf{n}$ a unit vector normal to the surface, $\Pi(c)$ the osmotic pressure just inside the membrane surface and P_0 the outside pressure. In the convection/diffusion equation the flux of the non–permeable species, N, normal to the membrane is imposed as $(\mathbf{n} \cdot N) = 0$. The flow geometry is a rectangular straight channel with an inlet and an outlet section without a membrane and a membrane part with one wall with the natural boundary condition (5) imposed.

RESULTS

In Figure 2 the developing secondary flow in a Quarterbend for $Re=500$ is shown. The mesh used has the inlet and outlet elements as shown in Figure 1. The number of elements in the bend is $N_{bend}=32$ and $N_y=16$. The coordinates have been transformed so that the new y–direction, y', is pointing towards the center of the bend.

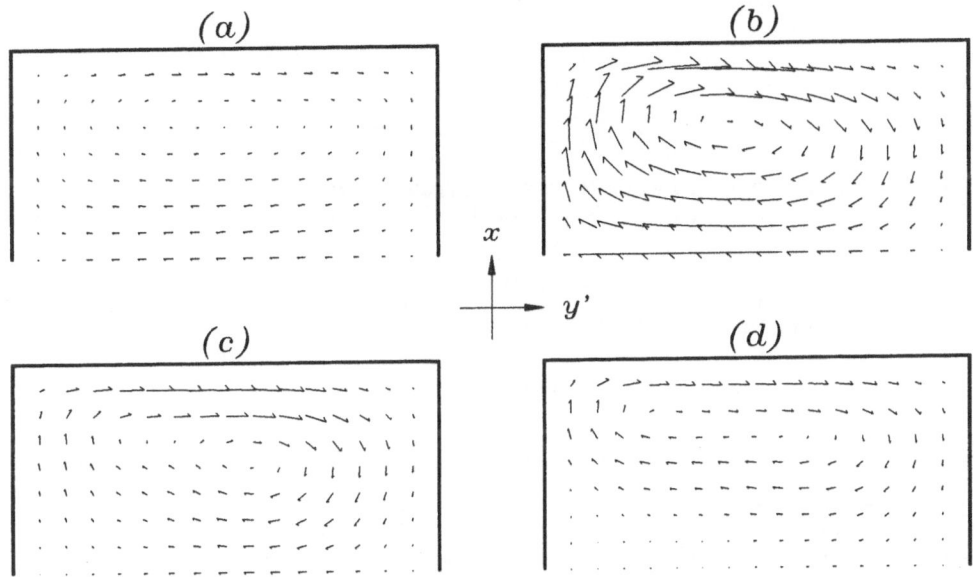

Figure 2. The developing of secondary flow. The upper half of the four xy'-planes shown is located at (a) $\pi/64$, (b) $8\pi/64$, (c) $15\pi/64$ and (d) $22\pi/64$ after the start of the bend. $N_{bend}=32$, $N_y=16$, $Re=500$.

The results shown are for different xy'-planes. The $22\pi/64$-plane is by a qualitative look on the remaining, not shown xy'-planes, close to the steady state. Shanthini and Nandakumar [2] found, with the same curvature, height/width relation and Reynolds number, a 4–vortex flow in the total cross section for a Newtonian fluid flowing in a 2–dimensional curved rectangular duct. This is also what is seen in Figure 2(d) although the weakest of the 2 vortices in the upper half of the cross section, positioned in the lower left of Figure 2(d), is not resolved to be visible with the given number of elements. Since the secondary flow is seen to be developing rather fast after the start of the bend(already in Figure 2(a) a vortex is evidently present) and is showing clear development through most of the bend, the proposed benchmark has plainly three dimensional effects obtainable by numerical simulations.

CONCLUSIONS
The 4–vortex flow found by Shanthini and Nandakumar [2] is also found in our calculations. The results show that the proposed 3D benchmark problem has a clear three dimensional behavior which allow for unambiguous comparisons between different numerical solvers.

The work was supported by the Danish Technical Research Council, Grant 16-4308.K.

REFERENCES
1. Happel, J. and Brenner, H., Low Reynolds number hydrodynamics, Noordhoff International Publishing, Leyden, 1973, pp. 38–39.
2. Shanthini, W. and Nandakumar, K., Bifurcation Phenomena of Generalized Newtonian Fluids in Curved Rectangular Ducts, J. of Non–New. Fluid Mech., 1986, 22, pp. 35–60.

THE USE OF PSEUDORANDOM TEST SIGNALS IN THE RHEOMETRY OF LINEAR-VISCOELASTIC MATERIALS

LUTZ HEYMANN
Academy of Sciences of the GDR/Institute of Mechanics
Postfach 408 / Karl-Marx-Stadt / DDR-9010

ABSTRACT

In the paper, disadvantages of conventional test signals in the rheometry are depicted. A method is proposed for the determination of the memory function of a linear-viscoelastic material. Some results of the numerical simulation of a rheometric experiment with a pseudorandom test signal are pointed out.

INTRODUCTION

One of the main tasks of the rheometry is the identification of a rheological model on the base of experimentally obtained data. For this, it is useful to have the data in a wide range of the test signal such as deformation, deformation rate, frequency or time.

Most commonly used experiments in the rheometry are characterized by a certain number of one-point-measurements at constant test signals. This applies both to stationary or dynamic and instationary experiments. A disadvantage consists in a considerable experimental and time expense, which affects especially the experiments with reacting fluids. A way to overcome this disadvantage consists in the application of multifrequent test signals in connection with methods of computerized processing of measured data. Some authors used such methods to calculate the dynamic properties of polymers in terms of the complex modulus G^* from experiments with transient stress growth and relaxation [1] or with a harmonic deformation signal composed of three different frequencies with a sum of the single amplitudes lying in the linear range of the material [2], respectively.

METHOD

A further increase in efficiency of the experiment can be attained by the application of test signals which have an infinite frequency contents such as the Gaussian white noise which is, nevertheless, only a theoretical signal model and can not be realized practically. A pseudorandom binary signal (PRBS) may be denoted as a useful deterministic approach of the white noise and can be described with statistical methods. The properties and the construction of the PRBS are described elsewhere [3]. The parameters of the PRBS (time period T, amplitude

Y_{max}, clock cycle Δt) are selected corresponding both to the expected rheological properties of the material (longest rela- xation time, linear range of material properties) and the dyna- mic properties of the used rheometer.

The generalized functions for periodic signals (like the PRBS), i.e. the correlation function (CF) R_{ik} and the power spectral density (SD) S_{ik} as the Fourier transform $\mathcal{F}\{.\}$ of R_{ik}, are defined by

$$R_{ik}(\tau) = \frac{1}{T} \int_O^T z_i(t)z_k(t+\tau)dt; \quad S_{ik}(\omega) = \mathcal{F}\{R_{ik}(\tau)\} \quad (1a,b)$$

where τ and ω are the time shift and the frequency, respecti- vely.

For i=k and $z_i(t)=x(t)$ eqns. (1) results in the autocorre- lation function (ACF) $R_{xx}(\tau)$ and the autospectral density (ASD) $S_{xx}(\omega)$ of the input signal. For $z_i(t)=y(t)$ and $z_k(t)=x(t)$ there are obtained the cross correlation function (CCF) $R_{xy}(\tau)$ and the cross spectral density (CSD) $S_{xy}(\omega)$.

The starting point of the method is the simple constitutive equation of a linear viscoelastic material

$$6(t) = \int_{-\infty}^t m(t-t') \, y(t') \, dt' = m(t)*y(t) \quad (2)$$

where $6(t)$ is the measured shear stress as response, $y(t)$ the deformation as input signal, $m(t)$ the unknown memory function (MF) and t, t' the time parameters. The asterisk denotes the convolution product.

Eq. (2) may be transformed into the equivalent relation

$$R_{6\gamma}(\tau) = \int_O^\infty m(\tau-\theta) \, R_{\gamma\gamma}(\theta) \, d\theta = m(\tau)*R_{\gamma\gamma}(\tau) \quad (3)$$

with θ as time shift. The ACF for a PRBS is given analytically by

$$R_{\gamma\gamma}(\tau) = \gamma_{max}^2 \begin{cases} (1- \frac{N+1}{N} \frac{|\tau|}{\Delta t}) & \text{for } 0 \leq |\tau| \leq \Delta t \\ -1/N & \text{for } \Delta t < |\tau| < (N-1) \, \Delta t \end{cases} \quad (4a)$$

$$R_{\gamma\gamma}(\tau) = R_{\gamma\gamma}(\tau+kT), \, k=0,1,2,3,\ldots \quad (4b)$$

with N as number of clock cycles per period [3]. $R_{6\gamma}$ is calcu- lated numerically from the data. The Fourier transformation of (3) leads to

$$S_{6\gamma}(\omega) = M(\omega) \cdot S_{\gamma\gamma}(\omega) \quad \text{or} \quad M(\omega) = S_{6\gamma}(\omega)/S_{\gamma\gamma}(\omega) \quad (5)$$

with $M(\omega)$ as the Fourier transform of the MF. The unknown MF, which may be denoted as a generalized material function, can be obtained by the inverse Fourier transformation of $M(\omega)$

$$m(t) = \mathcal{F}^{-1}\{M(\omega)\} \tag{6}$$

By the help of m(t) the linear material functions, i.e. G^*, can be calculated.

RESULTS

For the present, the method described above was implemented on a personal computer. The aim was the simulation of a rheological experiment. The computerized test signal construction was carried out by the help of the method of quadratic residua [3] on the base of the input parameters given by the operator. Using a special MF the algorithm was carried out and the calculated values of the MF were compared with those of the given one. There was found out a good agreement between the given and the calculated values.

CONCLUSIONS

The theoretical results show that the method is suitable for the determination of the memory function of a linear viscoelastic material. The ultimate object of the work is , firstly, the implementation of the method tested theoretically on a modern computerized rotational rheometer which has a free programmable drive with favourable dynamic properties, where the measured data have to be either smoothed and filtered or averaged over some periodes to avoid measuring errors and to decrease the signal-to-noise ratio, and, secondly, the extension of the method to nonlinear properties of viscoelastic materials.

REFERENCES

[1] Marin, G., Peyrelasse, J., Monge, Ph.
    Rheol. Acta, 1983, 22, 476-81.

[2] Holly, E. E., Venkataraman, S. K., Chambon, F. ,
    Winter, H. H.
    J. Non-Newt. Fluid Mech. 1988, 27, 17-26.

[3] Strobel, H., Experimentelle Systemanalyse, Akademie-Verlag
    Berlin 1975.

THE RHEOLOGICAL BEHAVIOUR OF WATER BORNE BASECOATS

R. Hingmann, H.M. Laun
BASF Aktiengesellschaft, 6700 Ludwigshafen, West Germany
W. Goering, T. Dirking
BASF Lacke und Farben AG, 4400 Muenster-Hiltrup, West Germany

ABSTRACT

The rheological characterization of water borne metallic basecoats is discussed with regard to a prediction of the application behaviour on the basis of laboratory data. Pertinent experimental techniques and the evaluation of some characteristic parameters are presented.

INTRODUCTION

At present efforts are made to replace solvent borne coatings by water borne systems. In the case of metallic basecoats for the automotive industry the latter systems are already developed and successfully applied. Coatings with up to 65 % of water and only 14 % of organic solvent are available. During spraying and flash off the loss of solvent and, as a consequence, the necessary viscosity increase is much smaller, compared to conventional systems. To compensate this phenomenon, the rheological properties of water borne coatings have to be more complex and systems with an optimized rheological profile and pronounced thixotropy are developed.

EXPERIMENTAL TECHNIQUES AND SAMPLES

The measurements were performed with a Rheometrics Fluids Spectrometer (RFS), a Carri-Med controlled stress instrument (CS 100), and a nitrogen capillary rheometer [1]. The RFS was used for oscillatory shear experiments and the detection of the viscosity recovery after pre-shearing at high shear rate. Creep measurements at various stresses were performed with the Carri-Med instrument. Viscosity functions at high shear rate were measured with the capillary rheometer. Coatings prepared for spraying as well as samples collected after application were studied.

EXPERIMENTAL RESULTS

Measurements of flow curves for different basecoats clearly show the thixotropy of water borne systems. Therefore, a well defined pre-shearing treatment has to be applied prior to each measurement in order to obtain reproducible results. The viscosity functions show a pronounced shear thinning behaviour below shear rates of 100 s^{-1}. Whereas distinct differences for the coatings of this study are observed below this value, at higher shear rates the viscosities tend to reach the same level. The dynamic moduli in oscillatory shear depend on the strain amplitude as expected for multiphase systems. As a consequence, the Cox-Merz rule does not hold for water borne basecoats. The ratio of the storage modulus G' and the loss modulus G" may be used to characterize the elastic properties. This ratio strongly depends on the solid content.

In the nozzle of the spray-gun the coatings are exposed to high shear rates. Therefore, the time scale of the recovery of the structure after pre-shearing has to be quantified. Fig. 1 shows the viscosity from small amplitude oscillatory measurements as a function of time after pre-shearing the coatings at a shear rate of 1000 s^{-1}. The recovery of the absolute value of the complex viscosity $|\eta^*|$ can be described by an exponential function governed by a characteristic time t_o.

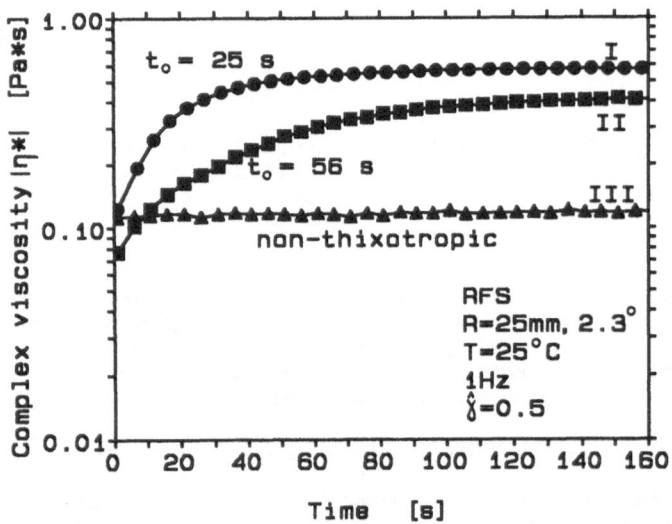

Figure 1. Recovery of the viscosity $|\eta^*|$ after 30 s pre-shearing at 1000 s^{-1} for three coatings.

Creep measurements were performed down to shear stresses of 0.1 Pa. Problems due to wall slip and concentration profiles near the wall could be avoided by using a vane geometry [2,3]. Fig. 2 shows the evaluation of experiments on three coatings.

The deformation for a given period of time is plotted as a function of the applied shear stress. Using this representation and the time constants for the viscosity recovery, the coatings can nicely be distinguished with respect to their sagging resistance.

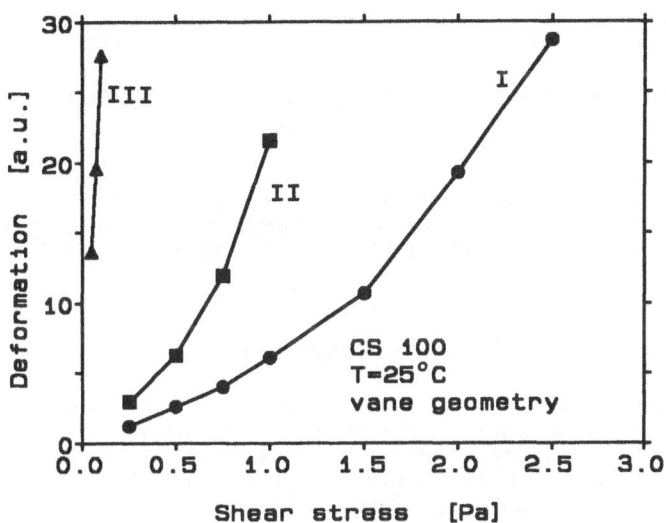

Figure 2. Deformation after a time of 20 s versus applied shear stress for the coatings from Fig. 1.

SUMMARY AND CONCLUSION

The rheological behaviour of water borne basecoats is demonstrated by experimental results obtained on various coatings by using different rheometers. Thixotropy, shear thinning, and viscoelasticity are the main characteristics of these fluids. The complex rheological behaviour can be reduced to a small number of relevant parameters which are suited for a correlation with the application behaviour. These parameters are a time constant for the recovery of the viscosity after pre-shearing, the deformation of the sample at a given level of stress after a given period of time, and the ratio of the storage modulus and the loss modulus. However, more quantitative information on the application behaviour of the coatings is needed in order to define a rheological profile for the coatings which ensures good application results.

REFERENCES

1. Laun, H.M. in H. Giesekus and M.F. Hilbert (Eds.), Progress and Trends in Rheology II, Suppl. to Rheol. Acta, 1988, **26**, 287
2. Dzuy, N.Q., Boger, D.V., J.Rheol., 1985, **29**, 335
3. Yoshimura, H.M., Prud´Homme, R.K., J. Rheol., 1987, 31, 699

RECOVERY OF STRAIN-INDUCED DAMAGE IN FILLED AND UNFILLED ELASTOMERS

Ch. Hübner and R.G. Stacer[*]
Fraunhofer-Institut für Chemische Technologie (ICT)
D - 7507 Pfinztal-Berghausen, FRG

ABSTRACT

This paper describes a preliminary investigation on the effects of elastomer composition on strain-induced damage and subsequent recovery. Damage/recovery data were obtained for a polybutadiene (cis:trans:vinyl=60:20:20) elastomer as functions of volume fraction glass bead filler, crosslinking level and plasticizer content. These data were collected from cyclic tensile tests conducted over a range of temperatures. Damage was observed as a pronounced decrease in strain-energy density E. Modeling of this behavior was accomplished using a power-law equation. Recovery as functions of time and temperature was monitored through measurements of both residual strain (set) and E. These data were found to be unamenable to the method of reduced variables without a vertical shift factor.

INTRODUCTION

Response properties of elastomers are extremely dependent on material deformation history. This history dependent behavior is generically termed the Mullins effect. Although a number of definitions and observed phenomena are associated with the Mullins effect, it is most often discussed for crosslinked materials in terms of strain-induced damage (stress-softening) and subsequent recovery. Of these two features, damage has received the bulk of the invesigative attention with only a handful of studies being devoted to time-dependent recovery.

EXPERIMENTAL

Damage/recovery was experimentally investigated in the series of elastomers given in table 1. A controlled deformation cycle

---

[*] Visiting scientist - US/FRG Scientific Exchange Program

illustrated in figure 1 was used to induce reproducible damage at temperatures ranging from -60°C to 20°C.

TABLE 1
Material Formulations

INGREDIENT	Formulation Designation (phr)								
	1	2	3	4	5	6	7	8	9
Polybutadiene	100	100	100	100	100	100	100	100	100
Dicumyl peroxide	3	3	3	3	3	3	2	1	–
Dioctyl adipate	–	–	–	50	100	–	–	–	–
Glass Beads 12 μm	0	25	50	75	100	75	50	50	50

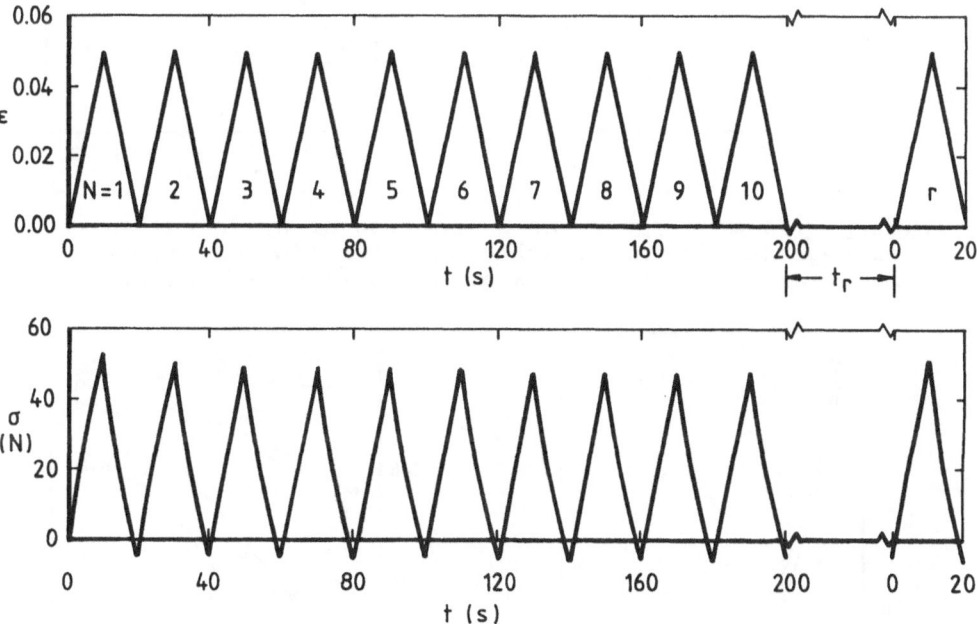

Figure 1: Cyclic strain (ϵ) input and typical corresponding measured stress (σ) response.

RESULTS AND DISCUSSION

Figure 2 presents some of the damage/recovery data obtained from material 2. Damage is shown as a ratio of strain-energy density during each cycle to its initial value E_o at the five temperatures considered. Subsequent recovery is also given in terms of the same ratio as a function of recovery time t_r.

Some comparisons between the different materials can be made from the data plotted in figure 3. A power-law exponent, obtained from log-log plots of the data in figure 3 (left), is used here to quantify damage versus temperature. Recovery data at -60°C are also given for the four materials.

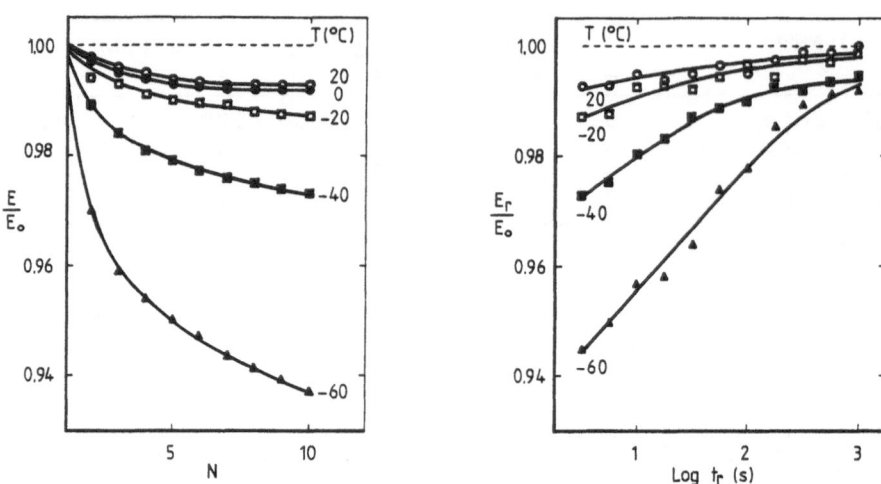

Figure 2: Damage (left) as a function of cycle number and subsequent recovery (right) as a function of time for material 2. Recovery data at 0°C omitted for clarity.

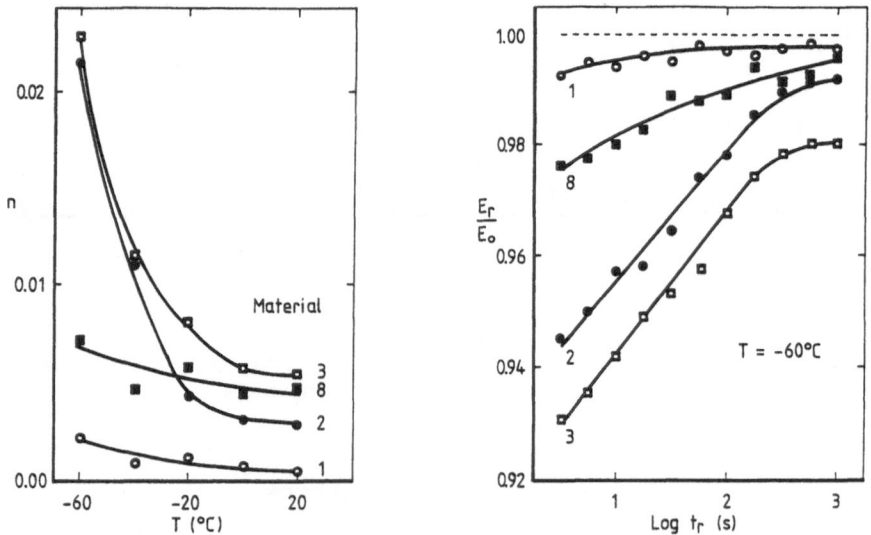

Figure 3: Damage power-law exponent (left) as a function of temperature and recovery versus time at -60°C for materials as indicated.

FLUID FAILURE MECHANISMS IN STRETCHING FLOWS: EXTENSIONAL VISCOSITY AND COMPLEXATION

N.E.HUDSON[1], J.FERGUSON[1], and B.C.H.WARREN[2].
[1]Dept of Pure and Applied Chemistry, University of Strathclyde,Scotland.
[2]Chemical Defence Establishment, Porton Down, Salisbury, England.

ABSTRACT

It is generally accepted that an elongating filament will fail by either a capillary or cohesive fracture process. This paper examines the effect of these mechanisms on the mass median diameter of the droplets formed when a solution is injected into a high velocity air stream. It will be shown that strain hardening and polymer/polymer interactions can play significant roles in determining droplet size.

INTRODUCTION

When a fluid is injected into a high velocity airstream, it will be dispersed into fine droplets. The rate at which these droplets settle is largely determined by their size. Obviously, there will be a size distribution, but the mass median diameter is usually used as a categorising parameter. Studies of the mass median diameter of droplets produced from Newtonian fluids have generated empirical formulae such as that of Weiss and Worsham [1] and Wigg [2]. From previous work by the authors, using Newtonian fluids, the MMD, X, was found to be related to the fluid properties and the system geometry by the equation

$$X = K (Q \mu / \rho \sigma)^{0.1} \sigma V^{1.5}$$

(1)

where K is a parameter dependent upon the system geometry and the dispersing fluid viscosity and density, Q is the volume injection rate of the test fluid, V is the dispersing fluid velocity, and μ, ρ, σ are the solution shear viscosity, density and surface tension.

The classical work of Rayleigh and Weber has shown that the Newtonian fluid breaks up into droplets by a capillary mechanism. Generalising the analysis to viscoelastic fluids produced the result that the elasticity in the fluid tended to delay the breakup, but the mechanism was still capillaric. It is now generally accepted that in such a dispersion situation, though there is shear deformation on the surface of the

injected liquid filament, the dominant deformation is extensional in
character. Hence, for a fluid with spinnability, the capillary mechanism
should be virtually eradicated, and the dominant mechanism for break up
will be cohesive fracture.

For a Newtonian fluid, the extensional viscosity is three times the shear
viscosity, so substituting the extensional viscosity (or a combination of
shear and extensional viscosities) for the shear viscosity in the
empirical formula will not change any exponents, only the constant
pre-factor.

However, for a non-Newtonian fluid, three problems arise.

a. The shear viscosity depends upon the shear rate, which is a function
   of injection velocity and injector diameter (and hence of mass
   injection rate).
b. The extensional viscosity can be many hundreds of times the shear
   viscosity, and is a function of the strain rate.
c. Recent work on extensional flow of polymers in a variety of solvents
   has indicated that polymer-polymer complexation can be induced in
   strong flow fields [3]. In the extreme case, phase changes have been
   shown to occur, both in shear and, more recently, in elongation [4].
   When a non-Newtonian fluid is extended, migration of polymer molecules
   from the surface may occur, thus changing the surface characteristics,
   including the surface tension.

It is therefore not surprising that empirical formulae such as the Weiss
and Worsham have not been successful in predicting the MMD of viscoelastic
solutions.

EXPERIMENTAL

Polymer Solutions
A series of polyacrylamides has been used as part of an EEC Science
Stimulation Project. Full details are available. These polyacrylamides
have been used in aqueous solutions, at concentrations ranging from 5ppm
to 1% (w/w).

Steady Shear Flow
The steady shear flow characteristics of these polyacrylamides were
determined using the Carrimed Controlled Stress Rheometer. To cover a
sufficient range of shear stress, both cone-and-plate and concentric
cylinder geometries were used. At each value of the shear stress, shearing
was continued until a steady state was achieved. For each polymer, at each
concentration, the data were fitted to a Kreiger-Dougherty model to
determine the viscosities at zero and at very high shear stress.

Though the graphs for all solutions have the same shape, there are
differences. In particular, the onset of non-Newtonian behaviour, at a
given concentration, occurs later in some polymers than in others.

Elongational Flow Studies
The Carri-Med Elongational Viscometer was used without a spinnaret, the
fluid issuing straight from the nozzle. Hence, shear rates at the nozzle
were between 100 and 300 s^{-1}, roughly comparable to those at the nozzle in

the wind tunnel. Take up was by means of the vacuum tube, which was necessary both to achieve high strain rates, and also because the low concentration solutions did not always adhere to the drum. The elongational viscosity of solutions of concentrations in the range 30 to 500 ppm were determined, and fitted to power law equations of the form

$$\mu_E = m \, \dot{\varepsilon}^t \tag{2}$$

Typical values of the parameter t ranged between 1.1 and 3.2, and both t and m were determined as functions of concentration. Hence it was possible to extrapolate down to the lower concentrations used in aerosol studies.

Wind Tunnel Experiments

The series of polyacrylamides have been used in wind tunnel experiments at CDE. Concentrations used were 5 to 150 ppm. Throughout the programme, the flow rate was kept constant at 20 ml/s, and the air velocity was found to vary between 90 and 96 m/s. Attempts to correlate the MMD with either the zero or high shear rate viscosities were unsuccessful. During any test, all variables except μ and σ are expected to remain constant. At this point, it does not seem feasible to determine surface tension changes during an extensional deformation, but it may be possible to monitor such changes during the weaker deformation of shear, where similar molecular migrations from the surface are known to occur. In a typical dispersion experiment, the strain rates experienced by the test fluid are expected to be initially very large, and remain so until fluid break up. The final strain rate can be approximated by $\dot{\varepsilon} = V/L$, where V is the relative velocity and L is the length to break up. By taking typical values of the fluid and air parameters, the formula for the MMD can be reduced to

$$X = 186(\mu_E)^{0.1} \tag{3}$$

where μ_E is the elongational viscosity. Therefore, using eqn (3), it was possible to determine elongational viscosities from the MMDs, and then, by using eqn (2), to estimate the strain rates involved. Though the ranges of strain rate at each concentration are, in many cases, quite wide, these ranges do overlap between concentrations. In the wind tunnel experiments, the strain rates to which a solution is subjected would initially be much higher than these values. The values determined would be achieved further downstream, and in some way represent maximum values that droplets could withstand without breaking further. It is interesting to note that higher strain rates than these were achieved in the elongational viscometer without the filament breaking.

REFERENCES

[1] M.A. Weiss & C.H. Worsham, ARS Journal, 1959, 29, 252-259
[2] L.D. Wigg, J. Inst. Fuel, 1964, 37, 500-505
[3] J. Ferguson, N.E. Hudson, B.C.H. Warren and A. Tomatarian, Nature, 1987, 325, 234-236
[4] J. Ferguson, N.E. Hudson and B.C.H. Warren, J. Non-Newtonian Fluid Mechanics, 1987, 23, 49-72

TRI-DIMENSIONAL EFFECTIVE PROPERTIES OF QUASI-PERIODIC POROUS ELASTIC ANISOTROPIC COMPOSITE MATERIALS THROUGH A NEW HOMOGENIZATION TECHNIQUE

Christian HUET, Parviz NAVI

Ecole Polytechnique Fédérale de Lausanne
Department of Materials, Laboratory for Building Materials
32, ch. de Bellerive, CH - 1007 Lausanne (Switzerland)

The existing analytical and numerical homogenization methods for periodic inhomogeneous materials suffer various limitations when applied to the real materials. To overcome to some extent to these limitations a geometrico-numerical method for estimating the effective properties of quasi-periodic inhomogeneous medium was presented in [1]. The principle of this method is laid on an application of the modification theorem established by Hill, in 1963 [2], which proves that in all elastic inhomogeneous materials, if the elastic modulus tensor of any constituants is increased (or decreased) in some region, its effective moduli is increased (or decreased). Based on this theorem the idea of this homogenization technique was proposed by one of us [3]. It consists to perform a monotonic transformation on the micro-structure of the real material in such a way that the Hill modification theorem can be applied, and the effective moduli of the modified materials "augmented materials" and "diminished" one, can be calculated analytically or numerically. Thus two computations are performed, on two modified materials and the results provide two bounds (an upper one and a lower one) on the real effective properties of the material. The real material effective moduli is obtained from the bounds through interpolation techniques. In the following sections two examples are performed. In the first one, we deal with a simple example, in order to show the numerical validity of the theorem in porous materials. In second example a quasi-periodic porous material is considered and its effective moduli is estimated from those "augmented" and "diminished material". The elastic effective properties of the periodic modified materials are calculated by a method developed by Navi [4].

Let us consider an unidirectional hollow periodic medium, called M_1. The basic unit cell of the medium in the plane perpendicular to the hollow direction is shown in Figure 1b. The matrix is considered as an isotropic material with elastic moduli in the Voigt representation is given in Table 1. In addition to this real medium, we consider two other periodic hollow medium M_2 and M_3 with the basic unit cells illustrated in Figure 1a and 1c respectively. Medium M_2 is obtained from M_1, by tranforming the real hollow cross-section into a largest included square, whereas medium M_3 is obtained from transforming the real hollow material into a smallest including one. Therefore, M_2 is the "augmented material" and M_3 the "diminished"

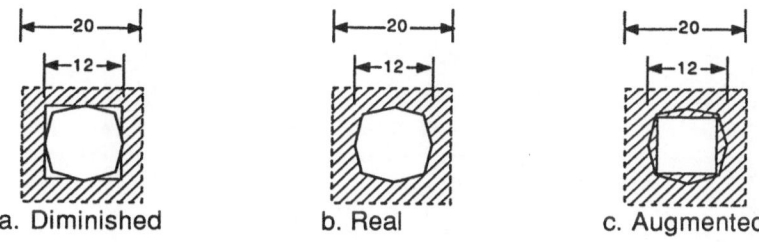

a. Diminished   b. Real   c. Augmented

Fig 1: Basic unit cells of the real and monotonically modified materials

one. The transverse effective elastic moduli of the three periodic medium are calculated and the results are given in Table 1.

<u>Table 1</u> : Elastic moduli of matrix and calculated transversal effective moduli in GP_a

$$\left(a = C_{ij} \text{ or } (1 - P); \ \delta C_{ij} = \left| C_{ij}^r - C_{ij}^1 \right| ; r = 2,3; \ P \text{ porosity} \right)$$

	Matrix	Disminished Medium M_3	Real Medium M_1	Augmented Medium M_2	$\dfrac{a^1 - a^3}{a^1}$ %	$\dfrac{a^2 - a^1}{a^1}$ %
1 - P		0.64	0.684	0.75	6.4	9.6
C_{11}	233	90.052	102.79	117.203	12.4	14.0
C_{22}	233	90.52	102.79	117.203	12.4	14.0
C_{12}	93	26.39	38.60	48.342	31.6	25.2
$\sqrt{C_{11} C_{22}}$	233	90.052	102.79	117.203		
δC_{12}		12.214	0	9.742		
$\sqrt{\delta C_{11} \delta C_{22}}$		12.270	0	14.413		
C_{66}	70	11.945	19.872	22.782	39.9	14.6

These results confirm numerically the Hill modified theorem showing that, the effective elastic moduli of real medium M_1 in the transverse plane is bounded from both sides as we expected.

In the following, the application of the proposed homogenization technique to estimating the effective moduli of a quasi-periodic porous medium like wood is illustrated. Wood is a highly porous, anisotropic and heterogeneous material and the early wood porosity can be as high as 60 % for soft woods. For late wood this value can reduce to some amount of 15%. The effective moduli of wood cell wall (the anisotropic matrix) also varies from early to late wood. This is strongly correlated to the microfibril angles which are turned helically inside the cell wall around the cell axis. To simplify all these complexities a constant microfibril angle of 40° is considered for both early and late wood and in addition it is assumed that, wood is constructed of long narrow hollow tubes of equal cross-sections of 50 by 50 µm where the porosity varies from 52 to 19 %. A schematic representation of such a wood model perpendicular to the cell direction is shown in Figure 2b and called M_b.

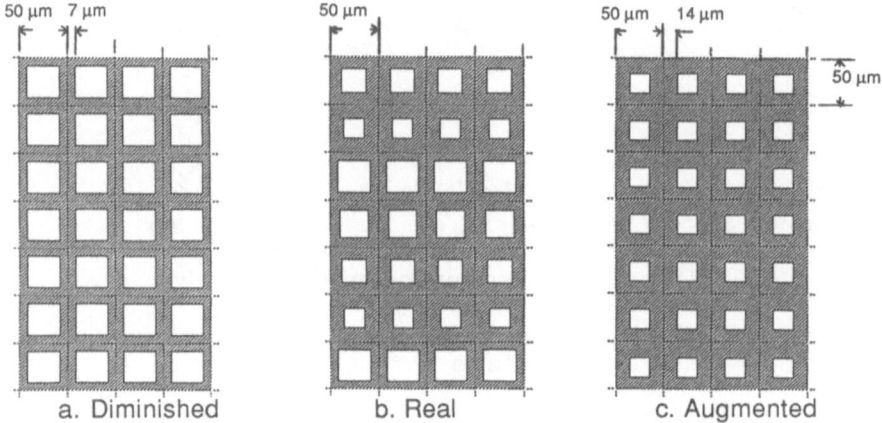

Fig. 2 : Segments of a real and monotonically modified materials.

Two additional porous periodic medium which are monotonically transformed from M_b are shown in Figures 2a and 2c. M_c and M_a are the "diminished" and "augmented" media with porosities 52 % and 19 % respectively. The 6x6 effective moduli of the wood cell wall in the direction of the microfibril principal axis is taken from litterature. The transversal effective elastic moduli of the periodic medium M_a and M_c are calculated by Navi's method [4] and the results shown in Table 2. Effective elastic moduli of the real quasi-periodic porous material M_b is calculated through linear interpolation in terms of (1-P), from the results obtained for M_a and M_c.

Table 2 : Calculated transversal effective moduli in GP_a

	Disminished Periodic Medium M_c	Interpolated Quasi-Periodic Medium M_a	Augmented Periodic Medium M_b
P	0.52	0.355	0.19
C_{11}	4.4008	5.86	7.3211
C_{22}	4.4008	5.86	7.3211
C_{12}	0.201	0.348	0.4858
C_{44}	0.231	0.907	1.58213

References:

1. C. Huet,P. Navi and P.E. Roelfstra, (1989) A homogenization technique based on Hill's modification theorem, 6th Symposium on Continuum Models and Discrete Systems, Dijon, June 26-29, unpublished.
2. R. Hill,. (1963) Elastic properties of reinforced solids : Some theoretical principles, J. Mech. Phys. Solids, 11, 357-372.
3. C. Huet, (1988) Unpublished technical note, ENPC-CERAM, Noisy-le Grand.
4. P. Navi, (1988) Evaluation des constantes élastiques d'un matériau hétérogène par analyse de la dispersion des ondes, Journées Physiques "Les méthodes d'auscultation et d'imagerie", Les Arcs, France, 135-140.

STABILITY OF MELT SPINNING PROCESSES

JAE CHUN HYUN
Department of Chemical Engineering,
Korea University
Anam-dong, Sungbuk-ku, Seoul, KOREA

ABSTRACT

The stability of melt spinning processes was theoretically studied on both isothermal and nonisothermal conditions. The governing equations of the system were based on a simple one-dimensional model which incorporated a strain-rate dependent material relaxation time into the convected Maxwell fluids. The instability behavior of the spinning, i.e., draw resonance and spin-line necking, produced by the model, corroborates well other research-ers' experimental findings. The results also show that the dichotomy of viscoelastic fluids depending on the White's parameter "a"(representing the degree of strain-rate dependency of the material relaxation time), turns out to be a very useful criterion for interpreting and predicting not only the spinning behavior but also other flow situations where extensional viscosity plays a key role.

INTRODUCTION

Among many instabilities encountered in polymer processing [1], the melt spinning instability is a major one which has attracted many researchers' attension around the world (e.g., [2][3][4]). In this study, we address the two instabilities of melt spinning, i.e., draw resonance and spin-line necking. The governing equations employed are the usual ones: continuity, momentum, energy and constitutive equation. For the last equation, we adopted a convected Maxwell model. We don't repeat them here, however, but refer to previous articles [4][5][6].

In addition to the above four equations, we add the White's variable relaxation time [7] as shown below.

$$\lambda = \frac{\lambda_o}{1 + a\sqrt{3}\lambda_o\dot{\varepsilon}} \tag{1}$$

where λ = strain-rate dependent material relaxation time
   λ_o = material relaxation time(constant for isothermal spinning and an exponential function of temperature for nonisother-mal spinning)

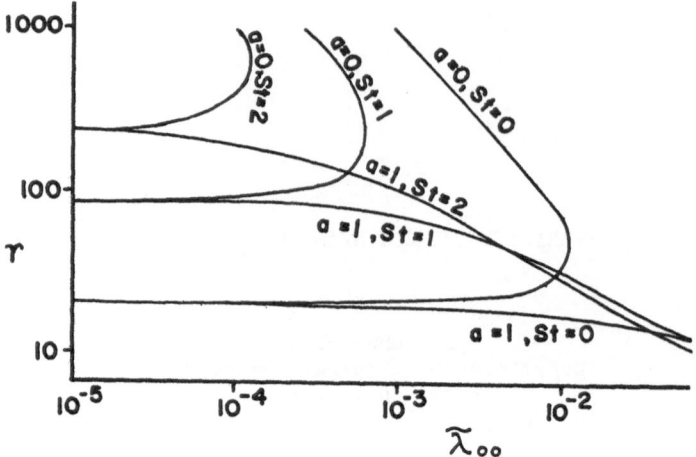

Figure 1. Critical Draw Ratio vs. The Relaxation Time with the Varying
Parameter "a" and Stanton Number.

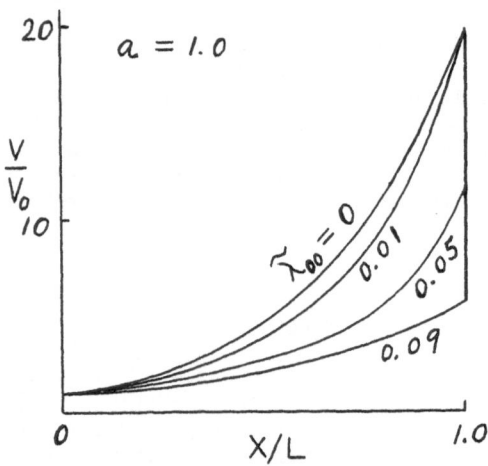

Figure 2. Spin-line Velocity Profile with the Varying Relaxation Time.

$\dot{\epsilon}$ = extensional strain-rate
"a"= White's parameter
The draw resonance instability can then be easily studied by solving the above five governing equations either numerically or analytically [5]. The new concept of draw resonance theory [5] based on the comparison of the two traveling times along the spin-line, i.e., one for throughput waves and another for the fluid elements, was recently confirmed [8] as pointed out by others [9] with some revisions after equestions were raised before [10].

RESULTS AND DISCUSSION

Fig.1 is a typical picture showing the critical draw rates, r, as a function of the dimensionless relaxation time, $\tilde{\lambda}_{oo}$, the parameter "a" and the cooling conditions represented here by the Stanton number, St. We can readily see that depending on the value of "a", there are fundamental differences between two fluids groups(one group of a < 1/$\sqrt{3}$ and the other group of a > 1/$\sqrt{3}$) in their behavior. For example, increasing $\tilde{\lambda}_{oo}$ stabilizes the former but destabilizes the latter. Cooling, in general, stabilizes the both groups. The results in Fig.1 agree well with others' experimental findings [11].
Regarding the spin-line necking [12], we have found that only the fluids group of a > 1/$\sqrt{3}$ can exhibit spin-line necking. Experimental results by others [11][13] corroborate our results. Fig.2 shows typical such results produced by our model. As the relaxation time,$\tilde{\lambda}_{oo}$,increases, the spinline stress increases and then a velocity jump(necking) occurs. If we specify the critical necking stress for a > 1/$\sqrt{3}$ materials,the velocity jump should occur at a mid-point of the spin-line as found by others [14].

REFERENCES

1. Petrie, C. J. S. and Denn, M. M., AIChE. J., 1976, 22, 209.
2. Ishihara, H. and Kase, S., J. Appl. Poly. Sci., 1975, 19, 557
3. Pearson, J. R. A. and Matovich, M. A., Ind. Eng. Chem. Fundamentals, 1969, 8, 605.
4. Fisher, R. J. and Denn, M. M., AIChE J., 1976, 22, 236.
5. Hyun, J.C., AIChE J., 1978, 24, 418.
6. Hyun, J.C. and Ballman, R. L., J. Rheology, 1978, 22, 349.
7. Ide, Y. and White, J. L., J. Non-Newtonian Fluid Mech., 1977, 2, 281.
8. Hyun, J. C., Oh, J. S. and Lee, S. J., Hwahak Konghak, 1989, 27, 556.
9. Mewis, J. and Petrie, C. J. S., Encyclopedia Fluid Mech. ed. by Cheremisinoff, 1987, 6, 111.
10. Denn, M.M., AIChE J., 1980, 26, 292.
11. Minoshima, W. and White, J. L., J. Non-Newtonian Fluid Mech., 1986, 19, 275.
12. Ziabicki, A., Fundamentals of Fiber Formation, John Siley & Sons, New York, 1976, p.69.
13. Wada, Y., J. Appl. Poly. Sci., 1971, 15, 183.
14. Kikutani, I., et. al., Proceedings of Polymer Processing Society, Kyoto, Japan, 1989, No. 06-02, 108.

RHEOLOGICAL PROPERTIES OF IRON-FREE KAOLIN/WATER SUSPENSIONS

YACHKO IVANOV, ALEXANDER POPOV[x], EKATERINA DIMITROVA
Central Laboratory of Physico-Chemical Mechanics, Bulg.Ac.of
Sci.,"Ac.G.Bonchev" Str., Bl.I, Sofia 1113, Bulgaria
[x]Institute of Applied Mineralogy, Bulg.Ac.of Sci.,"Rakovski"
Str. 92, Sofia 1000, Bulgaria

ABSTRACT

Rheological parameters are extremely sensitive to structural changes on microlevel. This has been demonstrated on the example of kaolin/water suspensions. Two types of kaolin were used: initial processed kaolin (K) and iron-free kaolin (IFK), obtained after a special procedure of iron leaching. The sharp increase of the rheological characteristics of IFK suspensions was not accompanied by a noticeable growth of the specific surface or by considerable changes in particle size distribution and electrophoretic mobility. Some peculiarities in the rheological behaviour of the kaolin/water dispersions were observed too. Most probably the structural changes are due to the predominating number of edge-to-edge and edge-to-face contacts, created by the newly formed active centres on the places of the leached iron ions.

INTRODUCTION

Rheology has found recently a very wide scope of application both in solving theoretical and technological scientific problems. As was stated by Mendeleev about a century ago in his "Fundamentals of Chemistry": "...The relationship existing between the viscosity and the other physical and chemical properties compels us to think, that this quantity, expressing the internal friction, will occupy an important place in molecular mechanics."[x] As a result from the classic works of a number of prominent scientists rheology has firmly found its proper place among the new fastly developing branches of science. Especially intensive investigations were carried out in the field of microrheology I2I. The present work is an illustration of the possibilities afforded by rheological methods, which turned out to be most sensitive to structural changes on microlevel in a practical case concerning kaolin/water suspensions.

---

[x] - recited from I1I, authors' translation

MATERIALS AND METHODS

Bulgarian processed kaolin was used in the investigation. The chemical and technical characteristics of the material are given elsewhere I3I. The method of iron leaching is described in detail in I4I. It consists of iron removal by sodium dithionate in a sulphuric acid medium with subsequent washing, settling and decantation of the liquid phase. This procedure lead to the increase of the rheological and mechanical characteristics. The measurements were performed on "Rheotron Brabender" viscometer by automatical recording of the flow curves in 40 s for the up- and the downcurve respectively. Electrophoretic mobility, BET surface area, sedimentation and pH measurements were performed too.

EXPERIMENTAL RESULTS AND DISCUSSION

Kaolinite has a 1:1 crystal structure, made up of tetrahedrically arranged silicate sheet and octahedrically arranged alumina sheet. The layers are held together by strong hydrogen bonding between the hydroxyl groups and the adjacent oxygen atoms of the layers, this being the reason for their nonswelling in water. Kaolinite particles are grouped in stacks of about 40 flat hexagonal platelets. Hence the dominating type of contacts the face-to-face (FF) one, while the edge-to-edge (EE) and edge-to-face (EF) contacts are weak and easily destroyed by salt presence and other factors.

The rheological behaviour of processed (K) and iron-free (IFK) kaolin/water suspensions was compared for samples from different deposits and at diverse regimens of testing. On Fig.1 are presented the flow curves for 33% solid mass suspensions. There are two phenomena that should be discussed in this case: 1) the dilatant behavior of the suspensions in a wide range of concentrations and 2) the considerable increase of the rheological parameters of the IFK suspensions. It should be noted that one of the classical examples of thixotropic behaviour - cement pastes, display thixotropy in viscometric tests for a relatively long period of time (usually 5-20 min) and later exhibit certain dilatant effect due to the enhanced chemical reaction on one hand and to the accelerated sedimentation in the thinned suspension under dynamic conditions on the other. Anyway, the particle size dis-

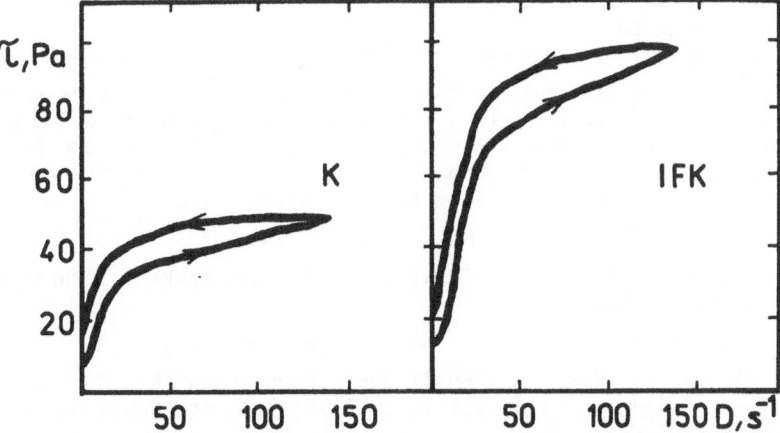

Figure 1. Flow curves of processed (K) and iron-free (IFK) kaolin/water suspensions (33 % solid mass content).

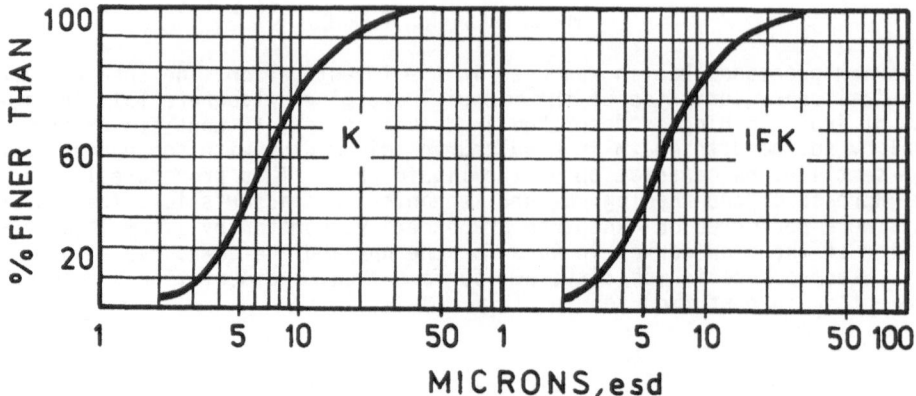

Figure 2. Particle size distribution for K and IFK samples

tribution for the K and IFK samples (Fig.2) testify to the presence of enough fine particles to prevent rapid sedimentation. Besides that the pastes,
especially the IFK ones were rather stable. When comparison is made with a
bentonite clay the difference is still more drastic. Kaolinite seems to display a somewhat retarded increase of the shear stress. We are inclined to
think that in this case the particles configuration and the type of contacts between them are the main reason for the observed peculiarity in the
flow behaviour.

As far as the comparison between K and IFK suspensions is concerned,
similarity is observed between the two samples as follows in Table 1:

TABLE 1

Parameter	Sample	
	K	IFK
Specific surface area, m^2/g	23	25
Particles finer than $10 \mu m$ esd, %	71	77
Electrophoretic mobility, $m^2V^{-1}s^{-1}$	$-1,47.10^{-8}$	$-1,48.10^{-8}$
Power of Hydrogen pH	~ 4	~ 4

Obviously the difference in the rheological characteristics should be
explained with changes on microlevel. Most probably the newly formed active
centres on the places of the leached iron ions contribute to the formation
of stronger EE and EF contacts and a firmer structure. Possibly a certain
breakdown of stacks occurred in the IFK suspensions too.

REFERENCES

1. Bezborodov, M.A., Viscosity of Silikate Glasses, Nauka i Technika,
   Minsk, 1975, (in Russian).

2. Ivanov, Ya., D.Sc.Thesis, 1988, Sofia, (in Bulgarian).

3. Popov, Al. et al., Proc. of the II World Congress on Non-Metallic Mine-
   rals, 1989, Beijing, China, Vol.I, 1071-74, Int. Ac. Publishers,
   Pergamon - CNPIEC.

4. Popov, Al. et al., Patent Requirement Nr 86928, 1989, Bulgaria.

EXTENSIONAL VISCOSITY, AN ELUSIVE PROPERTY OF MOBILE LIQUIDS

DAVID F. JAMES
Department of Mechanical Engineering
University of Toronto
Toronto, Canada M5S 1A4

Extensional viscosity is simply defined – it is the analogue of shear viscosity – but it is a much more difficult property to measure. A steady state can almost always be achieved in shear flow and thus it is straightforward to measure shear viscosity as a function of shear rate. In extensional flow measurements, however, steady-state conditions have been achieved only for a few melts and Newtonian fluids, not for any mobile viscoelastic fluids. A steady state should not be expected in extensional motion because the fluid microstructure – generally entangled polymer chains – generally will not reach an equilibrium configuration in finite time when the fluid and its microstructure are subjected to a constant axial tension. Because the microstructure evolves with time, the normal stresses and therefore the extensional viscosity are functions of the history of deformation. This complexity is often overlooked by rheologists, both theoretical and experimental, who report extensional viscosity η_E as a function of extensional rate $\dot{\varepsilon}$ only, in analogy with shear viscosity; η_E is a function of $\dot{\varepsilon}$ only when steady stresses are achieved during constant-$\dot{\varepsilon}$ motion. Since steady-state cases are rare, it is proposed that the term "ultimate extensional viscosity" and the symbol η_E^{∞} be reserved for these special cases, leaving the term extensional viscosity η_E to indicate a property which depends on strain history.

In many extensional rheometers, the strain history is difficult to determine and the flow field is often not well suited for the measurement of extensional viscosity. The remainder of this note deals with some of the significant but lesser-known weaknesses of current techniques to measure η_E.

One flow which seems obviously exploitable for η_E measurements is Fano flow (the tubeless siphon) because there is a free surface everywhere. Paradoxically, however, the zero shear stress at the free surface causes shearing within the flow. This curious result, originally pointed out by MacSporran [1] and demonstrated experimentally by Matthys [2], can easily be shown by a stress balance on a fluid element adjacent to the free surface. The induced shearing is not negligible, for Matthys's measurements show that the shear rate is up to twice the extensional rate.

Shearing induced by the free surface is also present in fibre-spinning flows, but a more serious defect is shearing in the issuing tube. The ratio of tube shear rate to filament extensional rate is about $8LD^2/D_o^3$ where D is the filament diameter after length L and D_o is the tube diameter. Typical values for these dimensions, taken from [3] for example, yield a ratio around 20. This significant prior deformation likely accounts for higher η_E values obtained by this technique than by other techniques.

Matta [4] has developed a new stretched-filament technique which avoids the problems of shear. His method is to pull out a fluid drop by a falling weight; the filament remains cylindrical as it is stretched and thus the motion is purely extensional. While this technique is similar to the stretched-cylinder technique for polymer melts, the strain rate cannot be controlled in Matta's device because the rate necessarily rises rapidly and then decreases inversely with time. A further problem is that the strain history cannot be made the same for different fluids.

Internal flows are also used for the measurement of extensional viscosity. Entry flows, into a tube or an orifice, or sudden-contraction flows can subject the fluid to high extensional rates and thus generate easily-measurable extensional effects. The difficulties with these flows are that strain rates must be found from velocity measurements (which is rarely done), strain histories are different for different fluids, and axial tension cannot be determined directly. However, Hasegawa and colleagues [5,6] found that axial tension calculated from measurements of jet thrust agreed with calculations based on measurements of excess pressure drop, which suggests that indirect measurements of axial tension are valid.

An extensional rheometer based on converging channel flow has been developed by the writer and colleagues [7]. The claim that constant-extensional-rate flow is generated in their channel has not been confirmed by velocity measurements. Also, shearing in the upstream connecting tube is not small, the shear rate being about 40% of the extensional rate in the channel.

Perhaps the most promising technique for the measurement of η_E is that developed by Fuller and co-workers [8]. Shear-free stagnation point flow is created by drawing fluid into opposed entry tubes and the dynamical reaction on one tube is measured. Rheometrics Inc. is developing an instrument based on this concept and intends to market the instrument in the summer of 1990. It will be the second commercial extensional rheometer for mobile liquids, the first being the fibre-spinning instrument sold by Carri-Med. The Fuller device is simple to operate and can test liquids having a wide range of viscosities. Its main drawback is that the residence times of fluid particles in the constant-extensional-rate flow field vary widely.

This brief critique of available techniques may help to explain why a proper measurement of extensional viscosity is so difficult. In essence, we are still in the formative stages of trying to measure this elusive fluid property.

REFERENCES

1. MacSporran, W.C., J. Non-Newt. Fluid Mech., 1981, **8**, 119-138.

2. Matthys, E.F., J. Rheology, 1988, **32**, 773-788.

3. Chang, J.C. and Denn, M.M., J. Non-Newt. Fluid Mech., 1979, **5**, 369-385.

4. Matta, J., "Liquid Stretching Using a Falling Cylinder", in press, J. Non-Newt. Fluid Mechanics.

5. Hasegawa, T. and Fukutoni, K., Proc. Xth Int. Congress on Rheology, Sydney, Australia, 1988, page 1.395.

6. Hasegawa, T. and Nakamura, H., "Experimental Study of the Elongational Stress of Dilute Solutions in Orifice Flow", 1989, private communication.

7. James, D.F., Chandler, G.M. and Armour, S.J., "A Converging Channel Rheometer for the Measurement of Extensional Viscosity", in press, J. Non-Newt. Fluid Mech.

8. Fuller, G.G., Cathey, C.A., Hubbard, B., and Zebrowski, B., J. Rheology, 1987, **31**, 235-249.

THE VISCOSITY OF A SMECTIC POLYMER

S.G. JAMES AND P. NAVARD
Ecole Nationale Supérieure des Mines de Paris
Centre de Mise en Forme des Matériaux
URA CNRS 1374
Sophia-Antipolis
06560 VALBONNE CEDEX - FRANCE

ABSTRACT

The rheology of a smectic main chain polymer shows that the activation energy and the viscosity are very high. The recoverable strain in the smectic phase is large due to the reformation of the smectic layers after cessation of shear.

INTRODUCTION

There is no report in the litterature on the rheology of a main-chain polymer in a smectic phase. The results published concerned either smectic low molar mass liquid crystals [1] or side-chain polymers [2]. The present work concerns measurements of viscosity and of recoverable strain of a smectic main chain polymer. The polymer used is described in reference 3, under the code name TO 11. It is in a smectic C phase above 125°C and changes to the smectic A phase between 190 and 210°C. Its smectic-isotropic transition is between 235 and 253°C. The rheological measurements were performed with a rheometrics RSR, with a cone and plate equipment and under nitrogen.

RESULTS AND DISCUSSION

The viscosity of the polymer as a function of temperature, at a shear rate of 0.03 s^{-1}, is shown in figure 1, along with the phase transition temperatures. One can see that in the smectic phase at low temperatures the viscosity is extremely high. If one assumes an Arrhenius type behaviour the activation energy in the smectic phase can be calculated as 275 kJ mol^{-1}. This value is very high when compared with the values for conventional thermoplastics (30 to 100 kJ mol^{-1}), but is within the range of values obtained for the side chain smectic polymers (2). The reason for this high value can be seen if one examines the structural model proposed by Noel et al. (3). During flow polymer chains tend to be aligned parallel to the flow direction, and so using the proposed structure the smectic layers will be perpendicular to the cone and plate and also to the flow direction. This is the situation for the highest viscosity. As the temperature is increased towards the smectic-isotropic transition temperature the perfection of the smectic layers decreases and the energy required to move one layer over another also decreases.

As the temperature is increased above 235°C the polymer becomes biphasic and the drop in viscosity with temperature decreases, until in the isotropic phase the slope of the viscosity/T^{-1} curve is similar to that of many conventional isotropic polymers.

The recoverable strain (s_r) has been measured for the polymer after shear at three different temperatures. At 229°C when the polymer is in the pure smectic phase the strain recovered after the cessation of shear is large s_r = (3.25) and even after 5 minutes the limiting value has not been reached. A tentative explanation is to relate this to the displacement of the rigid-units in one chain in relation to the rigid-units in an adjacent chain.

On increasing the temperature to 234° and 240°C where the polymer is biphasic the recoverable strain decreases (σ_r = 2) due to the presence of an isotropic component but still remains higher than in the purely isotropic phase (σ_r = 1.8).

References

1.   HORN R.G. and KLEMAN M., Ann. Phys. 1978, **3**, 229.

2.   ZENTEL R. and WU J., Makromol. Chem. 1986, **187**, 1727.

3.   NOEL C., FRIEDRICH C., BOSIO L. and STRAZIELLE C., Polymer 1984, **25**, 1281.

Figure 1 : Plot of log (viscosity) as a function of T^{-1} in the smectic, biphasic and isotropic states.

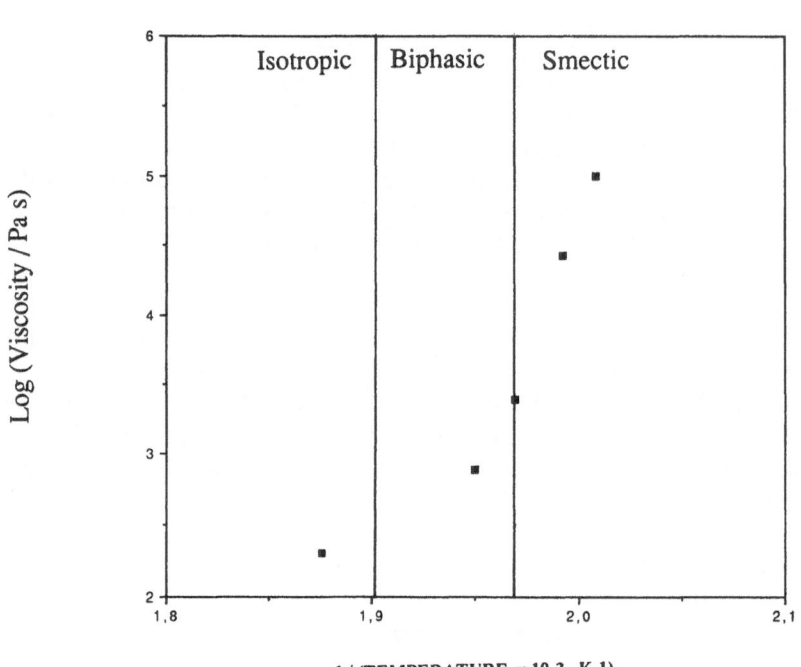

1 / (TEMPERATURE x 10-3 . K-1)

RHEOLOGICAL STRUCTURE IN BULK SEAWATER

IAN R. JENKINSON
Agency for Consultation and Research in Oceanography
Lavergne, F-19320 La Roch Canillac, France

ABSTRACT

Recent rheological meaurements on seawater samples indicate it to be a weak organic gel. Superimposed on the newtonian viscosity of the aqueous phase, η_w, is a non-newtonian excess viscosity, $\eta_E = k \cdot \dot{\gamma}^{-P}$, where P has been found to vary from 0.9 to 1.6. The non-newtonian properties appear to be due to polymeric material secreted primarily by algae and bacteria. Calculations made on bubbles trapped in an exceptional phytoplankton bloom indicate an excess total modulus, G_E^* of $\geqslant 0.2$ Pa. In parts of the sea $\eta_E > \eta_w$ at the low shear rates *in situ*, 3×10^{-4} to 1 s^{-1}, thereby damping turbulence. Some gel is flocculated into complex organic aggregates. It is proposed that seawater deserves the epithet "biomaterial".

INTRODUCTION

Surface films have been recognized as viscoelastic since at least 1872 [1]. Present models of oceanographic processes such as turbulent dispersion and energy dissipation, however, assume that seawater viscosity, η, depends only on temperature and salinity, varying from 0.8 to 1.9 mPa·s[2,3]. These values were obtained with a Morton (capillary) viscometer at shear rates, $\dot{\gamma}$, from 100 to 5000 s^{-1}. This is 2 to 7 orders of magnitude higher than ambient values of $\dot{\gamma}$, which range from only 1 s^{-1} in surface water stirred by a wind of $\approx$ 10 m s^{-1}, to 3×10^{-4} s^{-1} or lower in deepwater thermoclines [4].

Polymer content of seawater

Seawater is an organic solution, containing about 0.5 to 3 mg·kg^{-1} of organic carbon, at least half of which is in molecules larger than 20 000 daltons [5]. Most of these polymers are considered to be produced by phytoplankton, either directly or by secretion of small molecules later polymerized by bacteria [6].

The seawater surface film also consists of large, soluble, polymer-like materials [7], the amount of which has been found related to phytoplankton productivity in the underlying water [8]. Phytoplankton can also secrete material which modifies bulk rheological properties [4,9,10]. Based on detailed study of the way it reduces turbulent drag at high values of $\dot{\gamma}$, material secreted by algae has been categorized as consisting of covalent polyanions bound by weak forces into a hydrocolloid matrix [10].

The aim of the present paper is to show that seawater is at times functionally non-newtonian, and to briefly review progress in the young field of seawater rheology.

EVIDENCE FOR NON-NEWTONIAN PROPERTIES

Anecdotal reports of exceptional rheological properties

Reports exist of exceptional bulk rheological properties during phytoplankton blooms. Water sampled during one bloom sprang back out of sampling containers like white of egg [11], and in another, the water was so gelatinous that it kept clay in suspension, prevented weighted fishing nets from sinking, clogged nets so that they tore when attempts were made to recover them, and even made the steering of boats difficult [refs cited in 4,12].

Mechanical spectra in some phytoplankton cultures

In some phytoplankton cultures, serving as laboratory models of blooms, the mechanical spectra were determined with a <u>Contraves</u> LOW SHEAR 30 rheometer fitted with a Couette measuring system. Some of the cultures were shear-thinning liquids showing power-law behaviour over discrete ranges of $\dot{\gamma}$. It was found that viscosity could be regarded as consisting of two components, such that,

$$\eta = \eta_w + \eta_E \tag{1}$$

where η_w newtonian viscosity due to the aquatic phase and η_E, excess viscosity, that due to polymeric material. Over ranges of $\dot{\gamma}$ showing power-law behaviour, the excess viscosity due to polymers,

$$\eta_E = k \cdot \dot{\gamma}^{-P} \tag{2}$$

where k and P are constants. P was found to range from 0 to 1.3.

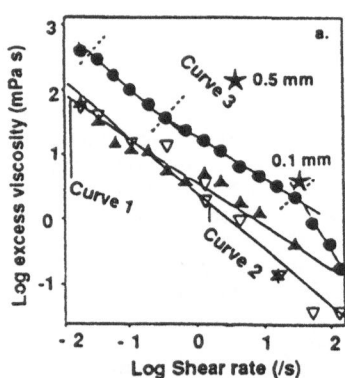 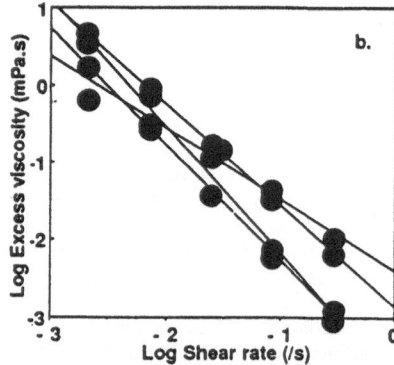

Figure 1. Mechanical spectra a. Culture of <u>Dunaliella marina</u> from different parts of culture vessel. (Asterisks show η_E–$\dot{\gamma}$ coordinates calculated from trapped bubbles observed rising in still seawater in a bloom of <u>Gyrodinium</u> cf <u>aureolum</u> at $\lessdot 1 mm \cdot s^{-1}$, the estimated limit of detection). b. Non-bloom seawater from the Mediterranean. Each point represents the mean of between 4 and 16 readings. From Jenkinson [13].

The highest viscosity measured, at $\dot{\gamma}$ = 0.017 s^{-1}, was ≈ 400 times that of seawater [13]. The spectrum incorporating this measurement is reproduced in Fig. 1a. Elasticity was also found in this culture.

Mechanical spectra of some non-bloom seawater samples

Thirty-two non-bloom seawater samples were taken at different depths and at four times of year from the Mediterranean off Villefranche-sur-Mer. They were measured using Couette geometry in oscillating mode at shear amplitude 240% and at $\dot\gamma$ from 0.002 to 0.29 s^{-1}. The excess viscous modulus, $G_E" = \eta_E/\dot\gamma$, averaged $\approx 5 \times 10^{-6}$ Pa.

Since no systematic effect of water depth could be detected, all the samples for each time of year were treated as replicates. The data are reproduced in Fig. 1b [13]. The water samples were shear thinning, with values of P for each time of year varying from 0.9 to 1.6. Interference from surface effects precluded determination of bulk elasticity.

Relationship found between seawater viscosity and biological factors

Viscosity, measured at $\dot\gamma$ = 0.072 s^{-1}, and various biological parameters were quantified during a fairly intense phytoplankton bloom in the German Bight. Of all the variables measured, the most correlated pair of parameters was chlorophyll concentration and η_E (r = 0.86, n = 12) [12]. This supports the contention that the principle determinant of η_E was phytoplankton biomass.

OCEANOGRAPHIC IMPLICATIONS OF VARIOUS RHEOLOGICAL PARAMETERS

Can exceptional thickening kill marine organisms?

On many temperate coasts, red tides of the phytoplankter _Gyrodinium_ cf _aureolum_ (= _Gymnodinium_ cf _nagasakiense_) are economically important. They are associated with mass mortality of wild and cultured animals such as fish and molluscs, both adult [14] and larval [15,16]. Because of this species, fish and mollusc culture is considered too risky on some coasts.

In a bloom of this species, bubbles were seen trapped in the water presumably by the yield stress of secreted polymers [13]. For a given rising speed of bubble, η-$\dot\gamma$ coordinates have been calculated (Fig. 1), which indicate an excess total modulus, G_E^* of $\geq$0.2 Pa. It was proposed that the fish are killed because the energy used in pumping this thickened water through the gills uses more oxygen than could be extracted from pumped water [13]. This idea is supported by failure to find significant quantities of toxin in _G._ cf _aureolum_, but adhesion of plankton mucus to the fish gills might also add to the effects of viscosity [17].

Reduction of turbulence at low shear rate

In the sea, increase in viscosity at low, ambient $\dot\gamma$ dampens turbulence and increases the size and duration of Kolmogorov eddies [4]. The mechanism is different from that operating in drag reduction by elastic polymers at high $\dot\gamma$ [10]. Since Kolmogorov eddies are considered to be the hydrographic units in which planktonic organisms live, it may suit them to gel their environment, thereby increasing spatial stability.

Marine organic aggregates

Some of the suspended organic matter is flocculated. These flocs, "marine snow", can contribute up to 1% of the seawater volume fraction. Sinking of marine snow is an important pathway for organic carbon fixed from atmospheric CO_2 to sink to the ocean bed [18]. Its mechanical properties are thus important inputs for models of global climatic control. Rheometry of marine snow, however, has only just begun [19]. Aggregates are a functionally solid, if elastic, pieces of seawater, and represent turbulence-free zones. The organisms living in them may be able to have more stable relationships with each other than those constantly subject to the vagaries of turbulence.

Influence of aggregates on inter-aggregate turbulence

The break up of aggregates in shear fields absorbs energy. Even wobbling aggregates that are sufficiently strong to avoid break-up may extract energy from turbulence by viscous loss. In particular, the presence of elasticity in a liquid decouples the relationship between energy dissipation and turbulent dispersion. This problem has not yet been addressed in oceanography.

Elongational viscosity

The measurements described so far in Couette flow have all measured only shear viscosity. The Trouton parameter, $T = \eta_T/\eta$, where η_T and η are respectively elongational and shear viscosity. In newtonian fluids $T = 3$, but in non-newtonian liquids, T may be as high as 1000 [20]. Since many authors believe that both elongational and shear flow are important in turbulence [20], it may be important to measure η_T as well as η in studies on non-newtonian damping of marine turbulence.

Many aggregates are deformed into "comet" shapes during sinking. This indicates elongational deformation. Modeling of aggregate break-up may thus also need to take into account their elongational rheology.

CONCLUSIONS

Measurements recently carried out with a commercially available instrument indicate that seawater behaves like a weak gel, and that the degree of thickening is positively related to the concentration of phytoplankton. Seawater may thus deserve the epithet, "biomaterial". It is hydrologically and biodynamically active, as well as lumpy. Determination of its rheological properties, and incorporation of these into physical, chemical, ecological and even climatalogical models thus requires intense interdisciplinary cooperation.

REFERENCES

1. Marangoni, C. Sul principio della viscosità superficiale dei liquidi stabili. Nuovo Cimento, Ser. 2, 1872, 5/6, 239-73.

2. Krümmel, O. and Ruppin, E. Über die innere Reibung des Seewassers. Wiss. Meeresunters. Kiel. Komm., 1905, 9, 27-36.

3. Miyake, Y. and Koizumi, M. The measurement of the viscosity coefficient of seawater. J. mar. Res., 1948, 7, 63-6.

4. Jenkinson, I.R. Oceanographic implications of non-newtonian properties found in phytoplankton cultures. Nature, Lond., 1986, 323, 435-7.

5. Toggweiler, J.R. Deep-sea carbon, a burning issue. Nature, Lond., 1988, 334, 468.

6. Biddanda, B.A. and Pomeroy, L.R. Microbial aggregation and degradation of phytoplankton-derived detritus in seawater. Mar. Ecol. Prog. Ser., 1988, 42, 79-88.

7. Goldman, J.C., Dennett, M.R. and Frew, N.M. Surfactant effects on air-sea gas exchange under turbulent conditions. Deep-Sea Res., 1988, 35, 1953-70.

8. Žutić, V., Ćosović, B., Marčenko, E., Bihari, N. and Kršinić, F. Surfactant production by marine phytoplankton. Mar. Chem., 1981, 10, 505-20

9. Hoyt, J.W. and, Soli, G. Algal cultures: ability to reduce turbulent friction in flow. Science, Wash. D.C., 1965, 149, 1509-11.

10. Ramus, J., Kenney, B.E. and, Shaughnessy, E.J. Drag reducing properties of microalgal exopolymers. Biotechnol. Bioeng., 1989, 33, 550-7.

11. Dreyfuss, R. Note biologique à propos des eaux rouges. Cah. Cent. Enseignem. Rech. Biol. Oceanogr. méditerr., Nice, 1962, N° 1, 14-15.

12. Jenkinson, I.R. and Biddanda, B.A. Microbial plankton and viscosity of bulk seawater in the German Bight. Oceanologica Acta, in the press.

13. Jenkinson, I.R. Increases in viscosity may kill fish in some blooms. In Red Tides: Biology, Environmental Science, and Toxicology, eds T. Okaichi, D.M. Anderson and T. Nemoto, Elsevier, New York, 1989, pp. 435—8.

14. Jenkinson, I.R. and Connors, P.P. The occurrence of the red-tide organism, Gyrodinium aureolum (Dinophyceae) around the south and south-west of Ireland in August and September 1979. J. Sherkin Isl., 1980, 1(1), 127-45.

15. Potts, G.W. and Edwards, J.M. The impact of Gyrodinium aureolum blooms on inshore young fish populations. J. mar. biol. Ass. U.K., 1987, 67, 293-7.

16. Erard-Le Denn, E. and Dao, J.C. Toxicity of Gyrodinium cf. aureolum towards Pecten maximus (post larvae, juveniles and adults). 4th int. Conf. on Toxic Marine Phytoplankton, held at Lund, Sweden, 26-30 June 1989.

17. Partensky, F., Le Boterff, J. and Verbist, J.-F. Does the fish-killing dinoflagellate Gymnodinium cf nagasakiense produce cytotoxins? J. mar. biol. Ass. U.K., 1989, 69, 501-9.

18. Alldredge, A.L. and Silver, M.W. Characteristics, dynamics and significance of marine snow. Progr. Oceanogr., 1988, 20, 41-82.

19. Jenkinson, I., Abreu, P., Biddanda, B., Ehlken, S., Riebesell, U. and Turley, C. Measurements of aggregate strength. Aggregate Methodology Workshop., Alfred-Wegener-Institut, Bremerhaven, June 1988, pp. 22-23.

20. Allsford, K. Elongational viscosity: its relevance and measurement. Appl. Rheol. News, Warren Spring Lab., April 1987, 1-2.

THE INFLUENCE OF CONCENTRATION AND PARTICLE SHAPE ON THE RHEOLOGY OF SILICA, FLYASH AND OTHER INDUSTRIAL SLURRIES

R.L. JONES* and R.R. HORSLEY**
*Department of Biochemistry, University of the Witwatersrand,
Johannesburg, South Africa
**Department of Mechanical Engineering, Curtin University of
Technology, Perth, Western Australia

ABSTRACT

Rheograms of aqueous suspensions of pure silica and commercial flyash of comparable particle size distribution but different particle shape, have been compared at different concentrations using a modified viscometer designed to prevent settling of the particles. Using the entropy approach to slurry viscosity, it was possible to show a relationship between the plastic viscosity (η) of the slurry and the vacancy fraction, where the latter is the fluid space available to the suspended particles. This relationship contains a factor which distinguishes between the rounded shape of the flyash particles and the irregular shape of the silica. Extrapolation of this relationship to zero solids concentration yields in both cases a value very close to that expected of water.

INTRODUCTION

Recently a new approach has been suggested which attempts to describe the rheograms of non-Newtonian shear thinning mineral slurries in terms of clearly defined material parameters [1,2]. One suggestion is that the plastic viscosity (η) of the slurry, that is the linear part of the rheogram where the shear rate becomes proportional to the shear stress, is directly related to the fluid space available to the particles, the so-called vacancy fraction. The vacancy fraction can be represented by

$$\left(1 - \frac{\Phi}{\Phi_M}\right)$$

where Φ is the volume fraction and Φ_M is the maximum packing density of the suspended particles.

Using the equation

$$\eta = \frac{T}{A} \exp\left(\frac{Q}{RT}\right)$$

to describe the plastic viscosity of the slurry, where T is the absolute temperature, R the gas constant, Q the activation energy for flow for the suspending fluid, and A a constant including such factors as the diffusion coefficient and particle size, the vacancy fraction can be incorporated as a configurational entropy term such that

$$\eta = \frac{T}{A} \exp\left(\frac{\Delta S}{R}\right) \exp\left(\frac{Q}{RT}\right)$$

where

$$\Delta S = R\ln\left(1 - \frac{\Phi}{\Phi_M}\right)^s$$

The factor s is a term introduced to compensate for the irregularity of the particles and varies from 1 for uniform spheres of similar size to 3 for irregularly shaped or sized particles since the latter can fit into vacant sites in three orientations. Thus the plastic viscosity can be written as

$$\eta = \left(\frac{T}{A}\right)\left(1 - \frac{\Phi}{\Phi_M}\right)^{-s} \exp\left(\frac{Q}{RT}\right)$$

Plots of

$$\log\eta \; vs \; \log\left(1 - \frac{\Phi}{\Phi_M}\right)^{-1}$$

should then give a straight line with an intercept value at zero solids concentration close to that expected of water and a slope which varies between 1 and 3 depending on the particle shape.

RESULTS

Materials used were industrial flyash obtained from power stations with an average particle diameter of 8 microns and pure (99.5%) silica obtained commercially with an average particle diameter of 12 microns, so chosen because the size distribution was the closest available to that of the flyash. Figure 1 illustrates the similarity between the particle sizes of the two materials and the distinct difference in particle shape.

 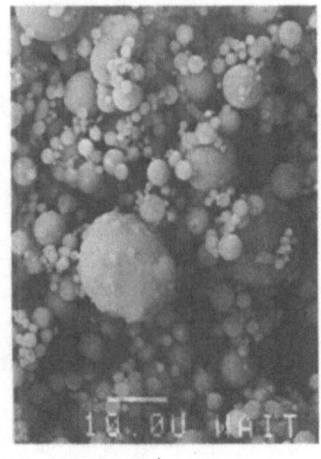

a   b

Figure 1. Electron micrograph of (a) silica and (b) flyash particles

Rheograms of aqueous suspensions of pure silica and flyash were obtained at 20°C with varying volume fraction concentrations in a modified Haake viscometer specially designed to prevent the solids from settling [3]. Maximum packing densities of the silica and flyash slurries were 0.60 and 0.59 respectively. Further samples were tested using mixtures of pure silica and flyash in the ratio of 1:1.

Plots of

$$\log \eta \ vs \ \log \left(1 - \frac{\Phi}{\Phi_M}\right)^{-1}$$

gave straight lines in all cases with a slope of approximately 3.0 in the case of silica, 1.7 in the case of flyash and intermediate values for silica-flyash mixtures. Extrapolation of the lines to zero solids volume fraction in all cases gave intercepts close to 10^{-3} Pa s, the viscosity of water. Data obtained using gold mine tailings which have irregularly shaped particles showed behaviour similar to silica. Thus the s factor in the model described can be identifed as a particle shape factor which is dependent on the regularity of the particle shape.

REFERENCES

1. Chandler, H.D. and Jones, R.L. An entropy approach to determining the viscosities of concentrated slurries. Philos. Mag. Letters, 1988, 57, 81-84.

2. Chandler, H.D. and Jones, R.L. FLow kinetics in concentrated suspensions of silica-iron(III) oxide-water. Philos. Mag. A, 1989, 59, 645-660.

3. Overend, I.J., Horsley, R.R., Jones, R.L. and Vinycomb, R.K. A new method for the measurement of rheological properties of settling slurries. Proc. IX Intl. Congress on Rheology, Mexico, 1984, 583-616.

EQUIVALENT OLDROYD MODELS

R S JONES
Department of Mathematics,
University College of Wales,
Aberystwyth, U.K.

ABSTRACT

In the numerical simulation of two-dimensional flows there is considerable simplification if one considers the co-rotational rather than the upper convected Maxwell model. This is because the co-rotational Maxwell is traceless and for two-dimensional flows only two of the extra stress components are independent. In this paper we consider other models which are traceless for two-dimensional flows but not for flows in three-dimensions. More generally we consider equivalent Oldroyd models, that is Oldroyd eight-constant models for which the stress components differ only by an arbitrary pressure. It is shown that in some circumstances it is possible to simplify the governing equations by choosing an appropriate equivalent model.

INTRODUCTION

The Oldroyd 8-constant model [1] can be written in the form

$$T+\lambda_1\overset{\circ}{T}-\mu_1(T.d+d.T)+\nu_1\text{tr}(T.d)I+\mu_0(\text{tr}T)d$$

$$= 2\eta_0\{d+\lambda_2\overset{\circ}{d}-2\mu_2d.d+\nu_2\text{tr}(d.d)I\}, \tag{1}$$

where T is the extra stress tensor, d the rate-of-strain tensor and the superfix • denotes the co-rotational derivative. This is the most general equation that is linear in T, bilinear in T and d and quadratic in d^2. The eight constants λ_1, μ_1, ν_1, μ_0, λ_2, μ_2, ν_2, η_0 are arbitrary. (Some restrictions, for physical reality, on the range of possible values for these constants are given in (1)). However the extra stress T is arbitrary to the extent of an added isotropic pressure and so we cannot distinguish between models for which the extra-stresses differ only by an arbitrary isotropic pressure. Such models we call equivalent rheological models.

EQUIVALENT MODELS

If we replace T by $T' + pI$ in (1) we obtain the equivalent model

$$T'+\lambda_1\overset{\circ}{T}-\mu_1(T'd+d.T')+(3\mu_0p+\mu_0(trT')-2\mu_1p)d+(p+\lambda_1\overset{\circ}{p}+\nu_1tr(T'.d))I$$
$$=2\eta_0\{d+\lambda_2\overset{\circ}{d}-2\mu_2d^2+\nu_2(trd^2)I\}. \tag{2}$$

We can choose the pressure to make the coefficients of d or I zero. In particular it is not difficult to prove the following:

(a) Provided $2\mu_1 \neq 3\mu_0$, p can be chosen such that $(3\mu_0-2\mu_1)p+\mu_0(trT')=0$, and it follows that (1) is equivalent to

$$T'+\lambda_1\overset{\circ}{T}-\mu_1(T'.d+d.T') + (\nu_1-\frac{\mu_0}{2\mu_1}(3\nu_1-2\mu_1)tr(T.d)I)$$
$$=2\eta_0\{d+\lambda_2\overset{\circ}{d}-\mu_2d^2+(\nu_2-\frac{\mu_0}{2\mu_1}(3\nu_2-2\mu_2)(trd^2)I,$$

i.e. a model without the μ_0 term involving trT'.

(b) If $\mu_1= \frac{3}{2}\mu_0$, then (1) is equivalent to a model with $\nu_1=\nu_2=0$.

(c) If $\mu_1/\mu_2=\nu_1/\nu_2$, then (1) is equivalent to a model with $\nu_1=\nu_2=0$ and with μ_0 replaced by $\mu_0+k(2\mu_1-3\mu_0)$, where $k=\mu_2/\mu_1$.

Two-dimensional flows
For two-dimensional flows the velocity vector v, rate-of-strain tensor d and extra-stress tensor T have the forms: $v = (u(x,y), v(x,y),0)$

$$d = \begin{pmatrix} d_{xx} & d_{xy} & 0 \\ d_{xy} & d_{yy} & 0 \\ 0 & 0 & 0 \end{pmatrix}, \quad T = \begin{pmatrix} T_{xx} & T_{xy} & 0 \\ T_{xy} & T_{yy} & 0 \\ 0 & 0 & T_{zz} \end{pmatrix} \tag{3}$$

when referred to orthogonal cartesian axes $0_{x,y,z}$. For the numerical simulation of these flows it is observed that the governing equations for a co-rotational Maxwell model are much simpler than those for the upper-convected Maxwell. This is because the co-rotational Maxwell is traceless, $T_{xx}+T_{yy}+T_{zz}=0$, and for two-dimensional flows $T_{zz}=0$. Hence only two of T_{xx}, T_{xy}, T_{yy} are independent. This is not generally regarded as being a very good model and the question arises whether it is possible to construct a model with 'better' viscometric functions but still maintaining the property $T_{xx}+T_{yy}=0$ for two-dimensional flows.

The upper-convected time derivative $\overset{\triangledown}{T}=\overset{\circ}{T}-(T.d+d.T)$. When T and d have the forms

(3), and writing $T_{xx}+T_{yy}=(\text{tr}T)'$,

$$\mathbf{T.d+d.T}=\text{tr}(\mathbf{T.d})\mathbf{I}+(\text{tr}T)'\mathbf{d}, \quad (\text{tr}\mathbf{d}=0).$$

It follows that the model

$$T+\lambda(\overset{\triangledown}{T}+(\text{tr}\mathbf{T.d})\mathbf{I})=2\eta_0\mathbf{d}, \tag{4}$$

has the property $T_{xx}+T_{yy}=0$ for two dimensional flows and will give the same equations for T_{xx}, T_{xy}, T_{yy} as the co-rotational Maxwell.

The condition for the Oldroyd 8-constant model to have the above property is $\mu_1=\nu_1$, $\mu_2=\nu_2$. We can now construct models equivalent to (1) with this property. For example if $\nu_1=2\mu_1(\mu_1-\mu_0)/(2\mu_1-3\mu_0)$ in (1) then it is equivalent to the model

$$\mathbf{T}'+\lambda_1\overset{\circ}{\mathbf{T}}'-\mu_1(\mathbf{T}'.\mathbf{d}+\mathbf{d}.\mathbf{T}')+\mu_1\,\text{tr}(\mathbf{T}'.\mathbf{d})\mathbf{I} = 2\eta_0\{\mathbf{d}+\lambda_2\overset{\circ}{\mathbf{d}}-2\mu_2\mathbf{d}.\mathbf{d}+\mu_2\text{tr}(\mathbf{d}.\mathbf{d})\mathbf{I}\}, \tag{5}$$

of which (4) is a special case. These models all have constant planar extensional viscosity $\eta_{EP} = 4\eta_0$. A variable η_{EP} can be introduced, somewhat artificially, by choosing $\eta_T=\eta_T(\text{tr}\,\mathbf{d}.\mathbf{d})$.

CONCLUSIONS

It has been shown, by choosing an appropriate isotropic pressure, that in some circumstances either the μ_0 or ν terms in the Oldroyd 8-constant model can be made redundant. In addition a class of models has been constructed which are traceless in two-dimensional flows but not generally so.

REFERENCES

1. Oldroyd, J.G., Non-Newtonian effects in steady motion of some idealized elastico-viscous liquids. Proc. Roy. Soc., 1958, A245, 278-297.

TRANSVERSE FLOW OF FIBRE-REINFORCED COMPOSITES

R S JONES AND A B WHEELER
Department of Mathematics,
University College of Wales,
Aberystwyth, U.K.

ABSTRACT

An experimental and theoretical investigation into the flow of continuous fibre-reinforced materials is presented. In particular a study is made of the flow mechanisms that take place during consolidation under constant load.

A theoretical model is considered in which the material is treated as an incompressible anisotropic fluid. Solutions to the governing equations are obtained for transverse flows and the predicted behaviour of the material is compared with that observed in the experiments.

INTRODUCTION

Thermoplastics reinforced with continuous inextensible carbon fibres have recently been introduced into the field of composite materials. During most processing operations, in particular consolidation, this composite material is subjected to a normal load. An experimental and theoretical investigation into the flow processes that occur during consolidation has been made in conjunction with ICI Wilton. The basic mechanisms of deformation of continuous fibre-reinforced composites have been outlined by Barnes and Cogswell [1].

A series of experiments were performed at ICI on carbon fibre reinforced PEEK and subsequently at Aberystwyth on a model system in which the samples were squeezed between parallel rectangular plates (Figure 1). The model system, supplied by ICI, could be used at room temperature and consisted of carbon fibres in a matrix of Golden Syrup. The experiments were carried out on laminates made up of a series of plies. The fibres were initially aligned both within the plies and from ply to ply. The laminate was therefore initially transversely isotropic with respect to the fibre direction. The experimental procedure and the results of the experiments at ICI have already been presented [2]. Subsequent experiments on the model system at Aberystwyth showed similar behaviour. Some *elastic* recovery was also observed on removing the load. In [2] the results were interpreted using analysis in which the following assumptions were made:

1) the fibres remain parallel;

2) the flow in the axial direction is negligible.

During the experiments a *barrelling* effect was observed and a small amount of resin was squeezed out along the fibre direction (Figure 2). In order to accommodate these, the analysis considered in [2] must be extended to three dimensions.

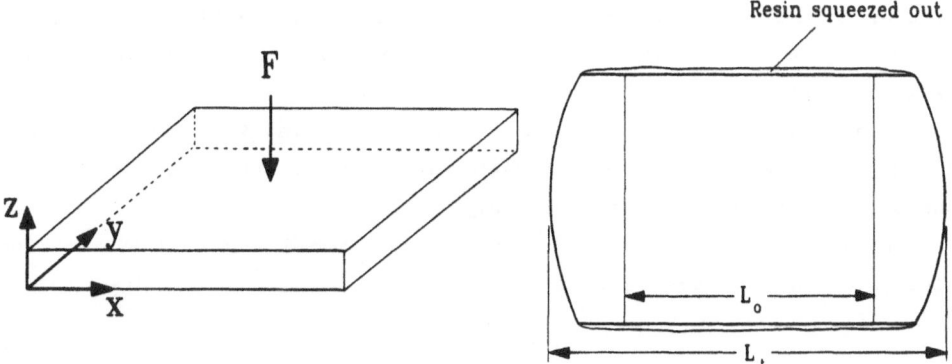

Figure 1. Geometry of sample.   Figure 2. Barrelling deformation.

THEORY

We represent the composite in its molten state as an incompressible anisotropic fluid in which, referred to Cartesian axes, the stress components σ_{ij} at any instant are related to the components of the rate-of-strain tensor d_{ij} by the constitutive equation

$$\sigma_{ij} = -p\delta_{ij} + 2\eta_T d_{ij} + 2(\eta_L - \eta_T)(a_i a_k d_{kj} + a_j a_k d_{ik}) + E_L a_i a_j a_m a_n d_{mn} \tag{1}$$

where p is the isotropic pressure and a is a unit vector representing the fibre direction in the composite; η_L and η_T represent shear viscosities in the axial and transverse directions and E_L represents an elongational viscosity in the axial direction. The vector a moves as a material line and satisfies

$$\frac{\partial a_i}{\partial t} + v_j \frac{\partial a_i}{\partial x_j} = a_j \frac{\partial v_i}{\partial x_j} - a_i a_j a_k \frac{\partial v_j}{\partial x_k} \qquad (2)$$

The material is consolidated between parallel rectangular plates. Initially the fibres are parallel and aligned with one pair of edges of the plates.

In order to solve the full equations numerical techniques are required and a finite difference solution is currently being investigated. Using the usual lubrication approximations it can be shown that the pressure p satisfies $\nabla^2 p = 0$ subject to zero pressure at the edges of the plates. A solution is now obtained in terms of a double Fourier series from which the velocity distribution is derived. Within this approximation it is shown that the vector **a** remains constant and, hence, all the fibres remain parallel. However, some indication of the changes in fibre direction predicted by the model can be obtained by considering the terms neglected in (2) under the lubrication approximation. It is shown that these are consistent with the observed *barrelling*.

In the limit as $\eta_T \rightarrow \infty$ we obtain the usual two-dimensional lubrication solution given by Rogers [3].

CONCLUSIONS

It is shown that some of the transverse flow properties of a fibre-reinforced composite can be predicted using a highly anisotropic continuum model.

REFERENCES

1. Barnes, J.A. and Cogswell, F.N., Transverse flow processes in continuous fibre-reinforced thermoplastic composites. Composites, 1989, **20**, 38-42.

2. Balasubramanyam, R., Jones, R.S. and Wheeler, A.B., Modelling transverse flows of reinforced thermoplastic materials. Composites, 1989, **20**, 33-37.

3. Rogers, T.G., Squeezing flow of fibre-reinforced viscous fluids. J. Engng. Math., 1989, **23**, 81-89.

THE RHEOMETRIC CHARACTERIZATION OF FLEXIBLE CHAIN AND LIQUID CRYSTAL POLYMERS

VINOD M. KAMATH and MALCOLM R. MACKLEY
Department of Chemical Engineering, University of Cambridge,
Pembroke Street, Cambridge CB2 3RA, UK

ABSTRACT

We have studied the linear and nonlinear rheological behaviour of mono and polydisperse flexible chain and polydisperse liquid crystal polymers (LCPs) in the molten state. A modified version of the Doi and Edwards (DE) constitutive equation, consisting of memory function, damping function and a nonlinear coupling function is used to analyse the data. Simple analytical approximations for these functions are found to satisfactorily describe the nonlinear step strain-stress relaxation and steady shear-stress relaxation at high shear rates.

INTRODUCTION

In this paper we characterize the flexible chain and liquid crystal polymers in terms of their response to linear and nonlinear deformations. In general, for steady state conditions, three basic functions are necessary a) a memory function, $M(s)$ or equivalently a relaxation modulus $G(s)$ where $s=t-t'$ with t the current and t' the past time, b) a nonlinear strain dependence described by a damping function, which is a function of strain γ, $h(\gamma)$ and c) a further coupling function $\mu(s,\gamma)$. By adopting the above strategy we are able to fully describe both the long time scale diffusional relaxation of polymer molecules and the short time scale convective chain stretching response. We find that the overall response of a material is a subtle combination of both the time and strain dependence of the materials. The approach has been inspired by the recent developments in molecular theories and hence can be directly related to the molecular mechanisms for monodisperse polymers that have been studied.

A CONSTITUTIVE EQUATION FOR COMPLEX FLOWS

In this work we limit our observations to simple shear deformation only and we adopt an integral formulation of the viscoelastic constitutive equation [1]. This is a direct generalisation of the DE constitutive equation for linear flexible polymers for transient deformations as originally formulated [2]. The constitutive equation is of a generalised K-BKZ type, relating stress $\tau(t)$ to the deformation gradient $E(t,t')$ by

$$\tau(t) = \int_{-\infty}^{t} \frac{\partial \widetilde{G}(t\text{-}t', \ \mathbf{E}(t,t'))}{\partial t'} D(\mathbf{E}(t,t')) \ dt' \tag{1}$$

where $D(\mathbf{E})$ is a generalised measure of deformation, and the first term is equivalent to a strain dependent memory function [2]. Under certain conditions, the nonlinear relaxation modulus $\widetilde{G}(t\text{-}t', \mathbf{E})$ can be written as product of linear relaxation modulus $G(t\text{-}t')$ and a nonlinear coupling function $\mu(t\text{-}t', \mathbf{E})$

$$\widetilde{G}(t\text{-}t', \ \mathbf{E}(t,t')) = G(t\text{-}t') \ \mu(t\text{-}t', \ \mathbf{E}(t,t')) \tag{2}$$

For steady shear deformation of a polydisperse polymer, the various functions can be written as

$$M(s) = -\frac{\partial G(s)}{\partial s} = \frac{G_i}{\lambda_i} \exp\left(\frac{-s}{\lambda_i}\right) \tag{3a}$$

$$\mu(s,\gamma) = \left[1 + (\alpha(\gamma)\text{-}1) \exp\left(\frac{-s}{\lambda_K}\right)\right]^2 \tag{3b}$$

$$D(\gamma) = h(\gamma) \ \gamma = \gamma \exp(-k\gamma) \tag{3c}$$

where we have assumed certain simplistic forms for all the functions which are justified below.

The justification for choosing the Maxwell type relaxation modulus with moduli G_i and time constants λ_i, is two-fold. First, in previous work, we have found experimentally that this is a satisfactory procedure when considering small strain oscillatory viscoelastic data [3]. Secondly, molecular theories derive analytic predictions for the relaxation modulus in terms of exponential functions. $G(s)$ is a measure of time constants present in the fluid.

The choice of an analytic form for damping $h(\gamma)$ is more subjective. From DE theory, this term $\{h(\gamma) \ \gamma\}$ is represented by their so called $\mathbf{Q}$ tensor. It has been shown that to first approximation a general deformation measure for polydisperse polymers can be represented as single decaying exponential type as considered above [4]. The constant k in the exponent is a parameter, which has to determined experimentally, if the measure of deformation is unknown. We have found that the damping function of the DE constitutive equation is in good agreement with single exponential behaviour and k approximates to 0.38. In view of its simple analytic form we generalise this to other polymers, and retain k to be an adjustable parameter to be determined experimentally for each polymeric system.

Finally, we choose the coupling function $\mu(s,\gamma)$ also to follow the DE theory. In their theory, this has the form of a Rouse type relaxation of polymer chain length, and the strain dependent term $\alpha(\gamma)$ can be approximated as $(2/15) \ \gamma^2$ for steady simple shear. This factor can be associated with the measure of deviations from viscoelastic behaviour when the material is sheared at a rate faster than the longest relaxation time λ_d of the polymer. For polydisperse polymers, we can assume that there exists a time, not necessarily the Rouse relaxation time of the polymer chain, before which the constitutive equation becomes non-separable. To avoid confusion we shall indicate this time by λ_K, in general based on observations in the literature [5].

When the above terms are introduced into the original constitutive equation (1) we can rewrite the constitutive equation as

$$\tau = \int_0^\infty \left\{ M(s)\left[1 + (\alpha(\gamma) - 1)\,e^{-s/\lambda_K}\right]^2 + \frac{2G(s)}{\lambda_K}(\alpha(\gamma) - 1)\,e^{-s/\lambda_K}\left[1 + (\alpha(\gamma) - 1)\,e^{-s/\lambda_K}\right] \right\} D(\gamma)\ ds \quad (4)$$

which will be used in analysing the experimental data.

RESULTS AND DISCUSSION

In the limit of linear viscoelasticity, the above equation reduces to a general constitutive equation for linear viscoelasticity, which can be analysed rigourously. Further, under linear viscoelastic conditions, the relaxation of both the mono and polydisperse polymers and LCPs in general requires a range of time constants to be specified. For monodisperse polymers, we can identify two dominant time constants, which can be associated with segmental Rouse relaxation time (not included in the above analysis of nonlinear viscoelasticity, as it becomes very small) and reptation time.

For large step strain-stress relaxation the nonlinear relaxation modulus, will be the same as that of DE constitutive equation which is

$$\widetilde{G}(s,\gamma) = G(s)\left[1 + (\alpha(\gamma) - 1)\exp\left(\frac{-s}{\lambda_K}\right)\right]^2 h(\gamma)\,\gamma \tag{5}$$

The above equation is tested by us for mono and polydisperse polymers and LCPs. For monodisperse polymers this is found to be satisfactory. Further, both the damping function and constant k are in agreement with DE theory, this agreement has also been reported by other researchers [4,5]. For polydisperse polymers, the coupling function characterized by the Rouse-like relaxation time λ_K, is not clearly defined. In order to obtain a value, a nonlinear optimization is adopted over the whole range of $\widetilde{G}(s,\gamma)$. In general, it was found possible to obtain a reasonable representation of the data.

In terms of LCPs three important features were established. Firstly, the small strain viscoelastic behaviour is characterized by a broad range of time constants. Secondly, the nonlinear response did not follow the DE flexible chain strain softening response characterized by constant k. For flexible molecules, k=0.38 and for the LCPs we examined k=0.20 which results in a substantially enhanced strain softening response. Finally, the effect of the coupling function became significant at short time scales.

For steady shear-stress relaxation, at high shear rates $\dot{\gamma}$, equation (4) can be integrated with appropriate limits to obtain the nonlinear stress decay coefficient $\bar{\eta}^-(t,\dot{\gamma})$ as

$$\widetilde{\eta}(t,\dot{\gamma}) = G_i\lambda_i\,e^{-t/\lambda_i}\left[\frac{1}{(1 + k\dot{\gamma}\lambda_i)^2} + \frac{24}{15}\frac{(\dot{\gamma}\lambda_i)^2 e^{-t/\lambda_R}\lambda_R^3(\lambda_R + \lambda_i)}{(\lambda_R + k\dot{\gamma}\lambda_i\lambda_R + \lambda_i)^4} + \frac{480}{225}\frac{(\dot{\gamma}\lambda_i)^4 e^{-2t/\lambda_R}\lambda_R^5(\lambda_R + 2\lambda_i)}{(\lambda_R + k\dot{\gamma}\lambda_i\lambda_R + 2\lambda_i)^6}\right]$$
$$(6)$$

Equation (6) clearly indicates the evolution of rheological behaviour for molten polymer systems where the steady flow can be characterized by linear (Newtonian), nonlinear (shear thinning) and full viscoelastic behaviour.

CONCLUSIONS

Linear viscoelastic response of flexible chain and liquid crystal polymers can be adequately described by the theory of linear viscoelasticity provided the data is analysed using a sound numerical procedure. So far as the the nonlinear response of polymers is concerned, a simplified form of Doi and Edwards constitutive equation is found to be useful and satisfactory. In general, for flexible polymers, as the polydispersity increases, the range of time constants present becomes broader. In addition, the ability to relate the coupling function time constant λ_K to a unique Rouse relaxation time becomes less clear. We find that LCPs can fit into the scheme described in this paper although certain rheological parameters are significantly different to flexible melts.

REFERENCES

1. Lin, Y. H., Explanation for slip-stick melt fracture in terms of molecular dynamics in polymer melts, J. Rheol., 1985, **29**, 605-37.

2. Doi, M. and Edwards, S.F., The theory of polymer dynamics, Clarendon, Oxford, 1986, pp. 239-66.

3. Kamath, V.M. and Mackley, M.R., The determination of polymer relaxation moduli and memory functions using integral transforms, J. Non-Newtonian Fluid Mech., 1989, **32**, 119-44.

4. Laun, H. M., Description of nonlinear shear behaviour of a low density polyethylene melt by means of an experimentally determined strain dependent memory function, Rheol. Acta, 1978, **17**, 1-15.

5. Osaki, K. and Kurata, M., Experimental appraisal of the Doi-Edwards theory for polymer rheology based on the data for polystyrene solutions, Macromolecules, 1980, **13**, 671-6.

ON-LINE MEASUREMENTS OF THE RHEOLOGICAL PROPERTIES OF FERMENTATION BROTH

Z. KEMBŁOWSKI, P. BUDZYNSKI, P. OWCZARZ
Institute of Chemical and Process Engineering
Technical University of Łódź, Poland

ABSTRACT

The paper is concerned with continuous measurements of the rheological properties of fermentation broth. An on-line rheo-meter "Rheohelix-1", based on the application of a helical screw impeller rotating in a draught tube, has been cons-tructed. The instrument was used for measurements of the rheo-logical parameters of fermentation broth of Aspergillus niger in a submerged fermentation process. The results of rheological and standard measurements have been compared.

INTRODUCTION

In an aerobic fermentation process the broth forms a three-phase system consisting of a suspension of microorganisms and air bubbles in the continuous liquid phase. In the course of the process changes in the biomass concentration, in the morphology of the microorganisms, and - in some cases - also changes in the concentration of extracellular metabolites occur. This may lead to significant changes of the rheological properties of the broth. Many mycelial fermentation broths, particularly those which contain microorganisms in filamentous form, exhibit highly non-Newtonian behaviour already at a relatively low biomass concentration.

The changes of rheological properties of fermentation broth influence the hydrodynamics of the bioreactor (bulk mixing and bubble coalescence), and therefore they influence mass and heat transfer. This has of course a profound effect on the fermentation process, and, as a consequence, also on the rheological properties of the broth. Hence, a complicated feed-back mechanism takes place in the bioreactor.

As it was pointed out in 1978 by Charles [1], "... in most cases where rheological properties are controlling factors they are also sensitive indicators of the state of the process and should be considered for purpose of monitoring and control".

Therefore, continuous measurements of the rheological properties of fermentation broth may contribute a lot to the improvement of process design, scale-up and operation.

EXPERIMENTAL

In one of our previous papers [2] the concept of a rotational
rheometer with helical screw impeller rotating in a draught
tube was presented, and a procedure for the flow curve determination was proposed. Contrary to the impeller instruments described in the literature, no calibration of the developed
measuring system is necessary. The experimental verification
proved the system to be fully applicable in the rheological
characterization of mycelial fermentation broths.
    Recently, the theory of the instrument has been improved
in comparison with the simplified approach presented in paper
[2]. This will be discussed in a separate paper.
    Taking advantage of the developed system we have con-
structed an on-line rheometer "Rheohelix-1" which is presented
schematically in Fig. 1. The instrument has been used for continuous measurements of the rheological properties of fermentation broth of Aspergillus niger during a submerged fermentation
carried out in a stirred tank bioreactor. A photograph of the
experimental set-up is shown in Fig. 2.

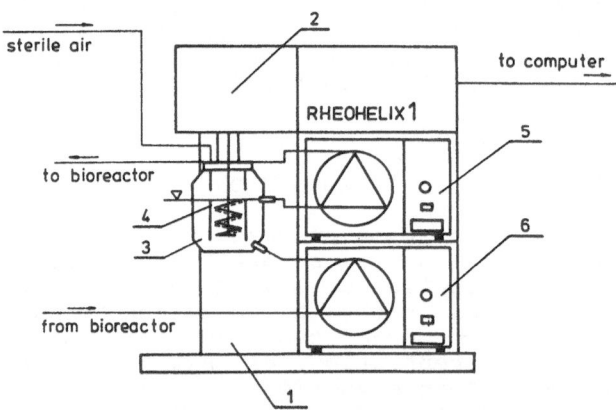

Figure 1. Diagram of the rotational rheometer "Rheohelix-1".
          1-electrical motor and rotational speed indicator,
          2-torque meter, 3-measuring tank, 4-helical screw
          impeller with draught tube, 5,6-peristaltic pumps.

The rheometer was equipped with a microprocessor which
acted as a sequential time controller. The microprocessor collected data on the rotational speed and torque of the helical
screw impeller, as well as data on the fermentation parameters:
temperature, rotational speed of the agitator, reading of the
oxygen- and pH-electrodes, and performance time of the foam
suppression system. Simultaneously, during the entire fermentation process, periodical analyses were carried out to determine
the concentration of biomass, citric acid and unfermented sugar.

The measurements were carried out during the whole fermentation cycle, i.e. ca. 140 hours. The rheometer worked at a preset rotational speed. After a chosen time interval of 40 minutes the sequential time controller started to change automatically the rotational speed of the helical screw impeller in 10 rising and 10 falling steps.

Figure 2. Experimental set-up.

RESULTS AND DISCUSSION

The obtained flow curves could be approximated by the power law, and the values of rheological parameters n and k were determined as a function of fermentation time. The parameters were compared with the results of standard measurements of the fermentation process and clear dependences were found. A detailed analysis of the results is outside the scope of this short report. The general conclusion is, however, that the constructed rheometer "Rheohelix-1" may become a sensitive tool which will enable monitoring of the fermentation process. It is also worth noting that the on-line measurements of rheological properties did not disturb the process.

REFERENCES

1. Charles, M., Technical aspects of the rheological properties of microbial cultures. Adv. Bioch. Eng., 1976, 8, 1-62.

2. Kembłowski, Z., Sęk, J., Budzyński, P., The concept of a rotational rheometer with helical screw impeller. Rheol. Acta, 1988, 27, 82-91.

SIMULATION OF VISCOELASTIC FLUID FLOW PAST A CYLINDER USING THE "METHOD OF LINES"

Riyaz Kharrat and Shapour Vossoughi
Department of Chemical and Petroleum Engineering
University of Kansas
Lawrence, Kansas, U.S.A.

ABSTRACT

An investigation is carried out into numerical techniques to solve the unsteady state flow of viscoelastic fluid past a circular cylinder rod. An implicit four-constant time dependent Oldroyd model is employed to separate shear-thinning and elastic effects.

The method of lines is found to produce stable solution for low as well as high elasticity number. However, in the case of high elasticity, finer mesh sizes and small time steps had to be used to reach convergence. A downstream shifting of the streamlines is found for low elasticity while an upstream shifting is observed for high elasticity. These predictions are in accordance with the available experimental evidence.

The pressure and vorticity distribution around the surface of the cylinder are found to alter in the presence of shear-thinning parameter.

INTRODUCTION

From the review of the literature [1-3], the following facts emerged :

I) Existing theoretical analyses predict a downstream shift for low Weissenberg number, We. However, no numerical simulations are available to evaluate any upstream shift associated with high We. In fact, divergence problems in numerical techniques have been reported for high We.

II) Many numerical schemes "diverge" for comparatively low elasticity values. In addition, it has been found very difficult to obtain solution for flows of highly elastic liquids, consequently, smoothing procedures and transient solution have been recommended.

For the transient solution of viscoelastic fluid past a cylinder, several finite difference methods may need to be tried, each of which has its own stability and convergence problems, as well as truncation errors. Hence, the purpose of this work is to find the suitable numerical method to solve the unsteady state flow of viscoelastic fluid past a cylinder.

BASIC EQUATIONS AND THEORY

The unsteady state flow of an incompressible viscoelastic liquid past a circular cylinder of radius "a" and of infinite length is considered. The equations of momentum, continuity, and the rheological equation of state are solved simultaneously. In addition, vorticity and stream function definitions are employed. The fluid is characterized by the four-constant time dependent Oldroyd model [4].

METHOD OF SOLUTION

Different numerical schemes and methods were considered in this work. However, the method of lines was found to produce the optimum results. Hence, the result of this method will be given only. The description and detail analysis of the other methods are given elsewhere [5]. In the method of lines, the partial differential equations are reduced to a system of ordinary differential equations. The resulting system of ordinary differential equations is then solved using Runge-Kutta-Gill method. The central difference is used in the space derivative, which gives a second-order truncation error while the time derivative is fourth-order using the Runge-Kutta method.

RESULT AND DISCUSSION

The numerical methods have been applied to Newtonian, elastic and shear-thinning fluid flow for Reynolds numbers equal to 0.25, 2.5, and 10, and various values of relaxation time, retardation time, and zero shear rate viscosity.

It was found that as the shear-thinning parameter μ_o increases, the maximum pressure at the surface of the cylinder decreases. However, the maximum vorticity value increased as the shear-thinning parameter was increased. Hence, the shear-thinning property might enhance the rotation of the fluid particles, which is manifested in the vorticity value.

Streamline profiles for Newtonian, low and high elastic (small and large relaxation time), and shear-thinning fluids were generated and compared for the sake of downstream and upstream shifting phenomena. They indicated that the small downstream shift due to elasticity increased when shear-thinning parameter was included. These results compare favorably with those of Pilate and Crochet [6], who used a constant second-order fluid model, and with those of Townsend [7], who used the Oldroyd model.

An upstream shifting for the highly elastic fluid was absorved, which is in agreement with the experimental observation of Manero and Mena [8]. The elasticity parameters used in this work are equivalent to the Weissenberg number used by Manero and Mena [8]. In the case of shear-thinning fluid flow the same streamline pattern was observed as in the case of the viscoelastic fluid flow, and little difference was found between them. It, thus could be argued, that shear-thinning properties might have been suppressed by the strong elastic effect.

CONCLUSIONS

A stable solution was obtained not only for low elasticity but also for highly elastic fluids. In addition, the upstream and downstream shifting of the steamlines have been predicted. Finally the presence of shear-thinning in the solution has been found to alter the pressure and vorticity profiles to some extend.

REFERENCES

1. Leslie, F.M., The Flow Of a Viscoelastic Liquid Past a Sphere. Q. J. Mech. Appl. Math., 1961, 14, 36.
2. Ultman, J.S. and Denn, M. M., Slow Viscoelastic Flow Past Submerged Objects. The Chemical Eng. J., 1971, 2, 81.
3. Townsend, P., A Numerical Simulation of Newtonian and Viscoelastic Flow Past Stationary and Rotary Cylinders. J. Non-Newtonian Fluid Mech., 1980, 6, 219.
4. Oldroyd, J. G., On the Formulation of Rheological Equations of State. Proc. R. Soc., 1950, A200, 523.
5. Kharrat R., Rheological Characterization of a Boger Fluid and the Numerical Simulation of its Flow Across a Cylinder. Doctoral Dissertation, The University of Kansas, 1989.
6. Pilate, G. and Crochet, M., J., Plane Flow of a Second-Order Fluid Past Submerged Boundaries. J. Non-Newtonian Fluid Mech., 1977, 2, 323.
7. Townsend, P., On the Numerical Simulation of Two-Dimensional Time-Dependent Flow of Oldroyd Fluids, part 1 : Basic Method and Preliminary Results. J. Non-Newtonian Fluid Mech., 1984, 14, 265.
8. Manero, O. and Mena B., On the Flow of Viscoelastic Fluid in Cross Flow Around a Circular Cylinder. J. Non-Newtonian Fluid Mech., 1981, 9, 379.

EXCESS THERMAL NOISE GENERATED DURING THE CAPILLARY FLOW OF POLY(ETHYLENE OXIDE) SOLUTIONS: THE INFLUENCE OF THE SOLVENT

CARL KLASON, JOSEF KUBÁT and OTAKAR QUADRAT*
Chalmers University of Technology, Department of Polymeric Materials,
S-412 96 Gothenburg, Sweden

*Institute of Macromolecular Chemistry, Prague 6. Czechoslovakia

ABSTRACT

The excess thermal noise generated during capillary flow of polymer solutions has been studied in detail for solutions of poly(ethylene oxide) of varying concentrations in water, and for constant concentration in various solvents (mixtures of water and isopropyl alcohol). The excess noise level, which increased with flow rate, was found to be a linear function of the dimensions of the polymer coils obtained from intrinsic viscosity data. Results obtained with different solvents and concentrations show that the excess noise depends on both the concentration and the solvent system. The noise power seems to be closely connected with the entanglement density in the system, expressed as the degree of overlapping of the macromolecules, or the correlation length, i.e., the mean distance between two points of entanglement.

INTRODUCTION

Measurements of electrical fluctuations (noise) associated with Poiseuille flow of various polymer solutions through capillaries have been the subject of a series of papers from our laboratory (1,2). The bulk of the experiments was carried out with aqueous solutions of poly(ethylene oxide), PEO.

Regarding the data gathered until now it appears natural to focus further work on the state of the molecular coils in the solution and its possible correlation with the spectral distribution of the noise signal. This is the aim of the present paper where we show data relating to the role played by the dimensions of the polymer coil, and the degree of overlapping and entanglement in the present context.

EXPERIMENTAL

A series of poly(ethylene oxide) solutions (WSR-301, Polyox, M_w 4×10^6) in water (conductivity 0.89 μS/cm) and its mixtures with 20, 40, 70 and 85 wt.% of isopropyl alcohol as a nonsolvent was used. Due to the preferential sorption of one component of the solvent on the polymer molecules, the dependence of the expansion factor varies with solvent composition and, consequently the intrinsic viscosity shows a maximum at a certain isopropyl alcohol content. The experimental set-up, was similar to that described in previous work (2).

RESULTS AND DISCUSSION

The spectral distribution of the noise generated by the capillary flow of all our samples showed a $1/f^\alpha$ type behaviour, with a white plateau at the low frequency end of the spectrum. The noise ratio U of the noise power obtained by the integration of the square of the noise voltage over the frequency range available and the corresponding noise power for the solution at rest (the noise ratio $U = U^2_{flow}/U^2_{rest}$) increased with the volumetric flow rate Q. It was found that the increase of U with Q depends strongly on the composition of the solvent. Fig. 1 shows the variation of U with Q for various solutions at constant polymer concentration. Surprisingly and significantly, the U-Q diagram consists of a family of straight lines going through the origin. The slopes of these lines depend on the composition of the solvent, although not in a monotonic manner.

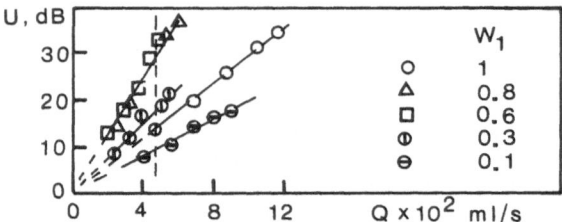

Figure 1. Noise ratio U vs. volumetric flow rate Q. w_1=weight fraction H_2O.

Fig. 2 shows a curve exhibiting a U-maximum for solutions containing about 55 % isopropyl alcohol. The data plotted in Fig. 2 relate to a particular value of Q (4.8×10^{-2} ml/s). The lower part of Fig. 2 is a corresponding plot showing the dependence of the intrinsic viscosity on the solvent composition. There is a striking similarity in the overall appearance of the two diagrams. The relation between the noise intensity U and the intrinsic viscosity was found to be linear. With regard to the role played by the polymer concentration c, it was found that the noise ratio U is approximately proportional to this quantity.

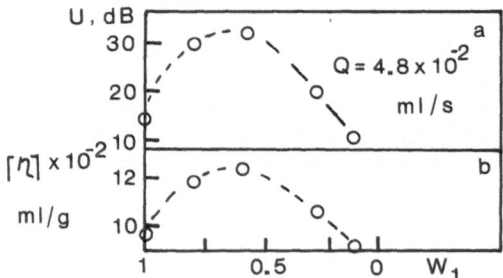

Figure 2. Noise ratio U (a) and intrinsic viscosity (b) vs. solvent composition. w_1 weight fraction of H_2O in mixture with isopropyl alcohol.

It is interesting to combine the roles played in this context by c and [η] into a more general relationship expressing the noise ratio in terms of parameters normally used to characterize semidilute polymer solutions. Among such parameters, the degree of overlapping of the polymer coils, c/c^*, and the correlation length ξ stand out as natural candidates for such a generalization. They are defined as follows (3)

$$c/c^* = (4\pi/3)6^{3/2} \ N \ R_h^3 \ c/M, \qquad (1)$$

$$\xi = 6^{1/2} < s^2 >^{1/2} (c/c^*)^{-3/4} \qquad (2)$$

where N denotes Avogadro's number, R_h the hydrodynamic radius of the polymer coils, M the molecular weight, $< s^2 >^{1/2}$ the mean square radius of gyration of the macromolecules, and c^* the coil-overlap critical concentration at which the space occupied by the solution is filled up by swollen macromolecular coils.

Fig. 3 shows that the data for different solvents and concentrations do not coincide and that the U values for a chosen c/c^* or ξ depend on both the concentration and the solvent system. The noise level increases with the intrinsic viscosity. Since the latter quantity is an increasing function of the solvent power we conclude that the noise effect is enhanced by the polymer-solvent interaction or - in other words - the solvation of the polymer segments.

However, a full elucidation of this problem will require additional experiments, in the first place with polymer samples having various molecular weights, in order to find the limiting conditions for the excess noise generation. This will be the subject of a coming investigation.

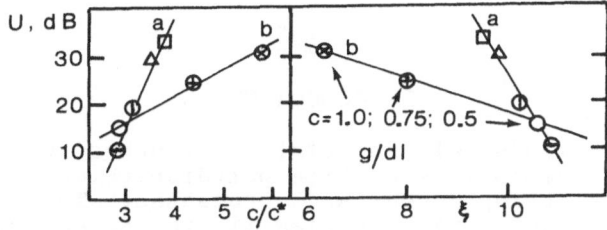

Figure 3. Noise ratio vs. c/c^* and correlation length ξ at a constant flow rate of 4.8×10^{-2} ml/s. Line 'a' relates to a constant polymer concentration (0.5 g/dl) in different solvent compositions (see Figure 1). Line 'b' relates to different polymer concentrations in one solvent (water).

REFERENCES

1. Hedman, K., Klason, C. and Kubát, J. J. Appl. Phys., 1979, 50, 8102.
2. Klason, C. and Kubát, J., J. Appl. Phys., 1985, 58, 2060.
3. Coviello, T., Bucrchard, W., Dnntini, M. and Crescenzi, V., Macromolecules, 1987, 20, 1102.

THE INFLUENCE OF MODIFICATION OF PARTICLE SIZE OF CERAMIC SUSPENSIONS UPON THEIR APPARENT VISCOSITY

JIŘÍ KONUPEK
Research Institute of Non-metallic materials
360 13 Karlovy Vary Czechoslovakia

ABSTRACT

The modifications of particle size of high concentrated suspensions influence expresively their rheological properties. It is demonstrated upon the example of kaolin suspensions how it is possible, by putting together two or three different size fractions, to compose the suspensions with various apparent viscosity. The new GV - method for this process is described here.

INTRODUCTION

In different kinds of industrial branches relatively concentrated suspensions of fine polydispersion non-metallic materials are being used and processed. To enable some operations without any troubles it is necessary for the concentracion and apparent viscosity of suspensions not to exceed preliminarily set limits at the defined range of shear stress or shear rate. The optimum grain size distribution can help to fulfile these requirements. It is possible to find it out by the original so - called GV - method /Granulometry - - Viscosity/ which is explained upon a concrete example. This method is also very effective in the theoretical research of rheological properties of all sorts of polydispersed suspensions.

Specification of the problem
To find out the influence of granularity upon the apparent viscosity of kaolin suspensions and determinate the optimum particle size distribution and its allowable fluctuation in order to fulfil the following conditions:
1. Range of particle size: 0 - 10 μm.
2. Concentracion of suspension: Minimaly 35,6 %vol. /57 %w./.
3. Limit apparent viscosity: 500 mPa·s in the shear rate range of 20 - 4860 s^{-1}.

MATERIALS AND METHODS

1. Fractions I - III were separated out from the original kaolin: I: 0,5-2 μm;   II: 2-5 μm;   III: 5-10 μm.

2. Suspensions with the concentracion of 35,6 %vol./57 %weight/ liquefied by the addition of 1 % of natriumhexametaphosphat were prepared by a uniform process from fractions I,II,III.

3. Apparent viscosities of ηz at 6 values inside of the given range of shear rate were measured by rotational viscosimeter.

4. A ternary diagram was drawn for each value of shear rate. To each point of ternary diagram whose coordinates correspond to the mixture proportions according to table 1 was subjoined the measured value of apparent viscosity.

5. Approximate coordinates of points that define the mixture proportions with apparent viscosity 500 mPa·s were derived in the each ternary diagram by the grafic linear interpolation of apparent viscosity. The connecting line of these points makes the boundary line of the so-called mono-shear rate existence areas of suspensions with apparent viscosity of $\eta z \gtrless 500$ mPa·s.

6. The boundary line of so-called poly-shear rate existence areas of suspensions with apparent viscosity of $\eta z \gtrless 500$ mPa·s may be drawn by overlaping all diagrams.

RESULTS

TABLE 1
The mixture proportions and corresponding apparent viscosities of ηz at given shear rates of D

No. of MIX	FRACTION CONTENT/%/			SHEAR RATE D /s^{-1}/					
	I	II	III	20	60	180	540	1620	4860
				APPARENT VISCOSITY ηz /mPa·s/					
01	100	0	0	13101	4174	1764	709	943	879
02	0	100	0	5722	1058	192	271	306	221
03	0	0	100	1724	226	118	91	80	77
04	75	25	0	4154	3422	1289	293	570	561
05	50	50	0	1254	8230	1301	222	344	387
06	25	75	0	157	26	74	113	238	242
07	0	75	25	8665	2524	1358	338	136	119
08	0	50	50	314	91	67	84	99	114
09	0	25	75	2704	784	122	77	72	80
10	25	0	75	7936	993	157	51	63	57
11	50	0	50	274	290	305	62	104	181
12	75	0	25	235	39	113	216	477	444
13	50	25	25	<500	314	44	151	243	168
14	25	50	25	274	91	70	104	145	133
15	25	25	50	745	1038	80	75	94	106

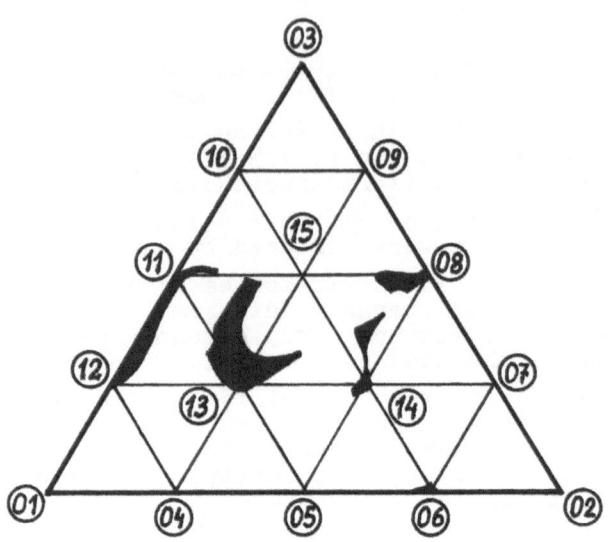

Figure 1. Poly-shear rate existence areas for suspensions with apparent viscosity of $\eta_z < 500$ mPa·s /black/ or $\eta_z > 500$ mPa·s /white/. It is valid for shear rates of D /20 - 4860/ s^{-1}. Inside rings are No. of MIX - see TABLE 1.

CONCLUSIONS

GV-method makes it possible to inquire into relationships between particle size and rheological properties of any suspension. The specification of conditions which characterise both properties quite exactly can be considerably heterogeneous. Not only apparent viscosity, but also other rheological characterizations of suspensions can be influenced by particle size distribution. However the generalization of results is not possible as a rule. In spite of the considerably complicated procedure, GV-method can well assert itself at work with many theoretical and practical problems.

REFERENCES

1. Heidiger, M., Messung rheologischer Eingeschaften. Bulletin 6704-652, Contraves AG 8052 Zürich.

2. Šatava, Vl., Úvod do fyzikální chemie silikátů, SNTL Praha, 1965.

3. Reiner, M., Deformation, Strain and Flow, London, 1969.

4. Batel, W., Einführung in die Korngrössenmesstechnik, Springer-Verlag, Berlin/Göttingen/Heidelberg, 1964.

RHEODYNAMICS AND HEAT TRANSFER OF DIELECTRIC SUSPENSIONS IN CONSTANT ELECTRIC FIELDS

E.V. KOROBKO
Heat & Mass Transfer Institute,Byelorussian
Academy of Sciences, P.Brovka, 15str.,220728
Minsk, USSR

ABSTRACT

Heat and momentum transfer processes in liquid-dispersed suspensions changing their structure in the presence of external electric fields are considered. Nonlinear, visco-elastic, thixotropic, relaxational regularities are revealed.

A phenomenon of a quick reversible change,in mechanical properties of dielectric suspensions (ERS) in the presence of strong electric fields has been referred to as an electrorheological effect. Being discovered for the first time in 1947 by American engineer W.Winslow (1) it has been registered as a change in the effective viscosity of suspensions. At the same time classical approach has been suggested to interpretation of the effect in terms of structural changes in sus - pensions. When an electric field is applied, solid particles in a condenser gap begin to oscillate between electrodes, to form separate aggregates, bridges and an anisotropically developed structural skeleton. A degree of internal structuri - zation dependent on electric intensity and ERS composition as well as the mechanism of its destruction due to mechanical load (continuous Couette shear, Poiseulle flow, oscillations) determine the specificity of transfer processes in such suspensions.

At the Heat and Mass Transfer Institute (Minsk,USSR) ERS effective compositions based on natural and polymer fil-

lers have been designed and investigated (2). It is shown
that all ERSs may be classified into three structural –rheological groups with respect to their behaviour specifying
the spheres of their application in modern technology.

Low–concentration (C ≤ 5%) ERSs with movable, relative
to a dispersed medium, structures behave like Newtonian liquids. Such media may be used as heat agents and in control-
lable optical devices.

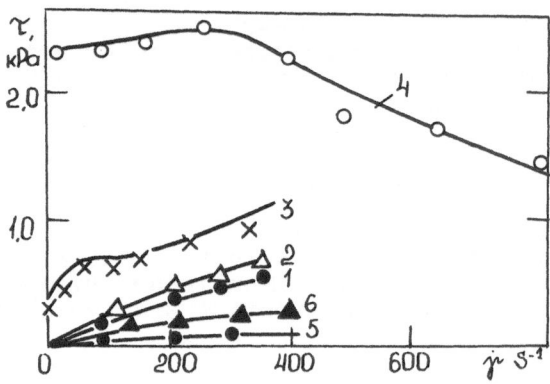

Figure 1. ERS flow curves at different electric intensities:
E=0(1); 0.2(2); 0.5(3); 1.0(4); C=40%; E=0(5),
1.3 (6) kV/mm, C=20%.

Suspensions (C>5)may be employed in different hydroautomated
systems with control of flow–rate–head characteristics, in
robotics, hydraulic power devices, viscous friction dampers,
etc.

High–concentration ERSs (C > 20%) are characterized by
a higher increment of the effective viscosity and manifestation of the solid properties. Besides, such ERSs possess
distinct elastic properties (Fig.2) (3), they may be used as
fixing pastes (e.g. in gripping devices in robotics) as well
as interlayers with variable acoustic properties.

Structural changes in ERS caused by an external electric
field give rise to anomalous effects noy only at momentum
transfer but also during its response to thermal impact.

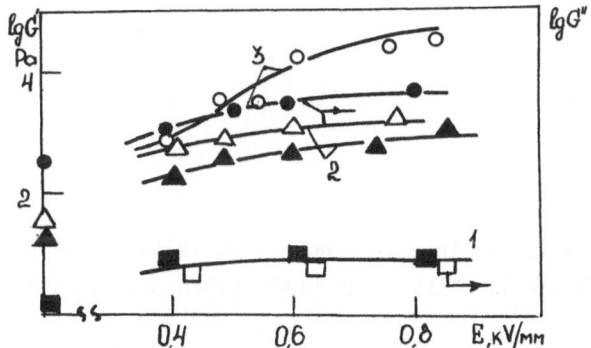

Figure 2. Log elasticity G' and loss G'' moduli at ERS concentrations C=10%(1), 20(2), 60(3) vs electric intensity.

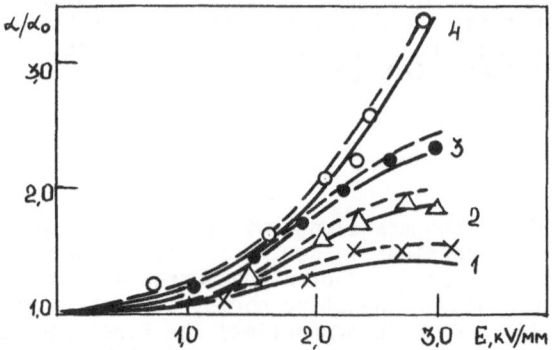

Figure 3. Relative convective heat transfer coefficient of transformer-oil -base silica suspensions at C=0(1), 1(2),3(3),5%(4) vs electric intensity in a coaxial-cylindrical channel-condenser. — ,theory;─── experiment:z_1/z_2=0.625.

Convective heat transfer in ERS in a coaxial-cylindrical channel-condenser has been studied experimentally at the 2nd kind boundary conditions(4). It is shown (Fig.3) that heat transfer enhances when electric intensity surpasses its threshold values.

REFERENCES

1. Winslow, W. J.Appl.Phys. ,1979, 20, 12, p.1137.

2. Luikov, A.V., Shulman,Z.P. Electrorheological Effect, Nauka i Tekhnika Publishers, 1976.

3. Shulman, Z.P., Korobko E.V., Yanovskii Y.G., J. of Non-Newtonian Fluid Mechanics, 1989, 33 , pp. 181-196.

4. Shulman, Z.P.; Korobko, E.V. J. Heat & Mass Transfer, 1977, 21 , pp. 543-548.

A STUDY OF TRANSIENT AND STEADY-STATE SHEAR
AND NORMAL STRESSES IN GLASS FIBER SUSPENSIONS

RUMIANA KOTSILKOVA, WOLFGANG GLEISSLE[x]

Central Lab.of Physico-Chemical Mechanics, BAS
Sofia 1113,Acad.G.Bonchev str.bl.I,Bulgaria
[x]Institut für MechanischeVerfahrenstechnik und
Mechanik,Uni.Karlsruhe, 7500 Karlsruhe,
Postfach 6980, BRD

ABSTRACT

An experimental study of the transient and steady,shear and normal stresses
of glass fiber model suspensions in Newtonian and non-Newtonian(viscoelastic) liquids is reported. Some important phenomena were observed (shear and
normal stresses overshoots,and negative normal stresses)although the basic
liquid have not manifested these properties. The first and second mirror
relations are applicable only in a narrow region of high shear rates and
short shear times,which may be due to the "structural effects" in fiber
suspensions.

INTRODUCTION

In the last several years some papersof the transient shear flow of fiber
filled materials gave an account of the importance of these investigations
for processing such materials I1,2I. It is considered that these properties
related to the structure developed by fibers or fiber orientation in the
polymer matrix. This paper presents the effect of glass fibers on the
transient and steady flow of concentrated fiber suspensions. The great
difference was found comparing the normal stresses of the suspensions in
Newtonian and non-Newtonian matrix.

MATERIALS AND METHODS

The samples used in this study are 11 v% (25 w%) glass fiber suspensions in silicone oil (η_o =600 Pas) and in epoxy resin (η =20 Pas).
Dimensions of the fibers are : the average fiber length of L_n=144,180 and.
215μm,diameter D=10μm . The transient viacosity and first normal stress
difference were measured by cone-plate rheometer (cone angle 6°, diameter

d=84.1 mm and gap h=158 μm).

RESULTS AND DISCUSSION

Fig.1 shows examples of shear and normal stresses growth functions, $\eta^+(t,\dot{\gamma})$; $\Psi^+_1(t,\dot{\gamma})$ of the glass fiber suspension in silicone oil(sample A). The basic matrix is a viscoelastic liquid and shows slight pseudoplastic bechavior. The transient curves of the fiber suspension show extremely high overshoot in contrast to the curves of the silicone oil. Moreover, the overshoots of the normal stress coefficient are much larger than the overshoots of the transient viscosity and depend strongly on the fiber length and aspect ratio.We found that the first and second modified Gleißle mirror relations for particle suspensions I3I are applicable also for fiber suspensions but only in the narrow range of high shear rates and short shear times,where,

$$\eta^+(t)=\eta_{st}(\dot{\gamma}) \quad at: B.\dot{\gamma}.t = 1 \qquad \Psi^+_1(t)=\Psi_{st}(\dot{\gamma}) \quad at: B.\dot{\gamma}.t=k \quad (1)$$

That curves strongly depart from eqs.(1) out of the upper range (see the dotted and full lines in Fig.1). This may be due to the "structural effects" in fiber suspensions which are generally ignored when dealing whith spheres.

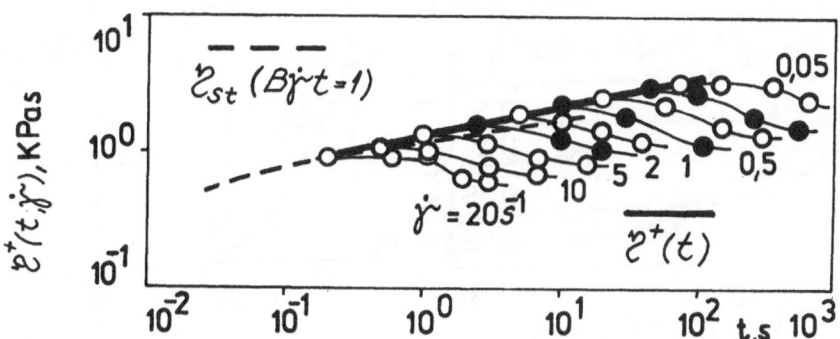

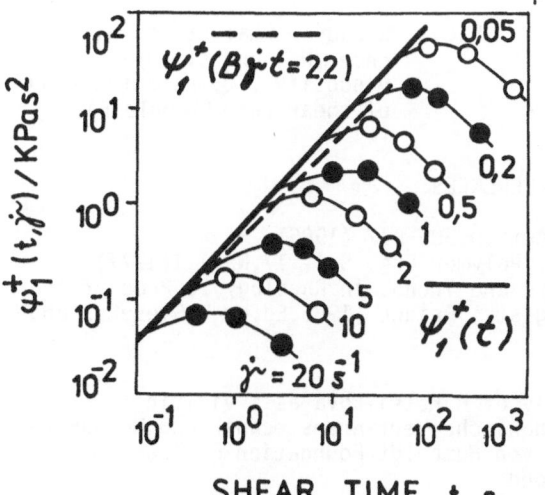

SHEAR TIME, t, s

Fig.1. Transient and mirrored steady-state viscosity and normal stress coefficient of a glass fiber suspension in silicone oil (sample A), at shear rate as a parameter.

$\eta^+(t)$-transient viscosity

$\Psi^+_1(t)$-transient normal stress coefficient

$\eta_{st}(B\dot{\gamma}t=1)$-steady state viscosity

$\Psi_{st}(B\dot{\gamma}t=2.2)$-steady state normal stress coefficient

The fiber dimensions:
L_n=215 μm ; a_r=22μm

$\dot{\gamma}$ Fig.2 shows the transient normal stress difference vs. shear strain $\gamma=\dot{\gamma}t$ of the glass fiber suspension in epoxy resin(sample B). The filler concentration and the fiber dimentions are the same like in the sample A. However, the basic matrix in the sample B is a Newtonian liquid. In the contrast to the results in Fig.1 extremely high negative first normal stresses were measured here,which depend significantly on the shear rate. If we plot the steady normal stress difference and the steady state viscosity against the shear rate (Fig.3) we see that N_1 oscillates,while η decrease monotonious with increasing the shear rate. The reason of the specific normal stress bechavior of fiber suspensions in a low viscosity Newtonian matrix is not quite clear and needs of following investigations.

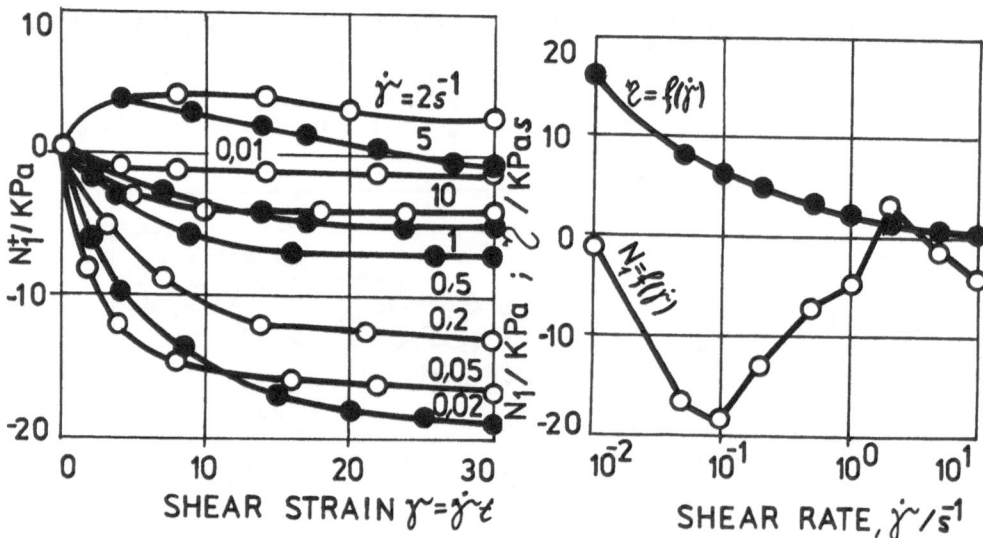

Fig.2. Transient normal stress difference vs. shear strain of glass fiber suspension in epoxy resin (sample B)

Fig.3. Steady normal stress difference ,N_1 and steady-state viscosity , η as a function of shear rate (sample B).

REFERENCES

1. Kitano T.,Funabashi M., Rheol.Acta,p.606-617 (1986)
2. Onogy S.,Mikami Y.,Mastumoto T.,Polymer Eng. Sci. 17,Nr.1,1(1977)
3. Gleißle W.,Mukherjee D.,Progress and Trends in Rheology II,Proc.of the 2nd Conference of Ruropean rheologists, Prague ,1986,Ed.by H.Giesekus and M.F.Hibberd,1988,379

[X]This work was done during the stay of R.Kotsilkova as a visiting scientist at the Institut für Mechanische Verfahrenstechnik und Mechanik-Uni.Karlsruhe, BRD. The Alexander von Humboldt Foundation Fellowship for this work is gratefully acknowledged.

RHEOLOGY OF SHORT FIBERS FILLED POLYMERS

RUMIANA KOTSILKOVA,WOLFGANG GLEISSLE[x], YACHKO IVANOV
Central Lab of Physico-Chemical Mechanics, BAS
Sofia 1113,Bulgaria
[x]Institut für Mechanische Verfahrenstechnik und
Mechanik,Universitat Karlsruhe, BRD

ABSTRACT

In the present paper an account is given of the effects of matrix rheology, fiber length and concentration on the viscous properties of suspensions. The flow properties of concentrated fiber suspensions of different silicone oils and glass fibers were investigated by capillary viscometer. The generalized flow bechavior at high shear rates was given independent of the oil viscosity and fiber concentration and it was compared with that of particle suspensions.

INTRODUCTION AND THEORY

The influence of short fibers on the rheological properties of polymers is of great scientific interest and has important industrial applications. There have been some studies of the influence of fibers on the viscosity, viscoelasticity,extrusion characteristics and orientation of fibers during the flow [1,2,3]. In this paper an experimental study of the effects of filler concentration,fiber length and rheological properties of basic matrix on the viscosity of concentrated fiber suspensions is reported.

For concentrated suspensions of particles the flow curves can be plotted independently of the solid concentration if one uses the"reduced shear rate",$B_\zeta \dot{\gamma}$,where B_ζ is the factor by which the shear rate of the matrix liquid increased by the filler [4]. In that case the shift procedure is given by:

$$\eta^*(\dot{\gamma})=B_\zeta \eta (B_\zeta \dot{\gamma}) \quad \text{and} \quad \tau^*(\dot{\gamma})= \tau (B_\zeta \dot{\gamma}) \tag{1}$$

On the other side Vinogradov's superposition [5] gives a generalized viscosity function of linear polymers independent of the molecular weight by reducing the shear rate with factor of η_0. Using these theoretical predictions we suggest the following equation of the shift procedure:

$$\eta^*(\dot{\gamma})=B_\zeta \eta_0 \eta (B_\zeta \dot{\gamma}_w \eta_0) \quad \text{and} \quad \tau_w^*(\dot{\gamma})= \tau_w (B_\zeta \dot{\gamma}_w \eta_0) \tag{2}$$

where $\eta^*(\dot{\gamma}),\tau^*(\dot{\gamma})$ -viscosity and shear stress of fiber suspension
$\eta, \tau_w (B_\zeta \dot{\gamma}_w \eta_0)$-viscosity and shear stress of silicone oil
B_ζ -shift factor ;η_0 -zero viscosity of the silicone oil

The eq.(2) presents the generalized flow bechavior of fiber filled silicone oils independent of the filler contain and the oil viscosity.

MATERIALS AND METHODS

The samples used in this study are suspensions of glass fibers in different molecular weight silicone oils. Table 1 summarized the results of the oil viscosity and fiber length distribution. The flow characteristics were measured by using the capillary viscometer. Correction of the end effects (Bagley plot) was determined.

Characteristics of fiber suspensions   Table 1

Code	Type silicone oil	Viscosity (Density) Pa.s g/cm^2	Weight fraction w%	Volume fraction v%	Average length L_n /mm	Average aspect ratio a_r	Shift factor B_c
S1	AK 5.10^3	5 (0.97)	0	0	-	-	1.0
S1-10	-	(2.55)	10	4.05	0.77	55	1.2
S2	AK 10^5	100 (0.98)	0	0	-	-	1.0
S2-10	-		10	4.09	0.70	50	1.2
S2-15	-		15	6.35	0.65	46	1.4
S3	AK 5.10^5	600 (0.98)	0	0	-	-	1.0
S3-15	-		15	6.35	0.50	36	1.4
S3-25	-		25	11.40	0.44	31	1.6
S3-25	Glass beads(GB)	(2.55)	25	11.40	-	1	1.6
S4	AK 10^6	1000 (0.98)	0	0	-	-	1.0
S4-15	-		15	6.35	0.50	36	1.4
S4-25	-		25	11.40	0.49	35	1.6

RESULTS AND DISCUSSION.

Figure 1 shows the shear stress $\mathcal{T}_w$ plotted vs.reduced shear rate $B_c \dot{\varepsilon}_0 \dot{r}_w$ using eq.(2),for all samples from Table 1. At constant shear stress the data for all suspensions and silicone oils are shifted along the shear rate axis with factor $B_c \dot{\varepsilon}_0$. The values of B_c are taken equal to that of glass bead suspensions at the same filler concentration. It can be seen that at high shear rates data af all suspensions coincide well with generalized flow curve of the silicone oils. However,at low shear rates a big deviation of the suspensions data from the master curve was found which depend significantly on the viscosity of the basic medium,e.g. the effect of fibers is very high at low viscosity silicone oils (S1-10, S2-10). We suppose that in this flow region the interaction between fibers are dominated on the hydrodynamic forces and very important factors here are fiber length and aspect ratio,and the viscosity of the basic medium adhered to the fiber surface. On the other side, in the hydrodynamically determined high shear rate flow region the flow curves of fiber suspensions coincide approximately well with these of glass bead suspensions at the same filler concentration. It may be connected with an orientation and convergation of random distributed fibers from reservoir into the highly oriented configurations in the capillary I6I.

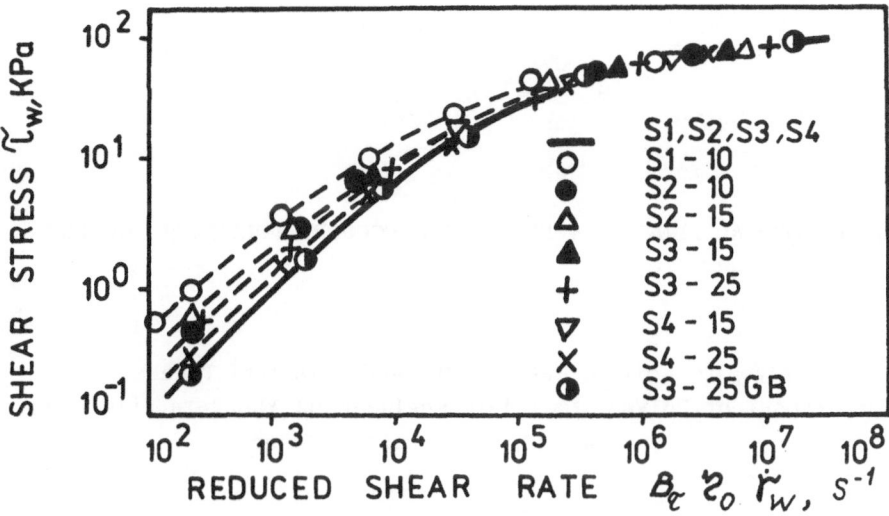

Figure 1.Master-curve $\mathcal{T}_w$ vs. $B_{\zeta}\, \mathcal{V}_0\, \dot{\mathcal{V}}_w$ of silicone oils,glass fiber suspensions and glass bead suspension (see Table 1),by ussing the eq.(2)

CONCLUSIONS

Addition of glass fibers to the silicone oils with different molecular
weight results an increase in shear stress and the influences of oil
viscosity, fiber length and aspect ratio are remarcable only at low shear
rates where the stresses generated by direct fiber-fiber interaction
dominate on the hydrodynamical effects. However, at high shear rates as a
result of fiber alignment the suspensions viscosity rise may be due solely
to hydrodynamic factors and depend only on the filler concentration.
Therefore, in this flow region fiber suspensions show the flow bechaviour
similar to that of glass bead suspensions at the same filler concentration.

REFERENCES

1. Maschmeyer R.O.,Ch.T.Hill. Trans.Soc.Rheol.,21,183,195 (1977)
2. Kitano T.,T.Kataoka,Rheol.Acta 20,390,403 (1981)
3. Crowson R.I.,M.Y.Folks,P.F.Bright,Polym.Eng.Sci.,2/,925,934 (1980)
4. Gleißle W.,Baloch M.K.,Proc.IX Int.Congress on Rheology,Acapulco,v.2,
   549,1984

   Gleißle W.,Mukherjee D.,Progress and Trends in Rheology II,1988,379,
   Proc. 2nd Conf.European Rheologists,Prague,1986
5. Vinogradov G.B.,Malkin A.Y.,Rheology Polymerov,M.1977
6. Chan Y.,Oyanagi Y.White J.,Polymer Eng.Sci.,v 18 (4),268 (1978)

[x]This work was attained with assistance of the Alexander von Humboldt
Foundation. The autors gratefully acknouledge this support. R.Kotsilkova
would like to thank Prof. H.Buggisch for the helpful comments.

ELONGATIONAL FLOW BEHAVIOR OF SUPERCOOLED POLYETHYLENE MELT

Kiyohito Koyama, Takashi Ota and Katsufumi Tanaka
Department of Polymer Materials Engineering, Yamagata University
Jonan 4-3-16, Yonezawa 992, Japan

ABSTRACT

The experimental study of elongational flow behavior under supercooled state has been carried out for high density polyethylene melt. A elongational rheometer was modified to measure the elongational viscosity under supercooled state. The elongational modes were a constant strain rate elongation and a dynamic strain rate elongation. In the dynamic strain rate elongation, the sinusoidal strain with small amplitude is superposed to the constant strain rate. The dynamic modulus of supercooled melt was calculated from the response of small strain amplitude. The crystallization process during elongation of supercooled melt was estimated from the dynamic modulus. The experimental results were compared and discussed with a orientation induced crystallization theory.

INTRODUCTION

The practical thermoplastic polymer processing contains the elongational flow under supercooled state. The elongational viscosity of molten polyethylene has been studied above melting temperature in the previous papers[1]. The flow of supercooled molten polymer has not been analyzed sufficiently. In the present paper the elongation and the crystallization behavior of supercooled high density polyethylene melt under isothermal state is discussed.

MATERIALS AND METHODS

The commercialized high density polyethylene (HDPE) with number average molecular weight of 8.1×10^3 and weight average molecular weight of 2.7×10^5 was used. The samples were extruded into a rod shaped form with the diameter of ca. 3mm. The melting point is 135.6°C and the crystallization temperature is 115.8°C by DSC measurement.

The experimental apparatus is essentially a rotating-cramps-stretching rheometer[1] and a melting bath is added to the rheometer. A sample was floated on the bath fluid of silicone oil and is melted at a constant temperature for 10 minutes. The molten sample is transported from the melting bath to the elongational bath through the narrow channel which exists between two baths and has a mobile partition. As to the dynamic strain rate, the sample is elongated as the following strain rate.

$$\dot{\gamma} = \dot{\gamma}_0 + \dot{\gamma}_1 \sin(2\pi t / T) \qquad (1)$$

The elongational forces were measured by a leaf spring and a differential transducer.

RESULTS AND DISCUSSIONS

The elongational viscosity of supercooled melt showed the same time dependence as that of molten state until a certain time at constant strain rate. The elongational viscosity increases smoothly by a critical strain of ca. 0.7 and then rapidly increases until breaking at strain of ca. 3. This is the same behavior for the molten state above melting temperature. The supercooled melt below melting temperature was elongated continuously up to large strain. The increase in elongational viscosity at large strain was steeper than that at small strain. The peculiar viscosity behavior can be correlated to the orientation induced crystallization.

The dynamic elongation experiments were carried out at a condition of $\dot{\gamma}_0 = 0.08(\text{sec}^{-1})$, $\dot{\gamma}_1 = 0.02(\text{sec}^{-1})$ and T=1.0(sec) in eq. 1. The amplitude of the elongational stress and the phase difference between stress and strain were estimated for the small superposed sinusoidal elongation. The amplitude was approximately constant and the phase difference increased gradually by ca. 20 seconds.

The storage modulus E' was calculated from the amplitude and the phase difference of various measuring temperatures with time as shown in Figure 1. The storage modulus increases remarkably beyond 15 seconds.

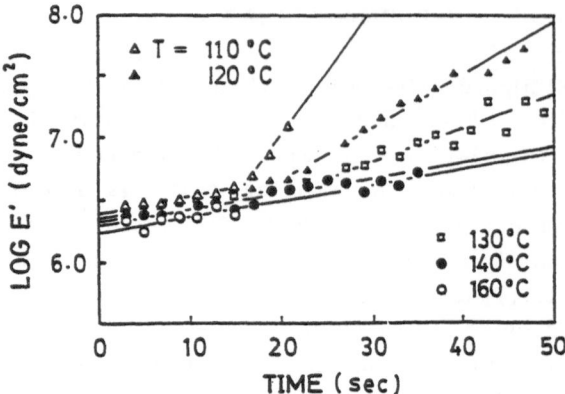

Figure 1. Variation of storage modulus with time (Temperature = 160, 140, 130, 120 and 110°C).

The crystallinity of supercooled melt was estimated by using E'. The assumption of the calculation is the simple series model of a crystal part and an amorphous part. The storage modulus of the crystal part is used to be $2 \times 10^{12}(dyn/cm^2)$. That of an amorphous part for each temperature is estimated by using the shift factor. The calculated crystallization curve corresponds to each measuring temperature. The results of crystallization behavior of supercooled melt was analyzed with a orientation induced crystallization theory. The qualitative agreement of both experiment and theory was obtained.

CONCLUSIONS

The supercooled HDPE melt showed strong nonlinearity compared with the molten HDPE. The variation of crystallinity of supercooled HDPE melt with time was be calculated qualitatively by assuming the supercooled sample to be a simple series model (crystal-amorphous). The prediction from orientation induced crystallization theory showed qualitative agreement with the experimental results.

REFERENCES

1. Koyama, K. and Ishizuka, O., Influence of molecular weight distribution on elongational flow of polyethylene. Sen-i gakkaishi, 1982, **6**, 239-245.

2. Koyama, K. and Ishizuka, O., Nonlinearity in elongational viscosity at a constant strain rate. J. Polymer processing engineering, 1983, **1**(1), 55-70.

PHASE SEGREGATION INFLUENCE ON OLIGOMERS RHEOLOGY

SERGEI G.KULICHIKHIN
Research Institute for Plastics,
35, Perovskii Proezd, 111112 Moscow, USSR

ABSTRACT

This paper considers the regularities of the rheological properties change at curing of the reactive oligomers. It is shown for a list of subjects that gel formation as a relaxation transition is preceeded by the formation in the reaction system of the fragments of a network structure of a colloid size. For such a structurization mechanism the main rheokinetics equation have been obtained.

In the general case, variations in the rheological characteristics of curing processes are conditioned by the whole complex of physical and chemical phenomena attending the chemical conversion of the oligomer into a non-fluid network polymer (1) . Specific character of curing is evidenced by the existence of the gelation point separating the whole process in two stages. The first one - gelation - ends in the formation of a single network structure of the whole bulk of the curing material.

Viscosity changes cannot be described from the viewpoint of homogeneous growth of the oligomer molecules as in a number of cases the build-up of viscosity is not characterized only by the change of the molecular characteristics of the oligomer but by a microphase segregation of the reaction system near the point of relaxation transition, which is gelation. The results of studies of curing rheokinetics of a number of urea-formaldehyde resins are most demonstrative from this point of view (2) . The time of

reaching gelation point does not change at varying resin
concentration in the solution and coincides with the time
of achieving the relaxation transition at block curing.
The steep rise in the optical density is observed long before
the gelation point on retention of the solution flowability.
Registrating jump of the optical density is associated with
the reaction system segregation and appearance in the
solution of a new phase composed of the fragments of
branched and cross-linked structures.

According to the theoretical calculations executed on
the basis of branching theory (3), as tridimensional
polycondensation goes on weight-average molecular mass of
the curing oligomer increases monotonically up to the
gelation point where turns to infinity. At the same point
of the reaction system gel fraction appears. A real picture
of the change in these characteristics obtained on the basis
of GPC analysis of epoxy resin curing by diamine has radical-
ly different character (4) . The molecular mass of the
formed polymer changes extremely with its maximum value
being observed at the point of microphase segregation. At
this very moment, rather that at the gelation point, the gel
fraction appears in the reaction system on retention of the
product flow in general.

Proceeding from the stated considerations, gelation
point from the viewpoint of the process physical picture
can be defined as a moment of reaching phase inversion in
the reaction system - fragments of the network structures
from the disperse phase become a continuous disperse medium
which, in turn, results in the change of the system relaxation state.

Then, viscosity change during the gel formation
providing the linear growth of the microgels concentrations
may be written as follows:

$$\eta (t) = \eta_m(t) \left(I - t/t^x \right)^{-b}$$

where $\eta (t)$ - the reaction system viscosity, $\eta_m(t)$ - function
responsible for the viscosity change in disperse medium,
t^x - gelation time, b - a constant.

Separation of the input of various components to the
change of viscosity of the epoxy oligomer in curing process
shows that at the initial stages viscosity change is in full
accord with the increase of the oligomer molecular mass.
After the reaction has attained a certain time, the viscosity
calculated on the basis of the change of the disperse medium
molecular mass becomes less than the viscosity of the
reaction system as a whole. At the general impetuous growth
of viscosity, a relative input of the molecular component
constantly decreases as well as its absolute value.

Thus, the change of the viscous properties at gel
formation of oligomers is conditioned by the whole complex
of dissimiliar physical and chemical phenomena displaying
in the process of chemical conversion of the initial oligomer
into a network polymer.

REFERENCES

1. **Kulichikhin**, S.G., Kinetics of the change of physical and
   chemical properties of adhesives at curing. Mekhanika
   kompozitnykh materialov (Mechanics of Composite Materials),
   1986, 6, 1087-92.

2. Kulichikhin, S.G., Abenova, Z.D., Bashta, N.I., Kozhina,
   V.A., Blinkova, O.P., Romanov, N.M., Matvelashvili, G.S.,
   Malkin, A.Ya. Rheological characteristics of the curable
   melamine-formaldehyde resins. Vysokomolekulyarnye Soedineniya (High-molecular Compounds), 1989, 31A, 2372-77.

3. Macosko, C.W., Miller, D.R., A new derivation of average
   molecular weights of non-linear polymers. Macromolecules,
   1976, 9, 199-206.

4. Kulichikhin, S.G., Nechitilo, L.G., Gerasimov, I.G.,
   Kozhina, V.A., Yarovaya, E.P., Zaitsev, Y.S. Rheokinetics
   of gel formation at interaction of epoxy oligomers with
   aromatic diamine. Vysokomolekulyarnye Soedineniya (High-
   molecular Compounds), 1989, 31A, 2538-43.

RHEOLOGY AND STRUCTURE OF POLYMER BLENDS CONTAINING LIQUID-CRYSTALLINE COMPONENTS IN MELT AND SOLID STATE

V.G KULICHIKHIN, N.A.PLATE
Institute of Petrochemical Sythesis, USSR Academy of Sciences
Moscow , 117912 , USSR

ABSTRACT
The rheological properties of four pairs of polymers containing
high-molecular mesophase components have been investigated. The
viscosity-composition curves were found to depend on the
interrelation between the initial polymer viscosities, stream
morphology and level of interaction on the interface . The
effects of the interphase interaction are determined from the
temperature changes of the relaxation transitions, and in
especial cases from their splitting. The localization of the
maxima of the mechanical properties of the solid state
extrudates in a specific region of the compositions is
accounted for by the peculiarities of the morphology and
adhesion between components.

INTRODUCTION
Apparently, one of the most popular application of LC-polymers
is their   use in blends with commercial thermoplastic
materials for modifying the rheological properties of melts and
physical-mechanical characteristics of the final   products
[1,2]. Unfortunately, though there are numerous articles covering LC-polymer applications, still there exist no scientifical-
ly-grounded principles for the selection of blend components
providing a maximal decrease in viscosity and an increase in
strength. These problems are to some extent mutually exclusive,
for the decrease in viscosity requires a low interphase interaction, whereas reinforcement require a high adhesion This
problem is discussed in the present work, mainly with the
framework of the rheological approaches for different polymer
couples involving the "condis" state polymers [3].

MATERIALS AND METHODS
The objects under investigation were the following polymer
couples: (I) aromatic polysulphone (PSPh) based on diphenylol-
propane and dichlordiphenylsulphone and LC-polyester Ultrax KR
4002 (dihydroxydiphenyl, terephtalic and isophtalic acids);
(II) the same PSPh and LC-polyester based on polyethylenetere-

phtalate and p-hydroxybenzoic acids (35:65); (III) polyethylene
and mesophase poly-bis-trifluorethoxyphosphazene(PPh); (IV) LC-
polyester containing alkylene-aromatic groups (PES-10) and
wholly aromatic polyester on a base of p-hydroxybenzoic and
terephtalic acids, phenylhydroquinone and resorcinol (PhHR-50).
The rheological properties of the blend melts were investigated
using the capillary viscosimeter. The extrudates of 1-2 mm
diameter were studied by the structural and mechanical methods.

RESULTS AND DISCUSSION

The mesophase components reveal a more pronounced viscosity
anomaly than the isotropic ones, moreover, PPh shows
appreciable yield stress values. In this case, practically no
extrudate swelling occurs, under certain conditions even
contraction takes place. The addition of the second component
leads to much steeper flow curves and a decrease in the
viscosity values (η). Here the role of the mesophase polymer is
most notable at high shear rates which is especially prominent
for composites containing more than 30% of LC-polymers. The
shape of the viscosity-concentration curves depends on the
temperature and constituents, though the blend viscosities are
as a rule lower than that of the initial polymers. An exception
is blend IV which at small PhHR-50 contents has a viscosity
which is higher than η PES-10. Apparently, the shape of the de-
pendecies $\eta(C_{LCp})$ is determined by the morphology of the stream
and the interaction between the constituents. As a rule, the
investigated polymers are non compatible on the macrolevel and
the blend melts look like emulsions. During the extrusion
process the dispersion phase drops are elongated in an
anisodiametric manner and a surface lubricant layer
simultaneously is formed which leads to distortion of the
parabolic velocities profile. However, such a consideration
does not involve the problem of interactions at the interface.
Moreover, in order to make an estimated approximation, it
appears possible to predict the viscous properties of the
blends operating only with ratio viscosity of the initial
components (η_2 / η_1). This fact is evidenced in Fig. 1 which
illustrates viscosity dependences of

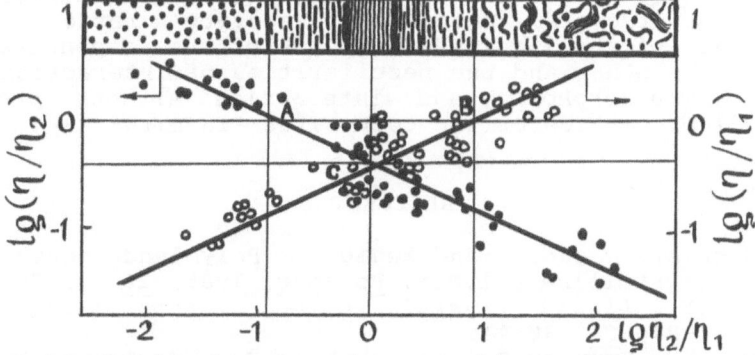

Fig. 1. Interelation between the normalized blend's viscosity
and the viscosity ratio of the initial components and
stream morphology.

13 different blends (30:70) normalized on the viscosity of the one component on η_2/η_1. Of special interest is the ABC triangle within the limits of which there is a sharp stream morphology change (schematically illustrated in the same figure). The increased scattering data in this region is due to the fact that the interphase interaction was not taken into account. Such an interaction exists as proved by the isolines map of the loss modulus (Fig.2). The temperatures of the γ- and α-relaxation transitions of one of the polymers "feel" the presence of the other; even splitting of T_g was registered for PSPh. The detected interaction, alongside with the homogeneity of the arming fibres, influences the mechanical properties of the solid extrudates (composites "in situ"), leading to the realization of the maximal elastisity modulus and strength values in the range of definite compositions.

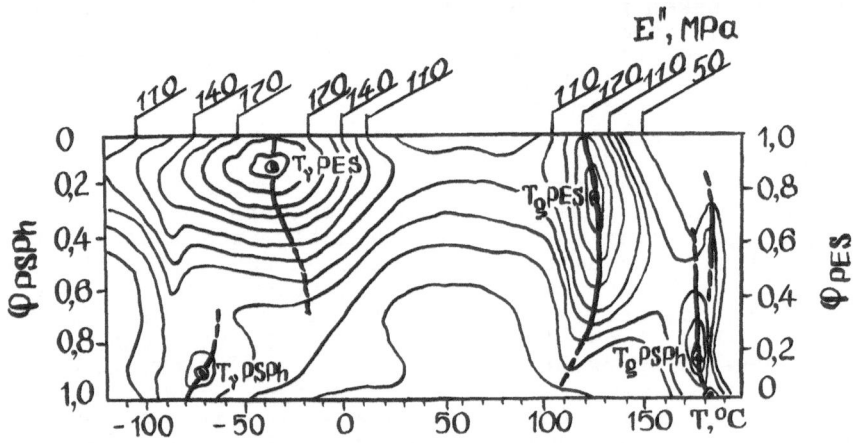

Fig. 2. The loss modulus isolines at the different temperatures and compositions for the blend I.

CONCLUSIONS

The rheological properties of polymer blends are determined by three major factors: component viscosity ratio, uniform distribution of the liquid jets of one of the components in the matrix of the other and the peculiarities of interaction at the interface. The morphology and interaction effects presuppose mechanical properties of the composites "in situ".

REFERENCES

1. Siegmann,A., Dagan,A. and Kenig.S., Polyblends containing a liquid crystalline polymer. Polymer, 1985, 26 , 1325-1330.
2. Oyanagi,Y., Liquid Crystal Polymer Blends (III). Techno Japan, 1989, 22, 30-41.
3. Antipov,E., Kuptsov,S., Kulichikhin,V., Tur,D. and Plate,N., Mesophase structure of flexible-chain polymers. Macromol. Chem., Macromol.Symp., 1989, 26, 69-89.

STRESS-STRAIN PROPERTIES OF POLY(BUTYL METHACRYLATE) AT UNIAXIAL EXTENSION IN LIQUID MEDIA AT TEMPERATURES ABOVE AND BELOW GLASS TRANSITION TEMPERATURE

M.K.KURBANALIEV[*], R.T.KADIROV[*], V.E.DREVAL[**], K.M.ABDULLAEV[*]

[*]Tadjik State University, Dushanbe, [**]Institute of Petrochemical Synthesis, USSR Academy of Sciences, Moscow, USSR.

ABSTRACT
The long term durability, breaking strain, viscosity and high-
elastic modulus of poly(butyl methacrylate) in different
alcohols at uniaxial extension at temperatures below and above
T_g are investigated. It is shown that the alcohols considerably reduce the strain-stress polymer parameters. This is due to
the fact that they act as surface active agents which reduce
surface tension at polymer-liquid interface or cause polymer
surface plasticizing.

INTRODUCTION
The active liquid media can dramatically influence stress-
strain polymer properties. These properties are usually studied
either below or above T_g using different liquid media at different specified temperatures [1]. This paper is devoted to the
study of stress-strain properties of amorphous linear polymers
both above and below their T_g using the same liquid media in
both cases.

MATERIALS AND METHODS
The investigations were carried out on amorphous poly(butyl
methacrylate)(PBMA) with MM=$5*10^4$, M_w/M_n=2 and T_g=300K. The
active liquid media used were normal alcohols: ethanol, propa-
nol, butanol, amyl alcohol. The non-active media were water and
air. The measurements of long-term durability (τ), (the time
from the beginning of deformation up to fracture of a polymer),
fracture strain, viscosity and high-elastic modulus were made
under uniaxial extension at different values of true stress (σ)

RESULTS AND DISCUSSION
Fig. 1A shows that in the region T<T_g the alcohols cause dramatic decrease in τ in comparision with its value in water or
air. This effect increases with decreasing of temperature and

transition from ethanol to amyl alcohol. Depending on the media
τ can change in a 5 to 6 decimal order. The function $\tau(\sigma)$ obeys
the well-known exponential equation for solid polymers.

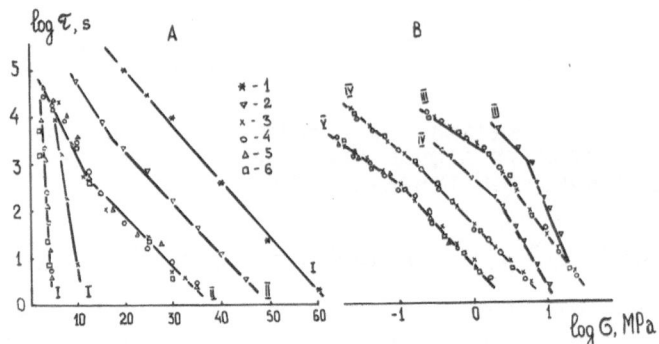

Figure.1A,B τ vs. σ 1-air,   2-water,   3-ethanol, 4-propanol,
5-butanol, 6-amyl alcohol.   T,K: 243(I), 283(II),
303(III),   323(IV), 343(V).

At $T \geqslant T_g$ all the alcohols reduce τ to an equal degree (Fig.1B).
In this case the influence of alcohols is much weaker, and the
values of τ in water or air differ from those in alcohols 10 to
15 times. This difference decreases with the increase in σ. The
function $\tau(\sigma)$ is described by two straight lines $\lg\tau - \lg\sigma$, whereas for other linear polymers it obeys the power law [1]. It
is obvious that the effects under consideration are due to
PBMA-alcohol interaction. Our own results complemented with
literature data show that the transition from ethanol to amyl
alcohol systematically decreases the surface tension at the
polymer-liquid interface $\backsim$ 2.5-times. The swelling mesurements
show that with ethanol the equilibrium degree of PBMA swelling
increases from 1% (at 283K) up to 15% (at 343K). As to other
alcohols at elevated temperatures, they produce not only volume
swelling but some surface dissolution of PBMA.
The decrease in surface tension correlates with the decrease in
τ in different alcohols at $T<T_g$ when PBMA swelling is negligible. This correlation is in qualitative agreement with
Griffith's theory predicting the decrease in solid body
strength with decreasing the body surface energy.
It is known that at $T>T_g$ τ and other stress-strain characteristics of linear polymers depend on their rheological parameters
[2]. The results of the rheological experiments at $T \geqslant T_g$ in
alcohols, water or air show that in some exception the liquid
media do not influence viscosity and high-elastic modulus of
PBMA during τ measurements. This allows us to conclude that
under the above conditions the active liquid media act as a
surface agent plasticizing the polymer sample surface which
decreases τ and polymer strength. In this case the increase in
the applied σ and, hence, the decrease in τ cause a decrease in

polymer surface swelling and diminish its influence on polymer stress-strain properties.

At $T \geqslant T_g$, the only exception is the case of low σ at elevated T when there occurs a noticeable volume PBMA swelling during experiments. This swelling leads to a noticeable decrease in viscosity,bringing about a decrease in τ and deviation from the straight line dependence $\lg\tau - \lg\sigma$ in the region $\sigma \leqslant 0.1$ MPa. A similar deviation takes place at $\sigma \geqslant 1-10$ MPa due to the decrease in PBMA viscosity under action of high σ [2].

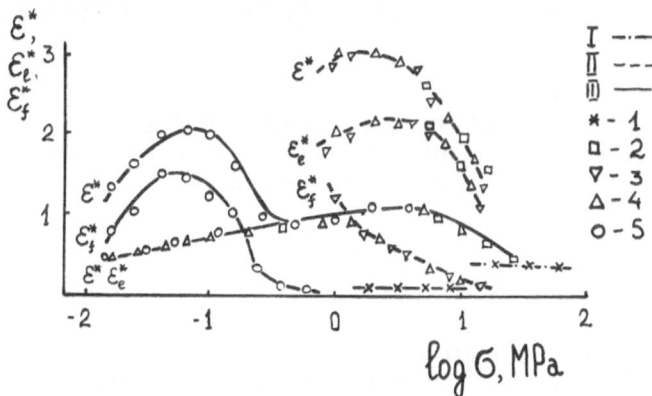

Figure.2 $\varepsilon^*, \varepsilon_e^*, \varepsilon_f^*$ vs. σ. I-air, II-water, III-alcohols. T,K: 243(1), 303(2), 313(3), 323(4), 343(5). The Henky strain measure is used.

Fig.2 demonstrates the influence of the alcohols on the breaking strain (ε^*) and its recoverable (ε_e^*) and non-recoverable (ε_f^*) components. It is seen that the alcohols decrease ε^*, ε_e^* and ε_f^* to an equal degree. At $T<T_g$, PBMA in alcohols shows brittle fracture with $\varepsilon^* \leqslant 10\%$. At $T \leqslant T_g$ and in the $\sigma \geqslant 10$ MPa region $\varepsilon^*(\sigma)$ and $\varepsilon_e^*(\sigma)$ in all the media are extremal functions with maximum. Such a behaviour is typical for ofher polymer melts and is due to their forced transition to high-elastic state under action of high stresses and high strain rates [1,2]. In the region of lower σ, the $\varepsilon^*(\sigma)$ of PBMA studied in alcohols shows another maximum whereas for polymers in non-active media, ε^* should pass through a minimum [1]. This maximum is due to volume swelling of PBMA inducing a decrease in τ and hence a decrease in time of deformation development prior to breaking.

REFERENCES
1. Kurbanaliev M.K.,Vinogradov G.V.,Dreval V.E.,Malkin A.Ya. Polymer 1982,v.23, 100
2. Vinogradov G.V.,Dreval V.E.,Borisenkova E.K.,Kurbanaliev M.K Zabugina M.P. J.Polymer.Sci. Polym.Phys.Ed., 1986, v.24,1971

NUMERICAL SIMULATION OF THE FLOW OF TWO NON-MIXING INELASTIC AND VISCOELASTIC FLUIDS WITH A FREE INTERFACE

JIANDI LAI , KARL STRAUSS
Energieprozesstechnik/Department of chemical engineering
University Dortmund

INTRODUCTION

In many technical applications one has to deal with the flow of two or more immisible liquids or melts where the different materials are separated by a continous free interface. Typical industrial processes which with flow field of such type are the productions of composite materials such as bicomponent fibers and multilayered films. One of the fundamental questions which arises, is the shape of the interface between the two components. The paper reports on a primitive variable algorithm combined with boundary fitted coordinates to analyse the flow of two non-mixing fluids. The position of the free interface between the fluids is calculated simultaneously with the solution of the governing equations. The constitutive equation used is the four-constant Oldroyd model. Finite differences are employed for the solution of the partial differential equations. The method is applied to the incompressible two-dimensional channel flow of two fluids with inealstic or viscoelastic feature.

MATHEMATIC MODEL

The governing equations for viscoelastic fluid flow are the equations of the conservation of mass and momentum and the constitutive equation. We consider in Fig.1 a typical interface between two non-mixing fluids 1 and 2. It is assumed that the velocity field is continuous across the interface i.e. $\underset{\sim}{u}_1 = \underset{\sim}{u}_2$

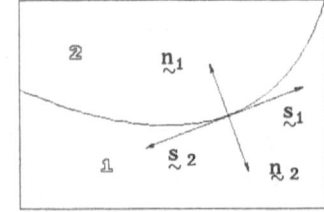

Fig.1. Typical interface

For steady flow we have $\underset{\sim}{u}_1 \cdot \underset{\sim}{n}_1 = \underset{\sim}{u}_2 \cdot \underset{\sim}{n}_2 = 0$

and $\underset{\sim}{u}_1 \cdot \underset{\sim}{s}_1 + \underset{\sim}{u}_2 \cdot \underset{\sim}{s}_2 = 0$

where $\underset{\sim}{n}$, $\underset{\sim}{s}$ are the unit normal and tangential vectors with $\underset{\sim}{n}_1 + \underset{\sim}{n}_2 = 0$ and $\underset{\sim}{s}_1 + \underset{\sim}{s}_2 = 0$.

In the absence of surface tension, the equilibrium of the interface requires that

$$(\underset{\approx}{S}_1 - P_1 \underset{\approx}{I}) \cdot \underset{\sim}{n}_1 + (\underset{\approx}{S}_2 - P_2 \underset{\approx}{I}) \cdot \underset{\sim}{n}_2 = 0$$

Since the interface boundary does not coincide with the cartesian coordinate line, a curvilinear coordinate system is introduced, in which the interface is described as a constant coordinate line. The transformation from the cartesian system to the curvilinear system is done using the relation:

$$\frac{\partial}{\partial x} = \xi_x \frac{\partial}{\partial \xi} + \eta_x \frac{\partial}{\partial \eta} \quad \text{and} \quad \frac{\partial}{\partial y} = \xi_y \frac{\partial}{\partial \xi} + \eta_y \frac{\partial}{\partial \eta}$$

The problem now is simplified in the sence that one has to solve the transformed equations within the rectangular domain for both fluids.

NUMERICAL ALGORITHM

The transformed equations are discretized by an averaging procedure over small control volumes surrounding the nodal points. Velocities, pressures and stresses are defined at the geometric centres of the control volumes. The notation is for a typical grid node P enclosed in its cell and surrounded by its neighbours N,S,E and W. The finite difference approximation to the governing equations can be performed by taking the intergral over the control volumes.

The general resulting discretized equtions become

$$A_P \Phi_P = A_N \Phi_N + A_S \Phi_S + A_E \Phi_E + A_W \Phi_W + S_\Phi \qquad (\Phi: u, v, p, s_{xx}, s_{xy}, s_{yy})$$

where S_Φ represents all the terms that cannot be approximated by value of Φ at the five points.

The discretized system of equations is solved iteratively as follows:

1. Generate adapting grid for a given interface line.
2. Solve the momentum and constitute equations with the given pressure field to obtain the velocity and the stress tensor.
3. Solve the continuity equation to get the pressure correction and update the pressure field.
4. Use the boundary conditions for the interface to get the velocities at the interface and update its position.
5. Return to 1 and repeat until convergence is reached.

RESULT

We now apply the method to caculate the flow of two non-mixing fluids in a channel with various material parameter. At he entrance fully developed flow conditions for both flows are assumed. For viscoelastic fluid the material constant λ_2 is always taken to be 1/6 of λ_1. The influence of the elastisity on the shape of the interface of newtonian-viscoelastic flow is to be seen in Fig.2. The system of viscoelastic-viscoelastic is more complicated. There is no clear tendency for the shape of the interface, wenn the We-number of the lower fluid flow increases (fig. 3).

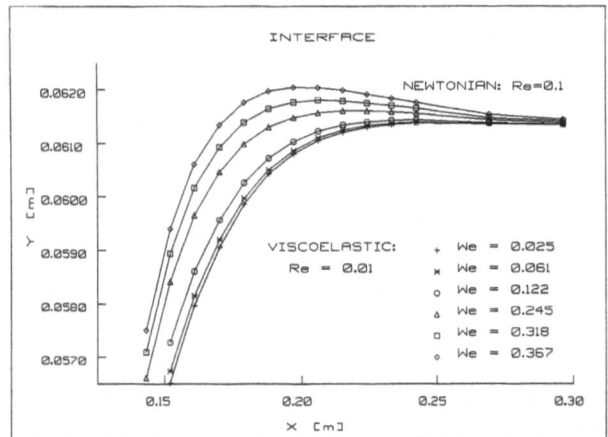

Fig. 2   Interface of newtonian-viscoelastic Flow

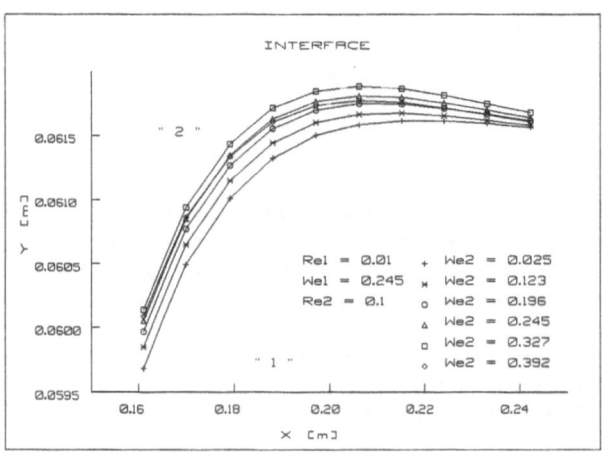

Fig. 3 Interface of viscoelastic-viscoelastic Flow

RHEOLOGICAL AND STRUCTURAL PROPERTIES OF MODEL COMPOSITE FOODS

KEITH R. LANGLEY, MARGARET L. GREEN AND ALAN MARTIN
AFRC institute of Food Research, Reading Laboratory,
(University of Reading), Shinfield, Reading, Berks. RG2 9AT, U.K.

ABSTRACT

Composites of solid, porous semi-solid and soft particles in a soft matrix have been modelled by using glass and cross-linked dextran beads and vegetable oil droplets suspended in a thermally-set protein gel.

We have compared the rheological and fracture properties with volume fractions of particles, particle size, porosity and interaction between particle and protein matrix. Fractured surfaces were examined by electron microscopy.

The measured mechanical properties depended on all the variables investigated and could be understood in terms of modification to existing theory and the interactions between the particles and the matrix.

INTRODUCTION

Food systems can be considered as complex mixtures of many components, a large group of them having one of the components as particulates. Particulate foods vary in composition and particle size and shape, with such diverse products as salad dressings, mayonnaise, pastes, pet foods, chocolates and cakes.

The rheological properties of food emulsions can be explained in terms of particulate suspensions and it would seem likely that the mechanical properties of solid food composites could be explained in terms of theories that have been developed for polymers and plastics.

Composites of solid, porous semi-solid and soft particles in a soft matrix have been modelled by using glass and cross-linked dextran beads and vegetable oil droplets suspended in a thermally-set protein gel.

MATERIALS AND METHODS

Preparation of Composites

The materials used, and the preparation of composites containing glass, corn oil and Sephadex particles, in a heat-denatured whey protein gel matrix, have been described previously [1,2].

Mechanical Tests

Most tests have been described elsewhere [1]. The stress and strain at failure, and elastic modulus were determined by loading 15 mm diam. x 15 mm long cylinders in compression at 0.83 mms^{-1}. The load and cross sectional area, assuming cylindrical shape were used to determine the stress, the natural strain was determined from the equation $\varepsilon = \ln^{ho}/\eta$ [3] and the modulus was determined from the slope of the initial linear stress-natural strain plot.

Scanning Electron Microscopy

Surfaces of the composites fractured at ambient temperature were examined after dehydration and critical point drying.

RESULTS AND DISCUSSION

1. Elastic Moduli

In general the elastic moduli of the composites increased with increase in particle phase volume (Table 1). This would be expected from Composite Theory [4], where the moduli $E \propto (1-\Phi)^{-5/2}$. Composite theory assumes that E is independent of particle size. Previously [1] we have shown that although the theory predicts the fail stress of hydrophobic coated glass particles, it does not for hydrophilic coated glass and oil droplets and Sephadex particles. This discrepancy was explained by an underestimation of particle volume due to adsorbed protein on the glass and oil droplets, and an overestimation due to protein absorption into the Sephadex particles. It seemed likely that this volume correction would again be required.

Only the moduli of composites containing 5 μm diam. oil droplets, decreased with increase in Φ. It proved difficult to produce stable composites with large droplets, oil droplet coalescence occurring during gel preparation, which would explain this moduli decrease.

2. Strain at Failure

Gels containing hydrophilic coated particles failed at low strains, indicative of brittle behaviour, whereas those containing hydrophobic coated surfaces tended to fail at higher strains indicative of ductile failure (Table 1). Composites containing large oil droplets, expected to be brittle, had ductile failure, and it would seem that at fracture oil coalescence occurred, which would act as an internal lubricant [5].

3. Electron Microscopy

Examination of the fractured surfaces showed that hydrophilic coated glass and oil, with the exception of 5 μm diam. droplets, were integral to the composite, failure occurring within the protein matrix. Composites containing hydrophobic coated particles failed adjacent to the particle surface indicating that there was little or no particle–matrix interaction. Gels containing porous Sephadex particles failed at or near the particle surface. Protein was adsorbed into the particle and it would seem likely that failure had occurred at the matrix protein–adsorbed protein interface.

TABLE 1

Moduli and natural strain at failure of particulate whey protein gel composites

Particle Type	Particle diam. (μm)	Relative elastic modulus			Relative natural strain at failure		
		Phase volume of particle					
		0.1	0.25	0.40	0.1	0.25	0.40
Corn oil	0.5	1.4	3.7	14.2	0.96	0.87	0.79
droplets	3.0	0.9	1.0	1.3	1.02	1.01	0.97
	5.0	0.4	0.2	0.1	1.02	1.07	1.15
Hydrophilic	30	2.5	7.4	20.4	0.97	0.90	0.82
coated glass	150	1.7	2.8	6.9	0.99	0.95	0.90
	600	1.1	1.5	2.5	1.01	0.98	0.94
Hydrophobic coated glass	(All sizes)	1.0	1.4	2.1	0.92	0.94	1.09
Sephadex gel	45	1.4	2.4	4.6	0.97	0.90	0.81
filtration	75	1.2	1.7	2.7	0.98	0.96	0.92
	125	1.1	1.4	1.8	0.98	0.94	0.91
(Theory)		(1.30)	(2.05)	(3.59)	(0.90) (0.63)	(0.78) (0.53)	(0.67)[1] (0.48)[2]

[1] Based on low particle–matrix adhesion
[2] Based on high particle–matrix adhesion

REFERENCES

1. Langley, K.R. and Green, M.L., Compression strength in relation to their microstructure and particle–matrix interactions. J. Text. Stud., 1989, 20, 191–207.

2. Green, M.L., Marshall, R.J. and Langley, K.R., Structure and mechanical properties of model composite foods. In Rheology of Food, Pharmaceutical and Biological Materials, Elsevier Science Publishers, London, 1990.

3. Bagley, E.B., Wolf, W.J. and Christianson, D.D., Effect of sample dimensions, lubrication and deformation rate on uniaxial compression of gelatin gels. Rheol. Acta., 1985, 24, 265–271.

4. Ross–Murphy, S.B. and Todd, S., Ultimate tensile measurements of filled gelatin gels. Polymer, 1983, 24, 481–486.

5. Marshall, R.J., Composition, structure, rheological properties, and sensory texture of processed cheese analogues. J. Sci. Food Agric., 1990, 50, 237–252.

RHEOLOGY OF FILLED SCLEROGLUCAN GELS

Romano Lapasin, Sabrina Pricl and Italo Colombo[o]
Institute of Applied and Industrial Chemistry University of
Trieste, I-34127 Trieste, Italy
[o]Vectorpharma Int. S.p.A., I-34148 Trieste, Italy

ABSTRACT

Two different gel matrices of scleroglucan were considered for examining the effects of particle addition on the continuous and oscillatory shear properties of these systems. The experimental results obtained suggest that a strong interaction exists between the gel matrix and the filler particles with substantial adsorption of scleroglucan onto the particle surfaces and the consequent reduction of polymer concentration and connectivity of the gel matrices.

INTRODUCTION

Scleroglucan, a neutral exocellular polysaccharide, can form gels in water at polymer concentration C_p greater than 0.25% (w/w) [1]. The shear-dependent behaviour of scleroglucan aqueous gels is of the plastic type and can be satisfactorily described by a modified Cross equation [1]. The time-dependent properties under continuous shear [1] as well as the mechanical spectra in oscillatory flow conditions [2] revealed a typical 'strong gel' behaviour. All these properties are nearly independent of temperature [3].

The gel behaviour of scleroglucan suggests its application in controlled release polymeric systems for drug delivery. To this purpose, experimental tests were performed on scleroglucan gels filled with cross-linked polymer particles in a concentration range of practical interest.

MATERIALS AND METHODS

Scleroglucan (SG, DP ~ 800, Ceca, France) and cross-linked povidone (PVP, Crospovidone[R], BASF, F.R.G.) were supplied by Vectorpharma Int. (Trieste, Italy). Two different series of filled systems were prepared, starting from SG gels with C_p = 0.5% and 1% (w/w).

Continuous shear tests (stepwise procedure) and dynamic measurements were carried out with a rheometer Rotovisco Haake RV100 (measuring system CV100, coaxial cylinder sensor system ZB15). The shear rate range explored was 0.3 - 300 s^{-1}; oscillatory measurements were performed in a frequency range from 0.134 up to 7.536 rad/s, at a constant strain of 2.4.

RESULTS AND DISCUSSION

Tests carried out under continuous shear conditions revealed that PVP additions do not modify substantially the shear dependent properties of the gel matrices. At low PVP concentration, the τ ($\dot{\gamma}$) profiles parallel those of the unfilled gel; at higher concentration, the curves diverge with increasing shear rate. Figure 1 reports the shear rate dependence of the reduced viscosity η_R, defined as the ratio between the corresponding shear stress values of filled and unfilled gels, for systems with C_p = 0.5%. The same behaviour can be observed for gels with C_p = 1% and for aqueous solutions with C_p = 0.15%. It must be emphasized here that the reduced viscosity η_R cannot be confused with the relative viscosity η_r, since the substantial adsorption of the scleroglucan molecules onto the particle surfaces and the consequent depletion of the continuous phase in C_p must be taken into account. In fact, η_R values are apparently low and their variation with particle concentration is quite different from the η_r dependence usually exhibited by particle suspensions in Newtonian and non-Newtonian media. From an analysis of filled scleroglucan solutions data were obtained which confirm the appreciable decrease of the continuous phase concentration because of the polymer transfer into the disperse phase. Similar effects are

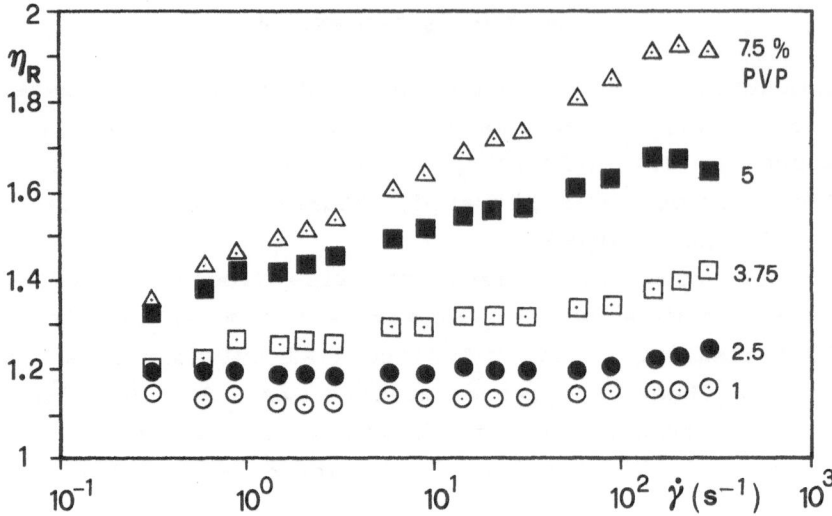

Figure 1. Reduced viscosity vs. shear rate for filled gels
with C_p = 0.5%

expected in the filled gels. The modifications of the time-
dependent properties induced by the addition of PVP particles
lead to the same conclusions concerning the decrease of concentration and connectivity of the gel matrices in the presence of suspended particles. The frequency variation of the re-
duced complex modulus G_R^* closely resembles that of $\eta_R(\dot{\gamma})$. On
the contrary, the phase lag δ is almost independent of filler
concentration and, for all the systems examined, it lies between 1.08 and 1.18 rad in all the frequency range examined.

REFERENCES

1. Lapasin, R., Pricl, S. and Esposito, P., Rheological
    behaviour of scleroglucan aqueous systems in the solution
    and gel domains. In Rheology of Food, Pharmaceutical and
    Biological Materials, ed. E.R. Carter, Elsevier Science
    Publishers, Barking, Essex, U.K., 1990, in press.

2. Lapasin, R., Pricl, S. and Tracanelli, P., Different
    behaviors of concentrated polysaccharide systems in large-
    amplitude oscillatory shear fields, Carbohydr. Polymers,
    1990, in press.

3. Rodgers, N.E., Scleroglucan. In Industrial Gums: Polysac-
    charides and Their Derivatives, eds. R.L. Whistler and
    J.N. BeMiller, Academic Press, New York and London, 2nd
    edition, 1973, pp. 499-511.

THE RELATIVE VISCOSITY OF DILUTE SUSPENSIONS IN GENERALIZED NEWTONIAN LIQUIDS

J.LAVEN, H.N.STEIN
Laboratory of Colloid Chemistry and Thermodynamics
University of Technology Eindhoven
P.O.Box 513, NL-5600 MB Eindhoven, The Netherlands

ABSTRACT

A new, theoretically more satisfactory, definition for the relative viscosities of suspensions in non-Newtonian media is presented, according to which viscosities of suspension and pure liquid should be compared at equal liquid phase viscosities. Theoretical implications of this definition have been investigated, especially for very dilute suspensions. Rheological data of suspensions in liquids with a wide range of degree of pseudoplasticity have been analysed. According to the new definition the Einstein viscosity coefficient of these suspensions is 2.5 independent of the amount of pseudoplasticity.

INTRODUCTION

The relative viscosity η_r of a suspension in a liquid is the ratio of the suspension and the pure liquid viscosities. These viscosities usually are evaluated at equal macroscopic shear rates $\dot{\gamma}_m$ (η_r^m) or at equal shear stresses τ (η_r^τ). With pseudo plastic liquid media, Einstein coefficients $K_E = \lim_{\phi \to 0} (\eta_r - 1)/\phi$, where ϕ is the volume fraction of solids, have been reported to be smaller than the Newtonian value 2.5 (K_E^r) or slightly larger (K_E^r).

We propose [1] to evaluate relative viscosities of suspensions under shear conditions at which the viscosities of the liquid phases of the suspension and of the pure liquid are equal (η_r^L). In case of generalized Newtonian liquids (non-elastic; viscosity is function of the squared shear rate $\dot{\gamma}^2$) this means that viscosities should be compared at equal averaged values (superscript $^{(A)}$) of the squared shear rates in the liquid phases $\dot{\gamma}_L^{2\,(A)}$.

THEORETICAL RESULTS

From theoretical arguments interrelations between the different K_E values of a suspension could be derived:

$$K_E^m = n\, K_E^\tau = [n+1]K_E^L/2 + [n-1]/2 \qquad (1)$$

where $n = d(\log \tau)/(d(\log \dot{\gamma}_m)$ at the shear rate involved.

The mean squared shear rate for a dilute suspension in a generalized Newtonian liquid can be expressed as $\dot{\gamma}_L^2(\text{susp})^{(A)}/\dot{\gamma}_L^2(\text{pure liquid})^{(A)}=1+\beta\phi$. When evaluating η_r^m, η_r^τ and η_r^L the values of β are 7/2, −3/2 and 0 respectively.

Evaluating η_r^m, η_r^τ and η_r^L implies comparing the viscosities found by cross sectioning the flow curves of suspension and pure liquid with a line with a value for α of ∞, −1 and −10/7 where $\alpha=d(\log \eta)/d(\log \dot{\gamma}_m)$ (see Figure 1).

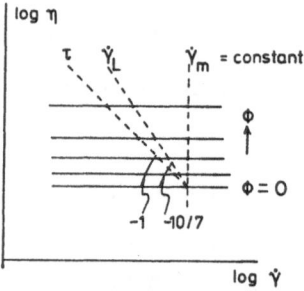

Figure 1. Shear rates involved in the determination of η_r^m, η_r^L and η_r^τ for dilute suspensions in Newtonian media.

EXPERIMENTAL RESULTS

Data of suspensions of glass spheres (with $0 \leq \phi \leq 0.4$) in polymer solutions and liquid crystals, from Newtonian to almost plastic, were analysed. The newly proposed, more objective, relative viscosity η_r^L is remarkably independent of the degree of pseudoplasticity of the liquid medium, K_E^L being 2.5 in the range $0.07 \leq n \leq 1$.

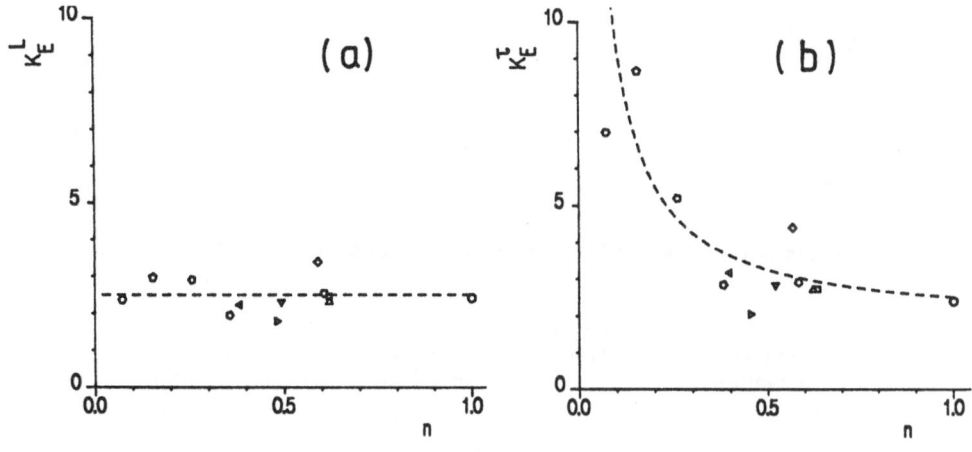

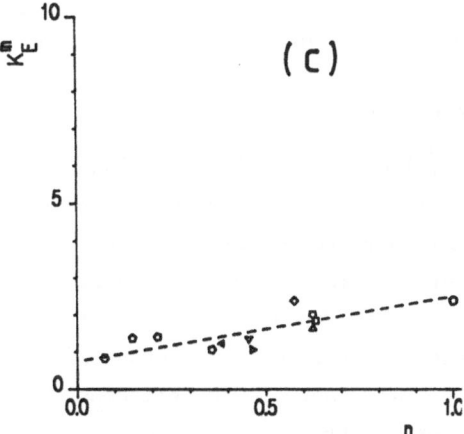

Figure 2. Einstein coefficients K_E^L (a), K_E^τ (b) and K_E^m (c) as functions of $d(\log \tau)/d(\log \dot{\gamma}_m)$. Dashed lines are the theoretical predictions (Eq.1) that best fit the data.

REFERENCES

1. J.Laven, H.N.Stein, The Einstein coefficient of suspensions in generalized Newtonian liquids, submitted to Journal of Rheology.

RHEOLOGICAL PROPERTIES OF BIAXIAL NEMATIC LIQUID CRYSTALS

J.S. LAVERTY and F.M. LESLIE
Department of Mathematics, University of Strathclyde,
Livingstone Tower, Richmond Street, Glasgow G1 1XH.

ABSTRACT

This paper presents a curtailed form of continuum theory for biaxial
nematics suitable for rheological studies, and examines aspects of interest,
particularly flow alignment and the measurement of viscous coefficients.

INTRODUCTION

In their simplest form termed nematics, liquid crystals are uniaxial or
transversely isotropic, and the successful continuum theory for these
anisotropic liquids is based on a single director model [1]. The
resultant equations contain a number of material parameters, basically
three elastic coefficients associated with the couple stress and five
viscosities, but it has proved possible to determine these constants
through experiments in which one essentially controls the alignment of the
anisotropic axis by use of a magnetic or electric field, or by imposed
surface alignments in the case of thin layers.

Given that nematics and their theory are now tolerably well understood, interests in thermotropic liquid crystals have largely moved to
other systems, primarily smectics, but also biaxial nematics given the
recent appreciation that such materials can occur in practice [2]. To
model these more complex anisotropic liquids satisfactorily one requires
more than a single director theory, and a first step is clearly to develop
a general biaxial theory. In this paper, therefore, we briefly present
continuum theory for biaxial nematics, but in a form suitable for
rheological studies, primarily flow alignment and viscosity measurements.

THEORY

A number of authors have derived continuum models for biaxial nematics from different standpoints, but there is now reasonable agreement as to the format of such theory (see, for example, [2,3,4]). As might be anticipated, the number of material coefficients rather increases in this more general theory to twelve elastic and also twelve viscous coefficients. However, at this stage it is not clear to what extent it is possible to control the alignment of both axes by suitable prior treatment of the bounding solid surfaces, this being essential for the determination of the elastic parameters. Here, therefore, we elect to ignore surface effects, and to concentrate solely upon the influence of flow and external fields.

With this simplification the equations for the velocity vector $\underline{v}$ and the two directors $\underline{n}$ and $\underline{m}$ reduce in Cartesian tensor notation to

(i) constraints of incompressibility and orthonormality

$$v_{i,i} = 0 \quad , \quad n_i n_i = m_i m_i = 1 \quad , \quad n_i m_i = 0 \quad , \tag{1}$$

(ii) the reduced balance laws for linear and angular momentum

$$\rho \dot{v}_i = \rho F_i + t_{ij,j} \quad , \quad 0 = \rho K_i + e_{ijk} t_{kj} \quad , \tag{2}$$

(iii) the constitutive relationship for the stress

$$\begin{aligned}
t_{ij} = &-p\delta_{ij} + \alpha_1 D_{kp} n_k n_p n_i n_j + \alpha_2 N_i n_j + \alpha_3 N_j n_i + \alpha_4 D_{ij} \\
&+ \alpha_5 D_{ik} n_k n_j + \alpha_6 D_{jk} n_k n_i + \beta_1 D_{kp} m_k m_p m_i m_j + \beta_2 M_i m_j \\
&+ \beta_3 M_j m_i + \beta_4 D_{ik} m_k m_j + \beta_5 D_{jk} m_k m_i + N_p m_p (\gamma_1 m_i n_j + \gamma_2 n_i m_j) \\
&+ D_{kp} n_k m_p (\gamma_3 m_i n_j + \gamma_4 n_i m_j),
\end{aligned} \tag{4}$$

where

$$2D_{ij} = v_{i,j} + v_{j,i} \quad , \quad 2W_{ij} = v_{i,j} - v_{j,i}, \tag{5}$$

$$N_i = \dot{n}_i - W_{ik} n_k \quad , \quad M_i = \dot{m}_i - W_{ik} m_k,$$

and (iv) the Onsager relations

$$\alpha_2 + \alpha_3 = \alpha_6 - \alpha_5, \ \beta_2 + \beta_3 = \beta_5 - \beta_4, \ \gamma_1 + \gamma_2 = \gamma_4 - \gamma_3. \tag{6}$$

In the above ρ denotes density, the superposed dot a material time derivative, $\underline{F}$ and $\underline{K}$ body force and body couple per unit mass, respectively, $\underline{t}$ the stress tensor, here asymmetric, e_{ijk} the alternator, δ_{ij} the Kronecker delta, and p pressure. Finally the body couple arising from a magnetic field $\underline{H}$ is given by

$$\rho K_i = e_{ijk} J_j H_k \quad , \quad J_i = (\chi^n n_i n_j + \chi^m m_i m_j) H_j \quad , \tag{7}$$

where χ^n and χ^m are constants. The couple due to an electric field is given by an analogous expression.

FLOW ALIGNMENT AND VISCOSITY MEASUREMENTS

The equations describing flow alignment follow from the asymmetric part of the stress and from equations (2), (4) and (6) prove to be

$$\underline{k}.(\tau_1 D\underline{n}+\underline{N}) = 0, \ \underline{k}.(\tau_2 D\underline{m}+\underline{M}) = 0, \ \underline{m}.(\tau_3 D\underline{n}+\underline{N}) = 0,$$

$$\underline{k} = \underline{n}_\wedge\underline{m} \ , \ \tau_1 = \frac{\alpha_3+\alpha_2}{\alpha_3-\alpha_2} \ , \ \tau_2 = \frac{\beta_3+\beta_2}{\beta_3-\beta_2} \ , \ \tau_3 = \frac{\alpha_3+\alpha_2-\beta_3-\beta_2+\gamma_1+\gamma_2}{\alpha_3-\alpha_2+\beta_3-\beta_2+\gamma_2-\gamma_1} \ .$$

(8)

For simple shear flow the above equations yield either six, four, two or no steady state solutions, these occurring in pairs with one of the three vectors $\underline{n}$, $\underline{m}$ or $\underline{k}$ perpendicular to the plane of shear. However, no matter the number of possibilities, consideration of time dependent perturbations selects at most one solution on stability grounds.

As Chandrasekhar has discussed [5] it may be possible to align both axes by a suitable choice of magnetic and electric fields. This being so, shear flow experiments appear to offer nine measurements, three for each of $\underline{n}$, $\underline{m}$ and $\underline{k}$ normal to the plane of shear. Detailed calculations confirm this conjecture. In addition three experiments of the type employed by Zwetkoff with a rotating magnetic field [1] seem possible, allowing a further three measurements. Thus, if Chandrasekhar's technique is effective, measurement of the twelve viscosities appears to be possible by a repetition of methods used for a uniaxial nematic.

REFERENCES

1. Leslie, F.M., Theory of flow phenomena in liquid crystals. Adv. Liq. Cryst., 1979, 4, 1-81.

2. Saupe, A., Elastic and flow properties of biaxial nematics. J. Chem. Phys., 1981, 75, 5118-5124.

3. Kini, U.D., Isothermal hydrodynamics of orthorhombic nematics. Mol. Cryst. Liq. Cryst., 1984, 108, 71-91.

4. Govers, E. and Vertogen, G., Fluid dynamics of biaxial nematics. Physica, 1985, 133A, 337-344.

5. Chandrasekhar, S., Paper presented at 8th Liquid Crystal Conference of Socialist Countries, Krakow, Poland, August 1989.

WALL SLIPPAGE AND THE FLOW OF RUBBER COMPOUNDS IN CONVERGING PATH

Jean L. LEBLANC
MONTEDISON, Belgium

ABSTRACT

Experiments were made to specifically study the converging flow behaviour of various rubber compounds. A set of cylindrical dies with 45° conical entrance was used to perform experiments with a capillary rheometer. The entrance pressure drop Pent, as derived according to the Bagley method, was compared with the extrusion pressure through zero-length dies $P_{(L/D=0)}$, and discrepancies were observed which vary with the level of filler or the amount of processing oil used in the compound. The hypothesis that such discrepancies are due to slippage effects within the converging path, leads to a discussion regarding the significance of the Bagley correction.

INTRODUCTION

Wall slippage is known to be a major component in the flow behaviour of rubber compounds (1) and is particularly relevant in a number of practical processing situations, such as the entrance flow in extrusion dies, the converging-diverging flow path in injection moulding, etc. Both qualitative and quantitative information is however needed, particularly in what concerns the flow of rubber compounds in converging geometries.

When a viscoelastic material flows in a converging path, the overall output not only depends on the true viscosity, but also on the pressure loss; the latter component, in addition to its obvious dependence upon the elastic properties, is strongly affected by wall slippage conditions. LEBLANC, VILLEMAIRE et al. (2) paid attention to the modification of wall slippage in converging path, using NR compounds with various levels of lubricant. They reported discrepancies between the entrance pressure drop, as determined using long die data and the Bagley method, and the extrusion pressure through zero-length die. They drew the hypothesis that such discrepancies reflect wall slippage effects in the converging entrance.

The purposes of the experiments reported here were to further document the discrepancies observed between $P_{(L/D=0)}$ and Pent, with rubber compounds.

EXPERIMENTAL

Rubber compounds were prepared using either Natural Rubber (SMR 5CV), or Ethylene-Propylene Rubber (DUTRAL CO 054) and various amounts of carbon black (N330), as well as several NR compounds with 50 phr carbon black and

increasing level of processing oil. Compounding was made in a 1.5 l.
Banbury mixer using mixing procedures monitored by energy, in such a manner
that an optimal dispersion of the ingredients was achieved.

Extrusion experiments at 100°C were performed with the Rheoplast, a
capillary rheometer with preshearing capabilities (3) whose commercial
version has been developed for our laboratory. The capillary rheometer was
driven in the constant piston rate mode, using a set of dies with diameter
D=1 mm, L/D= 0, 1, 5 and 10 and a conical entrance angle of 45°. The
experimental approach consisted in measuring the Output-Pressure curve with
each die. All experiments were performed at least two times, in order to
compensate for possible compound inhomogeneities and to assess the
experimental scatter. According to the Bagley method (4), the entrance
pressure drop were derived by plotting the extrusion pressure versus the
length-to-diameter ratio of the dies, at constant volumic output (or
constant shear rate). Note that the pressure data with the L/D=0 dies were
not used in deriving the Bagley entrance loss.

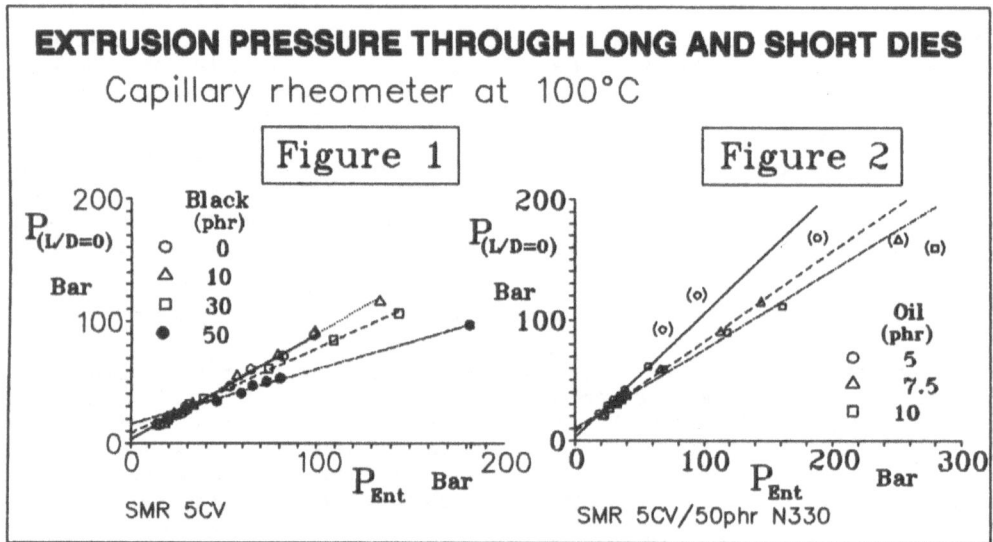

As illustrated in Figures 1 and 2, the extrusion pressure with the
zero-length die (i.e. $P_{(L/D=0)}$) is not equal to the entrance loss P_{Ent}, at
equal output, and the discrepancy increases either with increasing carbon
black level or with increasing oil level. Least square fittings of
$P_{(L/D=0)}$, P_{Ent} data yields straightlines with slope A and intercept B, which
are used to report all our experimental results (see Table). Note that
compounds 1 to 7 were mixed using an upside procedure; a normal mixing
procedure was used for compounds 8 to 10; this explain the differences
between the results obtained with compounds 4 and 8 (same formulation).

DISCUSSION

As shown in the Table, the higher the filler content the lower the slope A
and the higher the intercept B (experiments 1-4 and 5-7). The particular
values of A and B depend on the nature of the rubber and other compounding
ingredients, but in an homologous series of experiments, the results are
consistent. A similar effect is observed with increasing oil level.

MA, WHITE et al. (1) demonstrated that slip phenomena do appear in the flow of rubber compounds when the filler content is above 10%, and further increase in magnitude as the filler level increases. Our data show that gum compounds exhibit either zero or low intercept values and slope close to 1; with gum compounds wall slippage is likely to be indeed negligible. By reasonning on the likely morphology of (filled) rubber compounds, it has been demonstrated that no wall slippage condition is a physical impossibility with such materials, owing to interactions between the complex rubber-filler aggregates (5). Compounds with higher oil contents are exhibiting larger slippage effects through enhanced lubrication at the rubber-metal interface. The variation in slope and intercept with compounding variations are thus conform to the logics of enhanced slippage effects within the convergent.

Table : $P_{(L/D=0)}$ vs P_{Ent} for various rubber compounds

Exp.No	Rubber	Black	Oil	Intercept A	Slope B	R squared
1	SMR 5CV	–	5	1.740	0.858	0.997
2	SMR 5CV	10	5	3.242	0.850	0.998
3	SMR 5CV	30	5	7.661	0.693	0.998
4	SMR 5CV	50	5	16.803	0.443	0.995
5	CO 054	–	–	0.002	0.962	0.998
6	CO 054	10	–	1.320	0.958	0.998
7	CO 054	30	–	5.490	0.918	0.998
8	SMR 5CV	50	5	3.266	1.020	0.994
9	SMR 5CV	50	7.5	7.149	0.746	0.997
10	SMR 5CV	50	10	9.450	0.657	0.992

Such results obviously address the significance of the Bagley correction. As pointed out by HAN(6), the Bagley correction includes in fact contributions for both the entrance pressure drop and the residual exit pressure. The exit pressure can be assessed by extrapolating the straight line portion of wall pressure measurements to the exit of slit dies. HAN reported that P_{Exit} increases rapidly with decreasing L/D ratio; his observations were made with hd polyethylene and the smallest L/D considered was 4. It is obvious that a zero length die has very little in common with the flow situation achieved in a die system consisting of a conical entrance and a cylindrical tube; clearly with the very short dies used in practical rubber extrusion, the entrance flow is merged with the exit flow and in such case it is likely that the P_{Exit} contribution (which is assumed to exist with zero-length dies) is considerably larger than with longer dies, and therefore the discrepancies observed between $P_{(L/D=0)}$ and the Bagley entrance loss.

REFERENCES

(1) C.Y.MA, J.L.WHITE, F.C.WEISSERT, A.I.ISAYEV, N.NAKAJIMA and K.MIN – Rubb. Chem. Technol., 58, 815 (1985)
(2) J.L.LEBLANC, J.P.VILLEMAIRE, B.VERGNES and J.F.AGASSANT – Plast. Rubb. Proc. Applic., 11(1), 53 (1989)
(3) J.P.VILLEMAIRE and J.F.AGASSANT – Polym.Proc.Eng.,1,221 (1983)
(4) E.B.BAGLEY – J.Appl.Phys.,28, 624 (1957)
(5) J.L.LEBLANC – Eur.Rubb.J.,171(10), 33 (1989)
(6) C.D.HAN – "Rheology in Polymer Processing", Chap. 5 – Acad.Press, New-York (1976)

CHARACTERIZATION OF BREAD STEELING PROCESS BY MEANS OF AN INSTRON UNIVERSAL TESTING MACHINE OPERATING IN DYNAMIC MODE

D.M.B.LEITE, L. PIAZZA* AND P. MASI

Istituto di Ingegneria Chimico-Alimentare, Università degli Studi di Salerno, Via S. Allende, 84081 Baronissi - Italy.

* D.I.S.T.A.M., Università degli Studi di Milano, Via Celoria, 2, 20123 Milano - Italy.

ABSTRACT

This paper illustrates the application of the Nutting general low of deformation in combination with a simple experimental technique to characterize the rheological changes which accompain the steeling process of bread. Basically it consists in submitting a sample having regular geometry to a series of uniaxial compression tests during which a maximum constant stress or strain is reached. The reological behavior of bread in the course of the aging process is described by means of the Nutting's constitutive equation whose parameters are evaluated analyzing how the sample height, the extend of deformation and the stress vary in each test. Data produced according to this technique appear to be an usefull tool to characterize the evaluation of the rheological behaviour of solid food materials during ageing.

INTRODUCTION

Very often solid food materials during ageing undergo to extensive rheological changes. Because of that, commonly, the rheological analysis is used to investigate the food texture evolution during this stage. Dealing with food materials, however, is not simple. In fact, in contrast with many other materials, rheological evaluation of solids food, for practical reasons, is carried out by using techniques which are based on test procedure that involves large compressive deformations.

Since most solid food are viscoelastic this introduces the problem of taking into account non linear effects in characterizing their rheological behaviour. The literature presents many examples of non linear phenomenological rheological models which have been successfully used to characterize solid food materials (1-3). However in most cases a well balanced compromise between a demand for mathematical semplicity and predictive capability is not reached. From this point of view, Nutting general low of deformation (2) appears quite promising. According to this model the material strain is related to the applied stress and to the time during which the load is applied by a simple relationship:

$$\epsilon = K \ (\sigma)^a \ (t)^b \tag{1}$$

where K, a and b are material constants which are independent on testing procedure. Ideal elastic and liquid behaviour can be regarded as limiting cases of equation 1 which correspond to a=1 and b=0 or 1 respectively.

This paper illustrates the application of this phenomenological equation in bread steeling process characterization.

MATERIAL AND METHODS

Commercial french type bread was used in this work. Bread loafs having 12 hours age were purchased at once and stored in a confined room at constant relative humidity.

Cylindrical samples having 3 cm. diameter and 4 cm. height made from slices 4 cm thick, by means of a cork bore, were each submitted to twenty consecutive compressive tests by means of an INSTRON machine mod. 4301 equipped with computer device for data acquisition and elaboration.

During each test the sample was deformed at a constant crosshead speed equal to 10 mm/min up to a given fix load (150 g). The crosshead was then suddenly reported to the initial position (V = 500 mm/min) and the sample deformed again. At each run the sample height and the extend of deformation was detected via computer which assumes as initial samples height the distance between the dinamometer plates at which the detected load exceeds 1 gram.

RESULTS AND DISCUSSION

By applying equation 1 to two generic compressive run the following relationship hold:

$$\epsilon_0 = K \ (\sigma max_0)^a \ (t_0)^b$$
$$\epsilon_n = K \ (\sigma max_n)^a \ (t_n)^b \tag{2}$$

where 0 and n indicate a reference run and a generic successive one respectively. Being σmax constant in each run and K, a and b material constants one can write:

$$\epsilon_n/\epsilon_o = (t_n/t_o)^b \qquad\qquad (3)$$

which is equivalent to:

$$H_n/H_o = (\; \Delta H_n \; / \; \Delta H_o)^{1-b} \qquad\qquad (4)$$

in which H is the sample heigt in each run and H the correspondent extend of deformation.

By plotting in a log-log scale the experimental data according to equation 4 the value of the parameter b can be easily obtained. Figure 1 shows a typical log-log plot of Hn/Ho vs. ΔHn/ ΔHo. A straight line with a very high correlation (R=0.998) is obtained, whose slope provides the value of b.

Although not reported here, for sake of brevity, but extensively described elšewere (4), a similar procedure allows one to determine the value of the parameter a from constant strain tests. Once the parameter a and b are known, K can be extimated from a generic compressive test.

Figure 2 shows how the parameter b varies with varing the bread ageing. As the age of the bread increases the value of b decreases approaching asymptotically a value close to zero. This is in agreement with what one would expect as a consequense of steeling process. In fact the parameter b assume a value equal to zero in the case of an ideal elastic solid and the value equal to 1 in the case of a pure viscous liquid. Therefore the fact that b decreases with increasing the age of the bread indicates that the bread with increasing time loose progressively its viscoelastic characteristics and behaves in a way similar to an ideal elastic solid.

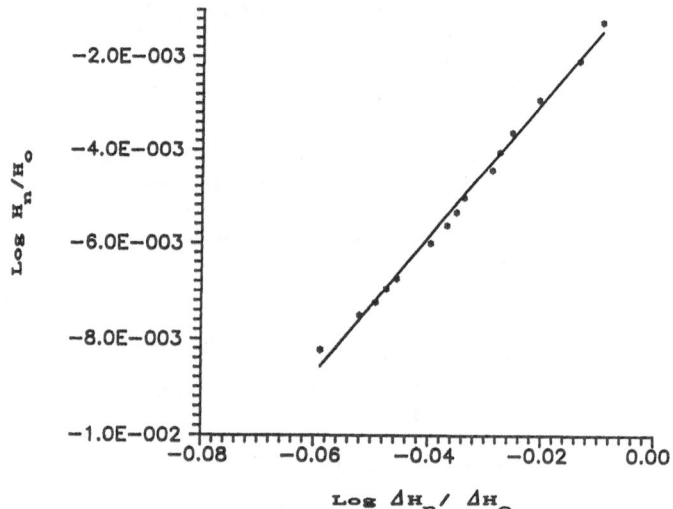

Figure 1. Typical log - log diagram obtained by plotting experimental data according to eq. 4.

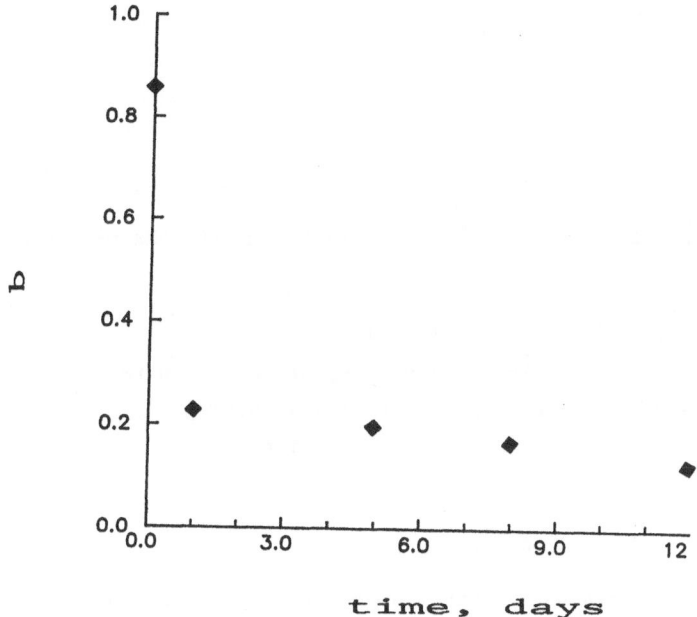

Figure 2. Evolution of the   parameter b of eq. 1 during bread
        ageing.

In conclusion, although more work is needed to elucidate the
relationship existing between food structure , composition and
model parameters, the Nutting general low of deformation can
represent a powerfull tool in characterizing the nonlinear
rheological behaviour of solid food materials.

REFERENCES

1. Reiner, M., Phenomenological macrorheology In Rheology -
   Theory and Applications , F.R. Eirich (ed.), Academic
   Press, New York, 1956, pp. 9-62.

2 Sherman, P., Industrial Rheology, Academic Press, London,
   1970.

3. Peleg, M. Application of non-linear phenomenological
   rheological models to solid food materials. J. of Texture
   Stud., 1984, 15, 1.

4. Leite, D.M.B., Piazza, L., and Masi, P., On the use of the
   Nutting general low of deformation to non linear
   rheological behaviour characterization of solid foods.
   Submitted for pubblication to J. Texture Stud., 1990.

A STUDY OF ELASTIC-VISCOPLASTIC CONSITITUTIVE EQUATION FOR TIAN-QING GEL

J.Li   T.Q.Jiang

Division of Rheology of Chemical Engineering Research Centre

East-China University of Chemical Technology

200237   Shanghai   P.R.China

ABSTRACT

An elastic-viscoplastic consititutive equation has been proposed to describe the rheological properties of Tian-Qing gel.   The experimental results show that the theoretical analysis give satisfactory agreement not only for Tian-Qing gel but also for Xanthan gum and HGP gels.

INTRODUCTION

Gelled materials are usually encounted as fracturing liquids in rude oil production. A suitable rheological model of gel is necessary for engineering applications. Unfortunately, because of the complication, there is, up to date, no one which can be used to describe them flawlessly. In this paper, we attempt to propose a rheological model which is suitable to predict the rheological properties of this visco-elastic fluid with yield stress.

THEORY

General speaking, the method of consititutive equation construction for an elastic-viscoplastic fluid can be divided into different ways as Hutton[1] and White[2] did respectively. On the basis of their methods, a new rheological model which is obedient to the Von Mises yield criterion is given by:

$$\mathbf{D} = \mathbf{O} \qquad\qquad tr\tau^2 \leqslant 2\tau_y^2 \qquad\qquad (1a)$$

$$\tau = \frac{\tau_y}{\sqrt{\frac{1}{2}tr\mathbf{D}^2}}\mathbf{D} + \mathbf{H} \qquad\qquad tr\tau^2 > 2\tau_y^2 \qquad\qquad (1b)$$

where: **D** is strain tensor, τ_y is yield stress, τ is deriatoric stress tensor and **H** is nonlinear memory function. We assume the build-up of stress only depand on the history of the rate-of-strain during flow, therefore, we have:

$$H = \int_{-\infty}^{t} G[(t-t'), \amalg_c] C^{-1} dt'$$

(2)

in which: $\amalg_c$ is the second invariant of right Cauchy-Green tensor and C^{-1} is the Finger strain tensor; G is the relaxtion modulus formulated in the following way:

$$G(t) = \frac{G_f}{\Gamma(1-\alpha)} t^{-n} exp(-t/\lambda)$$

(3)

where:

$$G_f = \frac{G_0}{(1 + \frac{1}{2}\lambda_0^2 \amalg_c)^{(1-n)/3}} \qquad \lambda = \frac{\lambda_0}{(1 + \frac{1}{2}\lambda_0^2 \amalg_c)^{(1-n)/6}}$$

and G_0, λ_0, n are empirical constants determined by experimential data.

MATERIALS AND METHODS

In the experiment, Tian-Qing gel which is widely used in China was made of the solution of Tian-Qing powder to which few cross-linking agents were added and the measurments of simple shear flow and die swell were performed in capillary viscometers, moreover the osillatory shear flow and part of simple shear flow were carried out in RMS-605.

RESULTS AND CONCLUSIONS

Fig. 1 and 3 show the experimential results and computations of equation in which the parameters are obtained by regressing the data of simple shear flow. Based on Metzner's a new suggstion(3), the X-axis is not used ω as usually but $\gamma_0 \omega$ in fig. 3.

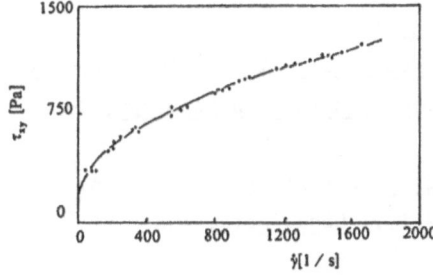

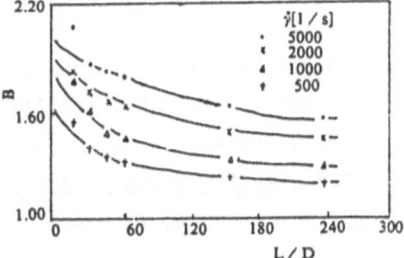

Fig. 1. Simple shear flow results in capillary viscometer(TQ gel)

Fig. 2. Experimential die swell ratio B and computation(TQ gel)

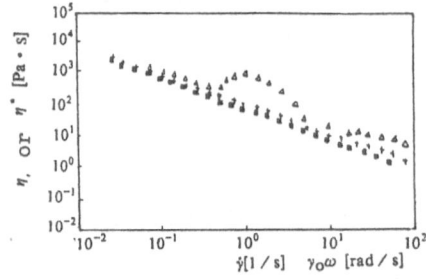

Fig. 3. Experimential results in
RMS-605 and theoretical results

Δ   simple shear flow

+   osillatory shear flow

■   theoretical computation

Fig. 2 shows the comparison between die swell data and theoretical predictions(4). In order to check whether the equation can be extended into other kind of gels, we also make the comparisons in which the data of fig. 4 and 5 are chosen from literatures(5)(6). It is evident that the equation developed by us is also suitable for Xanth gum and HGP gels as well as Tian-Qing gel.

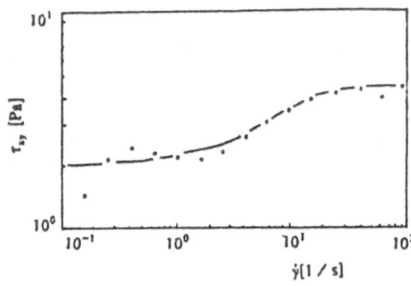

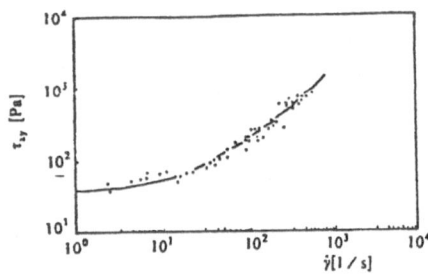

Fig. 4. Simple shear flow of Xanth gum(5)

Fig. 5. Simple shear flow of HGP gels(6)

REFERENCES

1. Hutton, J.F, Rheol. Acta, 1975, 14, 979.

2. White, J.L, and Tanka, H, JNNFM, 1981, 8, 1.

3. Metzner, A.B., etc., Xth Interl. Congr. on Rheol., Vol 1, 1988, Sydney.

4. Li, J. and Jiang, T.Q., The third National Congress on Rheology, 1990, Shanghai, China.

5. Xu, Z.F., Master Dissertation of Zhe-Jiang University, 1989, China.

6. Jiang, T.Q., Young, A.C. and Metzner, A.B., Rheol. Acta, 1986, 25, 397.

RHEOLOGICAL PROPERTIES
OF XANTHAN AND XANTHAN-GALACTOMANNAN GUM MIXTURES

Luyten, H., Kloek, W., Vliet, T. van
Dept. of Food Science
Wageningen Agricultural University
Wageningen, the Netherlands

Xanthan and other polysaccharides are often used as thickening agents e.g. for liquid-like food materials. For the rheological properties of these materials normally two major requirements are set:
- the material must behave like a gel or must have a high apparent viscosity at low stresses and strain rates in order to prevent sedimentation or creaming of dispersed particles,
- one must be able to drink the liquid, this means that the apparent viscosity at higher stresses and strain rates must be relatively low.

Xanthan gum dispersions as such, show 'Newtonian' behaviour at both low and high shear rates (figure 1) and cannot fulfil these requirements. However, a mixture of xanthan with a galactomannan may have a high viscosity or (apparent) gel structure at low stresses in combination with a low enough viscosity at higher stresses. The rheological properties of some of these materials depend, besides on the normal factors, also on the dimensions of the volume of material which is actually involved in the testing procedure.

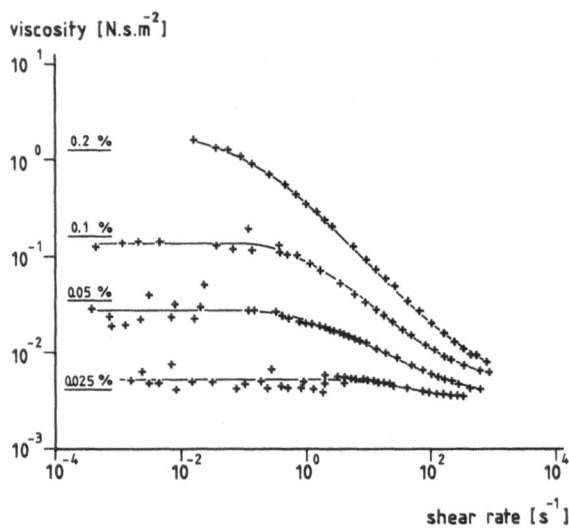

FIGURE 1

The apparent viscosity of xanthan gum dispersions of different concentrations (w/w %) in 0.1 % NaCl.

For these dispersions the apparent viscosity as calculated from the sedimentation velocity of glass beads was found to be higher than the apparent viscosity measured in an apparatus at the same stress applied (table 1). This difference probably may be explained by the volume of material, which is involved in testing, being smaller in a sedimentation experiment. Inhomogeneities in the material which are larger than this volume will not effect the sedimentation viscosity, but will lower the viscosity as determined in an apparatus with a larger testing volume.

TABLE 1
Some examples of the ratio of the sedimentation viscosity
to the apparatus viscosity for different dispersions as
measured at a stress of 3 to 4 N·m^{-2}.
lbg = locust bean gum

dispersion	$\eta^*_{sedimentation} / \eta^*_{apparatus}$
glycerol/water	1
0.1 % guar + 0.1 % NaCl	1
0.05 % xanthan + 0.1 % NaCl	1
0.05 % xanthan + 0.05 % guar + 0.1 % NaCl	10 to 16
0.05 % xanthan + 0.05 % guar + 0 % NaCl	100
0.05 % xanthan + 0.05 % lbg + 0.1 % NaCl	150
0.02 % xanthan + 0.02 % lbg + 0.1 % NaCl	280

Dispersions of xanthan with a galactomannan indeed
appeared to be inhomogeneous: a low-viscous liquid is
expelled from lumps which have a thicker consistency.
These lumps could be up to a few centimeter diameter. The
expelled liquid can cause slip to occur in a rheometer and
therefore lower the apparent viscosity measured in an
apparatus.

For more simpler and homogeneous systems the viscosity
as measured with sedimentation experiments and with a
rheometer were, at comparable shear stresses, equal within
the experimental accuracy (table 1).

ACKNOWLEDGEMENT

This work was supported by Suiker Unie and by the Dutch
Innovation Oriented Program Carbohydrates (IOP-k).

REFERENCES

(1)   Vliet, T. van, Walstra, P., Weak Particle
        Networks. In: Food Colloids, eds. R.D.Bee, P.
        Richmond and J. Minguins, Royal Soc. of Chem.,
        1989, pp 206 - 217.

FLOW OF POLYMER SOLUTIONS THROUGH BEDS OF SOLID PARTICLES

IVAN MACHAČ, JIŘÍ CAKL, PETR MIKULÁŠEK
The Institute of Chemical Technology Pardubice
Department of Chemical Engineering
nám. Legií 565, 532 10 Pardubice, Czechoslovakia

ABSTRACT

On the basis of analysis of an extensive set of experimental data, the possibility of exploitation of a modified capillary model of the bed is presented for description of a flow of polymer solutions through beds of solid particles.

The model gives good results for pressure drop prediction of pure viscous and weakly elastic liquids flowing through fixed beds of spherical and nonspherical particles.

In contrast to the flow through fixed beds, the capillary model cannot be used for reliable prediction of expansion of beds fluidized with polymer solutions in creeping flow region. In this case, a limited expansion of the beds was observed.

INTRODUCTION

A modified capillary model has been suggested [1] which enables the relations to be derived describing the flow of generalized Newtonian liquids (GNL) through fixed beds of particles, the wall effect being accounted for. In this model, the consistency variables have been defined as

$$\tau_s = \Delta p \, l_{ch}/L \qquad (1), \qquad D_s = 2 \, u_{ch}/l_{ch} \qquad (2)$$

Here

$$l_{ch} = \varepsilon / [a_p \varphi(1-\varepsilon) + a_w] \quad (3), \qquad u_{ch} = u/\varepsilon \qquad (4)$$

are the characteristic linear dimension of the bed and characteristic velocity of the liquid, Δp the pressure drop, L the height of the bed, ε the bed porosity, a_p the specific surface of particles, a_w the specific surface of the wall, φ the dynamical characteristic of the bed (bed factor) representing a ratio of the total and friction drag of the bed.

It is assumed that the relation between τ_s and D_s enabling the pressure drop calculation can be obtained by integrating the Poiseuille equation

$$D_s = (4/\tau_s^3) \int_0^{\tau_s} \tau^2 \ D(\tau) \ d\tau \qquad (5)$$

using a corresponding flow model $D = D(\tau)$ of GNL.

The form of eq. (5) makes possible the pressure drop calculated to be easily corrected on the surface effects [2] or elasticity effects [3].

After substitution

$$\Delta p/L = g \ (\rho_s - \rho) \ (1 - \varepsilon) \qquad (6)$$

where ρ_s and ρ are the densities of the solid particles and the liquid, the relations for description of fluidized beds expansion result from eq. (5).

EXPERIMENTAL RESULTS AND DISCUSSION

Fixed beds

In our research programme, the dependence of the pressure drop on the volume flow rate of Newtonian liquids (oil, water solutions of glycerol) and shear thinning non-Newtonian liquids (water solutions of different polymers) was measured in fixed beds of spherical and nonspherical particles (sand, cylinders, polyhedrons, cubes, lenses, and Raschig rings). In total, approximately 300 different fixed bed-liquid systems were tested.

It was found that the bed factor φ is independent on the rheological characteristics of GNL and can be expressed as

$$\varphi = \omega (1 + \alpha \psi) \qquad (7)$$

where ψ is the drag criterion given as a ratio of the form and friction drag of the spherical particles bed, ω the correction factor for the wetted surface, α the form factor ($\omega = 1, \alpha = 1$ for spherical particles). In the creeping flow region $\psi = 0.5$; in the transient region, the drag criterion is a function of Re_m [4].

The power law, Ellis, Tyabin, Sutterby, Carreau, and modified Eyring models were tested for expressing viscosity function of GNL. The pressure drop can be most simply calculated using the power law. If the parameters K and n of the power law are determined in the appropriate interval of the shear rate, the accuracy of the calculation is sufficient, and more complicated flow models need not be employed.

Fluidized beds

The expansion of spherical particles beds was measured in our fluidized beds experiments. Over 200 systems differing in particles composing the beds, liquid medium used, or column diameter were investigated. The beds were composed of monodispersed glass, steel, and lead balls. Water

solutions of glycerol and different polymers were used as
liquid tested. Viscosity function of the liquid was approximated with the power law $(0.35 \leq n \leq 1)$ or the Carreau flow
model.

   A limited expansion has been observed to compare with
the Newtonian liquids fluidization in our measurements with
polymer solutions in the creeping flow region. In this case,
the shear thinning effect was pronounced by an elastic
effect [5,6]. The relations based on a capillary model do
not express the limited expansion and therefore cannot be
used for correct prediction of the expansion in the fluidized
systems mentioned. To this purpose, the equation

$$\varepsilon = (\frac{u}{u_t})^{0.218 - 0.404 \, d/D} - 0.224 \, (1 + 4.75 \, \frac{d}{D})[\wedge_t^{0.5}(\frac{u}{u_t})^{1.55}]^m \quad (8)$$

has been evaluated from experiments based on the Carreau
flow model $(0.35 \leq m \leq 0.79, \quad 0 \leq \lambda \leq 39.3 \, s)$.
Here u_t is the terminal falling velocity of the particles,
$\wedge_t = \lambda \, u_t/d$ the dimensionless time parameter, λ and m are
parameters of the Carreau model.

   In the transient flow region, when the influence of
inertia forces and turbulence begin to be conspicuous, the
effect of the elasticity upon the retardation of the bed
expansion is weakening [7].

REFERENCES

1. Machač,I. and Dolejš, V., Flow of generalized Newtonian
   liquids through fixed beds of nonspherical patricles,
   Chem. Eng. Sci., 1981, 36, 1679-86

2. Machač,I., Cakl, J. and Lecjaks,Z., Anomalous effects
   in fixed bed flow of polymer solutions. 8th International
   CHISA Congress, Prague, 1984

3. Cakl,J., Machač,I. and Lecjaks,Z., Flow of viscoelastic
   liquids through fixed beds. Progress and Trends
   in Rheology II, Supplement to Rheol. Acta, 1988, 26, 266-8

4. Dolejš,V. and Machač,I., Pressure drop in the flow
   of a fluid through a fixed bed of particles.
   Int. Eng. Chem. 1987, 27, 730-6

5. Machač,I., Balcar,M.and Lecjaks,Z., Creeping flow of
   non-Newtonian liquids through fluidized beds of spherical
   particles. Chem. Eng. Sci., 1986, 41, 591-6

6. Machač,I., Mikulášek,P. and Lecjaks,Z., Flow of non-
   Newtonian liquids through fluidized beds of spherical
   particles. Progress and Trends in Rheology II, Supplement
   to Rheol. Acta, 1988, 26, 268-70

7. Machač,I., Mikulášek,P. and Ulbrichová,I., Flow of non-
   Newtonian liquids through a fluidized bed of spherical
   particles in transient flow region. 35th National CHISA
   Congress, Karlovy Vary, 1988 (in Czech)

RHEOLOGY OF SMALL SAMPLES

M.E. MACKAY
DEPARTMENT OF CHEMICAL ENGINEERING
THE UNIVERSITY OF QUEENSLAND
ST. LUCIA, QUEENSLAND 4072
AUSTRALIA

ABSTRACT

An attachment was developed for a standard torsional rheometer based on a
sliding plate rheometer. The attachment can measure the dynamic shear
rheological properties using samples of total volume on the order of $25mm^3$.
This is the volume of two pellets of polymer and eliminates the need to
press samples as is needed for the cone and plate geometry. The results
from this attachment agreed to within 4% of rheological properties measured
with cone and plate and parallel plate geometries for a variety of fluids:
a Brookfield oil, a high density polymer melt and a paint base. Surface
tension was found to influence the dynamic modulus particularly for the low
viscosity Brookfield oil. This attachment is most useful for thermally
sensitive materials as one thermal cycle is eliminated since a sample need
not be moulded.

INTRODUCTION

Measurement of the dynamic properties of small samples is in general
difficult, especially for polymer melts. The attachment developed here
consists of two plates connected to a central shaft via two opposing lever
arms. The shaft oscillates about a null point and the sample shears
between the oscillating plates and two matching plates connected to a
torque transducer. A standard rheometer can be used by making suitable
attachments. The software for a cone and plate can also be used by
assuming the attachment behaves as a phantom cone and plate.

MATERIALS AND METHODS

Three fluids were used: a Brookfield oil, a high density polyethylene
(HDPE) and a paint base (Paint). To eliminate curing of the paint a layer
of a fine oil was used to seal the edges. The attachment (hereafter
referred to as the prong) was made from stainless steel and used with a
Rheometrics Fluids Spectrometer. The radius and angle of the phantom cone
and plate were determined by equations derived elsewhere [1]. The results
from the prong were compared to standard cone and plate and parallel plate
geometries. A strain of 10% was used in all tests.

RESULTS AND DISCUSSION

An optimum gap between the plates was found for all the fluids used presumably because of the superposition of a twist on the translational motion which exists. The magnitude of the twist was fairly small due to the relatively small strain used. For 5mm by 5mm plates with a lever arm of 13.5mm to the plates' centre the optimum gap was found to be 0.5mm as shown in Figure 1.

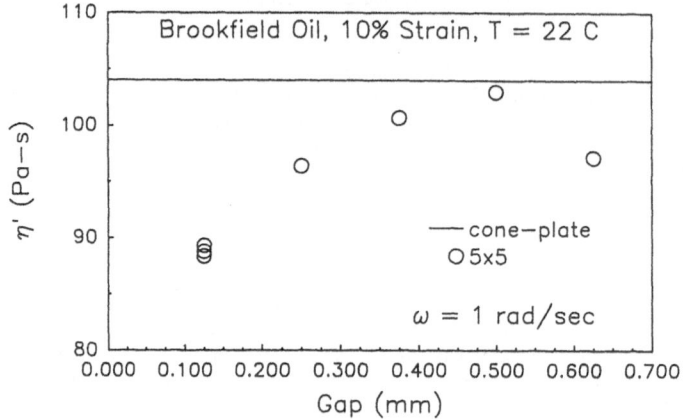

Figure 1. Procedure to find the optimum gap with the prong attachment. η' is the dynamic viscosity.

Overall the error was less than 4% for both η' and G' (storage modulus) as shown in Figure 2. It was found that surface tension affected the results particularly for the Brookfield oil. In fact, by taking account of surface tension in a force balance allowed its calculation. This analysis lead to a value of 30.2 mN/m which was within experimental error of the surface tension determined by plate detachment (27.7mN/m).

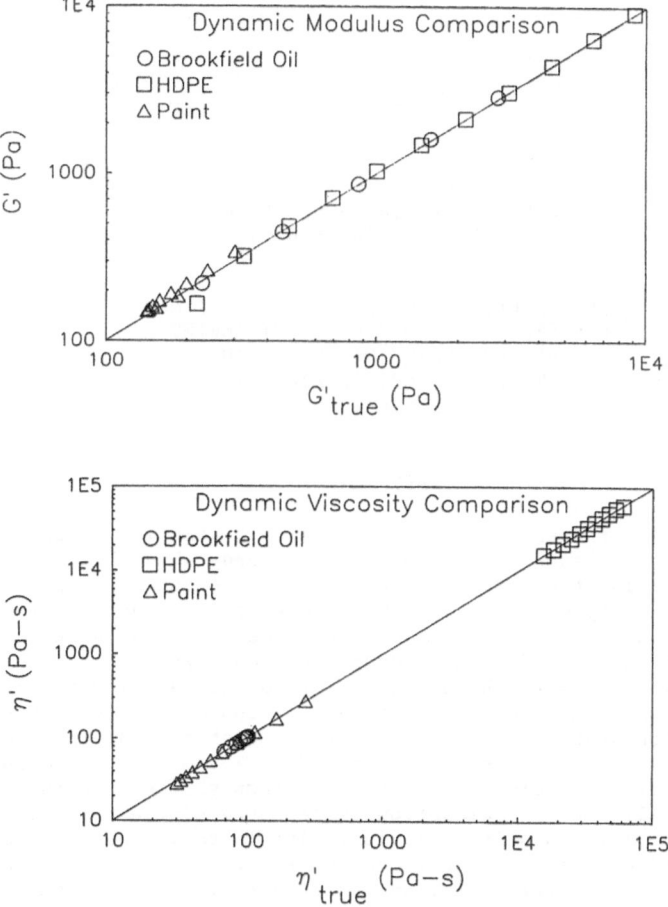

Figure 2. Comparison of the dynamic properties determined with the prong with a 0.5mm gap to those measured with standard geometries.

ACKNOWLEDGEMENT: This work was funded by the Australian Research Council. The author thanks Dr. C.A. Cathey for taking most of the data.

REFERENCE

1. Mackay, M.E., submitted to J. Rheo.

RHEOLOGY OF RIGID MACROMOLECULES
IN THE SEMI-DILUTE REGION

M.E. MACKAY, C.-H. LIANG AND C. JINAN
DEPARTMENT OF CHEMICAL ENGINEERING
THE UNIVERSITY OF QUEENSLAND
ST. LUCIA, QUEENSLAND 4072
AUSTRALIA

ABSTRACT

Various rheological properties were measured of a solution containing a
rigid macromolecule, Xanthan Gum, including: steady shear viscosity, stress
jump and stress relaxation after steady shear and exponential shear. The
steady shear viscosity was found to shear thin for all the solutions tested.
The stress jump on cessation of steady shear was found to increase with
increasing shear rate and depend on concentration. When the stress jump
was made dimensionless with the stress at steady state prior to cessation
the stress jump was found to be independent of concentration at higher
shear rates and solvent viscosity and increase with increasing shear rate.
This is consistent with predictions of dilute solution rigid molecule
theories. However, the relaxation after the jump showed that the stress
decayed with a multitude of relaxation times not a single relaxation time
as predicted by dilute solution theories. The instantaneous viscosity in
exponential shear exhibited a large overshoot and the stress jump and
subsequent relaxation after cessation depended on the time of shear.

INTRODUCTION

Xanthan Gum is a semi-rigid macromolecule and has been used in this work as
a test macromolecule to determine if stress jumps are experimentally
realisable. Stress jumps on cessation of shear flow have been predicted
theoretically for rigid molecules as well as molecules with internal
viscosity, they have never been observed experimentally. Because the
velocity profile requires a finite time to achieve equilibrium, stress
jumps on start-up are more difficult to observe. In this paper the shear
rheological properties of semi-dilute solutions of Xanthan Gum are presented.
The results for exponential shear will be presented separately due to space
limitations.

MATERIALS AND METHODS

Food grade Xanthan Gum was used and mixed with distilled water by slow
rotation of a vessel. Fructose was then added after one day and allowed to

mix for a further week. The rheological properties were measured with a modified Rheometrics Fluids Spectrometer. The internal filtering was shorted out since this would produce phase lags and eliminate visualisation of stress jumps. A PC-AT computer was used as a function generator to perform exponential shear.

RESULTS AND DISCUSSION

The steady shear viscosity of the solutions was measured and is presented in Figure 1. All the solutions shear thin.

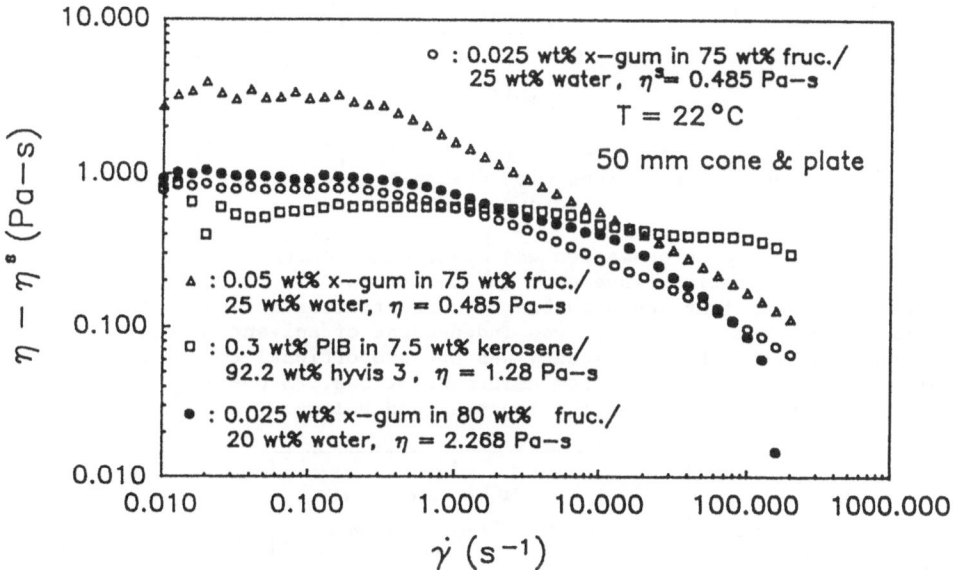

Figure 1. Viscosity minus the solvent viscosity versus shear rate for the various solutions.

The viscosity was divided by the zero shear viscosity and the shear rate multiplied by the square of the weight percent of Xanthan Gum in solution This scaling should produce a master plot if existing theories for semi-dilute/concentrated solutions hold true. This is shown in Figure 2 below for the shear rates examined in this study. The zero shear viscosity did not scale exactly with the concentration to the third power.

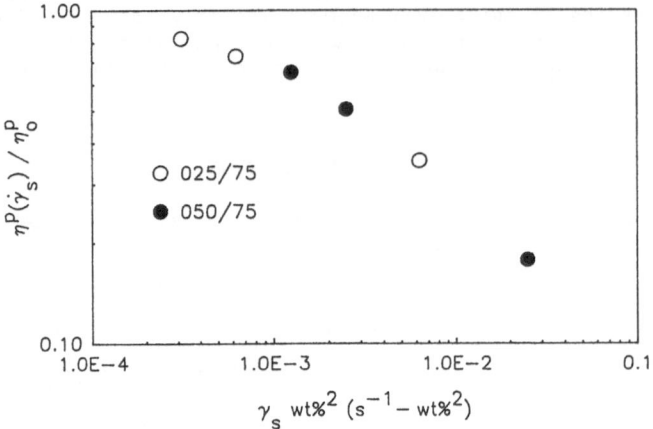

Figure 2. Scaling of the viscosity shear rate plot. 025/75 and 050/75 represent solutions with designated weight percent Xanthan Gum/fructose weight percent.

The stress jump on cessation was measured by extrapolation back to zero time over a small time range near the cessation time. At higher shear rates the stress jump tends to become concentration independent as shown in Figure 3. The stress jump was independent of solvent viscosity, however, the zero shear viscosity was not proportional to the solvent viscosity as should be true for dilute solutions. A chemical interaction may therefore exist between the fructose and polymer.

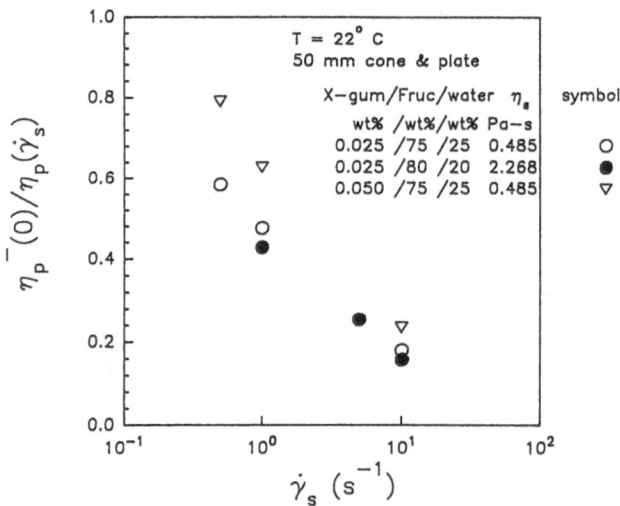

Figure 3. Stress jumps on cessation of shear flow for various Xanthan Gum solutions divided by the viscosity prior to cessation versus shear rate.

ACKNOWLEDGMENT: This work was supported by the Australian Research Council.

VISCOELASTICITY OF POLYMER GELATION

TOM McLEISH
Department of Physics, University of Sheffield
Sheffield S3 7RH

ABSTRACT

We contrast the theoretical relaxation behaviour of cross-linked polymer gels in unentangled and well-entangled cases. The former gives the observed power-law decay, but the latter case an "anomalous" relaxation modulus, based on a tube model calculation for the topological interactions. A transition to percolation statistics is predicted at long length scales. At the gel point itself the microscopic processes of reptation and fluctuation contribute equally at all timescales. Calculations beyond the gel point define new rheological criteria for the detection of gelation when the system is entangled.

INTRODUCTION

The dynamical properties of polymer gelation have enjoyed considerable recent interest both experimentally and theoretically [1-6]. Near the gel point, at which an infinite cluster first appears, static properties of the system take on a scaling form. The mass M and the size R of clusters are related via a "fractal dimension" d_f, $M \sim R^{d_f}$, and the molecular weight distribution as specified by the number density of clusters of mass M, $n(M) \sim M^{-\tau}$, is a power-law [6]. Dynamical quantities also scale: the zero shear rate viscosity η_0 and recoverable compliance J_e^0 diverge at the gel point and the zero shear rate modulus G_0 appears beyond it:

$$\eta_0 \sim (p_c - p)^{-s} \text{ for } p < p_c \; ; \quad J_e^0 \sim |p_c - p|^{-t} \; ; \quad G_0 \sim (p - p_c)^t \text{ for } p > p_c \tag{1}$$

Here p is the number density of bonds in the system, or an extent of reaction, and p_c is the critical density at percolation. The frequency-dependent modulus also exhibits scaling behaviour at the gel point [3-5]:

$$G^*(\omega) \sim (i\omega)^u \tag{2}$$

but there is no direct way of determining the dynamic exponents s, t and u from the static ones without further assumptions about the nature of the dynamic interaction [2,5].

One of the most important quantities affecting the dynamics of gelation clusters is their degree of overlap. This determines whether the hydrodynamic interaction between monomers is screened or not; ie whether the dynamical behaviour is Rouse-like (local effective friction) or Zimm-like (non-local) [7], as in solutions of linear polymers. Strong overlap will additionally mean that entanglements may dominate, though these become important at rather higher concentrations than the onset of hydrodynamic screening. This is seen in melts of linear chains where there is a critical molecular weight (which is usually several thousand) above which dynamical behaviour diverges from the phantom-chain predictions. For gelation clusters we therefore expect entanglements to be important only when a cluster is strongly overlapped with others as large or larger than itself [8].

GELATION OF LOW MOLECULAR WEIGHT POLYMER

This overlap is small in 3-dimensional "percolation clusters" which result from cross-linking low-molecular-weight polymers. This was used by Rubinstein et al. [5] to calculate the dynamical scaling of polymeric percolation clusters on the assumption that marginal overlap was effective in hydrodynamic screening but did not cause the clusters to entangle strongly, results which are confirmed by experiment [3, 5]. The key idea is that a cluster or sub-cluster of mass M has a terminal Rouse-like relaxation time T which is the time taken for the cluster to diffuse its own radius:

$$T(M) \sim R(M)^2/M \sim M^{2/d_f - 1} \tag{3}$$

Near the gel point we find that the dynamic modulus satisfies (2) with an exponent u which depends on the times T(M) and the concentration of clusters of mass M via the distribution exponent τ:

$$u = \frac{d_f(\tau - 1)}{d_f + 2} \tag{4}$$

The percolation values for the static exponents give a theoretical value for u of 0.67 which compares favourably with experimental values of 0.69 ± 0.02.

GELATION OF HIGH MOLECULAR WEIGHT POLYMER

A contrasting case is found in the cross-linking, or vulcanisation, of dense polymer chains when excluded volume is screened throughout the distribution except for the very largest clusters [9]. Now the classical theory of gelation [10] is valid and gives strong overlap of clusters. We must therefore account for entanglements. The viscoelasticity of entangled linear polymers is now understood using the tube model of topological interactions [7] and its extension to branched polymers has been an active area of interest recently [11]. The main results for entangled tree-like branched polymers (containing no closed loops) may be summarised as follows:

1)As for linear polymers, stress comes from the anisotropy of the polymer segments.

2)Stress relaxes by arm retraction via fluctuation.

3)Relaxation is hierarchical.The time taken for a branch-point to disentangle the segment connecting it to its cluster becomes the time for a single step in the diffusion of the next branch-point in "seniority" (defined precisely in ref.[12]).

4)The large separation of timescales on which different parts of a molecule relax means that all relaxed segments act as effective solvent for unrelaxed parts.

These results have recently been applied to well-entangled gelation clusters at the gel point [12]. In this case the relaxation time T of a segment is not controlled by the mass of its cluster, but only depends on its "seniority" m, or chemical distance to the cluster edge:

$$T(m) = \exp\left\{ \ln(T_\infty)\left(1 - \frac{1}{m}\right)\right\} \tag{5}$$

where T_∞ is a time depending exponentially on the ratio of the molecular weight of segments between cross-links M_x and the entanglement molecular weight M_e. This implies a relaxation modulus which has a logarithmic form before a transition at T_∞ to unentangled behaviour at longer times:

$$G(t) \sim \left[\ln\left(\frac{T_\infty}{t}\right)\right]^4 \text{ for } t < T_\infty; \quad G(t) \sim t^{-u} \text{ for } t > T_\infty \tag{6}$$

u=0.67 as before. Note that this is *not* a simple power-law. The calculation is complicated by some clusters which behave like linear polymers at long times and lose stress by reptation. A remarkable scaling of reptation and fluctuation exists at the gel point itself [12].

The calculation may be extended to cross-link densities close to but beyond the gel point. In this case the relaxation spectrum exhibits three regimes: 1) the logarithmic decay as at gel point at short times; 2) a power-law decay with a new exponent which depends on on the funtionality of the cross-links z, and M_x:

$$u_2 = \frac{32}{15}\left(\frac{M_e}{M_x}\right)\left(\frac{z-2}{z-1}\right) \tag{7}$$

at intermediate times; 3) a plateau corresponding to the finite gel fraction at long times. The intermediate power-law regime may dominate the spectrum not too far from the gel point and give a misleading indication of its rheological determination.

REFERENCES

1. Stauffer, D., Coniglio, A. and Adam, M., *Adv. Polym. Sci.* **44**, 103.
2. Cates, M.E., *J. Phys. (Paris)* , **46**, 1059.
3. Durand, D., Delsanti, M., Adam, M.and Luck, J.M., *Europhys Lett.*, **3**, 297.
4. Winter, H.H. and Chambon, F., *J. Rheol.*, **30**, 367 (1986); *J. Rheol.*, **31**, 683.
5. Rubinstein, M., Colby, R. H. and Gillmor, J.R. in *Space-Time Organisation in Macromolecular Fluids* eds. F. Tanaka, T. Ohta and M.Doi, Springer-Verlag, Berlin (1989); Polymer Preprints 30(1), 81.
6. Stauffer, D., *Introduction to Percolation Theory*, Taylor and Francis, London (1985).
7. Doi M. and Edwards S.F.E., *The Theory of Polymer Dynamics*, Oxford University Press, Oxford.
8. Cates M.E., *J. Physique Lett.*, **46**, L837.
9. de Gennes P.G., *J. Physique Lett.*, **38**, L355.
10. Stockmayer, W.H., J. Chem. Phys., **11**, 45; Flory P.J., *J. Am. Chem. Soc.*, **63**, 3083, 3091, 3096.
11. McLeish T.C.B., *Macromolecules*, **21**, 1062; McLeish T.C.B., *Europhys. Lett.*, **6**, 511; Ball R.C. and McLeish T.C.B., *Macromolecules*, **22**, 1911.
12. Rubinstein, M., Zurek, S., McLeish, T.C.B. and Ball, R.C., *J. Phys. (Paris)*, **15**, in press.

OSCILLATORY SHEAR FLOW OF INTERNAL VISCOSITY DUMBBELLS:
A NEW ANALYSIS EMPLOYING AN EXPANSION IN STRAIN

Tina P. McMahon[†] and Charles W. Manke[‡]
[†]Chemical Engineering and [‡]Materials Science and Engineering Departments
Wayne State University, Detroit, MI 48202, USA

ABSTRACT

A new formulation is presented for the oscillatory shear flow complex viscosity components $\eta'(\omega)$ and $\eta''(\omega)$ predicted by the dilute-solution internal viscosity model. For IV dumbbells (two beads connected by one submolecule), an expansion of configurational quantities in orders of strain is employed to avoid the pre-averaged angular velocity assumption that has been used in most previous IV model predictions. The new predictions for $\eta'(\omega)$ and $\eta''(\omega)$ retain most of the qualitative features of previous IV model formulations, including a non-zero limiting value for $\eta'(\omega)$ as frequency ω becomes infinite. Quantitative features of the complex viscosity components, however, are quite different from those obtained previously. Unlike previous formulations, the model presented here correctly predicts that $\eta'(\omega)$ and $\eta''(\omega)$ for the infinite-IV dumbbell approach the corresponding predictions of the rigid rod model.

INTRODUCTION

Accurate description of the complex viscosity of polymer solutions [1,2] has been one of the most successful features of internal viscosity models, which modify dilute solution bead-spring kinetic theories with an additional frictional resistance proportional to submolecule deformational velocity (the velocity of the end-to-end separation of the spring(s)). However almost all IV models have approximated the submolecule deformational velocity by subtracting a rotational velocity, calculated from the average angular velocity of the polymer coil, from the total velocity difference for the two beads of the submolecule. This approximation facilitates stress calculations by linearizing the IV force, which is highly nonlinear in the original IV model proposed by Kuhn and Kuhn [3]. Recently, however, doubt has been cast on the validity of this linearization approximation, especially in the high-IV limit, by discrepancies that have occurred, in certain cases [4,5], between stresses evaluated with the linearization approximation and those evaluated with the Kuhn and Kuhn nonlinear IV force expression. Because IV theory has been used most extensively, and most successfully, to describe small amplitude oscillatory shear flow of polymer solutions, investigation of the validity of the linearized rotational velocity (LRV) approximation for this case is clearly important. The dumbbell model examined here is an extremely useful theoretical tool for this purpose, despite the limitations imposed by its single relaxation time.

THEORY

We consider only the dumbbell IV model, with two frictional beads connected by a single, linear entropy spring. The internal viscosity force $\mathbf{F}^{IV}$ in the connector, which acts on both beads, is given by:

$$\mathbf{F}^{IV} = \phi \; [(\dot{\mathbf{Q}} \cdot \mathbf{Q})/Q^2] \; \mathbf{Q} \; , \tag{1}$$

where ϕ is the scalar IV coefficient, $\mathbf{Q}$ is the vector connecting the beads, and $\dot{\mathbf{Q}}$ is the velocity difference of the beads. By using Eq. 1, the LRV approximation is avoided.

Stresses are evaluated by Kramers' equation [6], using configurational moments of the bead configurational distribution function ψ and the bead force balance equations containing the IV force given by Eq. 1. Expressions for the configurational moments are developed from the diffusion equation for ψ. In general, evaluation of stress requires closed form solutions for the configurational moments, and the nonlinear IV force given by Eq. 1 becomes a formidable mathematical obstacle.

For small-amplitude oscillatory shear flow, however, only the linear stress response to strain amplitude is required for evaluation of $\eta'(\omega)$ and $\eta''(\omega)$. We take advantage of this special case by expressing all configurational moments appearing in the equation for shear stress as perturbation expansions about their equilibrium (rest state) values employing strain as the perturbation parameter. This version of the shear stress equation can then be truncated after the first order strain terms. Since the bead velocity difference $\dot{\mathbf{Q}}$ is proportional to strain amplitude, the equilibrium values for the configurational moments arising from $\mathbf{F}^{IV}$ are first order in strain. Higher-order IV terms are therefore not retained in the truncated shear stress equation. The equilibrium moments involving $\mathbf{F}^{IV}$ can be calculated readily and exactly because ψ is known for the rest state.

The nonlinear problem posed by Eq. 1 is thereby circumvented, and an analytical expression for the linear viscoelastic shear stress is obtained, yielding:

$$\eta'(\omega) = nkT\lambda \; [1 + 0.8\omega^2\lambda^2\phi/f/(1+2\phi/f)]/(1+\omega^2\lambda^2) \; , \tag{2}$$

and

$$\eta''(\omega) = nkT\lambda^2\omega \; [1 - 0.8\omega^2\lambda^2\phi/f/(1+2\phi/f)]/(1+\omega^2\lambda^2) \; , \tag{3}$$

where n is the number concentration of dumbbells, $\lambda \equiv fb^2/12kT$ is the dumbbell relaxation time, f is the bead friction factor, and b is the equilibrium RMS dumbbell length.

RESULTS

The dynamic viscosity predicted by Eq. 2 is shown in Fig. 1. The prediction for $\phi = 0$ exactly matches the prediction of the Hookean dumbbell model [6]. Similar behavior is predicted for finite ϕ, but then $\eta'(\omega)$ attains finite limiting values at infinite ω. For infinite IV, the predictions of Eq. 2 exactly match the $\eta'(\omega)$ predictions of the rigid rod model [7], and the infinite IV result $\eta'(\infty)-\eta_s = 0.4$, where η_s is solvent viscosity, is seen to be the largest possible value. Predictions obtained previously [2] with the LRV approximation at high ϕ are strikingly different. The LRV predictions successfully match the Hookean dumbbell when $\phi = 0$; but the response at $\phi/f = 1$ exhibits ω-thinning behavior at lower ω than Eq. 2, and the limiting $\eta'(\infty)-\eta_s$ value is much higher. At infinite IV the LRV model predicts $\eta' = 1$ for all ω.

The predictions of Eq. 3 for $\eta''(\omega)$ are shown in Fig. 2. These predictions also match the Hookean dumbbell for $\phi = 0$ and the rigid rod model when $\phi \to \infty$. The response for $\phi/f = 1$ is similar in shape, and falls between these two limiting cases. The LRV predictions, from [2], are quite different than Eq. 3 for the $\phi/f = 1$ case, and $\eta'' = 0$ is predicted for all ω when $\phi \to \infty$. However, the LRV predictions are identical to the Hookean dumbbell, and to Eq. 3, when $\phi = 0$.

Manke and Williams [4] have demonstrated generally that rheological responses of the IV dumbbell model in the limit $\phi \to \infty$ must be identical to those of the rigid rod model. Thus the success of Eq. 2 and Eq. 3 in matching rigid rod predictions for $\eta'(\omega)$ and $\eta''(\omega)$ is a confirmation of the validity of the IV model formulation presented here. Conversely, the failure of the LRV predictions to match the rigid rod when $\phi \to \infty$ would seem to indicate a fundamental flaw in the LRV approximation at high IV, at least for the case of small amplitude oscillatory shear flow. The evidence of Figs. 1 and 2 also suggests that, for the dumbbell, the LRV approximation may be in error even at more moderate IV values near $\phi/f = 1$.

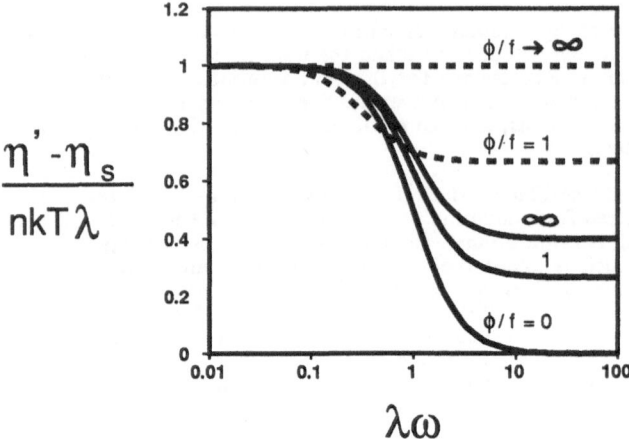

Figure 1. Dynamic viscosity predictions of Eq. 2, ▬▬, and the LRV approximation, ▪ ▪ ▪ .
Both formulations give the same prediction for $\phi = 0$.

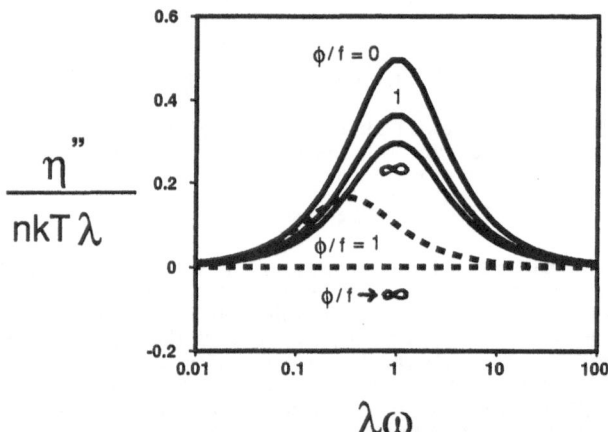

Figure 2. $\eta''(\omega)$ predictions of Eq. 3, ▬▬▬ , and the LRV approximation, ▪ ▪ ▪ ▪ .
Both formulations give the same prediction for $\phi = 0$.

REFERENCES

1. Massa, D.J., Schrag, J.L., and Ferry, J.D., Macromolecules, 4, 210 (1971).
2. Williams, M.C., AIChE Journal, 21, 1 (1975).
3. Kuhn, W. and Kuhn H., Helv. Chim. Acta , 29, 609, 830 (1946).
4. Manke, C.W. and Williams, M.C., J. Rheol., 30, 19 (1986).
5. Manke, C.W. and Williams, M.C. J Rheol., 31, 495 (1987), Errata, ibid, 32, 401 (1988).
6. Bird, R.B., Curtiss, C.F., Armstrong, R.C., and Hassager, O., Dynamics of Polymeric Liquids, Vol. II: Kinetic Theory, 2nd Ed., John Wiley and Sons, N.Y. (1987), pp 69, 74.
7. Bird, R.B., Warner, H.R., and Evans, D.C., Adv. Polymer Sci., 8, 1 (1971).

TRANSIENT STRESS RESPONSES PREDICTED BY THE
INTERNAL VISCOSITY MODEL IN ELONGATIONAL FLOW

Charles W. Manke
Materials Science and Engineering Department
Wayne State University, Detroit, MI 48202, USA
and
Michael C. Williams[*]
Chemical Engineering Department
University of California, Berkeley, CA 94720,USA
[*]Now at Ch.E. Dept., U. of Alberta, Edmonton, AB, Canada

ABSTRACT

Predictions are made for the elongational-flow viscosity growth function $\eta_e^+(t)$ of the dilute-solution internal viscosity (IV) model developed earlier by Bazua and Williams. For calculating η_e^+, a novel treatment of the initial rotation of chain submolecules is required; such rotation occurs in response to the macroscopic step change of elongational strain rate at $t = 0$. The major role of IV is to cause the following changes relative to the Rouse Model: (1) abrupt jump at $t = 0$ for η_e^+; (2) general time-retardance of response.

INTERNAL VISCOSITY MODEL

The internal viscosity (IV) concept in polymer chain dynamics was introduced years ago [1] to modify the bead-spring theories, correcting -- at least approximately -- their deficiency in presuming that all viscous effects can be accommodated merely through describing motions of the beads (mythical discrete frictional points joining the springs). A full mathematical treatment of our IV model formulation will appear elsewhere [2]; here we provide only a brief description of a few key features. We use only the Bazua-Williams (BW) IV model [3], wherein the net IV force imposed on a bead by two neighboring submolecules ("springs") is given by the product of a scalar IV coefficient ϕ and the difference of the deformational velocity components (i.e. end-to-end separation velocities) of the submolecules.

Customarily, high-N IV theories (N = number of submolecules) have assumed that deformational velocity can be evaluated as the difference between the full velocity and its rotational component, which is obtained from Ω, the average angular velocity of the coil. In pure elongational flow, however, $\Omega = 0$ when the macroscopic flow is fully developed. This would lead, unrealistically, to vanishing rotational velocity for all beads in the absence of a more detailed treatment of segmental rotations.

Evaluation of stress, employing Kramers' equation [2,4] together with the bead force balance equations, requires configurational moments of ψ, the bead configurational distribution function. In the present approach, the diffusion equation for ψ retains IV-dependent rotational velocity terms that would vanish if macroscopic dynamics ($\Omega = 0$) alone were employed. At $t = 0$, the "rotational moments" (configurational moments arising from the non-vanishing rotational velocity terms in the diffusion equation) can be calculated exactly because ψ is known for the rest state. For $t > 0$, we approximate these rotational moments by assuming that the known values at t = 0 decay to zero with the same time dependence as the evolution-to-steady-state of the cross-flow configurational moments (i.e. <xx> and <yy>) that supply the unbalanced torques for segmental rotation. This physical approximation gives analytical expressions for configurational moments and stresses, while preserving contributions from non-zero rotational moments that would otherwise be lost. We will demonstrate elsewhere [2] that substantial support for this approach is provided by comparison of the infinite-IV dumbbell with the rigid-rod model, for which rigorous solutions are known.

RESULTS

The stress growth function $\eta_e^+(t)$ is defined for a macroscopic deformation rate $\dot{\gamma}_e^o$ that is steady and uniform for $t \geq 0$ following equilibration of the fluid at rest and achievement of a zero-stress state for t < 0. Results, computed for the Rouse relaxation times $\{\tau_p\}$ and N = 100, are discussed with reference to a critical strain rate $\dot{\gamma}_e^c \equiv 1/2\tau_1$ which divides the long-term elongational responses into stable ($\dot{\gamma}_e^o < \dot{\gamma}_e^c$) and unstable ($\dot{\gamma}_e^o \geq \dot{\gamma}_e^c$) regimes for all finite values of ϕ, including the Rouse case of $\phi = 0$. For very short times, say $t/2\tau_1 < 1$, stability is not a factor and results are very similar for all $\dot{\gamma}_e^o$; thus, we present the short-time case in figure 1 only for $\dot{\gamma}_e^c$. The major effect of IV is seen to be the initial stress jump giving $\eta_e^+(0)$.

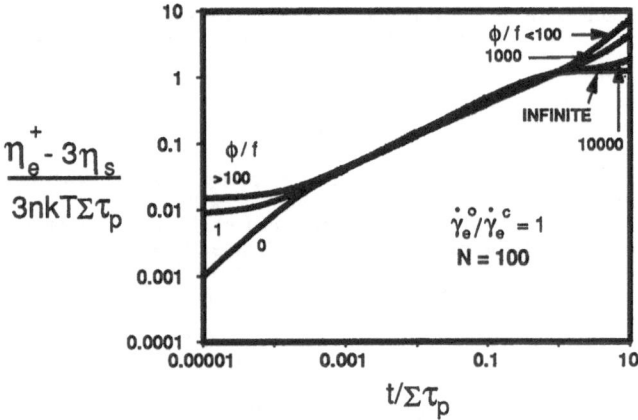

Figure 1. Short-time responses. (Internal viscosity ϕ is reduced by bead friction factor f.)

Long-time behavior is displayed in figure 2. Here, the stability problem is paramount, so results are displayed for a broad range of strain rates $0 \leq \dot{\gamma}_e^o/\dot{\gamma}_e^c \leq 100$. Behavior is qualitatively similar for the Rouse model and IV model, with the latter demonstrating primarily the expected time lag relative to the $\phi = 0$ responses, and this retardation increasing with ϕ. When $\dot{\gamma}_e^o < \dot{\gamma}_e^c$, a steady state is achieved for all ϕ and this level is entirely independent of ϕ. For $\dot{\gamma}_e^o > \dot{\gamma}_e^c$, $\eta_e^+(t)$ in figure 2 initially follows the curve for $\dot{\gamma}_e^c$ but breaks away (earlier at higher $\dot{\gamma}_e^o$) and grows exponentially with time. The physical limitation on validity of all supercritical $\eta_e^+(t)$ predictions at long times is the corresponding prediction (for all ϕ) that molecular extensions become impossibly large, whereupon the validity of the presumed Gaussian statistics of the coil is lost. This point is marked in figure 2 by dark circles on the η_e^+ curves.

Several predictions of $\eta_e^+(t)$ are displayed in figure 3 to highlight the contribution of rotational moments. First, $\eta_e^+(t)$ was evaluated for $\phi/f = 100$ and 1000 using the physical approximation for rotational moments described above. These two $\eta_e^+(t)$ functions proved to be virtually identical, represented by the solid line in figure 3, which begins at the true jump value computed from detailed submolecule mechanics and the known ψ at t = 0 [5]. Second, the calculations were repeated with rotational moments set equal to zero, as suggested by the purely macroscopic dynamics $\Omega = 0$, leading to the two broken-line curves in figure 3. Here, predicted jump values of the reduced η_e^+ function exceed the true ones by more than an order of magnitude, and the $\eta_e^+(t)$ curves exhibit a ϕ-dependence that seems unrealistically large. Figure 3 demonstrates clearly that the contribution of rotational moments cannot be ignored, especially during the short-time elongational dynamics.

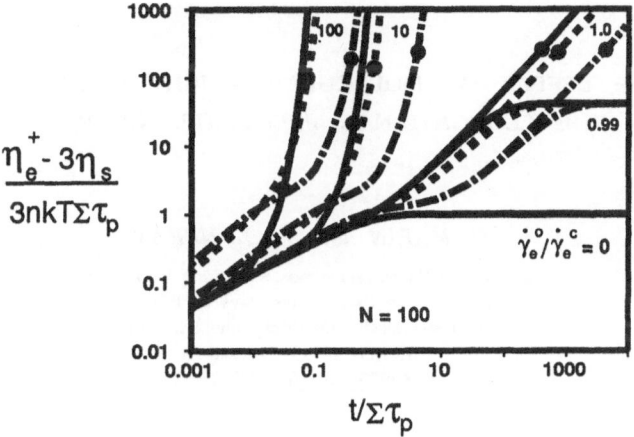

Figure 2. Long time responses: ———— , $\phi/f = 0$; ▪▪▪▪▪ , $\phi/f = 10^3$; ▪—▪—▪ , $\phi/f = 10^4$.

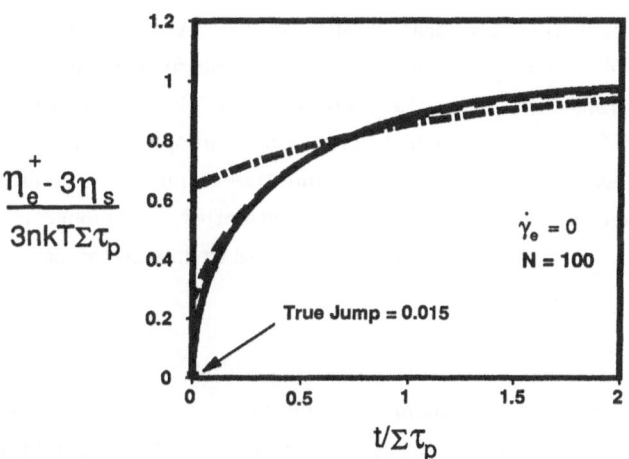

Figure 3. Predictions in the linear limit with and without rotational moments in the diffusion equation. With: ————($\phi/f = 100, 1000$). Without: ▪ ▪ ▪ ▪ ($\phi/f = 100$) and —▪— ($\phi/f = 1000$.)

REFERENCES

1. Kuhn, W. and Kuhn H., Helv. Chim. Acta , 29, 609, 830 (1946).
2. Manke, C.W. and Williams, M.C., Rheol. Acta, in press.
3. Bazua, E.R. and Williams, M.C., J Polym Sci (Phys), 12, 825 (1974).
4. Bird, R.B., Curtiss, C.F., Armstrong, R.C., and Hassager, O., Dynamics of Polymeric Liquids, Vol. II: Kinetic Theory, 2nd Ed., John Wiley and Sons, N.Y. (1987), pp 155-156.
5. Manke, C.W. and Williams, M.C. J Rheol. 31, 495 (1987), Errata, ibid, 32, 401 (1988).

THE EFFECT OF MOLECULAR WEIGHT DISTRIBUTION ON THE ELONGATIONAL PROPERTIES OF POLYMERS

G. MARIN and Y. LANFRAY

Laboratoire de Physique des Matériaux Industriels
Université de Pau et des Pays de l'Adour
Centre Universitaire de Recherche Scientifique
Avenue de l'Université
64000 PAU (FRANCE)

ABSTRACT

The transient properties of a large number of linear polymers in uniaxial elongational flows have been studied in a wide range of molecular weights and polydispersity indexes. The transient stress-strain curve at a given elongational rate may be related to the molecular weight distribution through a Wagner-like constitutive equation using as a kernel the memory function of linear viscoelasticity. The model allows to derive within experimental uncertainties the elongational properties of linear polymers from their molecular parameters at low and moderate rates of strain.

INTRODUCTION

The molecular models for linear viscoelasticity allow to explain and estimate quantitatively the viscoelastic properties of long chains as a function of molecular weight (model polymers) and Molecular Weight Distribution (commercial polymers). So it is possible to "predict" the viscoelastic effects of modifications in the molecular weight distribution. The picture is not so clear in the field on non linear viscoelasticity, where the continuum mechanics approach is maybe giving the best tools to relate quantitatively the non linear viscoelasticity to the molecular parameters. An important practical case, as far as polymer processing is concerned, is the elongational flow behaviour. We will present in this paper a model based on an integral constitutive equation that relates the molecular weight distribution to the elongational behaviour of commercial polymers.

EXPERIMENTAL

We have been using a series of monodisperse samples (polystyrene, polybutadiene), and samples having various polydispersity indexes and average molecular weights, along families (polystyrene, polyethylene, polypropylene). The linear viscoelastic properties in shear of the samples have been studied using an Instron 3250 and a Rheometrics RDA 700 rotary rheometers, in the frequency range 10^{-2} - $10^{2} s^{-1}$.

The uniaxial elongational flow measurements have been performed using a Metravib V.G.E. elongational rheometer at given strain rates ranging from 10^{-4} to $2 s^{-1}$ depending on the stress levels reached by the samples.

Samples have been vacuum-molded without residual stresses as circular discs (2mm diameter ; shearing) and parallelepipedic probes (90 x 25 x 5mm ; elongational).

A detailed description of the experimental procedures and data analysis methods may be found in ref. [1] and [2].

LINEAR VISCOELASTICITY

The memory function of linear viscoelasticity $\overset{o}{m}(t)$ is the kernel of the integral constitutive equation (see below) we are using for the description of the non linear viscoelastic behaviour.

$$\overset{o}{m}(t) = -\frac{dG(t)}{dt} = \int_{-\infty}^{+\infty} \frac{H(\tau)}{\tau} e^{-t/\tau} d\ell n \tau \qquad (1)$$

where $G(t)$ is the relaxation function and $H(\tau)$ the "Relaxation Spectrum" [3] that may be derived from the dynamic shear modulus through the usual equations of linear viscoelasticity [3,4]. These functions have been related to the molecular weight distribution through : (i) molecular models [5] or (ii) an analytic expression of the complex viscosity.

$$\eta^*(\omega) = \frac{\eta_o}{1 + (1\omega\tau_o)^{1-h}} \qquad (2)$$

η_o being the zero shear viscosity, τ_o the average relaxation time and h a dispersion parameter for the distribution of relaxation times. All three parameters may be related to the distribution of molecular weight.

Equation (2), although valid in a limited range of frequencies that covers the terminal region of relaxation, is a convenient tool to relate the viscoelastic properties with the structure of commercial polymers.

ELONGATIONAL FLOW

We have reported on Figure 1 the reduced transient elongational behaviour of a polypropylene sample at various strain rates (T = 200°C).

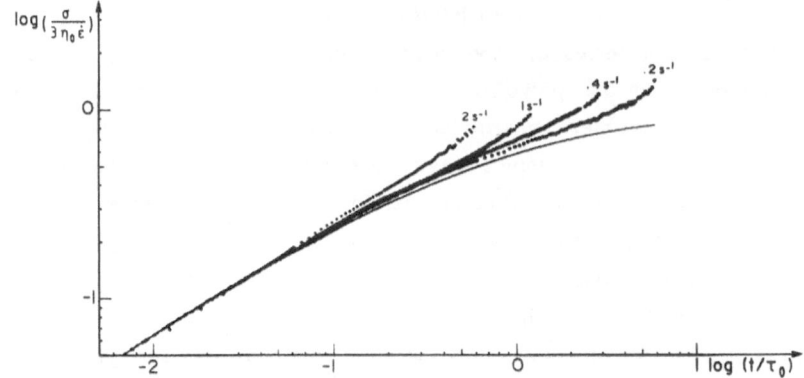

Figure 1 . Reduced stress-strain curve in elongation for a commercial polypropylene at various strain rates (- linear viscoelasticity).

The experimental behaviour may be fitted within experimental uncertainties by a modified Lodge-type constitutive equation :

$$\sigma_{11} - \sigma_{22} = \sigma(t) = \int_0^t \overset{\circ}{m}(t - t') \; h(t - t') \; (\exp 2 \dot{\varepsilon} t' - \exp - \dot{\varepsilon} t) \; dt' \\ + h(t) \; G(t) \; (\exp 2 \, \varepsilon t - \exp - \dot{\varepsilon} t) \tag{3}$$

The damping function $h(\varepsilon)$ has been numerically calculated from the experiments, and may be fitted by the equation :

$$h(\varepsilon) = \frac{a + 1}{a + \exp (1.4 \; \varepsilon)} \tag{4}$$

a is directly related to the molecular weight distribution, hence to the polydispersity index as a first approximation.

The model allows to get a very good estimate of the elongational behaviour of linear polymers simply from their molecular weight distribution (GPC curve).

REFERENCES

1 . Montfort, J.P., Marin, G. and Monge Ph, Macromolecules, 1984, 17, 1551.

2 . Lacaze, J.M., Marin, G. and Monge, Ph., Rheol. Acta, 1988, 540.

3 . Ferry, J.D., "Viscoelastic properties of polymers", 1980, Wiley.

4 . Marin, G., "Oscillatory rheometry", Ch. 10 in "Rheological Measurement", ed. A.A. Collyer and D.W. Clegg, Elsevier (London), 1988.

5 . Montfort, J.P., Marin, G. and Monge, Ph., Macromolecules, 1986, 19, 1979.

YIELD STRESS MEASUREMENTS

Christian D. Meier
Rhevisco AG, CH-8439 Wislikofen

ABSTRACT

Much has been written about the yield stress, specially in the field of blood rheology. The definition of the yield stress is very clear and says that this is the smallest load under which a substance will flow. Its measure is the yield stress value of the applied shear stress. The influence of the elasticity to the yield stress is automatically neglected by this definition with the assumption that this is also covered by the experimental conditions.

INTRODUCTION

The traditional representative for a material with a yield stress is the fluid known as a Bingham plastic. The shear stress/shear rate relationship is a straight line and the apparent viscosity becomes infinite at zero shear rate. With plastic substances we also should consider that they have at least two things in common: they have a yield stress value and they are in general empirical. But both offer generally a good fit for practical purposes in the range of shear stresses under consideration.

MEASURING METHODS

The results of practical rheological measurements depart always significantly from the ideal model. The plot of the flow curve is not linear, except over a very small range of strain rates and the yield stress corresponding to the yield value is not well defined.

The term "yield value" has been defined more in detail by Houwink (1). Corresponding to the measuring methods and also to the evaluation methods he distinguishes between different yield values. Most measuring methods used to determin the yield stress value are indirect. This means that the point of intersection of the flow curve with the shear stress axis has to be determined by extrapolating the linear portion of the curve; the intersection is called the extrapolated yield value. The shear stress value at which we can observe the beginning of linear flow is called the upper yield value.

By using a measuring method in which the shear stress is directely applied to the test material and is increased continuously, the yield stress will be measured directely by measuring the deformation of the test material. This direct measured yield value is the stress that produces the first non-zero shear rate value. The applications of shear stresses also make possible to determin the rheology of the yield stress region by using the creep test and the dynamic oscillation test methods.

CONCLUSIONS

A non-Newtonian fluid with a yield stress is not only a visco-elastic fluid but also a viscoelastic solid. Under these circumstances it is clear that it is quite difficult to describe the behaviour of the tested material by a simple rheological model assembled by fundamental elements. The most important elements are:
- The Hooke spring
- The Newtonian dashpot
- The St.Venant friction-element.

The St.Venant-element will flow only, if a certain limit stress is exceeded. Since no strain is recovered and the whole energy is converted into heat, the St.Venant-element has to be combined with spring- and dashpot-elements to describe the actual viscoelastic behaviour of a material with a yield value.

The behavior of real materials will never be accurately described by a mechanical model. But as soon as the rheological behaviors of the viscoelastic solid and the viscoelastic liquid are separately discussed, the description of the complex behavior of a material with plastic flow will be simplified. The simplest model with the characteristics of a viscoelastic solid is the Burgers model. Since the stress-strain diagram of the Burgers model depends on the strain rate, the time becomes a very significant parameter too and has to be controlled very carefully during yield stress measurements.

REFERENCES

1. Houwink, R., Elasticity, Plasticity, and Structure of Matter. Dover Publications, New York, 1958.

A RHEOLOGICAL STUDY OF THE STRUCTURE OF MICROEMULSIONS

J. Mellema, J.J. Beekwilder, C. Blom
Rheology Group, Faculty of Applied Physics,
University of Twente, Enschede, The Netherlands

ABSTRACT

Microemulsions are optically isotropic, stable mixtures of water, oil and surfactant(s) (mixture). Their internal structure can be characterized with length scales pertaining to the sizes of domains of the dispersed phase (e.g. oil) in the continuous phase (e.g. water). The dispersion occurs because of the presence of surfactant(s) at the oil-water interface.

The morphology and pertaining length scales are subject of research. As structure one can think of droplets, flexible rods and so-called bicontinuous structures.

Rheologically the microemulsions have been investigated with harmonic shear experiments to determine the complex viscosity η^*. Between 10^2 and $2 \cdot 10^5$ Hz all measurements of η^* show a transition of purely viscous to viscous and elastic behaviour. Each transition has been characterized with a relaxation time and relaxation strength. For the interpretation of these relaxation phenomena a few models are available. The measurements have been interpreted such that a consistent picture has been obtained for the various structures of the investigated microemulsions, in the present case a transition from spherical to rod-shaped droplets.

INTRODUCTION

Microemulsions are often defined as thermodynamically stable, transparent mixtures of oil, water (brine) and a surfactant (mixture). Here an example of a rheological study on microemulsions is presented. The investigated oil in water microemulsion consisted of n-hexane, deionized water and NNP7 (polyoxyethylene(7)-nonylphenol ether), a nonionic surfactant. A striking phenomenon observed was the remarkably increasing steady state viscosity and real and imaginary parts of the complex viscosity in a certain temperature range in the single phase region. The present work also shows the linear rheological phenomena in a frequency range between 10^2 and $2 \cdot 10^5$ Hz at a temperature such that the viscoelastic relaxation phenomena can be accurately analyzed as a function of the volume fraction of the combination of surfactant and oil. With help of the results of the analysis at this particular temprature a structure picture for the entire investigated temperature range can be given. The driving force behind the changing rheological behaviour as a function of temperature is known. The solubility of poly(ethyleneoxide) in water decreases rapidly when the temperature increases. Consequently the poly(ethyleneoxide) tails of the surfactant gradually effectively attract each other with increasing temperature. A few potential consequences can be anticipated. The droplets can attract each other, thus forming clusters, and at the same time the interfacial tension, the curvature moduli and the natural radius of the droplet surface increase. We assume that if these are spherical droplets dispersed at lower temperatures the clustering of spheres at higher temperatures does not reduce the energy sufficienlty to equilibrate the systems at the higher temperatures (only point contacts) but that instead shape change will do so.

The linear rheological measurements at the lowest temperature were consistent with the picture of dispersed spherical droplets. At higher temperature this is certainly not the case. The two structure pictures we will investigate here as candidates for an explanation of the rheological phenomena at higher temperature are: firstly a rigid rod model, such as originally described by Doi and Edwards and later refined by others: secondly a number of scaling laws from a so-called living polymer model, based on work by Safran et al. and Cates. Both models concern nonspherical, elongated domains.

Using the rigid rod model here presupposes that the rods suspension is dilutable (i.e. the rods do not change in size with volume fraction), interact only via infinitley steep repulsion at collisions or hydrodynamically at a distance and are rigid (at least for the forces present in our experiment).

In the living polymer model domains may join ends with each other or break up in two smaller domains, thus continuously changing their length. This model predicts a relation between the average domain length and volume fraction, governed by the elastic curvature energy present in the domain surface.

In both models the relaxation time, the viscosity and the infinite frequency limit of the complex viscosity are related to the volume fraction of the domains. The volume fraction dependencies can also be found experimentally and thus confrontation between theory and experiment is possible.

RESULTS

In figure 1 the complex viscosity as a function of angular frequency is shown for three volume fractions ϕ (top to bottom, ϕ = 0.20, 0.15 and 0.10) at temperature T = 19$°$C. Drawn lines are fits with which the longest relaxation time τ_1 and relaxation strength G_1 are determined.

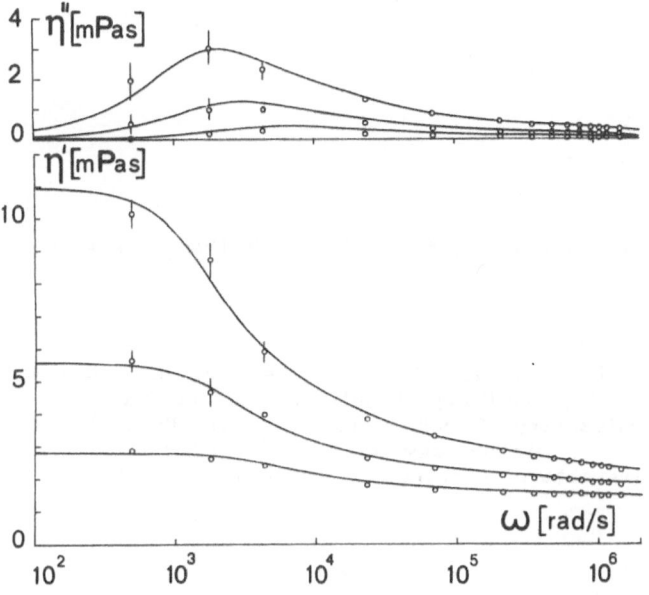

Figure 1.

In table 1 results are shown together with theoretical predictions. Obviously the rigid rods parameters show ϕ dependencies the closest to the experimental dependencies.

TABLE 1

	Theory		Experiment
	Rigid rods	Living polymers	
τ	$\propto (1+K\phi)^2$ K: constant	$\propto \phi^{1.4}$ (Cates) $\propto \phi^{1.8}$ (Safran)	$\propto \phi^{1.6\pm0.15}$
G	$\propto \Phi$	$\propto \phi^{2.3}$	$\propto \phi^{0.95\pm0.5}$

Apparently the rheological data are similar to those stemming from a rigid rod dispersion. Modelling the rods as cylinders (length L, diameter d) with spherical end caps a geometrical model, additional light scattering measurements and the rheological measurements led consistently to L $\approx$ 200nm and L/d $\approx$ 13.

Assuming that with increasing temperature T a transition occurs from spheres to rigid rods, the length L(T) could be deduced.

RHEOLOGY OF HUMAN SALIVA AND SALIVA SUBSTITUTES

J. Mellema, H.J. Holterman, E.J. 's-Gravenmade, H.A. Waterman, C. Blom
Rheology Group, Faculty of Applied Physics
University of Twente, Enschede, The Netherlands
*Laboratory for Materia Technica
University of Groningen, Groningen, The Netherlands

ABSTRACT

Saliva at rest separates in a bulk solution and a protein layer on top. The layer has been characterized ellipsometrically giving its thickness and density. Rheologically the layer and bulk were studied separately with linear and non-linear methods: harmonic shear experiments and flow curves. Thus a rheological framework resulted which was used for the study of existing saliva substitutes and the development of rheologically improved substitutes. The improvement of saliva substitutes was searched in mixtures of mucins and bovine serum albumin. These mixtures were studied with the same rheological means as the saliva samples. It turns out that mixtures exist that mimic saliva rheologically quite well.

INTRODUCTION

The rheology of saliva was the subject of several studies. However, the phenomenon of a protein layer adsorbed at the air-liquid interface necessitated reinterpretation of older results. Recently [1] a research was carried out on this surface layer, which is often clearly visible as a thin rigid film and the bulk below the layer. In these studies three types of human saliva were studied: submandibular, parotid and whole saliva. Firstly the samples were characterized with non rheological means. The bulk below the adsorbed layer is labeled by the acidity, density, index of refraction and electrical conductivity. An ellipsometric set up was built to investigate the development of the thickness and the index of refraction of the adsorbed layer in time. In addition the density of the layer resulted. Secondly rheological linear and non-linear behaviour has been investigated for the layer. The complex surface shear modulus at 66 Hz was used as representative of linear behaviour. An apparent yield stress and (subsequently) flow behaviour (shear viscosity of the surface as a function of the rate shear between 10^{-4} and $10^{-1}s^{-1}$) demonstrated non-linear behaviour. Thirdly the complex shear viscosity of the bulk below the adsorbed layer has been measured also at 66 Hz. In addition it has been measured between 0.03 and 1 Hz for whole and submandibular saliva to complete the rheological picture. We were not able to determine possible non-linear behaviour of the bulk underneath the layer. Using the results of these studies as a framework for the development of saliva substitutes that mimic saliva rheologically it is plausible to look for high molecular protein solutions as substitutes for saliva that show rapid protein adsorption at the air-solution interface and form layers with complex shear moduli in the same range as determined for real saliva.

Considering the composition of saliva mucins are good protein candidates to play this role. Here we report studies of solutions of porcine gastric mucin (PGM), bovine submandibular mucin (BSM), bovine serum albumin (BSA) and combinations of them. For comparison also commercially available saliva substitutes, often containing non biological polymers, have been studied with the same methods.

RESULTS

In fig. 1. surface layer shear viscosities are shown as a function of the rate of shear. Within the areas confined by the broken lines the curves of saliva fell. The path of the upper and lower broken boundary lines are typical for the $\tilde{\eta}_{so}(\dot{\gamma}_s)$ behaviour of saliva samples. The numbered lines

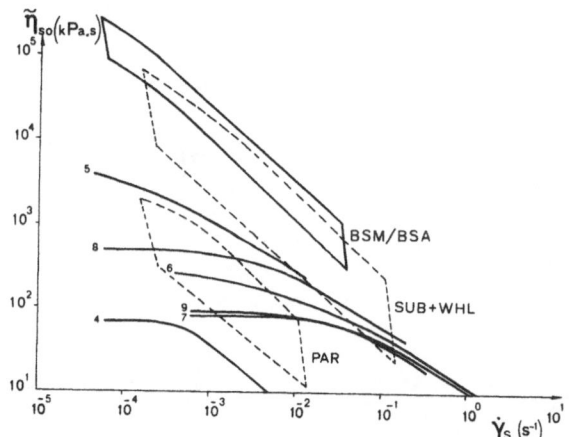

Figure 1.

stem from BSM (4), BSA (5, 6) and PGM/BSA (7, 8, 9) which are solutions of different compositions and pH. In the upper area confined by the solid lines the $\tilde{\eta}_{so}(\dot{\gamma}_s)$ curves of the BSM/BSA mixed solutions fell. The conclusion drawn from this figure is that BSM/BSA surface layer flow curves are similar to those of whole and submandibular saliva.

In fig 2. surface shear moduli (real parts primed and imaginary parts doubled primed) at 66 Hz are shown as a function of time. A', A" and B', B" are the measured upper and lower moduli values of saliva samples

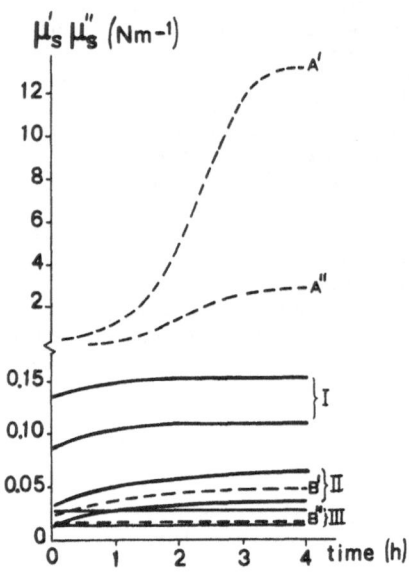

Figure 2.

respectively. In contrast to the $\tilde{\eta}_{so}(\dot{\gamma}_s)$ behaviour it is interesting that parotid and submandibular saliva samples have both as well high and low shear moduli within biological variation. The curves of μ_s' from BSM/BSA samples were found in region I. BSA and PGM/BSA samples had curves of μ_s' (time) in region II. Curves of μ_s'' (time) for all BSA containing samples fell region III. Clearly the BSM/BSA samples are linearly rheologically closer to low moduli saliva samples than to high moduli saliva samples.

In fig. 3. complex viscosities of the bulk below an adsorbed layer as a function of frequency is shown. In the hatched region the curves of human whole saliva fell. Lines marked O, S, L and G are commercial substitutes which had no measurable rheological surface

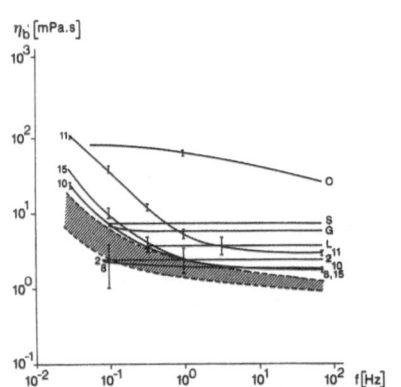

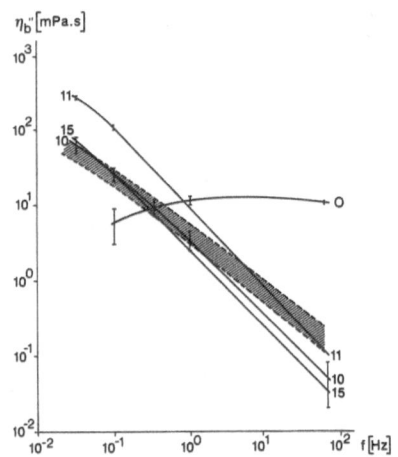

Figure 3.

properties and no elastic bulk properties. For the PGM (2) solution and a PGM/BSA (8) solution this was also the case. Only the bulk BSM/BSA (10, 11, 15) solutions behaved rheologically similarly as whole saliva.

Summerizing we conclude that BSM/BSA solutions are good candidates for substitutes that mimic saliva rheologically well.

REFERENCES

1. Holterman, H.J., On the rheology of human saliva and its substitutes thesis, University of Twente, 1989.

A SIMPLE MEASUREMENT OF ELONGATIONAL VISCOSITY

B. MENA
Instituto de Investigaciones en Materiales
National University of Mexico
Apdo. Post. 70-360, Coyoacan 04510, Mexico,D.F.

ABSTRACT

A very simple device is presented for measuring the elongational viscosity of dilute and semi-dilute polymer solutions. Basically it examines the flow between aligned orifices of different diameter separated by a variable distance. The flow field is generated by applying suction through the lower orifice. Once a steady state is achieved the average tensile stress is measured and the flow is photographed. The flow field is free of unwanted effects such as surface tension and gravity. Comparison with spinline rheometer and open-syphon measurements are in very good agreement

INTRODUCTION

In this article, we suggest a simple method as an attempt to measure "some kind of elongational property" for dilute and semi-dilute polymer solutions over a wide range of rate of stretch. In addition, the method eliminates surface tension and gravity corrections.

FIG I.

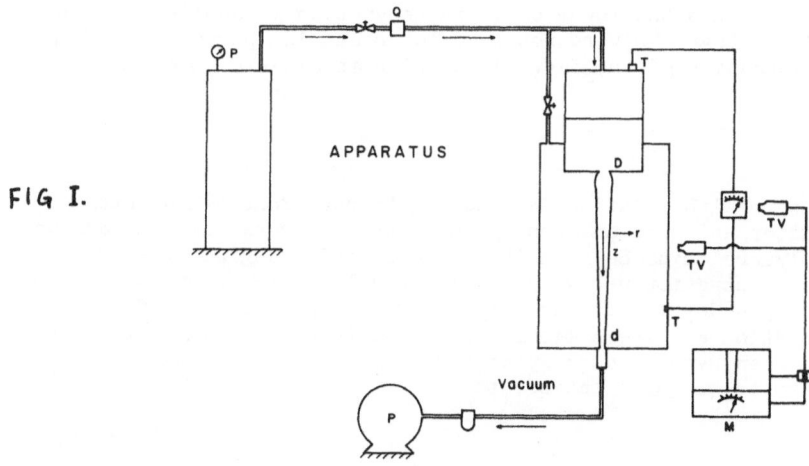

EXPERIMENTAL APPARATUS

The experimental arrangement is depicted schematically in Figure I. A
constant pressurized reservoir feeds the test fluid, through a mass flow-
meter Q into a cylindrical vessel of diameter D. The latter is partially
inmersed in a larger tank of square cross-section containing the same test
fluid but of a slightly different colour. The cylindrical vessel may vary
its position by sliding in or out of the tank. The test fluid exits from
the vessel, into the tank, through a small orifice of diameter d. The fluid
filament traverses the tank and exits through a smaller orifice of diameter
d located at the bottom. The pressure corresponding to the given initial
flow rate is registered by a miniature pressure transducer T located in the
cylindrical vessel. Suction is then applied at the bottom orifice through
a vacuum pump P. This creates a pseudo-extensional flow on the filament
which modifies the pressure reading of the pressure transducer. This
difference in pressure is approximately equal to a tensile stress upon the
filament.
Once the flow situation is stable and a steady is reached, the process s
filmed with two video cameras. One camera detects the change in filament
diameter and a second camera is focused on the reading of the pressure
scanner. Both cameras are linked through a splitters S to a video recorder
whose imagine appears in a monitor screen. A second pressure tranducer is
located in the bottom tank for reference purposes.
Orifices of various diamters may be used an since the separation between
them can be varied, a large variety of rates of elongation are easily
achieved.
The elongation rate $\dot{e}$ is calculated from the filament profile as $\dot{e} = \frac{dv}{dz}$
and the tensile   stress is measured directly as the difference between
the "stretched" pressure reading (applying vacumm) and the unstretched one
(without vacumm).
The obvious advantages of the measuring system described above over other
methods are the complete elimination of surface tension and gravity effects.
A disadvantage would be the presence of secondary flows in the tank of
square cross-section and a certain difficulty in achieving a steady state.
Nevertheless, these disadvantages may be easily overcome.

FLOW VISUALIZATION

Plate I shows the time sequence of the filament as it exists from the upper
orifice until it reaches the lower one, whese suction is applied. Although
these particular photographs were not used for measuring purposes, they
illustrate the formation of the flow field which is clearly an extensional
flow.

DATA COMPARISON

In order to compare with existing data on dilute solutions taken with a
spin-line rheometer and with an open-syphon system, several dilute aqueous
solutions of polyacrylamide Separan AP-250 were used, ranging from 0.1 %
to 2% polymer concentration per weight. The results are shown in Figure
II as an average pseudo-elongational viscosity n_e plotted as a function
of rate of elongation $\dot{e}$. The solid lines are average data taken from other
systems and the open symbols correspond to our experimental data. The
agreement appears to be quite satisfactory.

PRELIMINARY CONCLUSIONS

A relatively simple minded method for measuring an elongational property
which may be defined as a "transient elongational vicosity" has been illustrated.   This method, using a double orifice flow provides results which
are comparable to other methods without using any assumptions regarding
surface tension and gavity correction effects.

ACKNOWLEDGMENTS

The initial part of the experiments were performed during a short visit to
the Department of Mathematics of the University College of Wales,
Aberystwyth.   Financial and moral support by Prof. K. Walters   and the
skilfull technical help of Mr. R. Evans are greatfully acknowledged.
The sponsorship from Programa de Apoyo a Proyectos de Investigación y de
Innovación Docente (PAPIID-UNAM) is greatfully acknowledged.

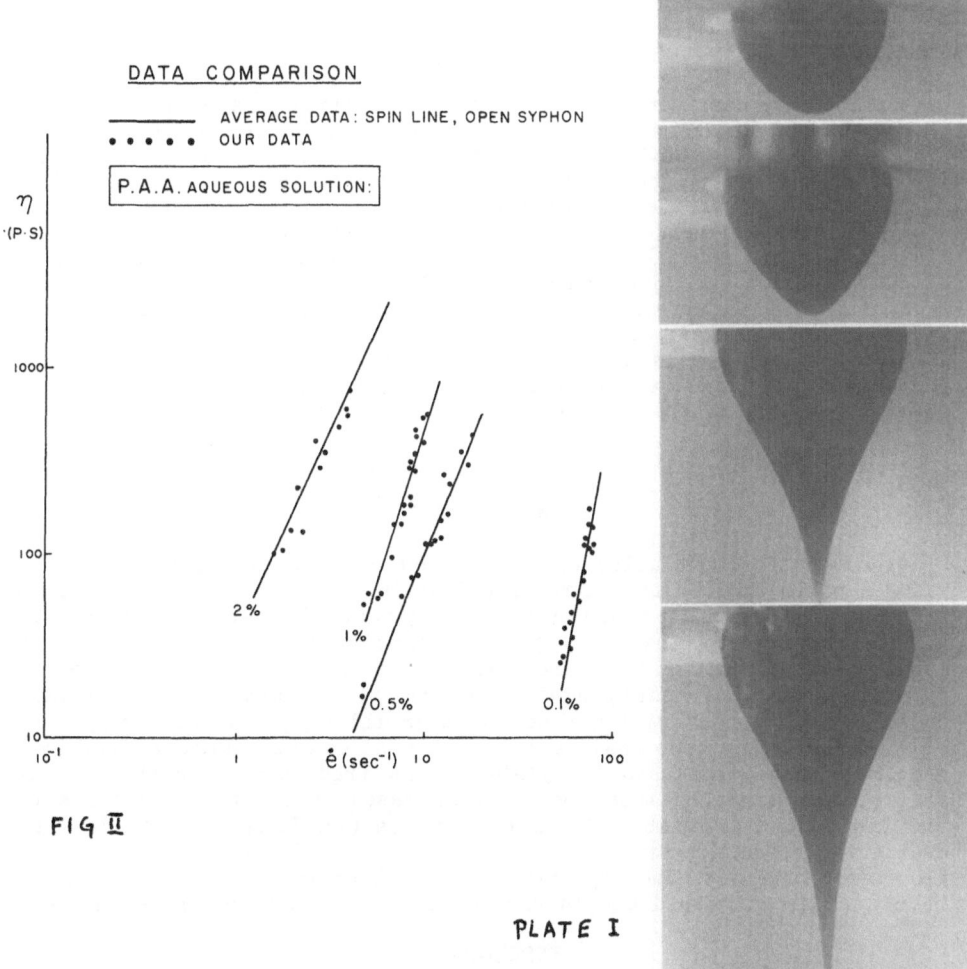

DATA COMPARISON

———— AVERAGE DATA: SPIN LINE, OPEN SYPHON
• • • • • OUR DATA

P.A.A. AQUEOUS SOLUTION:

FIG II

PLATE I

EQUILIBRIUM PROPERTIES OF REVERSIBLY FLOCCULATED DISPERSIONS

JAN MEWIS and ERIC VAN DER AERSCHOT*
Dept. Chemical Engineering, K.U.Leuven, 3030 Leuven,Belgium
*) present address: Du Pont de Nemours, 2800 Mechelen, Belgium

ABSTRACT

Dispersions are considered in which the floc structure changes nearly
reversibly upon shearing. The general rheological characteristics of
such systems are studied, using non-aqueous suspensions of fumed silica.
The water content of the particles is altered to investigate the effect
of floc strength. For stronger flocs the equilibrium viscosities are
shown to drop over nine orders of magnitude in a narrow stress region.
Yield stress and equilibrium storage modulus are measured as a function
of concentration. The dependence is best described by a power law relation, the power law being identical for modulus and yield stress. Its
value depends on floc strength. The results agree qualitatively with
some percolation models. However, yielding is often dominated by kinetic
phenomena, contrary to the assumptions of the models. The discrepancy
shows up clearly in the erronous predictions of the critical strain.

INTRODUCTION

Flow-induced deflocculation and subsequent reflocculation at rest provides a powerful tool to control rheology in colloidal suspensions. This
procedure is often used in industrial systems. The resulting variable
microstructure and the corresponding rheological behaviour are very
complex. At rest a solid-like response can be observed, which is caused
by a space-filling, particulate network. This breaks down gradually
under shear, thereby giving rise to shear thinning and thixotropy. The
equilibrium structure that develops at rest can be characterized by a
plateau storage modulus and a yield stress. Predictions for the concentration dependence of these properties, based on fractal properties of
the flocs, are available for strong flocs (1). Experiments are also
available for such materials (2) but this is not the case for reversibly
flocculated systems. The objective of this work is to test the applicability of the published models for the systems under consideration here.

EXPERIMENTAL

Nearly spherical particles of amorphous fumed silica (Aerosil A300 from

DEGUSSA) are used. The diameter of the elementary particles is approximately 7nm. As a dispersing medium methyllaurate (Janssen Chimica) is selected. It gives the desired structure, intermediate between a perfect dispersion and a strong gel, while avoiding the formation of two phases. Furthermore the chosen medium has a low volatility, which is essential to avoid changes in concentration during the long equilibrating periods.

The silica is dried for 1 to 2 weeks at 110°C until a constant weight is reached. The powder is then mixed with fluid and dispersed on a laboratory roller mill until the largest aggregates are smaller than 5 mm. In order to alter the floc strength the particles can be conditioned in an atmosphere with controlled humidity (81% RH, using $(NH_4)_2SO_4$) before being dispersed in the same medium .

On a Rheometrics 705F measurements are performed under controlled kinematics. For the stress controlled experiments a Rheometrics SR 8600 is available. Cone and plate geometry is used on both instruments, the temperature is kept at 20.0°C.

RESULTS AND DISCUSSION

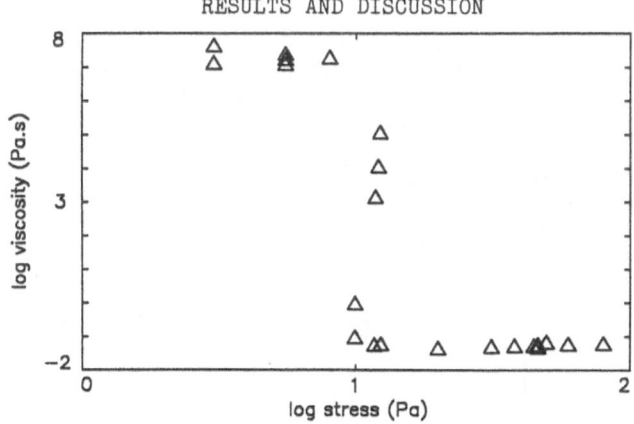

Fig. 1. Viscosity curve for a 2.5% SiO_2 suspension in methyllaurate

Combining data from the two instruments an extensive viscosity curve could be generated, covering the shear rate range from 10^{-7} to 10^3 s^{-1}. The viscosity-shear stress plot is shown in fig. 1. A spectacular drop of the viscosity, over 9 orders of magnitude, occurs in a narrow stress range. There is clearly a drastic change in flow mechanism at stresses around 10 Pa, even if there is some flow at lower stresses. In this latter region creep occurs through a gradual readjustment of the particulate network because of the external stress. At higher stress levels there is a macroscopic flow throughout the sample. In many practical cases the efffect of the creep can be ignored and an apparent yield stress can be introduced.

Next, the equilibrium moduli are measured as a function of concentration for dried and "wet" (i.e. conditioned as mentioned above) silica particles. After filling the instrument the sample is sheared. After stopping the flow, the particulate network is left to build up until the moduli reach equilibrium. This can take 4 to 5 days for each sample. At that stage the moduli are also independent of frequency.

The data indicate that modulus and yield stress have identical power law indices for dried particles. For wet particles (fig. 2) the power is higher and becomes slightly larger for the yield stress than

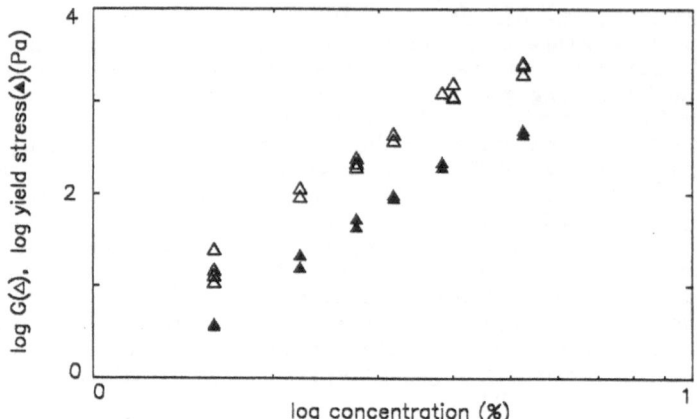

Figure 2. Moduli and yield stresses for "wet" silica particles.

for the modulus. Qualitatively these results agree surprisingly well with the model suggested by Patel and Russel (1). In fig. 3 the experimental values of the exponents for modulus and yield stress are compared with the model predictions. Quantitatively the data are somewhat outside the region of possible Poisson ratios. The data of Buscall (2) (point B in fig. 3) for strongly flocculated systems diverge further from the model predictions although they should actually be closer.

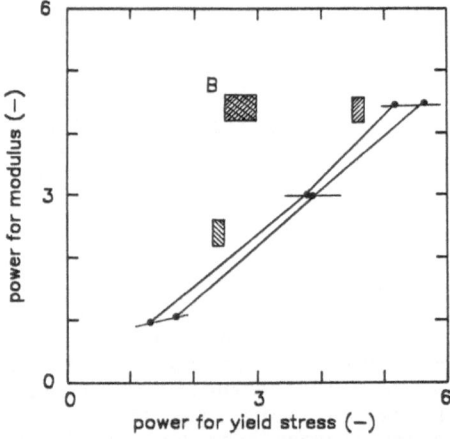

Figure 3. Comparison of experimental data with the Patel-Russel model

In the experiments yielding is clearly a kinetic phenomenon, whereas it is assumed to be static in the model. Furthermore the predicted value for yield strain are not very accurate.

REFERENCES

1. Patel, P.D. and Russel, W.B., A mean field theory for the rheology of phase separated or flocculated dispersions. Colloids & Surfaces, 1988, 31, 355-383.
2. Buscall, R. et al., The rheology of strongly-flocculated suspensions, J. Non-Newtonian Fluid Mech., 1987, 24, 183-202.

CONVERGENCE STUDIES IN THE NUMERICAL SIMULATION OF VISCOELASTIC MELTS USING THE K-BKZ MODEL

E. MITSOULIS and X.-L. LUO
Department of Chemical Engineering
University of Ottawa
Ottawa, Ontario, K1N 9B4, CANADA

ABSTRACT

A recently developed numerical scheme for flows with closed streamlines (such as entry flows with recirculating regions) for integral constitutive equations of the K-BKZ type is fully investigated with respect to its accuracy and convergence behaviour in entry flows of an LDPE melt. In particular, a parametric study is undertaken on the influence of mesh size on the results and the performance of the Picard iteration necessary for integral-type equations. An incremental loading technique is discussed for achieving solutions at high flow rates. Computational times are given for a series of runs with different finite element grids, and the minimum requirements for acceptable solutions are established.

INTRODUCTION

In recent years, the K-BKZ integral model has been used quite successfully in several attempts to quantitatively predict the behaviour of non-Newtonian fluids of commercial importance. These include prediction of difference in entry flow patterns between LDPE and HDPE melts [1].

Due to the complexity of the stress tensor for the K-BKZ integral model, the simulations have used a direct (successive) substitution scheme (Picard method). This scheme although robust and with a wider range of convergence, requires a great number of iterations to achieve relative errors of 10^{-4} or less, which a Newton-Raphson method can achieve within 3-5 iterations. Thus, the accuracy and convergence of viscoelastic simulations using the integral K-BKZ model have been questioned. In the present work, we make an attempt to answer these questions carrying out a thorough parametric investigation of accuracy and convergence using different density finite element meshes and different iterative strategies.

RESULTS AND DISCUSSION

We consider here the flow of an LDPE melt in a 4:1 abrupt circular contraction, which has been simulated previously [1]. The K-BKZ model is used to fit shear and elongational viscosity and normal stress data. The values of the constants can be found elsewhere [1]. A recently developed numerical scheme for flows with closed streamlines (recirculating regions) is used [2]. Figure 1 shows a partial view of 3 finite element meshes, named M1, M2 and M3, as well as a full view of mesh M2. Useful information regarding the number of elements, nodes and degrees of freedom (d.o.f.) for each mesh are given in Table 1. All meshes extend upstream to $-16R_O$ and downstream to $+30R_O$ to guarantee correct imposition of a fully-developed flow at inlet and outlet.

The solution features in terms of stresses, velocities and streamlines were found to be not very different for the 3 meshes used. This is evidenced in Figure 2, where the streamline patterns are shown for a selected run at $\dot{\gamma}_w = 104 \text{ s}^{-1}$, corresponding to a dimensionless stress ratio $S_R = N_1/2\tau_w$ of 2.5. Figure 3 shows the corresponding predictions for the stress components τ_{zz} and τ_{rz} along the $r = 1.0$ line. Quantitatively, Table 2 shows important global quantities obtained from the 3 meshes. Obviously, the agreement is very good. In terms of number of iterations N and the corresponding norm-of-the-error ε, again not much difference was found for the 3 meshes. Figure 4 plots the relative error as a function of the number of iterations. Note that the number of iterations was counted after the full elasticity level was reached, using the *elasticity increment scheme* with meshes M1 and M3, and the *flow rate increment scheme* with mesh M2. This figure clearly shows that the convergence speed and accuracy are largely mesh-independent. The relatively steep decrease in the value of ε , occurring once in each of the 3 curves in Figure 4, is due to underrelaxation with the introduction of a relaxation factor Ω from 0 to 0.5, after which the iterations were continued with $\Omega = 0.5$. Thus, a relaxation scheme used with the Picard method sped up the convergence and decreased the error.

All the comparisons made in the present work by using 3 different meshes to study the convergence of a K-BKZ model show good accuracy and very satisfactory mesh independence of the calculations. This is particularly encouraging, since it proves that a mesh with 200 elements such as M1 is adequate for obtaining reasonably accurate solutions without spending too much CPU time.

The present results show that it is now possible to study complex flows with or without recirculating regions using realistic integral-type constitutive equations. The range of simulations is not restricted to low shear rates but pretty well covers the practical range of operations, and the cost is moderate as well.

REFERENCES

1. Luo, X.-L. and Mitsoulis, E., A Numerical Study of the Effect of Elongational Viscosity on Vortex Growth in Contraction Flows of Polyethylene Melts. J. Rheol., 1990, **34**, 312-346.

2. Luo, X.-L. and Mitsoulis, E., An Efficient Algorithm for Strain History Tracking in Finite Element Computations of Non-Newtonian Fluids with Integral Constitutive Equations. Int. J. Num. Meth. Fluids, 1990, in press.

TABLE 1

Mesh characteristics of the 3 different finite element meshes used in the computations

MESH	No. of elements	No. of nodes	No of d.o.f.	Size of corner element (r*z)	CPU time* (secs/iter.)
M1	200	689	1623	0.18*0.22	320
M2	300	1007	2368	0.04*0.09	1000
M3	400	1317	3093	0.01*0.07	1650

* on a VAX-11/780 computer.

TABLE 2

Global quantities obtained from the 3 different finite element meshes

$$(\dot{\gamma}_w = 104 \text{ s}^{-1}, S_R = 2.5)$$

Mesh	No. of elements	Opening vortex angle, ϕ (deg.)	Relative vortex intensity, $-\Psi_{v,max}^*$ (%)	Total pressure drop, ΔP (MPa)
M1	200	49	13.7	5.61
M2	300	49	13.4	5.62
M3	400	48	13.2	5.64

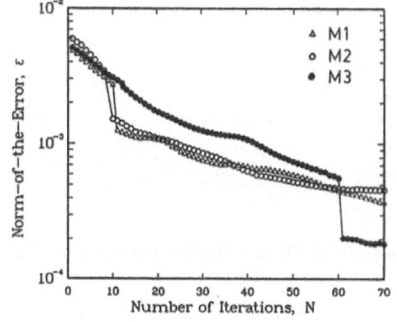

Figure 4. Convergence of solution with Picard method for the three meshes at $\dot{\gamma}_w = 104 \text{ s}^{-1}$, $(S_R = 2.5)$. The steep descent occurs when underrelaxation is introduced with a relaxation factor $\Omega = 0.5$.

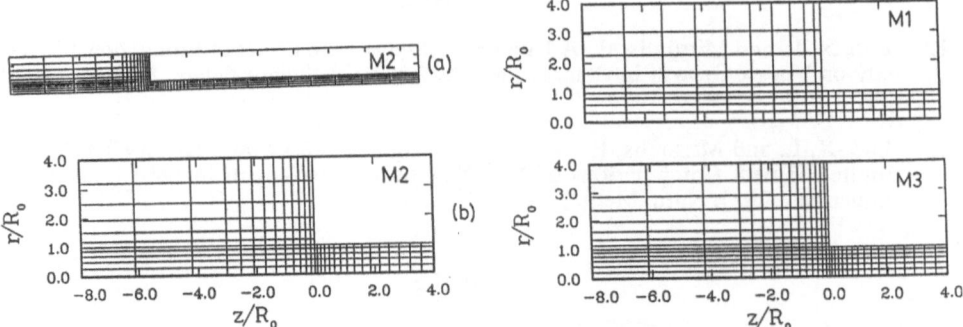

Figure 1. (a)Full finite element mesh containing 300 elements (M2) originally used in the 4:1 circular abrupt contraction computation; (b) Partial view of the finite element meshes near the contraction entrance (the meshes extend upstream to -16 R_O and downstream to +30 R_O; see also data given in Table 1).

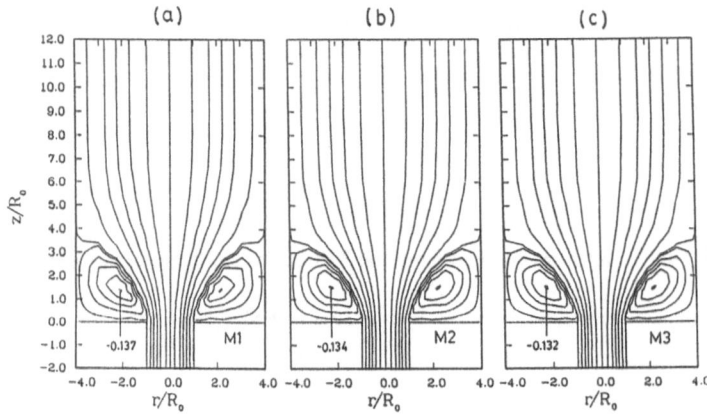

Figure 2. Streamline patterns obtained from the three meshes at $\dot{\gamma}_w = 104$ s^{-1}, ($S_R = 2.5$): (a) 200 elements, (b) 300 elements, (c) 400 elements.

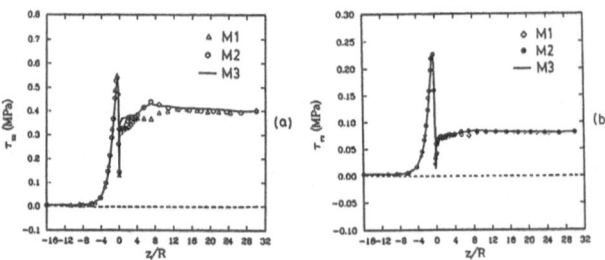

Figure 3. Predictions by all three meshes of the non-Newtonian stress components τ_{zz} and τ_{rz} along the r = 1.0 line: (a) τ_{zz}, (b) τ_{rz}.

INTERMITTENT SHEAR FLOW IN LIQUID CRYSTALLINE POLYMERS: RHEO-OPTICAL BEHAVIOUR

P. MOLDENAERS, H. YANASE[*] AND J. MEWIS
Department of Chemical Engineering, K.U. Leuven
de Croylaan 46, B-3030 Leuven (Belgium)
* Current address: Department of Poymer Chemistry, Kyoto University
Kyoto 606 (Japan)

ABSTRACT

Complementary rheological and rheo-optical measurements are used to investigate the behaviour of a lyotropic liquid crystalline sample under steady state shear and transient conditions. By comparing the transient stress during a step-wise increase in shear rate with that during flow reversal, the flow-induced anisotropy of the material is studied. If shearing stops, slow structural changes continue long after the stress is totally relaxed. This phenomenon is investigated using intermittent shear flow. With each start-up in shear flow corresponds an oscillating transient of the shear stress and the dichroism, both having the same period. This result fits in with existing calculations based on the Leslie-Ericksen theory, assuming director tumbling. The rheo-optical response changes with rest time for over 3 hours.

MATERIALS AND METHODS

The sample under investigation consists of a liquid crystalline solution of Poly(γ-benzyl-D-glutamate) (PBDG) in m-cresol (concentration 25% by weight, MW = 310000). The concentration is high enough to ensure a fully liquid crystalline phase. This model system is cholesteric at rest but develops into a nematic structure under shear. Its basic rheological behaviour has already been reported elsewhere [1].

The rheological experiments were performed on a Rheometics Mechanical Spectrometer RMS 705F with cone and plate geometry. For the conservative linear dichroism measurements a Rheometrics Optical Analyser, equipped with a He-Ne laser, was used. The technique is based on a polarization modulation technique to measure the optical anisotropy [2]. The geometry consists of two parallel quartz platens, separated by a gap of 460 nm. As parallel platens are used, the (1-3) plane is probed in the rheo-optical experiments. All experiments have been performed at a temperature of 293 K.

STEADY STATE SHEAR BEHAVIOUR

Figure 1 shows the steady shear viscosity and first normal stress difference for the PBDG sample under investigation. The value of the equilibrium dichroism as a function of shear rate is also displayed. The viscosity in figure 1 reaches a Newtonian plateau at small shear rates, perhaps preceeded by a shear thinning region at shear rates smaller than 0.03 (1/s). From a shear rate of 1 (1/s) on, the sample exhibits a second shear thinning region. The first normal stress difference shows the by now well established, although not fully understood, phenomenon of negative values in a limited shear rate region.

The equilibrium value of the linear dichroism ($\Delta n''$) decreases continuously over the entire shear rate range studied for the sample under investigation. This suggests that the structure of the sample is shear rate dependent, even in the region with constant viscosity. The existence of a variable structure with a constant viscosity has also been established by transient rheological experiments [1] and SALS experiments [3].

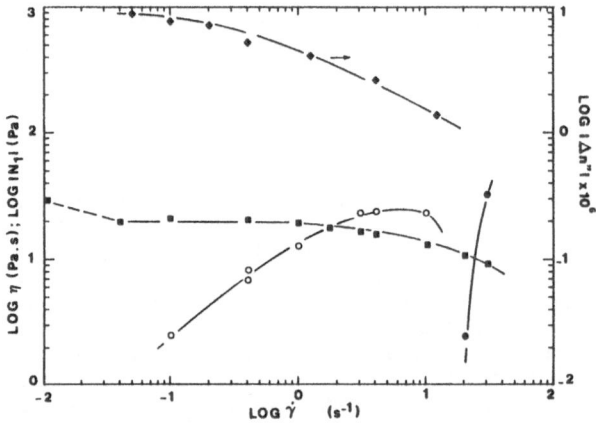

Figure 1: Steady state shear behaviour (■: viscosity; o: positive N1; ●: negative N1; ◆: |$\Delta n''$|).

FLOW INDUCED ANISOTROPY

Polymeric liquid crystals are assumed to be anisotropic. Comparing the stress response during a stepwise increase in shear rate and during flow reversal could provide means to assess anisotopy during normal flow conditions. Figure 2 shows the stress transients for flow reversal experiments at two shear rates in the Newtonian region. The curves could be superimposed approximately by scaling them with strain. The response for a stepwise increase in shear rate (figure 2) has a similar period as the flow reversal experiment. A phase shift close to 180°is however observed when comparing the two curves, indicating the presence of a flow—induced anisotropic structure.

The absolute values of the transient linear dichroism following a flow reversal is also represented in figure 2. Curves of |$\Delta n''$| at different shear rates could also be superimposed by representing them versus strain. It is clear from figure 2 that the transients have very similar patterns. Both describe a damped oscillation with the same period. This suggests a relation between the mechanical and the optical measurements, which was not evident from the steady shear measurements.

EVOLUTION UPON CESSATION OF FLOW

Various techniques can be used to probe the evolution of the structure upon cessation of shear. The relaxation of the shear stress, after shearing until equilibrium, is compared here with the transients of shear stress and dichroism when the shear is started again (intermittent shear flow). The relaxation of the shear stress is fast. For example, after shearing at a rate of 0.4 (1/s) the stress has decayed to 10% of its initial value after 20 seconds. The structure however evolves over a much longer time. This can be illustrated by comparing the mechanical and optical start–up transients after longer rest periods, when the shear stress has completely vanished. The profile of the oscillating transients clearly depends on rest time up to rest periods of three hours. Moreover, the instances at wich the curves of the transient shear stress and the transient dichroism reach their extrema turn out to coincide within experimental error. Hence the rheological and optical properties might be governed by the same physical mechanism.

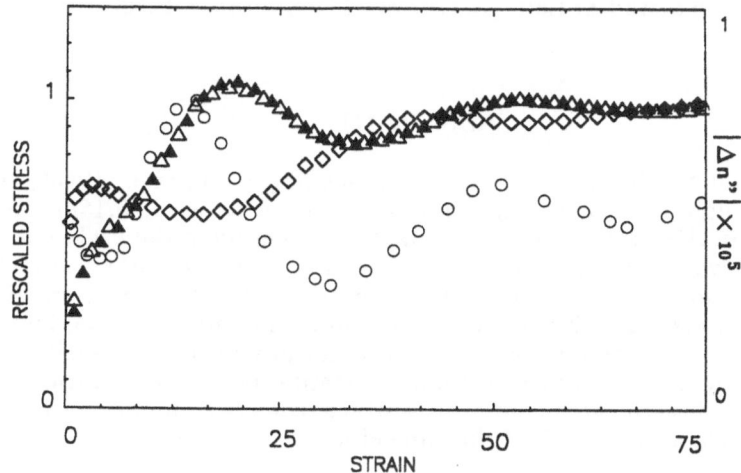

Figure 2: Comparison of scaled shear stress and the absolute value of the dichroism after flow reversal and scaled shear stress after stepwise increase in shear rate (stress flow reversal: ▲: $\dot{\gamma} = 1(1/s)$; △: $\dot{\gamma} = 0.4(1/s)$; dichroism flow reversal $\dot{\gamma} = 0.4(1/s)$: ○; stepwise increase in shear rate, $\dot{\gamma}_i = 0.1(1/s)$, $\dot{\gamma}_f = 1(1/s)$: ◇).

REFERENCES

1. Moldenaers, P., Yanase, H. and Mewis, J., Effect of the shear history on the rheological behaviour of lyotropic liquid crystals. In ACS Symposium Series, Advances in Liquid Crystalline Polymers, eds R.A. Weiss and C. Ober, 1990 (in press)
2. Fuller, G.G. and Mikkelsen, K.J., Optical rheometry using a rotary modulator. J. Rheol., 1989, 33, 761–769.
3. Takebe, T., Hashimoto, T., Ernst, B. and Navard, P., Small–angle light scattering of polymeric liquid crystals under shear flow. J. Chem. Phys., 1990, 92, 1386–1396.

368

EXPERIMENTAL DETERMINATION OF SLIP LAYER PROPERTIES IN POLYMER SOLUTION FLOW

H.Müller-Mohnssen and A.Tippe
Abt.Physiologie, GSF,
Ingolstädter Landstr.1, D-8042 Neuherberg, FRG

ABSTRACT

The apparent slip in ducted flow of aqueous polyacrylamide (PAM) solutions was studied by measuring velocity profiles as close as 0.15μm to the channel wall.The width δ of the slip layer as function of the PAM concentration c_∞ was determined to vary between δ = 0.36μm at c_∞ =0.005% and δ <0.15μm at c_∞ > 0.05%. The average PAM concentration within the slip layer proved to be smaller than that of the bulk solution. Donnan potential and slip velocity of the PAM solutions decrease parallel with increasing electrolyte concentration, suggesting that electrostatic forces between PAM molecules and wall surface (electric double layer) are responsible for at least 80% of the slip velocity.

INTRODUCTION

In laminar tube flow of certain polymer solutions the fluid velocity is known to be abnormally increased with respect to predictions based on viscometric data. This phenomenon was termed "apparent slip" [1], postulating the formation of a "thin depleted fluid layer"(slip layer) of finite thickness δ near the channel wall (TDFL-hypothesis)..According to this hypothesis δ should increase with decreasing polymer concentration.At sufficient low concentrations δ might,therefore, be accessible to a direct measurement.The decrease of the slip velocity Vs with increasing electrolyte concentration (Na,K) [2] suggests a slip mechanism based on electrostatic forces between PAM molecules and wall surface.

To prove these predictions, velocity profiles of a ducted flow of aqueous polyacrylamide (PAM) of different polymer concentrations c_∞ were measured down to a distance of 0.15μm from the wall.In addition the effect of electrolyte content in the PAM solution on Vs was compared with the corresponding changes of the Donnan potential E generated by the same PAM solutions.

MATERIAL AND METHODS

The velocity profiles of laminar PAM flow were measured in glass ducts of 0.1mmx1mm cross-section for different volume flow rates and PAM concentrations between 0.005% and 0.8% (weight percent) .Applying the laser-differential-microanemometer (LMA) and the total-reflection-microscope-anemometer (TRA;[2]) the velocity components of gold tracer particles parallel to the channel axis, as a function of the distance d from the wall were determined (LMA for $d > 1\mu m$; TRA for $d < 1\mu m$). The utmost spacial resolution is determined by the average diameter of the particles ($0.15\mu m$).

The Donnan potential was determined according to [3] in the concentration range 0-200 mM Na,K using dialysis tubes as semipermeable membranes.

RESULTS

The width δ of the slip layer, is defined by the point of intersection of the velocity profile within the slip layer and that of the bulk profile. An experimental determination of the profile within the slip layer could only be conducted for PAM concentrations $c_\infty < 0.05\%$, for in this region the width δ proved to be larger than $0.15\mu m$, the spatial resolution of the method ($\delta = .36\mu m$ for $c_\infty = .005\%$). For $c_\infty > 0.05\%$ the width δ was determined to be equal or smaller than $0.15\mu m$.

From the measured wall shear stress and the experimentally determined velocity gradient within the slip layer the fluid viscosity η_δ in this layer was calculated. The values obtained are smaller than the bulk viscosities by a factor of about five.

The addition of strong electrolytes as Na ,K (0-200mM) to the PAM solutions resulted in a reduction of the slip velocity Vs to about 20% of its value for electrolyte free solutions For higher salt concentrations the slope of this reduction in Vs became significantly smaller by a factor of about eight. The Donnan potential E of a PAM solution likewise decreased with increasing salt concentration, getting immeasurably small for concentrations > 200 mM (Fig.1).

CONCLUSIONS

The results strongly support the TDFL-hypothesis of apparent slip and establish experimental values of the slip layer width δ . The fluid viscosity η_δ within this layer proved to be smaller than predicted by shear thinning, thus indicating a depletion of polymer molecules in the slip layer.

The parallel decrease of E and Vs with increasing electrolyte concentration support the hypothesis that the dominant component of the slip phenomenon arises from electrostatic forces between the charged PAM-molecules and the wall surface.

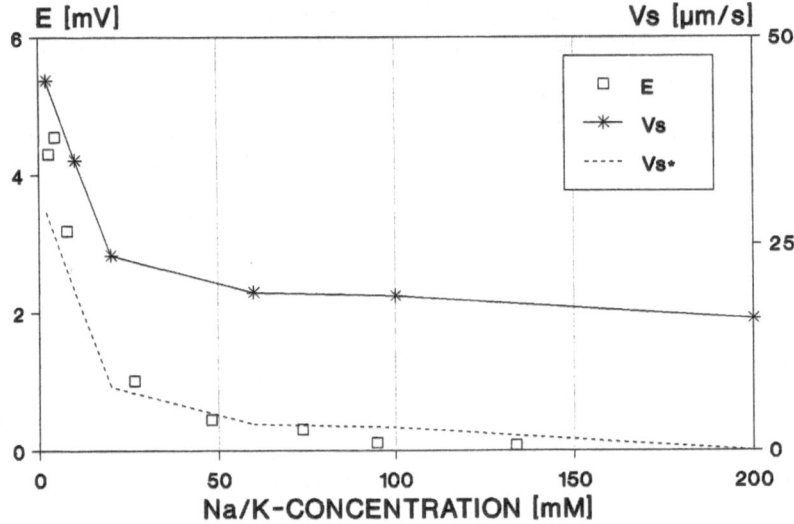

Figure 1.   Dependence of Donnan-potential E and slip velocity Vs of a 0.2% PAM solution on Na or K concentration.The dashed curve Vs * is identical to Vs but shifted by 15.9 μm/s, the value at 200mM Na. Note the correspondence in course between Vs * and E.

REFERENCES

1.   Cohen, Y. and Metzner, A.B., Apparent slip flow of polymer solutions. J. Rheol., 1985, **29**, 67-102.

2.   Müller-Mohnssen, H., Weiss, D. and Tippe, A., Concentration dependent changes of apparent slip in polymer solution flow. J. Rheol., 1990, in press.

3.   Overbeek, J.Th.G., The Donnan equilibrium. In Progress in Biophysics, ed. J.A.V. Butler, Pergamon Press, 1956, pp. 57-84.

VISCOELASTIC PROPERTIES OF A HEXAGONAL SURFACTANT LIQUID CRYSTAL

J. MUÑOZ; C. GALLEGOS
Departamento de Ingeniería Química. Facultad de Química.
Universidad de Sevilla. 41012 Sevilla (SPAIN).
A. SANTAMARIA
Departamento de Ciencia y Tecnología de Polímeros.
Universidad del País Vasco. 20080 San Sebastián (SPAIN).

ABSTRACT

This paper deals with the influence of surfactant concentration and temperature on the viscoelastic behaviour of a direct hexagonal liquid crystal. A nonionic surfactant, octylphenolpolyoxyethylene with 9.5 ethylene oxide groups, was used. This surfactant exhibited a direct hexagonal liquid-crystalline region within a composition range comprised between 40 and 55 wt% in surfactant.
From the experimental results, it can be deduced that only temperature has a significant influence on the linear viscoelastic functions of this kind of mesophase. On the contrary, both variables significantly influence the shear stress transient response at steady shear.

INTRODUCTION

Polyoxyethylene alcohols are a kind of nonionic surfactants which are widely used in detergency and as emulsifying agents (1). The stability of their viscoelastic emulsions has been attributed to the formation of three-dimensional association structures of liquid crystals yielding such rheological properties in the continuous phase so that coalescence of droplets is avoided (2).

Upon studying the nonionic/water phase diagrams the presence of different liquid crystal regions is noted (3). Normal hexagonal mesophase is composed of long cylindrical micelles packed in a hexagonal array, where the hydrophilic groups occupy the cylinder surfaces (4). This mesophase shows shear-thinning behaviour in steady-state shear, with the presence of an apparent yield stress (5).

In this paper, both the linear dynamic viscoelasticity and the transient response at steady shear of a direct hexagonal liquid crystal are studied. The first test permits the response of the mesophase in conditions close to the undisturbed state to be studied. The influence of surfactant concentration and temperature on the viscoelastic behaviour is analyzed. The transient response shows the influence of shear on the structural breakdown (reversible or not) of the hexagonal array of micelles.

EXPERIMENTAL

Triton X-100, octylphenolpolyoxyethylene with 9.5 ethylene oxide groups was used. This surfactant showed a hexagonal mesophase in a concentration range comprised between 40.0 and 57.5 wt% of surfactant, at temperatures up to 25°C. The existence of this mesophase was proved by means of polarising microscopy.

The transient (shear stress) experiments were carried out by a HAAKE Rotovisco RV-100 (measuring head M-150). Cone and plate geometry was used with a cone radius of 1,4 cm and a cone angle of 1°. A dynamic viscoelastomer (Polymer Laboratories DMTA) was used in parallel plate shear mode to measure the complex viscosity. The thickness of the samples was 2.75 mm and the amplitude of the oscillatory displacement was 16 µm, which corresponds to a maximum strain amplitude of 0.5%. The frequency was varied between 0.3 and 20 Hz. Measurements were carried out at 5°C, 10°C, 15°C and 20°C.

RESULTS AND DISCUSSION

a) Dynamic shear

The results obtained show a gel-like behaviour. Thus, the plot of η'' against η' (Cole-Cole diagram) exhibits a linear relationship within the frequency range studied.

In Table I, parameters of the power law equation:

$$|\eta^*|=|\eta^*_1|(\omega/\omega_1)^{-b} \qquad (1)$$

where $\omega_1=1$ rad/s, are shown.

TABLE I

Parameters derived from equation (1) at 10°C and 20°C.

		40%	42.5%	45%	47.5%	50%	55%
$\eta^*_1/5°C$	(Pa. s)	16857	–	13021	11797	11586	15325
$b_5°C$		0.83	–	0.83	0.83	0.82	0.84
$\eta^*_1/10°C$	(Pa. s)	13456	9404	12190	9984	10131	18051
$b_{10}°C$		0.81	0.79	0.81	0.79	0.78	0.85
$\eta^*_1/15°C$	(Pa. s)	10795	6492	11301	7413	7162	12689
$b_{15}°C$		0.77	0.78	0.80	0.76	0.76	0.80
$\eta^*_1/20°C$	(Pa. s)	8024	4608	7345	4642	5463	6822
$b_{20}°C$		0.76	0.77	0.75	0.73	0.75	0.74

373

The analysis of variance for the experimental dynamic values revealed that temperature provokes a decrease in the absolute magnitude of the complex viscosity. Furthermore, it is noteworthy that temperature has a more significant influence on the elastic component than on the viscous one.

Composition does not show any significant influence on the dynamic viscoelastic parameters.

$|\eta*|\cdot w$ was plotted against frequency in order to search for a possible apparent yield value. However, the dubious existence of a plastic behaviour for this kind of liquid crystals was verified.

b) Steady shear

Transient stress experiments at steady shear show an overshoot, which is produced at shorter time as the shear rate rises. Either an increase in surfactant concentration or a decrease in temperature significantly increase both the peak and the equilibrium viscosities. However, the equilibrium values lower for compositions close to the boundary of the hexagonal region. This is associated with a greater structural shear destruction in the sample.

From the results obtained, it can be deduced that there may be an orientation in the shear direction of the cylindrical aggregates, since a relatively fast and shear dependent recovery of the overshoot is always noticed. However, as such a recovery is not always complete, a mechanism based on the destruction of the "crystallites" structure must also be considered, above all at high shear rates and long shear times (5).

A comparison between the steady viscosities and the dynamic viscosities revealed that the Cox-Merz law is not followed.

REFERENCES

1. SAGITANI, H.; HAYASHI, Y.; OCHIAI, M. "Solution properties of homogeneous polyglycerol dodecyl ether nonionic surfactants". J.A.O.C.S., 1989, 66, 146-152.

2. PILPEL, N.; RABBANI, M.E. "Formation of liquid crystals in sunflower oil in water emulsions". J. Colloid Interface Sci., 1987, 119, 550-558.

3. GALLEGOS, C.; MUÑOZ, J.; FLORES, V. "Estudio reológico de diagramas de fases de sistemas alcohol graso polietoxilado/agua. I". Jorn. Com. Esp. Deterg., 1987, 18, 357-373.

4. TIDDY, G.J.T. "Surfactant-water liquid crystal phases". Physics Reports, 1980, 57, 1-46.

5. GALLEGOS, C.; MUÑOZ, J.; RODRIGUEZ, J.; FLORES, V. "Rheology of lyotropic liquid crystals for binary polyoxyethylene fatty alcohol/water systems. II". Progress and Trends in Rheology II, ed. H. Giesekus, Steinkopff. Darmstadt, 1988, 282-284.

6. VIOLA, G.G.; BAIRD, D.G. "Studies of the transient shear flow behaviour of liquid crystalline polymers". J. Rheol., 1986, 30, 601-628.

Measurements of the Rheological Properties of Suspensions based on Viscoelastic Liquids by means of a Cone-and-Plate- and a Capillary-Rheometer

N. OHL, W. GLEISSLE
Institut für Mechanische Verfahrenstechnik und Mechanik, Universität Karlsruhe (TH)
Postfach 6980, D-7500 Karlsruhe, Federal Republic of Germany

ABSTRACT

The shear stress function $\tau(\dot{\gamma})$, the first normal stress difference $N_1(\tau)$ and the pressure loss at a sudden diameter reduction $p_B(\tau)$ was measured using a cone and plate rheometer and a capillary rheometer. The investigated suspensions consisted of a limestone fraction (average particle size 8 µm), suspended in polyisobutylene and three viscoelastic silicone oils of different zero shear rate viscosity. By introducing two shift factors B_τ and B_N the influences of the liquid and the filler could be separated hence allowing the shear and normal stresses to be predicted for suspensions of viscoelastic liquids. The prediction of a typical process variable such as the Bagley-pressure p_B from these rheological functions, however, only partially succeeded.

INTRODUCTION

The flow behaviour of viscoelastic liquids is characterized by a shear-rate dependent viscosity, significant normal stress differences and pronounced memory effects. If solid fillers are suspended in these liquids, highly complicated rheological properties are to be expected from the resulting system. The list of parameters which influence the rheological properties of the suspension, ranges from those of the suspending liquid (see above), the characteristics of the filler (such as volume concentration, particle size, particle size distribution, shape etc.) up to the particle-particle- and particle-fluid-interactions (such as adsorption behaviour, surface charge etc.). In order to seperate the influences of the filler and the suspending liquid on the flow behaviour of these systems, Gleißle and Baloch |2| formulated the "concept of reduced shear rate", which is a generalization of the results of Kitano, Kataoka e.a. |1|.

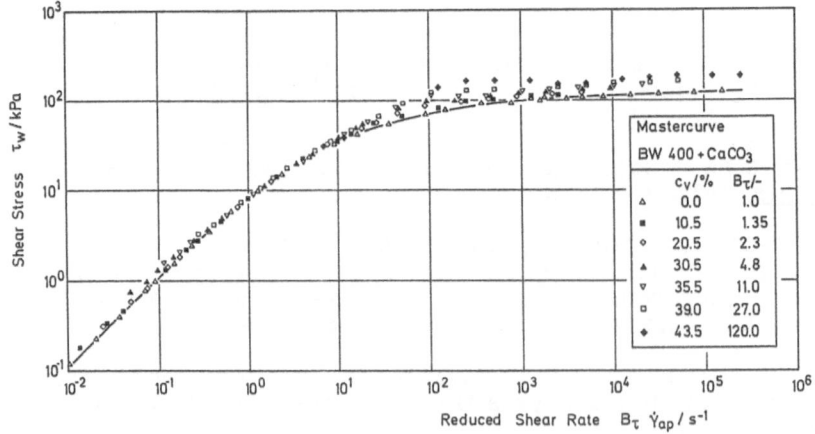

Fig 1: Mastercurve of BW 400 filled with limestone (average particle size $x_{50} = 8\mu m$)

STEADY STATE SHEAR BEHAVIOUR

According to this concept, the shear stress function of a suspension of fairly spherical particles (i.e. non fibers) can be predicted from the shear stress function of the pure liquid even if it is viscoelastic. The shear stress functions of the pure liquid and the suspensions plotted against the reduced shear rate $B_\tau \cdot \dot{\gamma}$ are represented by a single mastercurve. This is depicted in Fig. 1 for a system comprising BW 400, a silicone oil with a zero shear rate viscosity $\eta_o = 11$ kPas, filled with limestone particles of 8 μm average size. Thus the form of the pure liquid´s flow function governs that of the suspension´s flow curve. Whilst the shift factor B_τ is, in a first approximation, independent of the liquid, nevertheless it is a function of the filler and its properties. At higher stresses exceeding 50 kPa, however, the reduced flow curve of the suspension eventually deviates from the mastercurve as the filler concentration increases. A suspension containing 43.5 % limestone for instance possesses a flow curve maximum about 60 % higher than the shear stress function of the pure liquid at the same reduced shear rate $B_\tau \cdot \dot{\gamma}$.

The shear stress functions of the highly filled suspensions appear to possess a shear stress maximum. The decrease of shear stress at increasing shear rates can be explained by slip-stick effects.

The concentration dependency of the flow curve can hence be described by the shift factor B_τ, depicted in Fig 2 (open symbols) for three silicone oils of different η_o. Log B_τ can be represented by the same linear function of c_v^2 for the three liquids.

Generalizing the concept of reduced shear rates, one would expect the first normal stress difference N_1 as a function of the shear stress τ to be independent of the filler concentration. This is obviously a false conclusion as Fig. 3 demonstrates. Contrary to the concept, N_1 decreases as the limestone concentration is increased. By introducing a second shift factor B_N, however, the normal stress behaviour of the filled polymers could be adequately described. Similiar to the shift factor B_τ, which describes the shear stress behaviour, B_N is, in a first approximation, also independent of the suspending liquid (Fig 2, closed symbols). Lower viscosity polymers, however, no longer possess a constant slope of $N_1(\tau)$ but it decreases as the filler concentration is increased -i.e. η_o of AK 10^5 is 100 Pas; a factor of 100 lower than that of BW 400. Therefore for this liquid, B_N is an average value in the experimentally accessible shear stresse range. Nevertheless, for practical approximations this seems to be a powerful tool to describe the first normal stress behaviour of suspensions of viscoelastic liquids.

A very complex flow behaviour is revealed upon transforming from one flow pattern to another: Whereas the torque is not so much affected, the normal thrust either passes through a pronounced maximum or a minimum depending on the flow pattern. Even negative thrusts are reached.

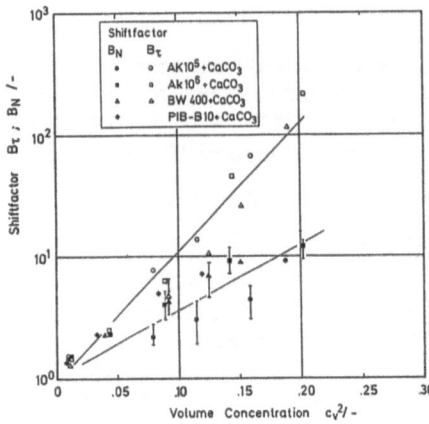

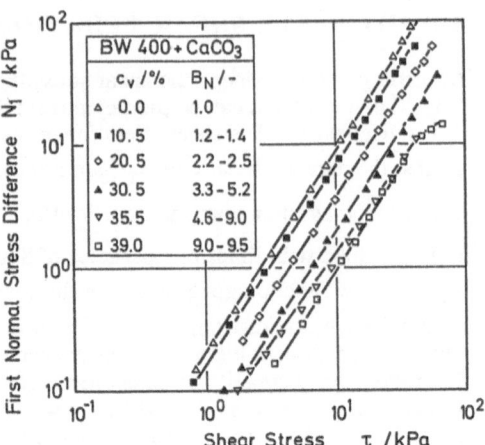

Fig 2: Shift Factors B_τ and B_N as functions of the volume concentration c_V for different viscoelastic liquids

Fig. 3: The first normal stress difference as a function of the shear stress; BW 400, with different concentrations of $CaCO_3$

PRESSURE LOSS AT A SUDDEN DIAMETER REDUCTION

By means of a capillary rheometer, the pressure loss at a sudden diameter reduction can be measured according to the well known Bagley procedure, using capillaries of different lengths but equal diameter. In a wide range of shear rates the Bagley pressure loss p_B was found to be proportional to the first normal stress difference at the capillary wall for pure viscoelastic liquids |3|.

This empirical result can for instrance be used to correct the viscosity function of an unknown viscoelastic liquid for the entrance pressure despite the use of just one capillary length. The question arises if this correlation between the Bagley-pressure and the normal stress, (to date only validated for pure liquids), might be applicable for suspensions. Measurements of the Bagley-pressure as a function of the shear stress at the capillary wall revealed a relatively sophisticated behaviour: p_B behaves differently at low and high shear stresses. At low shear stresses it is enhanced with increasing volume concentration compared to the entrance pressure loss of the pure liquid. In contrast it decreases with increasing filler concentration in the high shear stress range.In view of the proportionality of p_B and N_1 in pure liquids and the experimentally derived shifting procedure of $N_1(\tau)$, this latter result would have been expected throughout the accessible shear stress range.

LITERATURE

|1| Kataoka, T.; Kitano, T.; Sasahara, M.; Nishijima, K.: Rheol. Acta 17(1978) 149-155

|2| Gleißle, W.; Baloch, MK: Proc. IXth Intern. Congress. on Rheol., Mexico (1984) 549-556

|3| Gleißle, W: Proc. Xth Intern. Congress on Rheol., Sydney (1988) Vol. 1, 350-352

LOAD ENHANCEMENT EFFECTS IN LOOSE-FITTING BEARINGS CAUSED BY THE PRESENCE OF POLYMERIC ADDITIVES IN THE LUBRICANT

D R OLIVER
Department of Chemical Engineering
University of Birmingham, UK

ABSTRACT

A lubricated rotating cylinder fits into an outer cylinder which can be moved into eccentric positions; normal force and torque are measured. Inner cylinder diameters are typically 10mm to 20mm and clearance, $95\mu m$ to $500\mu m$. Load and friction are compared for Newtonian and polymer-thickened lubricants of similar viscosities.

Geometrical factors increase load enhancement effects due to polymeric additives - notably large clearances and short bearings. Short bearings are common on the crankshaft of modern car engines. Rapidly-changing loads also highlight load enhancement effects; this is achieved in our tests by having a rotor which is eccentric on its own shaft, thus simulating the time-dependent effects occurring in real bearings.

Load factors are enhanced by factors up to 18.0 for aqueous polymer solutions and by factors up to 2.70 for polymer-thickened oils. Corresponding values for friction reduction are 0.07 and 0.55 respectively.

INTRODUCTION

The journal bearing geometry is an important one; crankshaft bearings are normally of this type. Hutton et al [1] provided clear evidence of load enhancement effects for polymeric solutions in this geometry, but the solutions were of especially high elasticity. Later, Bates et al [2] showed that certain types of multigrade oil gave enhanced film thicknesses in an operating V-6 engine. Moreover, the effect was shown to be related to elastic properties possessed by these oils. Oliver [3] used a rotor eccentric on its own shaft and relatively large load enhancement effects were shown to be related to elastic properties possessed by the solutions. The present paper summarises further work in this field.

APPARATUS

This has been described previously [3]. The inner cylinder is driven at
speeds of up to 985 RPM by a Haake RV3 viscometer, which also measures
torque. The outer cylinder is attached to an Instron load cell mounted on
a carriage which permits accurate movement and alignment of cylinder
position. Normal force in the direction of eccentric movement is measured
on the load cell. The inner cylinder runs either "true" or eccentric on
its own shaft, the latter being adjustable in some cases.

The experimental procedure was to change the centreline eccentricity
ratio at constant rotational speed, making simultaneous measurements of
normal load and torque, which is then converted into frictional force.

Cylinder dimensions are quoted in Table 1.

TABLE 1
Cylinder dimensions

Test	Cylinder Diam(mm) inner(D)	outer	Shaft eccen-tricity e_R(μm)	Clearance c(μm)	Length L(mm)	D/L
A	10.00	10.20	30	100	10.0	1.0
B	9.99	10.18	15	95	3.30	3.0
C	20.00	20.38	35 or 90	190	6.60	3.0
D	20.00	20.96	0 or 200	480	6.60	3.0
E	20.00	20.96	0 or 200	480	1.75	12.0

Bearing "E" was "grooved". Dimensions of one of six areas of land are
quoted.

LIQUIDS USED

Aqueous solutions were an 80/20 v/v glycerol-water mixture of viscosity
56.0 mPas (23°C) and a 66.6/33.3 v/v glycerol-water mixture containing 500
ppm of Magnafloc E10. This is an anionic polyacrylamide of molecular
weight 20-25 million supplied by Allied Colloids, of Bradford, UK. The
viscosity is 46.0 mPas (23°C) at a shear rate of $2x10^4s^{-1}$.

Oil A is a Newtonian base oil of viscosity 79.0 mPas at 23°C. Oil B
is an oil containing 0.95 per cent of an olefin copolymer additive. This
5W/20 oil, supplied by Esso Research, is not fully formulated. The
viscosity is 62.0 mPas at a shear rate of $2x10^4s^{-1}$ (23°C).

RESULTS AND DISCUSSION

The theoretical load carried by a short journal bearing is defined in
reference 3. The load factor β is the measured load divided by the
theoretical load. For an eccentric rotor, a form of mean eccentricity

ratio is used in calculating theoretical load [3]. The coefficient of
friction μ is the tangential force divided by the measured load. The
dimensionless eccentricity ratio ε is the ratio between the movement of the
shaft centreline (e) and clearance (c). For eccentric rotors, this
movement is limited. Table 2 and Figure 1 show results; e_R is the
eccentricity of the rotor on its own shaft.

TABLE 2

Typical load enhancement (β^+) and friction reduction (μ^+) ratios due to the
presence of polymeric additives

Test	RPM range	$e_R(\mu m)$	ε	μ^+	β^+
Aqueous Solutions					
A1	500-700	30	0.15	0.94	1.32
A2	500-700	30	0.40	0.97	1.20
B1	800-985	15	0.65	0.64	2.00
B2	800-985	15	0.75	1.27	0.76
C1	500-700	35	0.50	0.48	2.40
C2	500-700	90	0.30	0.68	1.65
D1	600-800	20	0.80	0.10	12.00
D2	600-800	200	0.48	0.16	17.00
E1	800-985	0	0.82	0.10	6.50
E2	800-985	200	0.44	0.07	18.00
Oils					
A3	500-700	30	0.15	1.13	1.01
A4	500-700	30	0.40	1.06	1.01
B3	800-985	15	0.66	1.06	0.93
B4	800-985	15	0.80	0.75	1.23
C3	500-700	35	0.55	1.12	0.86
C4	500-700	90	0.30	1.05	1.03
D3	600-800	20	0.80	1.13	0.93
D4	600-800	200	0.50	1.09	1.10
E3	800-985	0	0.82	0.63	1.80
E4	800-985	200	0.50	0.55	2.70

$\beta^+ = \beta_P/\beta_N$ $\mu^+ = \mu_P/\mu_N$

Subscript "P" means polymer-thickened and "N" means Newtonian. In tests A
and B, the rotor was of fixed shaft eccentricity (30 and 15μm
respectively). Values of μ^+ and β^+ given for two values of ε.

The table shows the value of using highly elastic liquids (Ws $\doteq$ 20) in
addition to polymer-thickened oils (Ws $\doteq$ 2); changes in β^+ and μ^+ are
highlighted. Thus for runs A1, A2, B1 and B2, for close-fitting bearings
and small shaft eccentricity, load enhancement and friction reduction
effects are small. These effects are negligible in the equivalent runs for
oils (A3, A4, B3 and B4). As the clearance is increased (runs C, D and E)
the load enhancement and friction reduction effects become pronounced for
aqueous solutions, especially when the rotor is itself eccentric. For the

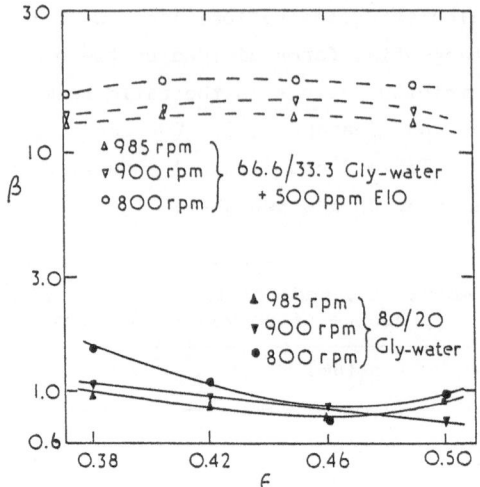

Figure 1. β vs ε for aqueous solutions (run E2)

oils, definitive effects occur in runs E3 and E4, the eccentric rotor again
giving the biggest effects. Runs E1-E4 were appropriate to a very <u>short</u>
bearing, of D/L ratio 12. Load enhancement effects due to polymeric
additives appear to increase with clearance, D/L ratio and shaft
eccentricity (ie time-dependent loading).

     For runs using oils, excluding E3 and E4, there is a surprising
tendency for friction to be slightly <u>increased</u> due to the additive. The
reason for this is unknown. Also notable in runs B1-B4 is the reversed
effect of centreline eccentricity (ε) on load enhancement for oils and
aqueous solutions. For the former, high eccentricity is an advantage,
whereas for the latter, high eccentricity is a disadvantage. Possibly, the
highly elastic solution is being rejected from the nip region under these
conditions. The results are obtained for a model bearing under
lightly-loaded conditions. The main conclusions are given in the Abstract.

ACKNOWLEDGEMENT

The author wishes to thank Esso Research, of Abingdon, UK, for their
generous help during this research programme.

REFERENCES

1.   Hutton, J.F., Jackson, K.P., and Williamson, B.P., ASLE/ASME Conf. San
     Diego Ca. Paper 84-LC-IC-1, 1984.
2.   Bates, T.W., Williamson, B.P., Spearot, J.A. and Murphy, C.K., Soc.
     Auto. Engrs. Int. Cong. Detroit, Mi. Paper 860376, 1986.
3.   Oliver, D.R., <u>J. Non-Newt. Fl. Mech</u>., 1988, <u>30</u>, 185-196.

ELASTIC RECOIL PREDICTIONS FROM THE CURTISS-BIRD MODEL

Hans Christian Öttinger
Institut für Polymere
Eidgenössische Technische Hochschule Zürich
ETH-Zentrum, CH-8092 Zürich, Switzerland

ABSTRACT

Computer simulations of the Curtiss-Bird model for the rheological behavior of concentrated polymer solutions and melts were carried out to determine the elastic recoil after steady shear flow. It is found that small but non-zero values of the link-tension coefficient are required to obtain consistency with typical experimental results.

INTRODUCTION

A most remarkable property of polymer melts is the large elastic recovery observed when the stresses causing shear or elongational flows are suddenly removed. For a molecular model to qualify as acceptable or useful, it should be capable of predicting such large recovery effects. In this note, we discuss the constrained elastic recovery after steady shear flow for the Curtiss-Bird (CB) model [1] which contains the Doi-Edwards (DE) model as a special case. A computer-simulation algorithm recently developed for a wide class of reptation theories is found to constitute a very convenient tool for discussing the time-dependent flow occurring during elastic recoil.

SIMULATION ALGORITHM

Computer simulations of reptation theories may be based on a rigorous reformulation of the underlying polymer dynamics in terms of a stochastic process [2], the so-called *reptation process* $\big(U(t), S(t)\big)$. The time-dependent random unit vector $U(t)$ then represents the direction of a given polymer in a melt at the position $S(t)$ within the polymer [$S(t)$ varies from 0 to 1 in going from one end of the polymer to the other]. In the presence of an applied flow field, the process $U(t)$ satisfies the deterministic differential equation $dU/dt = (\mathbf{1} - UU) \cdot \kappa \cdot U$, whereas $S(t)$ is a diffusion process in the interval $[0, 1]$ representing reptational motions and hence evolving according to a stochastic differential equation. The only coupling between $U(t)$ and $S(t)$ results from the boundary conditions which state that whenever $S(t)$ is reflected at one of the boundaries of its range $[0,1]$, $U(t)$ has to be replaced by a random unit vector (physically, this boundary condition represents the exploration of a new environment by the chain ends).

When the polymer dynamics is described by the reptation process, the stress tensor can be evaluated as an ensemble average (indicated by $\langle \cdots \rangle$) over the trajectories of $\big(U(t), S(t)\big)$; the effect of the flow history is fully contained in the current configuration of the reptation process, and hence no memory integrals need to be evaluated [2]. For example, the condition of vanishing shear stress to be imposed in order to determine the constrained elastic recoil after shear flow for the CB model reads as follows

$$\langle U_x U_y \rangle + \epsilon \lambda \dot{\gamma} \left\langle S(1-S) U_x^2 U_y^2 \right\rangle = 0, \tag{1}$$

where the flow field is assumed to be of the form $v_x = \dot{\gamma} y$, $v_y = v_z = 0$ with the time-dependent shear rate $\dot{\gamma}$ [to be determined from Eq. (1)], and the time constant λ and the link-tension coefficient ϵ are fundamental parameters of the CB model.

Starting from equilibrium, 2×10^5 trajectories of the reptation process were propagated in time according to the previously suggested simulation algorithm [2] with time steps of size $\Delta t = 0.01 \lambda$. This first stage of the simulation with constant shear rate $\dot{\gamma}_0$ was carried out until a steady state was reached (this can safely be assumed after the time 1.5λ). The steady-state configurations served as initial conditions for the determination of elastic recoil. In the second stage of the simulation, the shear rate $\dot{\gamma}(t)$ was determined such that Eq. (1) was satisfied. This way of proceeding is very similar to the experimental procedure in so far as the shear rate $\dot{\gamma}(t)$ is controlled such that the shear stress vanishes (experimentally, this is done with a servo drive [3]). Since during recoil $\dot{\gamma}(t)$ changes very rapidly in time, for this second part of the simulations we used a variable time-step width. From Eq. (1) the time-step width was determined such that in each time step the shear strain decreased by a given, fixed amount $\Delta \gamma$ (in all cases, $\Delta \gamma$ was less than 1% of the total recovery). For $\epsilon = 0$ (DE model), this method obviously needs to be modified; in particular, there occurs an instantaneous recoil which can be calculated directly from the steady-state initial conditions. For several values of ϵ, the entire procedure was carried out four times in order to improve precision and to estimate the statistical errors for the final results (which were typically of the order of 1%). Simulations with various numbers of trajectories for $\lambda \dot{\gamma}_0 = 1$ and 10 indicated that for $N = 2 \times 10^5$ trajectories the systematic error due to the finite number of trajectories in evaluating ensemble averages is smaller than the statistical error (systematic errors of order $N^{-1/2}$ should be expected). The effect of discrete time steps, in particular, the effect of adjustable time-step width in integrating a stochastic differential equation remains to be investigated in more detail.

RESULTS AND CONCLUSIONS

With the above simulation algorithm, the elastic recoil was calculated as a function of time; the simulation was stopped when a maximum in the recovery occurred which was then assumed to be the total recovery, $\gamma_{r,s}$ (due to fluctuations in a simulation with a finite number of trajectories, such a maximum must occur when eventually the strain rate almost vanishes).

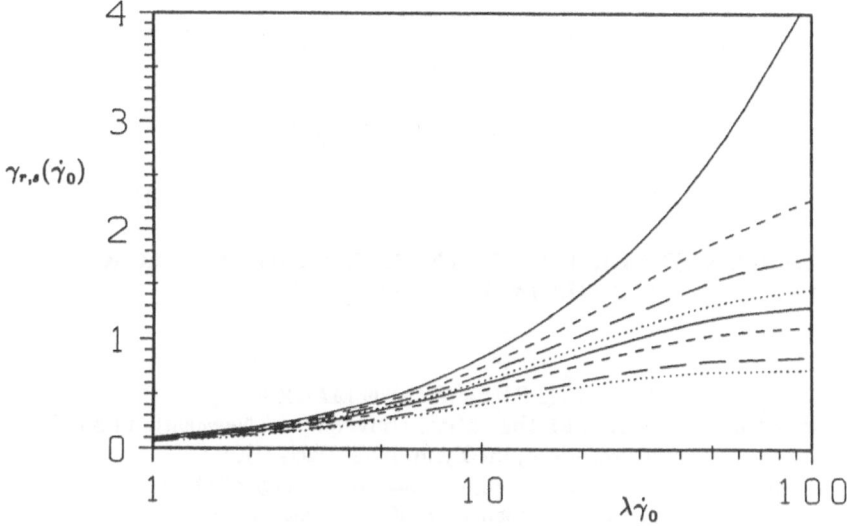

Figure 1. The total recovery after steady shear flow, $\gamma_{r,s}$, as a function of the steady-state shear rate, $\dot{\gamma}_0$. From top to bottom, the curves correspond to the link-tension coefficients 0 (DE model), 0.1, 0.2, 0.3, 0.375, 0.5, 0.8, and 1.0.

For the DE model ($\epsilon = 0$), at high shear rates one finds a total recovery which is much larger than typically observed experimental values ($\gamma_{r,s}$ of the order of 2 or 3 at high shear rates). For $\epsilon > 0$, the curves in Fig. 1 reach a plateau for $\lambda\dot{\gamma}_0$ of order 100 (recent, preliminary simulations indicate that the total recoil might begin to decrease at even higher shear rates). For small link-tension coefficients, the maximum recovery decreases very rapidly with ϵ; reasonable predictions are obtained for link-tension coefficients of the order $\epsilon \approx 0.1$. In conclusion, our preliminary simulation results clearly indicate that the experimental investigation of elastic recoil constitutes a very useful means for determining the link-tension coefficient of the CB model; typical experimental results suggest small values of ϵ, however, they appear to be inconsistent with the DE model ($\epsilon = 0$).

REFERENCES

1. Bird, R. B., Curtiss, C. F., Armstrong, R. C. and Hassager, O., Dynamics of Polymeric Liquids, Vol.2, Kinetic Theory, 2nd Edition, Wiley-Interscience, New York, 1987.

2. Öttinger, H. C., Computer simulation of reptation theories. Parts I and II. J. Chem. Phys., 1989, **91**, 6455-62 and 1990, **92**, 4540-47.

3. Meißner, J., Neue Meßmöglichkeiten mit einem zur Untersuchung von Kunststoff-Schmelzen geeigneten modifizierten Weissenberg-Rheogoniometer, Rheol. Acta, 1975, **14**, 201-218.

THOUGHT-EXPERIMENTS IN EXTENSIONAL FLOW OF POLYMER SOLUTIONS

C J S PETRIE and †M E MACKAY

Department of Engineering Mathematics, University of Newcastle upon Tyne,
Newcastle upon Tyne, NE1 7RU, UK
† Department of Chemical Engineering, University of Queensland,
St Lucia, Queensland 4067, Australia

ABSTRACT

Results of spinning simulations using the *FENE-P* dumbbell model are discussed in connection with extensional viscosity and "spinning viscosity" interpretations.

INTRODUCTION

One problem which is occupying a great deal of attention among rheologists at present is the estimation of extensional viscosities of mobile liquids. There is, for example, no experiment which gives a genuine value of the elongational viscosity of a dilute polymer solution because it is not feasible to carry out the steady, spatially uniform, axisymmetric extension of a sample of a mobile liquid.

Fibre spinning is probably the most useful of the flows which are predominantly extensional and may be used to obtain some sort of estimate of extensional properties. The major problem is that the flow is not spatially uniform, and in fact the variation of rate of strain along the spinline is not something which the experimenter can control. The question of how to interpret measurements made on fibre spinning is a crucial one, and theoretical analysis has an important role to play. Theoretical investigation of idealized experiments involving idealized materials may be used to compare estimates which might be obtained from actual experimental data with true extensional flow properties and to show how parameters like the Deborah number and features of the flow such as the upstream history of the material may affect the results. It is the detailed analysis of such "thought-experiments" which this paper seeks to introduce.

We have compared [1] the true elongational viscosity of a *FENE-P* dumbbell model with estimates of a "spinning viscosity" (as a function of an estimate of a rate of strain). We considered both "point values" (ignoring the fact that the ratio of stress to rate of strain at a point in a non-uniform flow is generally not a material property) and "average values", and from the comparison have made a proposal [1] for a better averaging formula for the "spinning viscosity" than one previously used [2]. Obviously such results depend on the model chosen; we have, of necessity, to abandon a quest for generality, at least for the present. However we believe that the *FENE-P* dumbbell model is both moderately realistic and is capable of modelling several important features of observed behaviour and so is a good first choice.

THE RHEOLOGICAL MODEL

We make quantities dimensionless using the velocity V_0 of the fibre at the start of the spinline ($x = 0$), the spinline length L and a stress FV_0/Q (where F is the applied force along the spinline and Q is the volume flow rate). Then the constitutive equation is

$$\mathbf{S} = -p\mathbf{I} + \mathbf{S}^{(p)} + 2(1 - \beta)\varsigma\mathbf{D} \tag{1}$$

in which p is the constitutively undetermined isotropic pressure and $\mathbf{S}^{(p)}$ is the polymer contribution to the total stress, $\mathbf{S}$. The solvent contribution, $2\eta_s\mathbf{D}v_0/L$, leads to the last term. The parameters here are $\beta = \eta_p/(\eta_s + \eta_p)$, the fraction of the total viscosity due to the polymer, and $\varsigma = (\eta_s + \eta_p)Q/FL$ which is a dimensionless reciprocal of the applied force. The equation for the polymer stress is

$$Z\mathbf{S}^{(p)} + \alpha\overset{\triangledown}{\mathbf{S}}^{(p)} - (\alpha\mathbf{S}^{(p)} + \beta\varsigma\mathbf{I})\frac{d\log Z}{dt} = 2\beta\mathbf{D} \tag{2}$$

in which $\alpha = \lambda V_0/L$, a Deborah number, and

$$Z = 1 + \frac{3}{b} + \frac{\alpha T}{b\beta\varsigma}. \tag{3}$$

T is the trace of $\mathbf{S}^{(p)}$, b is the finite extensibility parameter and Z is related to the dumbbell stretching by

$$Z^{-1} = 1 - \langle\mathbf{R}\mathbf{R}\rangle/R_m^2 \tag{4}$$

so that $Z = 1$ when there is no stretching and $Z \to \infty$ as the dumbbell (the polymer molecule) is fully extended with end-to-end length R_m. We recover the upper convected (Oldroyd B) models by letting $b \to \infty$ so that $Z \equiv 1$ (infinitely extensible dumbbells, $R_m \to \infty$). If there is no solvent contribution we have a Maxwell model, $\beta = 1$.

THE MODEL EXPERIMENT

We consider the idealized spinning experiment in which gravity, inertia, surface tension and air drag are neglected. We then have a constant force, independent of x, which balances the rheological force given by $S_{zz} - S_{rr}$ (see [3] for details). From this, defining $N = S_{zz}^{(p)} - S_{rr}^{(p)}$, we have

$$N + 3(1 - \beta)\varsigma u' = u \tag{5}$$

and the constitutive equation (2) gives us the two further equations

$$\alpha u T' = 2\alpha N Z u' - Z^2 T, \tag{6}$$

$$\alpha u N' = \alpha(N + T)u' + 3\beta\varsigma u' - ZN + \frac{\alpha^2 N u T'}{b\beta\varsigma Z}. \tag{7}$$

This is a system of three first order differential equations for u, N and T (as functions of distance x down the spinline). We solve using the initial conditions

$$u(0) = 1, \quad N(0) = 1, \quad T(0) = 1 \tag{8}$$

for any set of values of the parameters α, β, b and ς, with Z given by equation (3).

DISCUSSION OF RESULTS

In Figure 1 we plot draw ratio, $(u(1))$, against force, $(1/\varsigma)$, and note the following features of the model's behaviour. At small forces the behaviour is Newtonian. If the Deborah number, α, is small and also we have an appreciable solvent contribution, $(1 - \beta)$, the Newtonian behaviour persists up to large draw ratios (and "small" forces can be as large as $10(\eta_s + \eta_p)Q/L$). At some stage, however, the polymer contribution becomes important and there is a marked increase in the force over the Newtonian value. In this regime polymer molecules (dumbbells) are being stretched and lined up with the flow. In some situations this molecular stretching appears to require so much work that the draw ratio passes through a maximum and decreases as the force increases. Finally we see the effect of finite extensibility (except, of course, in the case $b \to \infty$ – the Oldroyd fluid B, with infinitely extensible dumbbells). When the molecules are all fully extended and lined up with the flow we have essentially Newtonian behaviour once again, with a higher viscosity.

This pattern can be observed in some of the viscosity against rate of strain graphs in [1], Figure 4 of that paper, for example, which is reproduced here as Figure 2 with its caption corrected. The S-shape of the graphs, with three values of force for a given draw ratio in some situations, has implications for the stability of spinning in such situations. If we control the take-up force there is no problem, but if we control the take-up speed or draw ratio we shall find a sudden jump in force, and the middle part of the curve would not be reachable.

Another aspect of the results obtained to date is seen in Figure 2, where a very large range of values of α has been considered. For $\alpha = 1$ and $\alpha = 10$ the limits of "spinning viscosity", $\bar{\eta}$, as "rate of strain" becomes small ($\dot{\Gamma}_e \to 0$) are clearly different from $3 - 9\beta/(b+3)$ which gives a Trouton ratio of exactly 3 for this model in the limit of small rate of strain. An asymptotic analysis of equations (5), (6) and (7) for small force ($\varsigma^{-1} \to 0$) leads to the explanation of this: the values of stress being calculated in this situation are being influenced almost exclusively by the initial values chosen for the simulation, equation (8). Thus we are not seeing a stress arising from the extensional flow, but the decay (proportionally to $\exp(-x/\alpha)$) of the initial stress which depends not on the extensional flow but on the prior history of the material. Choice of a suitable value of $N(0)$ can eliminate this effect, which clearly is important only for large Deborah numbers.

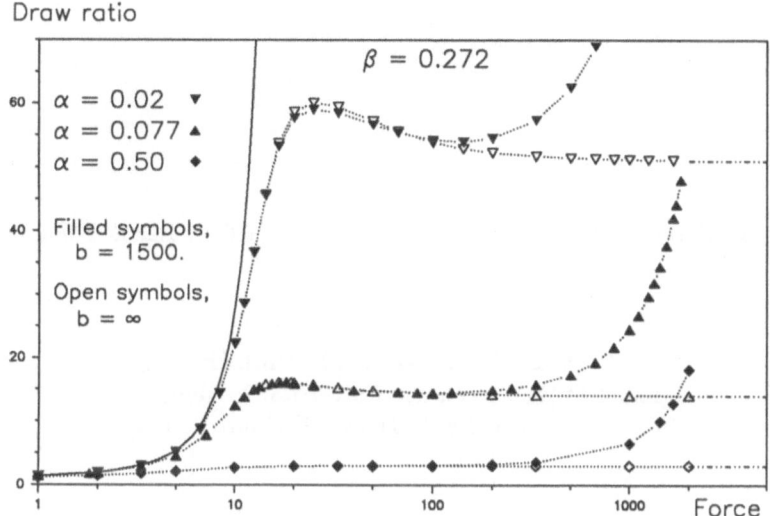

Figure 1. Typical results for the *FENE-P* dumbbell and Oldroyd B ($b = \infty$) models.

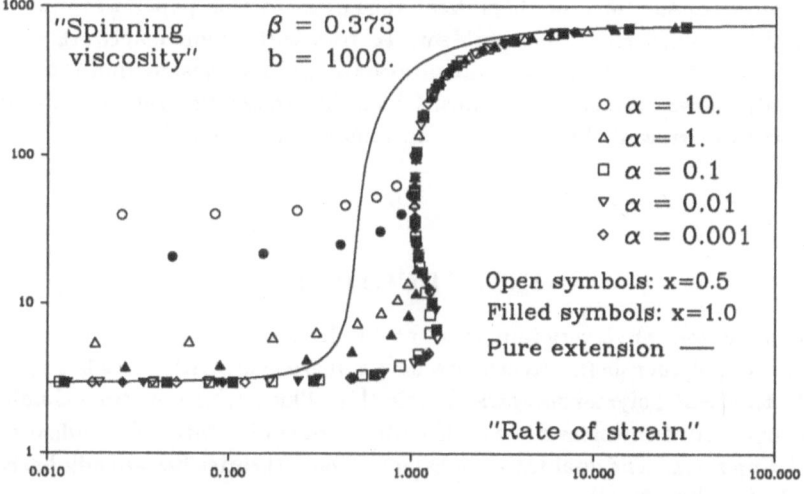

Figure 2. Point values of "spinning viscosity" and true elongational viscosity [1].

REFERENCES

1. M E Mackay and C J S Petrie, Rheol. Acta, 1989 **28** 281-293.

2. D M Jones, K Walters and P R Williams , Rheol. Acta, 1987 **26** 20-30.

3. C J S Petrie, **Elongational Flows**, Pitman, London, 1979.

CONTINUOUS TIME SIMULATION OF POLYMER MELTS

F. PETRUCCIONE and P. BILLER
Fakultät für Physik der Universität Freiburg
Hermann-Herder-Str. 3, D-7800 Freiburg , FRG

ABSTRACT

Predicting the nonlinear behavior of polymer melts from their known linear viscoelastic properties is one of the aims of theoretical rheology. In this paper we want to indicate a promising approach to this problem. In transient polymer network theories it is possible to derive an appropriate configuration dependent destruction rate from the experimantally known relaxation modulus. Given this destruction rate one can simulate the nonlinear properties of the melt by a continuous time simulation.

INTRODUCTION

Computer simulations are a very important tool to improve the mesoscopic modelling of the dynamics of polymer melts. Recently a new numerical algorithm has been presented to simulate transient polymer network theories [1]. This continuous time simulation is a very effective test of transient network theories with configuration dependent creation and destruction rates. The qualitative behavior of such theories has already been shown to be satisfactory [2], [3], [4].

In this paper we show how one can use this numerical tool to make quantitative predictions of material functions. The linear viscoelastic behavior of a melt is fully described by the relaxation modulus. In transient polymer network theories with configuration dependent rates a relation exists between the relaxation modulus and the destruction rate. One can thus extract the appropriate destruction rate from the experimental data of the relaxation modulus. With the help of this destruction rate one can then perform continuous time simulations to determine the material functions in typical flow situations and compare them with experimental data.

THEORY

In most transient polymer network theories the dynamics of the strands that form the network is described by the convection equation [5]

$$\frac{\partial}{\partial t}\Psi + \frac{\partial}{\partial \boldsymbol{Q}} \cdot (\boldsymbol{\kappa} \cdot \boldsymbol{Q}\Psi) = h(\boldsymbol{Q})[\Psi_0(\boldsymbol{Q}) - \Psi(\boldsymbol{Q}, t)] \quad . \tag{1}$$

Here $\Psi(\boldsymbol{Q}, t)d^3Q$ is proportional to the probability of finding a strand's configuration vector between $\boldsymbol{Q}$ and $\boldsymbol{Q} + d\boldsymbol{Q}$ at time t. The number density of strands $n(t)$ is defined as

$$n(t) = \int d^3Q \ \Psi(\boldsymbol{Q}, t) \quad , \tag{2}$$

and is generally a function of time. The tensor $\boldsymbol{\kappa}$ specifies the velocity gradient of the flow under study. We assume here that the strands obey a linear internal force law with spring constant H and that they are deformed affinely. The first term on the right hand-side describes the generation of new strands from a Gaussian equilibrium ensemble

$$\Psi_0(\boldsymbol{Q}) = \frac{n_0}{(2\pi\sigma^2)^{3/2}} \exp\left(-\frac{Q^2}{2\sigma^2}\right) \quad , \tag{3}$$

which occurs with a configuration-dependent rate $h(\boldsymbol{Q})$. This rate is assumed to depend on the length $Q = |\boldsymbol{Q}|$ and not on the orientation of the strand [2], [3], [4]. In the above equation n_0 is the equilibrium number density of strands and $3\sigma^2 = 3kT/H$ is the temperature dependent mean square equilibrium length of a strand. Finally, the second term on the right hand side of Eq. (1) describes the loss of strands from the actual ensemble, which also occurs with the rate $h(Q)$.

In general Eq. (1) cannot be solved analytically. Nevertheless it is possible to get results for the viscoelastic regime by an expansion for small velocity gradients. In this regime the stress tensor $\boldsymbol{r}(t) = H\langle\boldsymbol{QQ}\rangle$ [5] can be written as [6]

$$\boldsymbol{r}(t) = \int dt' G(t - t')\frac{1}{2}(\boldsymbol{\kappa} + \boldsymbol{\kappa}^T). \tag{4}$$

The linear visoelastic relaxation modulus $G(t - t')$ contains all of the information about the stress tensor in the fluid in the linear visoelastic regime. Following Wiegel and De Bats one can deduce that for our model the following relation between the destruction rate $h(Q)$ and the relaxation function $G(t - t')$ holds:

$$G(t - t') = n_0 kT \frac{2}{15\sqrt{2\pi}} \int_0^\infty d\rho \ \rho^6 \ \exp\left(-\frac{\rho^2}{2}\right) \exp\left(-h(\rho)(t - t')\right) \tag{5}$$

with $\rho = Q/\sigma$. Given the experimental data for the relaxation modulus one can extract the destruction rate by solving the above inverse problem.

NUMERICAL METHODS

The determination of the function $h(Q)$ is a typical inverse problem. To solve this problem we made a polynomial ansatz for $h(Q)$ and determined the parameters through the minimization of an appropriate χ^2 function. To minimize this function we made use of a Kalman-filter [7] and of a Levenberg-Marquardt routine [8].

Once the function $h(Q)$ has been determined from the linear viscoelastic properties, one can study the nonlinear predictions of the so specified transient network model. This is done most efficiently with the help of a continuous time simulation. The details of this technique have already been given elsewhere [1].

RESULTS AND OUTLOOK

A satisfactory destruction rate, reproducing the correct linear viscoelastic properties, can be easily obtained. With this destrution rate we then simulated simple flows and compared the simulated material functions with experimental data. The agreement was found to be good.

Work is in progress to improve the numerical efficiency of this approach, both for the minimization routines and for the simulation.

REFERENCES

[1] P. Biller and F. Petruccione, Continuous Time Simulations of Transient Polymer Network Theories, to be published in J. Chem. Phys., 1990.

[2] F. Petruccione and P. Biller, J. Chem. Phys., 1988, **85**, 577.

[3] F. Petruccione and P. Biller, Rheol. Acta, 1988, **27**, 559.

[4] F. Petruccione, Continuum Mech. Thermodyn., 1989, **1**, 97.

[5] R.B. Bird, C.F. Curtiss, R.C. Armstrong, and O. Hassager, Dynamics of Polymeric Liquids, Volume 2, Kinetic Theory, 2nd ed., Wiley, New York, 1987.

[6] F. W. Wiegel and F. Th. De Bats, Physica, 1969, **43**, 33.

[7] J. Honerkamp, Stochastische Dynamische Systeme, VCH Verlag, Weinheim, 1989.

[8] W. H. Press, B. F. Flannery, S. A. Teukolsky, and W. T. Vetterling, Numerical Recipes, Cambridge University, Cambridge, 1986.

BANDS TEXTURES IN LIQUID CRYSTALLINE POLYMERS SUBJECTED TO ELONGATIONAL FLOWS

EDITH PEUVREL AND PATRICK NAVARD
Ecole Nationale Supérieure des Mines de Paris
Centre de Mise en Forme des Matériaux
URA CNRS 1374
Sophia-Antipolis
06560 VALBONNE CEDEX – FRANCE

ABSTRACT

Band textures are known to occur after cessation of shear for lyotropic and thermotropic polymers. In order to gain more information on their behaviours we submitted an elongational flow to a lyotropic polymer with a rheo-optical system. The results are:
- elongation is very efficient for orienting the polymer,
- after cessation of elongation, a band texture occurs,
- the band spacing is the same as the one obtained after shear,
- there is a critical elongation rate below which the bands do not form.

INTRODUCTION

Band textures are known to occur after cessation of shear when the fluid is a polymeric liquid crystal. Both lyotropic and thermotropic polymers are subjected to this phenomenon [1-3]. It consists of long, dark, black lines when seen between crossed polarisers, the extinction direction of one of the polarisers being parallel to the previously applied shear direction. It has been seen in the past that dark bands seem to occur on drawing a fibre. In this case, the flow is mainly elongational. It is extremely difficult to obtain a pure elongational flow with a polymer, and even more difficult to study optically the relaxation.

During the course of another study, we fitted out our transparent cone-and-plate rheometer with a round glass fixed obstacle [4]. This gives an elongation after the obstacle, and a compression before. Such a device allows a correct exploration of the behavior of liquid crystalline polymers during and after elongation.

MATERIALS

The experimental equipment has been described elsewhere [3]. An aqueous solution of 55 %, of hydroxypropylcellulose [Mw = 100.000], kindly provided by Aqualon Hercules, was used as the liquid crystalline fluid.

RESULTS AND DISCUSSION

The first crude way of obtaining an elongational flow is to draw a fibre from a solution batch. The water evaporates in air and this "freezes" in the textures obtained during the relaxation. This gives a nice band texture, with all the optical characteristics of the one obtained after shear [2,3].

If we compare the band spacing obtained by this method with the one obtained after shear followed by evaporation in air, one finds nearly the same one:
- after shear : 1,70 μm,
- after elongation : 1,55 μm.

This suggests that the mecanisms can be similar.

Let's now look at the results obtained in the wake of the obstacle. An experi-mental determination of elongational rates α was performed by measuring the change in velocity of tracers.

The values of α were found to be well predicted by a finite element analysis model using a power law constitutive equation. The fluid accelerates from the obstacle to attain the mean velocity, and elongation occurs in the wake, behind the obstacle. Conversely, compression occurs before the obstacle.

The results are:
- elongation is very efficient for orientating the polymer,
- after cessation of elongation a band texture occurs, with the same optical features as the one obtained after shear (Figure 1),
- the band spacing after elongation is the same as the one obtained after shear,
- there is a critical elongational rate α_C ($\alpha_C = 0.04$ s^{-1}) below which no band texture occurs. This has to be compared to the critical shear rate $\gamma_C = 1,1$s^{-1}. The ratio of the two is 28. This is the efficiency ratio for orienting a liquid crystalline polymer: elongation is about 30 times more efficient than shear,
- compression does not give a band texture at low compression rates. Only the highest achievable velocities generate bands.

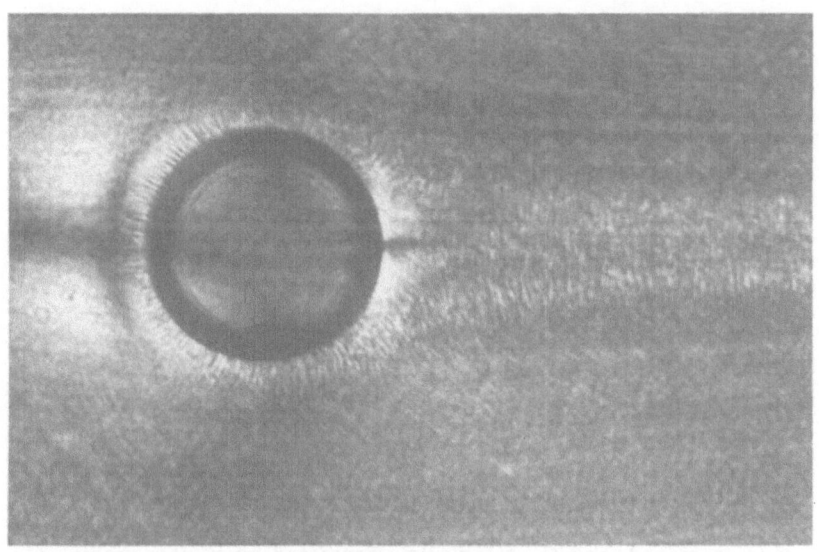

Figure 1. Band textures observed behind the obstacle after cessation of elongation.

References

1 C. VINEY, A.M. DONALD, A.H. WINDLE, _Polymer_ , 1985, **26**, 870.

2. P. NAVARD, _J. Polym. Sci., Polym. Phys. Ed._, 1986, **24**, 435.

3. B. ERNST, P. NAVARD, _Macromolecules_, 1989, **22**, 1419.

4. E. PEUVREL, P. NAVARD, _Liq. Cryst._, 1990, **7**, n° 1, 95.

Acknowledgments

This work was performed during the course of work sponsored by the DRET.

THE FLOW OF A VISCOELASTIC FLUID PAST A SPHERE

NHAN PHAN–THIEN and ROGER I. TANNER
Department of Mechanical Engineering
The University of Sydney, NSW 2006 Australia

ABSTRACT

The paper is concerned with the steady flow generated by a sphere falling along the centreline of a cylindrical tube containing a viscoelastic fluid which is modelled by the Oldroyd–B constitutive relation. By exploiting the similarity solution in the neighbourhood of the centreline of the tube it is found that there is a limiting Weissenberg number above which no steady state axisymmetric solution can exist. The full numerical solution to the problem using a boundary element method is reported and compared with results obtained by other numerical methods. We find an overall agreement between different sets of results pointing to the existence of the limiting Weissenberg number.

INTRODUCTION & RESULTS

The flow of a fluid past a sphere is a classical problem in fluid mechanics. The underlying theory for Newtonian fluids is well established [1]. For non–Newtonian fluids, however, the problem is considerably more complex and the analysis depends on the particular fluid model adopted.

The most commonly used constitutive equations in the numerical calculations are the upper–convected Maxwell (UCM) and the Oldroyd–B fluid models [2]. The flow of the UCM or Oldroyd–B fluids past a sphere placed on the centreline of a cylindrical tube was nominated as one of the benchmark problems for the comparison of different numerical techniques in the 5th Workshop on Numerical Methods in Non–Newtonian Flows (1987). The interesting quantity to calculate is the non–dimensional drag $\chi \equiv F/(6\pi\eta Ua)$ as a function of the Weissenberg number $Wi \equiv \lambda U/a$, where F is the drag on the sphere, a is the radius of the sphere, U is the relative velocity of the sphere with respect to the cylinder and λ is the relaxation time of the particular viscoelastic fluid model; η is the viscosity at zero–shear rate. In the agreed benchmark problem a/R = 0.5.

Unfortunately until now most numerical techniques used to solve this problem for the UCM or Oldroyd–B fluids have failed to obtain convergent results for values of Wi higher than 1, with very few exceptions [3,4]. Table 1 shows some previous investigations of the problem.

TABLE 1
Summary of previous numerical work on the flow of the
Maxwell fluid past a sphere in a cylinder (a/R = 0.5)

Investigator	Method	Limit of Wi
Hassager & Bisgaard (1983) [5]	Lagrangian FEM	0.57
Sugeng & Tanner (1986) [6]	BEM	0.70
Sugeng & Tanner (1986) [6]	FEM	0.30
Luo & Tanner (1986) [7]	Streamline FEM	0.40
Crochet (1988) [3]	Mixed FEM (Upwinded)	No limit
Carew & Townsend (1988) [8]	FEM	0.6
Debbaut & Crochet(1989) [4]	Mixed FEM (Upwinded)	No limit
Lunsmann, Brown & Armstrong (1989) [9]	EEME	~ 1.5
This paper.	BEM	0.4 − 0.7

The present work uses analytical and numerical methods to study the problem of viscoelastic flow past a sphere. A similarity solution for the velocities and stresses along the centreline is presented, and it is found that there is a limiting Weissenberg number above which no steady–state solution exists. This may yield a better understanding of the numerical breakdown problem. The numerical technique used in this work is the Boundary Element Method, which has certain advantages and its applicability in solving complex viscoelastic flows needs further investigations. Attention is paid to the effects of mesh size. The effects of elasticity on the drag, velocity and stress fields are discussed and wall effects are noted.

The Newtonian drag coefficient is found to be about 5.947. Comparing the result with those obtained by other authors, we found that the prediction of the Newtonian drag is confirmed to within 0.01%. The value differs from the exact theory of Haberman [1] by 0.385%. We have also found that the agreement between our numerical predictions and Haberman's theory becomes better as the ratio a/R becomes smaller. For example, for a/R = 0.2, we obtained χ = 1.67972, using a mesh with 90 boundary elements, in excellent agreement with Haberman's theory which predicts the value to be 1.680. For a/R = 0.8, our code predicts that χ = 72.97242, using a mesh with 242 boundary elements; while the value reported by Haberman [1] is 73.555.

The viscoelastic results are shown in Fig. 1 for the various investigations listed in Table 1. There is clearly considerable scatter in the results due to the various meshes and computational methods used. Considerable mesh refinement is needed to obtain results for this problem; in particular, the stresses near the sphere on the axis become very high as the Weissenberg number increases. In a more physically realistic non–linear model such as the PTT model, there is no limiting Weissenberg number in agreement with some previous work by others. Solutions indicate the presence of thin stress "boundary layers" fore and aft of the sphere and the resolution of these is a major numerical task.

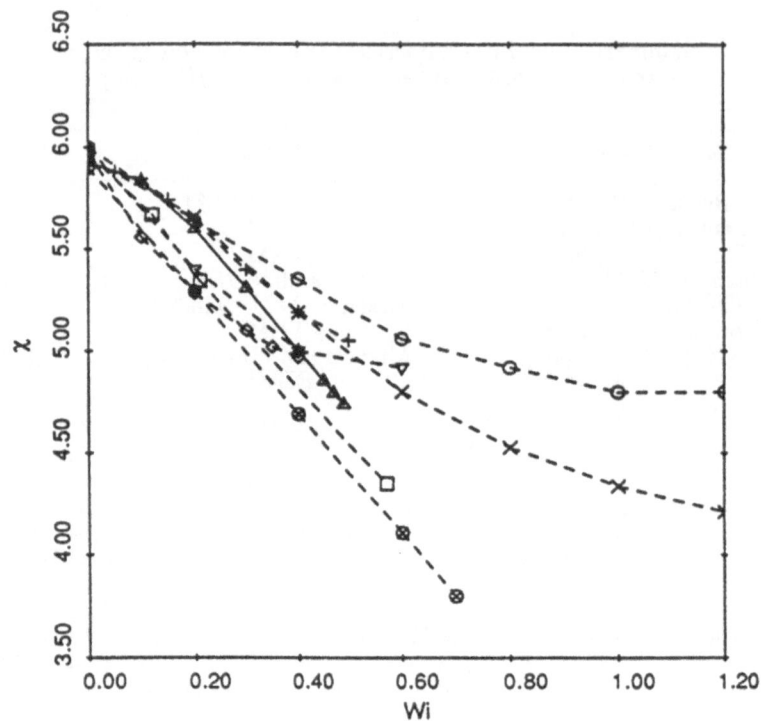

Figure 1. Dimensionless drag results.

Δ:	This work	○:	Crochet [3]
□:	Hasssager & Bisgaard [5]	∇:	Carew & Townsend [8]
⊗:	Sugeng & Tanner [6]	+:	Debbaut & Crochet [4]
◇:	Luo & Tanner [7]	×:	Lunsmann, Brown & Armstrong [9]

REFERENCES

1. Happel, J. and Brenner, H., Low Reynolds Number Hydrodynamics, Noordhoff, Leiden, 1973.

2. Tanner, R.I., Engineering Rheology, Oxford Univ. Press, Revised Edn., 1988.

3. Crochet, M.J., Numerical simulation of highly viscoelastic flows, in Proc. 10th Int. Congr. on Rheol., ed. P. Uhlherr, Sydney, Aug. 1988, pp. 1.19–1.24.

4. Debbaut, B. and Crochet, M.J., Int. J. Num. Meth. Fluids, 1988, 30, 169.

5. Hassager, O. and Bisgaard, C., J. Non–Newt. Fluid Mech., 1983, 12, 153.

6. Sugeng, F. and Tanner, R.I., J. Non–Newt. Fluid Mech., 1986, 20, 281.

7. Luo, X–L and Tanner, R.I., J. Non–Newt. Fluid Mech., 1986, 21, 179.

8. Carew, E. and Townsend, P., Rheol. Acta., 1988, 27, 125.

9. Lunsmann, W.J., Brown, R.A. and Armstrong, R.C., paper presented at 6th Workshop on Numerical Methods in Non–Newtonian Flows, Denmark, June, 1989.

THE RHEOLOGY OF WATERBORNE METALLIC BASECOATS FOR THE AUTOMOTIVE INDUSTRY AND ITS INFLUENCE ON THE APPLICATION PROPERTIES

Susanne Piontek, Theodora Dirking
BASF Lacke und Farben AG, 4400 Muenster, West Germany

ABSTRACT

The rheology of waterborne metallic basecoats was investigated and is discussed with regard to their application behaviour. Especially, the results of dynamic experiments reveal a correlation between the viscoelasticity of a basecoat on the one side and its metallic appearance on the other.

INTRODUCTION

More than 50% of the cars produced in the US and in Europe are coated with a metallic finish. This special effect is obtained by using a basecoat/clearcoat combination with the basecoat containing aluminium-flakes. The light reflected by the flakes gives these paints their characteristic luster and sparkle. Moreover, depending on the angle of observation, the lightness and frequently the colour of the paint changes. This is commonly referred to as "flip-flop"-effect. The optical quality of the metallic effect is highly correlated to the parallel orientation of the flakes. With conventional basecoats containing a lot of organic solvents, the parallel orientation is obtained by a high volume-shrinkage of the wet film. Immediately after the application, the solvents begin to evaporate very fast. Therefore, the flakes are forced to a position parallel to the substrate-surface. Simultaneously, the viscosity of the coating material increases steeply and prevents the flakes from reorientation. Because of environmental problems basecoats have been developed during the last few years containing water instead of organic solvents. Water evaporates very slowly and film shrinkage is not fast enough. Therefore, the rheology of the coating material plays a very important role for the flake-orientation in the finished film. Comprehensive rheological studies were done and some of the results are presented in this paper.

EXPERIMENTAL TECHNIQUES AND SAMPLES

The rheology of a waterborne basecoat is determined by its resin-composition and by special thickeners or so-called rheology control agents. Therefore, a first series of four basecoats was investigated which differ in their resin-compositions (A, M, AP, MP) and contain the same thickener. In a second series three systems were formulated with the same resin but with varying thickeners (VVG, HV, NV). For all measurements, the basecoats were adjusted to the solid-content that is used in practice for spray-application.

The rheological measurements were performed with a Carri-Med controlled stress instrument (CS 100), using a cone/plate geometry. For characterizing the metallic effect the reflection of light is measured at different angles with a goniophotometer instrument. The ratio of the face-brightness to the so-called "flop"-brightness is a measure for the optical quality: the higher this dimensionless "flop-value" the better the effect.

RESULTS

First, the flow behaviour of the basecoats was investigated. In principle, the viscosity/shear stress profiles of all samples are very similar. They are more or less pseudoplastic. Regarding only these curves, occurring differences in the metallic effects cannot be explained. But dynamic experiments measuring the storage modulus G' and the loss modulus G'' provided interesting results.

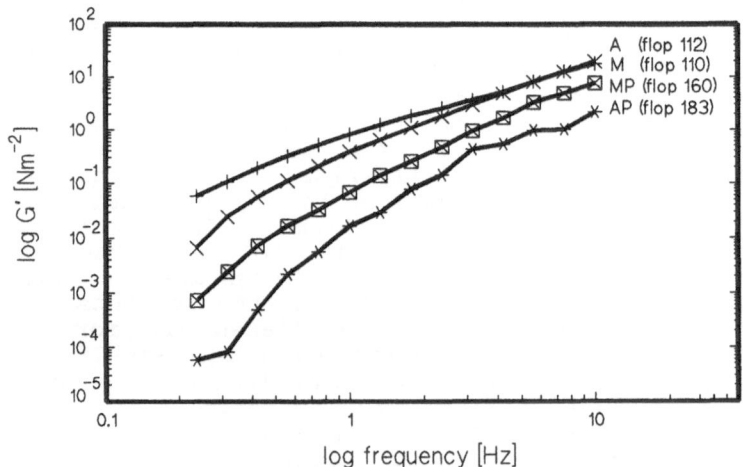

Figure 1. Storage Moduli and metallic effects of the basecoats containing different resins

In fig.1 G' is shown as a function of the frequency of

oscillation for the basecoats of the first series. The flop-
values are plotted in the diagram. The results elucidate that
the elastic part of a basecoat plays an important role for the
application. The more elastic the sample is, the worse is its
metallic effect.

The same correlation was found for the second series of basecoats. The thickeners which were used to formulate these
samples are chemically identical but with different molecular
weights. Measuring the dynamic moduli of the pure thickeners
resulted in pronounced differences concerning the elasticity.
These differences propagate in the basecoats (fig.2).

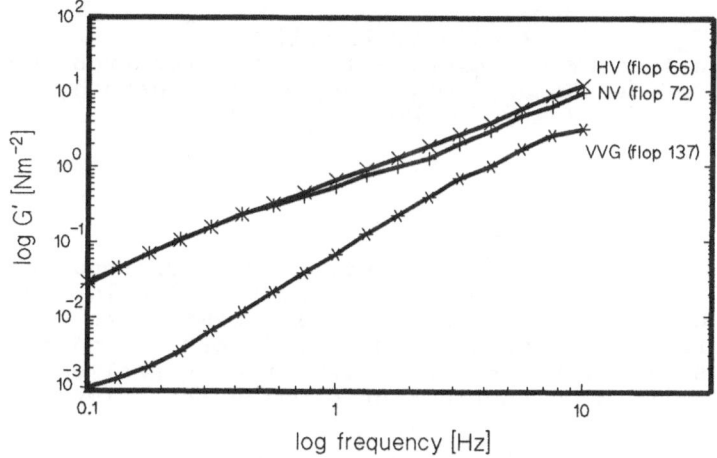

Figure 2. Storage moduli and metallic effects of the basecoats
containing different thickeners

DISCUSSION

The presented results indicate a correlation between the
elastic part of a waterborne basecoat and its metallic effect.
It has been stated in the literature (1) that elasticity has a
stabilizing influence on coatings subjected to the spray-
process. This can result in larger droplet size. The more
elastic the material is, the more energy has to be put in to
reach the same degree of atomization. It is known that the
finer the atomization the better the metallic effect (2).

REFERENCES

1. Golding et al, <u>Chem. Engng.</u>, 1972, **8**
2. Inkpen, Melcher, <u>Ind. Eng. Chem. Res.</u>, 1987, **26**, Vol.8,
   1645

ON SOME RHEOLOGICAL EFFECTS ARISING UNDER
DURABLE STRESS OF POLYMERIC LIQUIDS

ALEXANDER PROKUNIN
USSR Academy of Sciences Institute for Problems
of Mechanics, 101, Prospekt Vernadskogo,
117526, Moscow, USSR

ABSTRACT

The following effects: breaking away of the deformed polymeric liquid from the walls by a low-molecular liquid, self-heating of viscoelastic liquids and their foaming are considered. These effects appear upon durable deforming and are considered by us as related to the creation of a sealing of a rotating shaft with a polymeric liquid using the Weissenberg
effect.

BREAKING AWAY OF A POLYMER FROM A SURFACE

It is known that a flow in a polymer melt could not develop
under an intensive shear deformation, and it could break away
from solid surfaces during the period of time less than the
relaxation time of a polymer. Under a durable deformation,
another way of breaking away of a polymer from a wall was observed, which occurs under action of a film flow of a liquid
of low viscosity. For example, in the case of deforming of a
low- molecular polyisobutylene melt between steel disks (when
there was water on the rims of the disks, see Fig. 1a) during
several tens of minutes, the torque measured at the shaft decreased practically to zero. The disk surface was then wet
with water. In the case when there was air on the rims of the
disks, under the same conditions, the breaking away was not
observed.

THE VISCOSITY TYPES AND THEIR INFLUENCE ON
SHEAR STABILITY UPON POLYMER SELF-HEATING

Which of elastisity types is being ascribed to a polymeric medium under consideration, is essential in studies of motions

with self-heating for a chosen rheological equation. The type, first of all, influences the heat equation.

$$\rho \; c \; \frac{dT}{dt} = \mathscr{x} \nabla^2 T + D_k \tag{1}$$

The two limiting cases are being considered: the case of entropic elasticity caused by changes in macromolecules conformations and the case when elasticity is determined by changes in macromolecules lengths. We have correspondingly [1,2]

$$D_1 = \text{tr} \; (\sigma \, e), \quad D_2 = \text{tr} \; (\sigma \, e_p) \tag{2}$$

In [1], for cases [2], the heat capacity c depends only on temperature T, σ is the stress tensor, e and e_p are tensors of the total and irreversible shear rates, is the density, t, time, $\mathscr{x}$ is the heat conductivity coefficient.

The simple shear flow in the linear viscoelasticity model is being considered as an example of influence of the elasticity type on motion. In the D_2 case, the shear deformation from the rest state, when one of the plates moves with a constant speed, head is being removed from the plates, and t the adhesion condition is satisfied on them, could lead to autooscillations [3]. In the case D_1, under similar conditions, the stationary flow is being reached, as was observed in polyisobutylene experiments.

STATIONARY SHEAR FLOW ON POLYMER SELF-HEATING

First, simple shear flow between two parallel plates with the distance 2h between them and whose temperatures $T = T_0$ is being considered. In isothermic flow, the deformation rate γ is constant. But in flow with self-heating, the dimensionless deformation rate, $\Gamma = \gamma \theta$, (θ is the relaxation time) is constant. Using the heat equation, the exponential dependence of viscosity on temperature, the constancy of elasticity modulus ($\Delta T / T_0 \ll 1$), and the principle of temperature-time superposition, a system of equations with respect to temperature T and the medium velocity, V, is obtained, whose form does not depend on the form of the rheological equation for a viscoelastic medium. This system, which has an analytical solution, was studied in combustion theory, it also discribes the motion of non-Newtonian non-elastic liquids [4]. In particular, from this system, the dependence $\Gamma = \Gamma(\Gamma_{is})$, where $\Gamma_{is} = V_0 \theta_0 / 2h$, V_0 is the speed of the moving plate, $\theta_0 = \theta(T_0)$, was obtained. The dependence $\Gamma \; (\Gamma_{is})$ passes its maximum. The normal stresses exist in the system, they are fully determined by the dimensionless shear rate Γ. Based on the solution for simple shear, the solution of problem of stationary flow between disks (see Fig.1a) on self-heating was being obtained ($R \gg h$). It could be assumed that at a distance from both the centre and rim of the disks ($\sim h$), the temperature of the polymeric liquid changes considerably across the gap and in a small degree, along the radius. In this case, the problem is reduced to the former one, on simple shear, with accounting for the fact that $\Gamma_{is} = \Omega \, \theta_0 \, r / 2h$. The influences of the central and

the boundary regions could be accounted for by averaging the
equation over the gap width and by matching the obtained solution and the solution dependending on the radius. Fig.1b
shows typical dependencies of the pressure P measured with a
manometer as shown in Fig.1a on the angular speed, Ω. Cipher
1 denotes the isothermic case, [2], the flow with self-heating.
These dependencies for the torque on the disk are similar.
The non-linear model of a viscoelastic medium 2 was used in
calculations. A qualitative agreement with the experiment was
obtained.

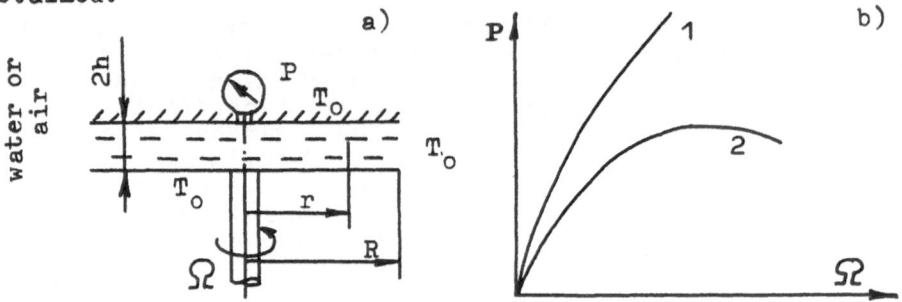

Figure 1. Deforming of the polymer between the disks: a) sche-
          matic of the experiment; b) dependence of the pres-
          sure on the speed of disk rotation.

REFERENCES

1. Astarita G., Sarti G.C. The dissipative mechanism in flowing polymers: theory and experiments. J.Non-Newtonian Fluid Mech., 1976, 1, pp. 39-50.

2. Prokunin A.N. On the description of viscoelastic flows of
   polymer fluids. Rheol. Acta, 1989, 28, pp. 38-48.

3. Stolin A.M., Khudyayev S.N. Non-isothermic unstability of
   viscoelastic medium flows. Soviet Phys. Dokl., 1972, 207,
   pp. 60-63.

4. Stolin A.M., Bostandzhian S.A., Plotnikova N.V. Critical
   conditions of a hydrodynamical heat explosion in power liquid flows. Heat-mass exchange in rheologically complex
   systems. Coll. of articles, ITMO of Byelorussian Acad. of
   Sciences, Minsk, 1976, 7, pp. 261-269.

USE OF AMPLITUDE AND FREQUENCY SPECTROMETRY FOR
DETERMINING THE SOL-GEL PHASE TRANSITION

E.-O. Reher, R. Schnabel, V. Schulze
Technical University "Carl Schorlemmer" Leuna-Merseburg
Department of Materials Science and Polymer Process
Engineering
Otto-Nuschke-Straße, Merseburg, GDR-4200

Abstract

The paper presents a method for determining the sol-gel
phase transition of gelatine-containing aqueous solutions.
Dynamic measurements carried out at varying amplitudes give
the possibility of drawing conclusions about structures of
gel formation from the sudden cooling regime by assessing
the validity of the COX-MERZ rule. The stationary and
dynamic solidification isotherms are quantified.

1. Problem statement/ methods

Determination of the temperatures of phase transition and
description of the sol-gel phase transition is necessary
for producing gelatine-containing photographical information
recording materials (IAM). The aim is to quantify the deformation properties during phase transition, using rheological
methods. Explanation of the formation of the structure and
its kinetics as a function of time and cooling gradient is
necessary for modifying these properties. In this respect,
the short-time behaviour within the time range of seconds
and gel formation at a strong cooling gradient of about
2 K/s is of great interest.

A review of the literature on this subject and results
have been presented in papers published by the authors /1/.
A promising method of thermorheology is amplitude-frequency
spectrometry. This method on the one hand allows determination of the viscous and elastic properties of the sol and
the forming and solidifying gel at small deformation amplitudes. On the other hand, the boundary of linear viscoelastic behaviour can be determined (GLV). Information on
structure concepts concerning the phase transition can
be obtained by comparing the information from stationary
measurements, i.e. the viscosity function $\eta(\dot{\gamma})$ with the
dynamic characteristics, e.g. the dynamic viscosity function, or by comparing normal stress measurements and the
dynamic shear modulus data G' (ω).
To this purpose, the Cox-Merz rule is analysed:

$$\eta(\dot{\gamma}) = |\eta^*(\omega)| \quad for \quad \dot{\gamma} = \omega$$

Conclusions concerning the types of interactions in the
gelatin-water system during phase transition can be drawn
from the position of stationary and dynamic characteristics,
using the following equations:

$\eta(\dot{\gamma}) \geq |\eta^*(\omega)|$ - type of solution, or network

$\eta(\dot{\gamma}) = |\eta^*(\omega)|$ - Cox-Merz rule

$\eta(\dot{\gamma}) \leq |\eta^*(\omega)|$ - type of gel, or specific type

of network (supramolecular structure, special interactions,
including formation of helices).
The experiments are carried out under a temperature regime
similar to that used in the determination of solidification
isotherms. The samples molten at 323 K are cooled to a
processing temperature of 313,3 K. Sudden cooling to this
solidification temperature is achieved in the rheometer.

2. Measuring cell for the Weissenberg rheogoniometer (R 18)

Following Gonzalez and Macosko /2/, a cone-plate apparatus
has been constructed, which makes it possible to inject
the cooled sample centrally and directly at the point of
the cone. Using the normal stress measuring equipment of
the R 18, the volume changes occurring during solidification can be compensated for. This ensures wall adhesion
of the gel during the period of experiments. The measurements showed that compensation of the volume shrinkage
requires changes of the slit to be made in the μm-range,
which does not affect the measuring results to an
appreciable degree. The measuring cell is equipped with
a temperature transmitter. Cleaning of the measuring cell
does not present any difficulties.

3. Experimental program and results

The dynamic characteristics of 10 mass % gelatine-water
solutions were determined as "solidification isotherms"
$\eta'(\omega, A)$ and G' (ω, A), respectively. ω is the
angular frequency, A is the amplitude of deformation.
The values varied were the solidification temperature,
the annular frequency ω and the amplitude. Furthermore,
amorphous silicon dioxide (Aerosil, e.g. A200) was used
in concentrations up to 5 mass % relative to the gelatine
concentration in order to determine its effects on the
sol-gel phase transition. The boundaries of linear viscoelasticity (GLV) were found to be A = 0,01 - 0,25 $\equiv A_{crit}$.
This confirms the boundaries found by Nagaslaeva and
Trapeznikov /3/. Above the critical amplitude, both η'
and G' decrease. The forming gel can only store part of
the energy of the vibrations. The stationary and the
dynamic characteristics are compared in the form

$$\eta(\dot{\gamma}) - |\eta^*(\dot{\gamma}_{dyn})|$$

where $\dot{\gamma}_{dyn} = \omega A$ is the maximum deformation rate of sinusoidal shear stress. For example, the solidification kinetics in the short-time range is described by the equation

$$\eta = \eta^{\pm} e^{bt^c} \mathcal{L}_0 c^{-E/RT}$$

it applies analogously to the change of the proportion of the shear modulus $G'(t)$ and to $\eta'(t)$ and $|\eta^*(t)|$ respectively. The shear stress and the solidification temperature are kept constant.

References

/1/ Schnabel, R.; Schulze, V.; Reher, E.-O.: Rheologische Aspekte der experimentellen und theoretischen Unter- suchung von Verfestigungsprozessen, Teil I: Überblick und das Konzept der Erstarrungskinetik bei Verfesti- gungsprozessen, Acta Polym. 37, 1986, 5; S. 268 - 272; Teil II: Rheometrische Untersuchungen, Acta Polym. 38, 1987, 2; S. 122 - 125, Zum Sol-Gel-Phasenübergang gelatinehaltiger Lösungen, IV, Kolloquium "Rheologie und Textur der Lebensmittel" 16. und 17. Nov. 1988, Dresden

/2/ Gonzalez, V.M. Macosko, C.W.: Viscosity rise during free radical cross-linking polymerization with inhibition, ANTEC '84, S. 1024 - 1028

/3/ Nagaslaeva, S.D.; Trapeznikov, A.A.: Vysokomolekuljar- nye soedinenija 21, 1979, 4; S. 836 - 841

THE EFFECT OF THE PARTICLE AGGREGATION ON THE RHEOLOGICAL AND THERMAL PROPERTIES OF SHEARED SUSPENSIONS

P. RIHA and A. PRUCHA
Institute of Hydrodynamics
Czechoslovak Academy of Sciences
Podbabska 13, 166 12 Prague, Czechoslovakia

ABSTRACT

The aggregation of the suspended particles owing to the linking inter-particle forces influences the mechanisms of the rheological behaviour and heat transfer. The results show that the reduction in size of the aggregates under shear manifests as thixotropy and/or shear thinning. At the same time, the breaking of links between particles by shear decreases the contribution of particle motion to the effective thermal conductivity.

INTRODUCTION

The fact that the suspended particles play a decisive role in the rheological behaviour of fluid suspensions is very known and had been the subject of many rheological studies. On the other hand, the investigation of the influence of the suspended particles on the heat transfer in fluid suspensions did not raise so far a comparable interest although the experimental results [1] give evidence of the similarly important particles role as in the case of the rheological behaviour.

The mechanisms of the influence of the particles on the rheological and on the thermal properties of fluid suspensions are among others dependent on the particle motion, the particle properties, the particle arrangement as well as on the particle aggregability. The latter influence is analyzed in the following sections.

RHEOLOGICAL PROPERTIES

The aggregates increase the local volume concentration of particles in comparison with the mean concentration. Describing this fact, the volume concentration ϕ of the particles in a particular position in the suspension volume element containing a large collection of particles may be related to the mean volume fraction $\bar{\phi}$ of the volume element as

$$\phi = \overline{\phi} + \phi_{,k} \xi_k + \phi_{kl} \xi_k \xi_l + \cdots , \tag{1}$$

where ξ_k is the position vector with respect to the mass centre of the volume element and $\phi_{,k}$, ϕ_{kl}, ... represent the variation of the volume concentration of particles.

Apparently, if $\phi_{,k} = 0$ (for the sake of a simplicity only the vector $\phi_{,k}$ is considered here), the distribution of the particles in the volume element is uniform. If $\phi_{,k} > 0$ then, due to the formation of the aggregates, the volume concentration ϕ in the particular position in the volume element is higher than the mean concentration in the mass centre.

Since the partial density of the particles is $\rho = \phi \gamma$, where γ is the constant particle density, the corresponding balance equations for $\overline{\phi}$ and ϕ_k may be obtained in the form [2]

$$\partial \overline{\phi} / \partial t + \overline{\phi} v_{k,k} = 0 , \quad \partial \phi_k / \partial t + \phi_k v_{k,k} + m_k = 0 , \tag{2}$$

where m_k is the growth of the aggregates of the suspended particles.

The appropriate constitutive equation which includes also the influence of the particles and aggregates rotation Ω_k on the rheological behaviour may be considered in the form suggested by several authors (e.g. [3])

$$t_{kl} = \mu(v_{k,l} + v_{l,k}) + \zeta(v_{l,k} - \varepsilon_{klm} \Omega_m) . \tag{3}$$

The constitutive coefficients μ, ζ are generally dependent both on the mean volume concentration $\overline{\phi}$ and ϕ_k which are balanced by Eqs. (2).

Considering $\overline{\phi}$ constant for a particular suspension then only the growth m_k in (2) has to be specified by a suitable kinetic equation, e.g.,

$$m_k = k_A ({}_o \phi_k - \phi_k) - k_D \phi_k , \tag{4}$$

where ${}_o \phi_k$ denotes the value under stationary conditions and k_A, k_D are rate constants for the aggregation and disaggregation, respectively.

Rheological properties of fluid suspensions with aggregating structure are thus quantified by the system of Eqs. (2) and (3). If the ϕ_k changes are time dependent, the system describes thixotropic properties. In the opposite case, the system describes only both the shear-thinning and shear-thickening behaviour depending on the relation of the rate constants k_A and k_D to the shear rate $\dot{\gamma}$. If this relation is expressed in agreement with the suggestions [2,3,4]

$$k_D / k_A = a \dot{\gamma}^b , \quad \Omega_k = -1/2 \, F(v_{l,m}) \, \varepsilon_{klm} v_{l,m} , \tag{5}$$

where $F(v_{l,m})$ is a scalar proportion function of the particle angular velocity to the vorticity of the suspension, the derived equation for the viscosity from Eq. (3) has the form

$$\eta = \eta_\infty + (\eta_0 - \eta_\infty)/(1 + a\dot{\gamma}^b) , \tag{6}$$

i.e. has the form of the Cross's viscosity equation [5].

THERMAL PROPERTIES

On the basis of the phenomenological constitutive theory, the heat flux q_k through the sheared suspensions, having generally anisotropic thermal properties, may be written in the form

$$q_k = k_{kl} \, \theta_{,l} \, , \tag{7}$$

where the bulk thermal coefficient k_{kl} is with respect to the Eq. (3)
a function of the shear rate deformation $v_{k,l} + v_{l,k}$, the relative particle
rotation $v_{l,k} - \varepsilon_{klm} \Omega_m$, the particle volume concentration ϕ and of the
variation of the volume concentration $\phi_{,k}$. For a stationary case at the same
arrangement of the arrays of suspended particles or aggregates, the bulk
thermal conductivity for the constant fluid and particle conductivities is
changing only with the volume concentration of particles (see Fig. 1). We
have confirmed this fact by the numerical computation of the bulk conductivities for the possible cubic symmetries of the sphere particles or aggregates arrangements (simple cubic and body centered cubic lattices).
The finite element method (FEM) was used for the computations.

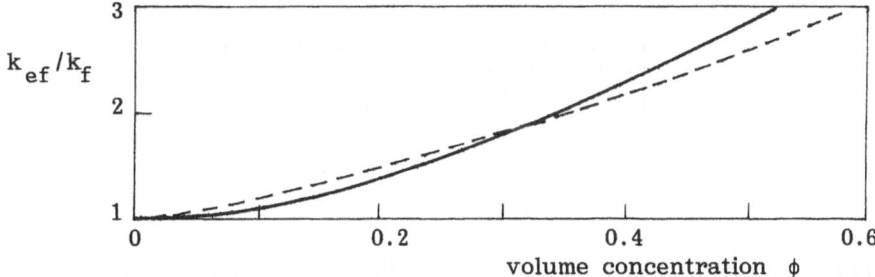

Figure 1. Predicted bulk thermal conductivities of a stationary suspension
of spheres. A solid line - FEM numerical results for
a simple cubic lattice arrangement of particles or
aggregates; a dashed line - Maxwell's eq. [1] . The ratio of
the particle to fluid thermal conductivity $k_p/k_f = 10$.

Under shear, the bulk thermal conductivity is influenced by the inner
structure motion [1]. With the shear increase, the size of the aggregates
reduces. An additional local convective heat flux dependent on the size of
rotating elements decreases as reveal preliminary numerical results. In this
way, the particle aggregation contributes to the mechanism of heat transfer
in sheared suspensions.

REFERENCES

1. Chung, Y.C. and Leal, L.G., An experimental study of the effective
   thermal conductivity of a sheared suspension of rigid spheres.
   Int. J. Multiphase Flow, 1982, 8, 605-25.

2. Riha, P., The influence of the RBC aggregation on the flow in large
   vessels. In Hémorhéologie et Agrégation Érythrocytaire, eds. J. F.
   Stoltz and M. Donner, Editions Médicales Internationales, Paris, 1990.

3. Brenner, H., Rheology of two-phase systems, In Annual Review of
   Fluid Mechanics, Vol. 2, eds. M. van Dyke and W.G. Vincenti, Annual
   Reviews Inc., Palo Alto, 1970, pp. 137-177.

4. Quemada, D., Rheology of concentrated disperse systems. Rheol. Acta,
   1978, 17, 632-42.

5. Cross, M.M., Kinetic interpretation of non-Newtonian flow. J. Colloid
   Interface Sci., 1970, 33, 30-35.

YIELD STRESS MEASUREMENTS ON WAXY NORTH SEA CRUDE OILS WITH CONTROLLED STRESS RHEOMETER AND MODEL PIPELINE

HANS PETTER RØNNINGSEN
Statoil a.s., Production Laboratories,
Forus, N-4001 Stavanger, Norway

ABSTRACT

A controlled stress rheometer (CSR) equipped with cone and plate geometry has been applied for yield stress measurements of gelled North Sea crude oils. The yield values measured in programmed shear stress mode, were found to be highly dependent on the cooling rate down to 5 °C and the stress loading rate while more or less independent of the aging time at 5 °C (within 6 hours). Similar, but less pronounced dependences were observed in a model pipeline test. Both tests had about ±20% repeatability. CSR results comparable to the model pipeline results were obtained by using higher cooling rate and stress loading rate in the CSR test. It seems to represent a useful alternative or supplement to the model pipeline test for assessing restart behaviour of gelled crude oils.

INTRODUCTION

A waxy crude oil which is allowed to cool down statically during a pipeline shutdown, develops a gel structure with a characteristic mechanical strength, which depends on the current pipeline temperature as well as the thermal and mechanical history of the crude oil. A measure of this gel strength, is the static yield stress of the gel. The restart behaviour of a crude oil is closely related to this material property. The common way of performing static yield stress measurements, is by a model pipeline test (1,2). Not considering complicating factors such as scale-up, autodestruction etc., the applied wall shear stress on yielding is easily converted to the required restart differential pressure in a full-scale pipeline. The development of computerized controlled stress rheometers (CSR) in recent years has suggested an alternative, far less sample consuming and possibly faster technique. In this presentation the effect of some experimental parameters in CSR measurements are discussed, and a comparison is made with model pipeline results. Four waxy crude oils were studied.

EXPERIMENTAL

Rheometer: Carri-Med CS100 with a 4 cm, 1 ° cone and plate geometry. Sample (415 μl) was transferred with pipette from a sealed pretreatment cell at 30 °C and cooled statically to 5 °C. The shear stress was programmed from zero to an appropriate upper value.

Model pipeline: A 14.7 m x 6 mm i.d. copper coil immersed in a thermostatted bath. Sample (415 ml) was transferred directly from the pretreatment cell at 30 °C, cooled statically to 5 °C and aged prior to measurement. N_2-pressure was applied to the pipeline inlet via the pretreatment cell in steps of 0.069 bar (15 min at each step). Yield criterion: 1 cm^3 flow in 15 min, i.e. wall shear rate 0.052 s^{-1}.

RESULTS

In the controlled stress rheometer test the static as well as the extrapolated, dynamic yield stress were highly dependent on cooling rate and stress loading rate. Going from 5 to 0.069 °C/min raised the static yield value by a factor 4 (see Figure 1A). The major part of the increase took place at the very low cooling rates, i.e. below about 0.2 °C/min. Increasing the stress loading rate from 0.021 Pa/min to 10.0 Pa/min, raised the static yield value of oil no. 3 from 2.5 to 11.3 Pa. The effect was largest at the lowest loading rates (see Figure 1B). The aging time at 5 °C had no significant effect on neither the static nor the dynamic yield stress within 0.5 min to 6 hrs. The repeatability of the yield stress measurements was about $\pm$ 20% which is similar to the model pipeline test.

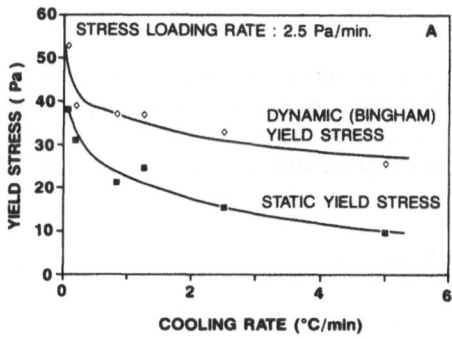 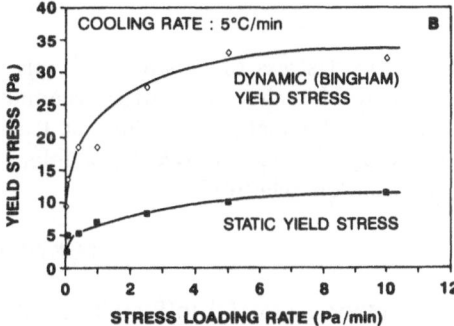

Figure 1. Effect of A) cooling rate and B) stress loading rate on yield stress at 5 °C in the controlled stress rheometer test.

The model pipeline test was generally carried out with 6 °C/hr cooling rate, 10 hrs aging at 5 °C and 0.047 Pa/min average stress loading rate. Very low cooling rates (0.1-1 °C/hr) and high stress loading rate (0.47 Pa/min) raised yield stress by about 20% and 50% respectively compared to the standard procedure. Aging times within 1 to 44 hrs gave no significant difference. At realistic (i.e. very low) cooling rates, extremely low stress loading rates and hence long analysis times had to be used in the CSR test to get results comparable to the model pipeline results. However, in general, good agreement was achieved by applying rapid cooling (5 °C/min) and a high stress loading rate (2.5 Pa/min) in the CSR test. This also gave flow curves that agreed well at the higher rates of shear, i.e. above about 200 s^{-1}, with controlled shear (concentric cylinder geometry) flow curves obtained with a Haake viscometer. Table 1 compares model pipeline (standard procedure) and CSR results of four representative, waxy North Sea crude oils which cover a wide range of yield stresses at 5 °C. Agreement within a few percent is seen for oils no. 2, 3 and 4, while oil no.1, which had the highest wax content and yield stress, was somewhat overestimated in the CSR test.

TABLE 1

Comparison of model pipeline and controlled stress rheometer (CSR) yield stress data.

| Oil no. | Static yield stress (Pa) at 5 °C | |
	Model pipeline[a]	CSR[b]
1	60.2	78.7[c]
2	20.3; 23.1; 30.1	23.5[d]
3	9.1; 7.0	9.8; 8.0; 7.0
4	7.0; 5.6	6.3; 6.5; 6.8

[a]6 °C/hr, 10 hrs aging, 0.047 Pa/min. [b]5 °C/min, no aging, 2.5 Pa/min. [c]6 measurements; SDEV = 10.6 Pa (13.4%). [d]21 measurements; SDEV = 3.9 Pa (16.6%).

CONCLUSION

Static yield stress of gelled crude oils comparable to model pipeline results can be obtained with controlled stress rheometer in less than one hour using only 0.4 ml sample. The method provides at least good early estimates and makes e.g. a fast screening of flow improvers possible. A broader discussion of this subject will be published elsewhere (3).

ACKNOWLEDGEMENT

The author is grateful to Statoil for allowing this presentation.

REFERENCES

1. Smith, P.B. and Ramsden, R.M.J., The prediction of oil gelation in submarine pipelines and the pressure required for restarting flow, Paper EUR 35, European Offshore Petroleum Conf. and Exhib., London, 1978.

2. Verschuur, E., Verheul, C.M. and den Hartog, A.P., Pilot-scale studies on re-starting pipelines containing gelled waxy crude, J. Inst. Pet., 1971, **57**, no. 555, 139-146.

3. Rønningsen, H.P., Rheological characterization of North Sea crude oils: Yield stress measurement, in preparation.

FINITE ELEMENT ANALYSIS
OF SHEAR–THINNING FLOW PROBLEMS IN MIXING VESSELS

Lothar Rubart

Institut für Strömungslehre und Strömungsmaschinen,
Universität der Bundeswehr Hamburg, Holstenhofweg 85, D–2000 Hamburg 70,
Federal Republic of Germany

ABSTRACT

A mixed finite element method for steady incompressible flow problems of generalized Newtonian fluids is presented. The Newton–Raphson scheme is used to treat the nonlinear terms in the resulting system of nonlinear algebraic equations and is combined with Uzawa's algorithm to solve the linear systems at each iteration step. Applications to the numerical analysis of various mixing problems in cylindrical unbaffled vessels are discussed. Computational results of the power consumption are compared with experimental data.

BASIC CONCEPTS

The theoretical analysis of isothermal, steady incompressible fluid flow with velocity $\mathbf{v}$ is based on the continuity equation

$$\text{div } \mathbf{v} = 0 \tag{1}$$

and the equation of motion

$$\rho \, \mathbf{a} = -\text{grad } p + \text{div } \mathbf{T} + \mathbf{f} \tag{2}$$

together with suitable boundary conditions. Here ρ is the density, $\mathbf{a}$ the acceleration vector, $\mathbf{f}$ the vector of the body forces and p the pressure. In the case of a generalized Newtonian fluid, the extra–stress tensor $\mathbf{T}$ is connected with the strain rate tensor $\mathbf{D}$ by [1]:

$$\mathbf{T} = 2\eta(\dot{\gamma}^2)\,\mathbf{D}\,, \quad \dot{\gamma}^2 = 2 \text{ tr } \mathbf{D}^2 \tag{3}$$

The viscosity η is considered to be an arbitrary function of the second invariant of $\mathbf{D}$.

NUMERICAL METHOD

In view of an approximate solution of the mentioned boundary value problem by finite elements the Ritz–Galerkin method is applied to obtain the corresponding weak formulation. A mixed finite element method is used to reduce the functional to a set of nonlinear algebraic equations in terms of coefficients representing the velocity vector and pressure values at the nodal points of the finite element mesh. An isoparametric element is proposed where the velocities are interpolated using bi–quadratic shape functions and the pressure is interpolated using linear shape functions defined on a triangular element, which is contained inside the quadratic element [2]. The numerical solution of the nonlinear equations is performed by the Newton–Raphson scheme where the dissipation term is treated in a way without specifying a special model for the viscosity function [3]. The linear subproblems at each iteration step are solved by Uzawa's algorithm, which is an efficient tool to treat the continuity constraint [4].

RESULTS

The method is applied to the numerical simulation of various mixing problems in cylindrical unbaffled vessels: the flow in the horizontal plane induced by rotating anchor or blade impellers and the three–dimensional axisymmetric flow due to a rotating disk, mounted coaxially on a shaft. The computations were performed for a real polymer liquid (CMC in water), where the viscosity data were found experimentally and were fitted to a suitable mathematical model. Two dimensionless groups are involved; the Reynolds number Re and a parameter α, defined by

$$\mathrm{Re} = \frac{\rho \; \mathrm{n} \; \mathrm{d}^2}{\eta_0} \; , \qquad \alpha = \frac{\eta_0^2}{\rho \; \tau^* \; \mathrm{d}^2} \; , \tag{4}$$

where α indicates the non–Newtonian viscosity behavior of the fluid. (η_0: zero–shear viscosity, τ^*: reference stress, n: number of revolutions, d: diameter of the impeller). It turns out that the computed flow fields agree with the real situation. This is approved by Figure 1 where the power characteristics, represented by the Newton number Ne, is shown as function of Re and α. Theoretical results are in accordance with experimental data even at high Reynolds number and for highly shear–thinning fluids. So the presented method is of practical importance.

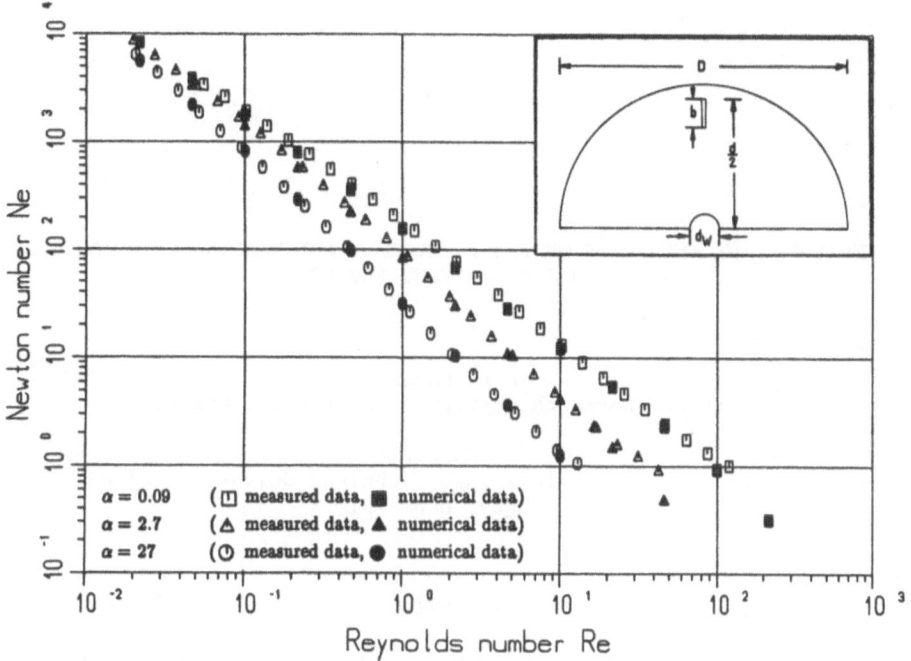

Figure 1.   Power characteristics of different CMC–liquids,
agitated in a vessel with anchor impeller

REFERENCES

[1]   Böhme, G., Non–Newtonian Fluid Mechanics. North–Holland, Amsterdam 1987

[2]   Kim, S.–W., A fine grid finite element computation of two–dimensional high
Reynolds number flows, Computers of Fluids, Vol. 16, (1988), pp 429–444

[3]   Böhme, G. and Rubart, L., Non–Newtonian flow analysis by finite elements,
Fluid Dynamics Research, Vol. 5, (1989), pp 147–158

[4]   Fortin, A. and Fortin, M., A Generalization of Uzawa's algorithm for the
solution of the Navier–Stokes equations, Comm. Appl. Numer. Methods,
Vol. 1, (1985), pp 205–208

INFLUENCE OF TIME CONSTANTS ON PHASE VELOCITY AND DAMPING RATES OF GRAVITY WAVES ON JEFFREY FLUIDS

ARILD SAASEN
Rogaland Research Institute, Stavanger, Norway

OLE HASSAGER
Department of Chemical Engineering, Technical University of Denmark,
Lyngby, Denmark.

ABSTRACT

The dispersion relation for gravity waves on a semi-infinite incompressible fluid layer satisfying the Jeffrey model was studied theoretically. The wave phase velocity and damping rates were studied as a fuction of wave numbers and time constants. Rayleigh wave modes can exist on a Maxwell fluid. For a Jeffrey fluid, any non-zero retardation time constants prohibit the existence of Rayleigh waves. However, elastic propagating wave modes may exist for a limited range of wave numbers.

INTRODUCTION

Linear gravity waves on Newtonian liquids have been thoroughly studied in the past. The effects of viscosity was studied in detail by Chandrasekhar [1]. Little attention has been given to gravity waves on viscoelastic liquids. Dispersion relations has been established as a function of wave numbers and relaxation time constants [2,3] . These works conclude that gravity wave modes on a Maxwell fluid become damped Rayleigh-waves in the limit of infinite wave numbers.

The present study covers investigations on dispersion relations for gravity waves on a semi-infinite fluid layer bounded above by a free surface. Interfacial tension effects are not studied.

The integral representation of the Jeffrey model is used since normal mode solutions are sought. The development of the characteristic equation is similar to the work reported earlier [2]. The dispersion relation is presented in a non-dimensional form using the Maxwell model non-dimensional time number, θ [2], and a Jeffrey model time number ratio, Λ. The time numbers, θ and Λ, are defined in Equations 1, and the non-dimensional velocity, c, and wave number, k, are given by the transformations in Equations 2, where the tilde, $\sim$, denotes dimensional quantities. g is the gravity, μ the viscous parameter, λ_1 the relaxation time constant and λ_2 the retardation time constant.

$$\theta = \left[\frac{g^2 \, \lambda_1^{\,3} \, \rho}{\mu} \right]^{\frac{1}{3}} \qquad \Lambda = \frac{\lambda_2}{\lambda_1} \qquad\qquad (1)$$

$$c = \left[g \, \frac{\mu}{\rho} \right]^{\frac{1}{3}} \tilde{c} \qquad k = \left[g \, \frac{\rho^2}{\mu^2} \right]^{\frac{1}{3}} \tilde{k} \qquad\qquad (2)$$

DISCUSSION AND RESULTS

The dispersion relation that is valid for the Maxwell fluid time number $\theta=0.25$ with zero retardation time is shown on Figure 1. The requirement for convergence of the Maxwell or Jeffrey fluid integrals sets an upper asymptote for the allowed damping rate. This requires a damping rate less than the inverse of the non-dimensaional time number θ; being 4 in the present case. The viscoelastic decay mode which exists for small wave numbers converges towards this value as the wave number $k \to 0$. The co-existing propagating mode converges towards the familiar deep water results, $Re(c)=(g/k)^{0.5}$. In a very short range of wave numbers the viscous and creeping modes of Chandrasekhar exist. The viscous mode and the viscoelastic decay merge into a branchpoint where the propagating Rayleigh type wave mode starts. In the limit of infinite wave numbers this becomes a damped Rayleigh wave mode [3] with a phase velocity of $Re(c)=0.9127\theta^{0.5}$.

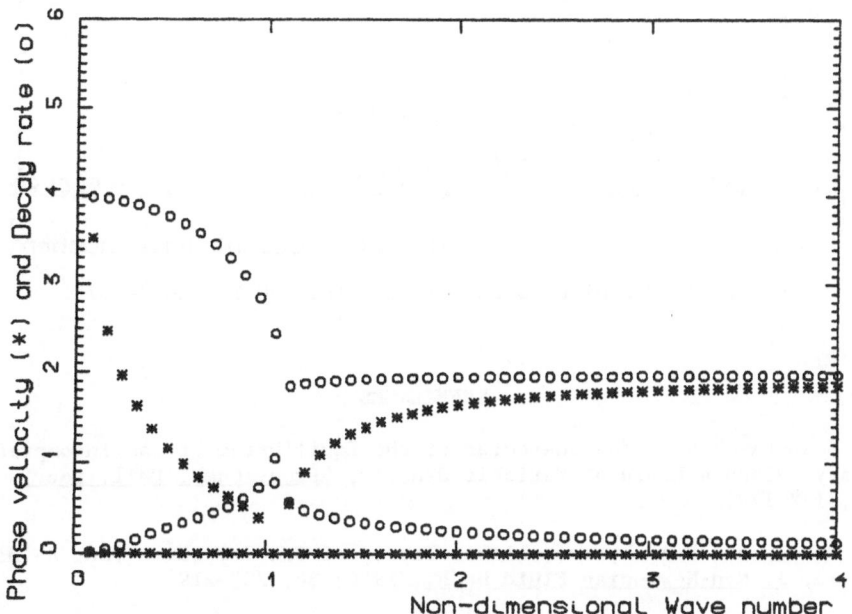

Figure 1. The dispersion relation for $\theta=0.25$ and $\Lambda=0$.

Rayleigh waves do not exist on a Jeffrey-fluid. Any non-zero time number ratio, Λ, allows only a zero phase velocity in the limit of infinite wave numbers. That is, only the creaping mode of Chandrasekhar exists. This is to be expected, as $\Lambda=0$ represents the Maxwell fluid and $\Lambda=1$ represents Newtonian fluids. On a Newtonian fluid, the phase velocity is equal to 0 for non-dimensional wave numbers larger than 1.1981 [1]. It is also seen from the governing equation for large wave numbers: In case of a Maxwell fluid this equation is a polynomial in $\theta^{0.5}c$, while in case of a Jeffrey fluid it will only contain a single term. Thus a constant phase velocity is not possible in the limit of infinite wave numbers. This is exemplified by the results shown in Figure 2 where $\theta=0.25$, and $\Lambda=0.75$.

To allow the damping rate to be arbitrarily large, θ must equal 0, or equivalently $\Lambda=1$. Thus, the viscous mode of Chandrasekhar will only exist for $\Lambda=1$ (or $\theta=0$) for large wave numbers.

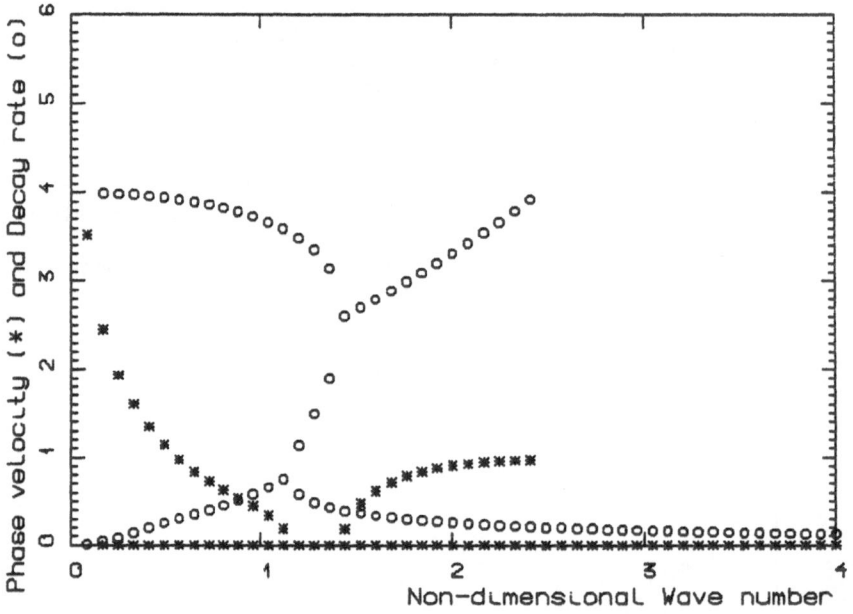

Figure 2. The dispersion relation for $\theta=0.25$ and $\Lambda=0.75$.

REFERENCES

1. S. Chandrasekhar, The Character of the Equilibrium of an Incompressible Heavy Viscous Fluid of Variable Density, Proc. Camb. Phil. Soc., 1955, 51, 162-178.

2. A. Saasen and P.A. Tyvand, Linear Theory of Gravity Waves on a Maxwell Fluid, J. Non-Newtonian Fluid Mech., 1990, 34, 207-219.

3. A. Saasen and P.A. Tyvand, Rayleigh-Taylor Instability and Rayleigh-type Waves on a Maxwell-fluid, J. Applied Math. and Physics (ZAMP), 1990, 41, 284-293.

UNIAXIAL TENSILE STRENGTH OF POLYMER FLUIDS

O. Yu. Sabsai, E. K. Borisenkova, V. A. Optov
NPO Plastik, Berezhkovskaya nab., 20; Institute
of Petrochemical Syntheses (Acad. of Sci.),
Leninsky prosp., 20; Institute of Chemical Physics
(Acad. of Sci.), Kosygina st., 4; Moscow, USSR

ABSTRACT

In dependance of the volume concentration of a filler there
are two different rupture mechanisms that are realising for
dispersion-filled high polymer melts.

INTRODUCTION

At high deformation rates polymer fluids may loos their
ability to flow as long as necessery and can break according
to cured rubber long-term durability mechanism. In [1] for
the ferst time the condition of fluid high polymers" rupture
at uniaxial extention was determined as follows:

$$\sigma/(\varepsilon - \varepsilon^*) = const > 0; \quad \varepsilon > \varepsilon^* \qquad (1),$$

where: σ - the true break stress, ε^* - the rupture recoverable
deformation (Hencky measure), ε^* -the critical recoverable
deformation, below which the rupture probability is equal to
zero. The rupture of cured high polymers may occur at any
value of deformation and thus for them $\varepsilon^* = 0$ [2], but in case
of high polymer melts" rupture $\varepsilon^* \neq 0$ and it does not depend
on the loading conditions and temperature [1,3].

MATERIALS AND METHODS

On Fig. 1-3 there are represented the results of investigation
of dispersion-filled high polymer melts" rupture. In a
regime of constant rate of uniaxial deformation at
temperature 293K series of experiments were carried on the

base of 1,2-polybutadiene (PB) with MM =1.8 x10^5 and
$\overline{M}_w/\overline{M}_n$=1.3. It containd 84% 1,2-units, its glass temperature
was equal to 256K. As despersed fillers were used shale-ash
and radiation-cured up to the glass and then milled 1,2-PB.
An average size of particles in both cases was equal to
20mkm.

RESULTS

As shown in Fig.1 the condition of durability (1) stays the
same, but the value ε^* falls linearly with the increase of
filler concentration φ_{vol} and does not depend on type or
nature of a filler. There is a critical concentration of a
filler φ_{cr}=25%$_{vol}$ -such, that at $\varphi > \varphi_{cr}$ ε^*=0 (Fig.2).

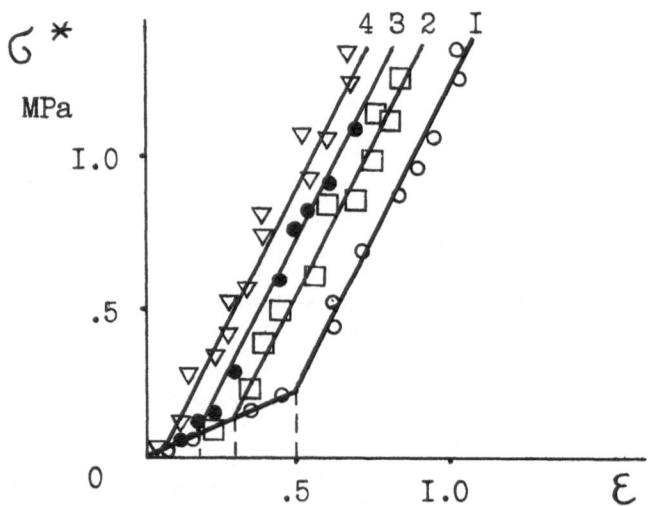

Figure 1. Rupture stress σ^* versus rupture recoverable
deformation ε. 1.-PB; 2.-(PB+10%$_{vol}$ of a shale-ash);
3.-(PB+15%$_{vol}$ of a cured PB); 4.-(PB+20%$_{vol}$ of a shale-ash).

At $\varphi > \varphi_{cr}$ the filled polymer melts may break at any
recoverable deformations, but in such cases the kind of the
dependance σ - ε changes qualitativly (Fig.3). As shown in
Fig.3 there are limited maximally achievable rupture
stresses and recoverable deformations at uniaxial extention
for highly-filled polymer melts.

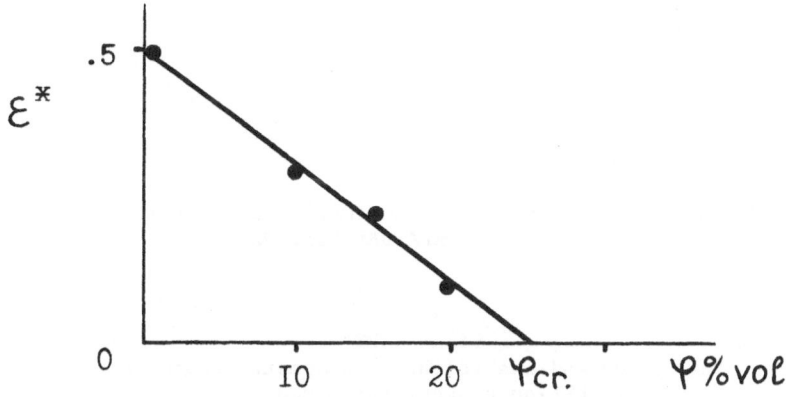

Figure 2. Critical recoverable deformation ε^* versus volume concentration of a filler φ.

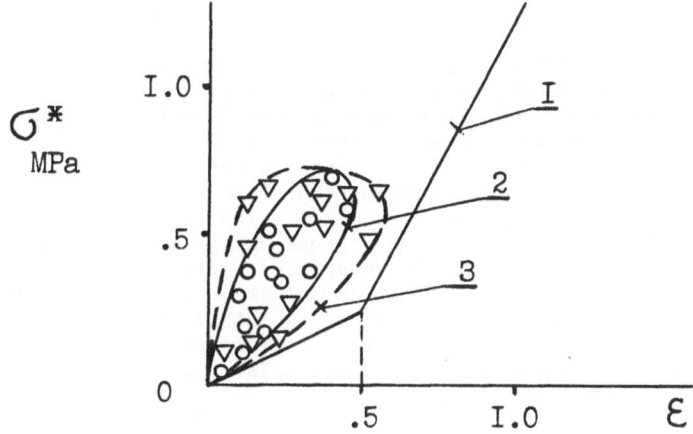

Figure 3. Rupture stress σ^* versus rupture recoverable deformation ε. 1. -PB; 2. -(PB+25%$_{vol}$ of a cured PB); 3. -(PB+30%$_{vol}$ of a shale-ash).

REFERENCES

1. E. K. Borisenkova, O. Yu. Sabsai, M. K. Kurbanaliev, V. E. Dreval, G. V. Vinogradov, Polymer, 1978, Vol.19, Desember, p.1473.
2. Smith T., in Rheology (Ed. F. R. Eirich), Acad. Press, N.-Y., 1969, 5, p. 143.
3. Chalaya N. M., Sabsai O. Yu., Abramov V. V., Prosessing of filled composit materials, Moscow, NPO Plastic, 1982, p.80.

THE USE OF NEWTONIAN VISCOSITY TO PREDICT SEDIMENTATION RATE IN TIME-TEMPERATURE MONITORS

SARAH SALT

Formulation and Analytical Development Department,
SmithKline Beecham Pharmaceuticals,
Clarendon Road, Worthing.

ABSTRACT

i-Point (TM) Time-Temperature Monitors (TTM) consist of enzyme coated PVC granules suspended in an aqueous Xanthan gum base. An enzyme mediated irreversible colour change occurs on storage which is time and temperature dependent, and which may be used as an indicator of end of shelf life.

Problems have been experienced with some batches of monitor in which the granules undergo sedimentation resulting in an uneven colour change.

Three batches of TTM with known sedimentation characteristics have been examined under flow and creep conditions using the Carri-Med CS Rheometer. The size and shape of the suspended particles was also investigated by light microscopy. Continuous shear measurements at shear rates up to 1000/sec showed small differences but creep results at shear rates of 0.001-0.1/sec gave Newtonian viscosity figures which correlated well with the observed sedimentation characteristics.

INTRODUCTION

i-Point (TM) Time-Temperature Monitors (TTM) consist of enzyme coated PVC granules suspended in an aqueous Xanthan gum base. These potentially have a widespread application in the food and pharmaceutical industries [1] where the time and temperature dependent enzyme mediated irreversible colour change could be used, as an easy to read indicator of product shelf life.

The manufacturers of the i-Point TTMs use 1-2% aqueous Xanthan gum, a high molecular weight polysaccharide extracted from seaweed, as suspending agent for the PVC particles [2]. The gel base is sterilized by autoclaving which markedly affects the viscosity. A series of batches of TTM having variable Xanthan gum concentration (1-2%) and variable autoclave cycles, were produced for evaluation by SmithKline Beecham. The relationship between rates of sedimentation, as determined by the evenness of the colour change upon storage, and the viscosity of the gel, as determined using the Carri-Med CS Rheometer, was investigated, with the aim of preparing a viscosity specification for the TTMs.

METHOD

Three batches of TTM with different sedimentation characteristics were examined using the Carri-Med CS Rheometer in both Creep and Flow mode (Version 4.3). A 4cm, 4 degree stainless steel cone and a temperature of 20°C ± 0.3°C, was used for all determinations. The contents of nine TTM were used for each measurement, as each sachet contained only about 200mg of gel. Each sample was allowed to equilibrate for 5 minutes after loading and then a series of six creep curves were produced at increasing torque (1-50µNm). Each sample was allowed to 'creep' for 5 minutes at each torque. Finally, a flow curve was produced for each of the gels, using the same sample as for the creep runs.

Newtonian viscosities for the three samples were compared at $0.001sec^{-1}$, in the linear visco-elastic region. Apparent viscosities at $100sec^{-1}$ and $200sec^{-1}$ were obtained from the rheograms and also compared.

Each of the batches was also examined by light microscopy to examine the shape and size of the suspended particles.

RESULTS AND DISCUSSION

TABLE 1

Batch	Sedimentation Characteristics	Particle Size	Particle Shape	Newtonian Visc. @ 0.001/sec* (Pa.S)
1	Poor	3-5µm	Spherical	57.1
2	Intermediate	3-5µm	Spherical	98.8
3	Good	40-170µm	Irregular	350.0
2% Keltrol (autoclaved)	−	−	−	52.7

*Newtonian Viscosity defined as the viscosity of the straight portion of the creep curve, in the linear visco-elastic region.

Flow data confirmed the rank order above, but the differences were much less marked at high shear rate.

It is possible to apply Stokes' Law to such systems to predict sedimentation rates [3], but assumptions such as sphericity of suspended particles and uniform particle size do not hold. When Stokes' Law is applied, it predicts that Batch 3 would have the fastest sedimentation rate and this does not occur in practice.

CONCLUSION

Newtonian Viscosity, a simple easy-to-determine parameter, has been used as the basis for a viscosity specification which defines the desired rheological characteristics of a gel-based TTM. As a result of this, the problem of uneven colour change has been eliminated.

REFERENCES

1. Taoukis, P.S. and Labuza, T.P. Applicability of Time-Temperature Indicators as Shelf Life Monitors of Food Products. Journal of Food Science, 1989, 54, No.4.

2. Kelco Technical Bulletin DB33. Keltrol T and Keltrol TF - Xanthan Gums for Food, Pharmaceutical and Cosmetic Application.

3. Power, M. An Approximate Method of Predicting the Shelf Life of Suspensions. Laboratory News, 12 June 1989.

A NEW MODEL FOR POLYMER MELTS AND CONCENTRATED SOLUTIONS

JAY D. SCHIEBER, P. BILLER, and F. PETRUCCIONE
Fakultät für Physik der Universität Freiburg
Hermann-Herder-Str. 3, D-7800 Freiburg , FRG

ABSTRACT

A simple mesoscopic model for polymer chains in melts and concentrated solutions at equilibrium is presented. The chain is modeled as a freely jointed bead-rod Kramers chain that makes jumps of the Orwoll-Stockmayer type. The elemental jumps are postulated to have a fractal waiting time distribution. The kinematics of the chain is investigated by a continuous time simulation in which the diffusion properties and the autocorrelation function for the end-to-end vector of the chain are followed.

INTRODUCTION

Concentrated polymer solutions and melts are dense systems with complicated many particle interactions. The motion of a chain is thus strongly hindered due to the presence of the other chains. We therefore propose here that a given polymer chain is frozen in space until it meets with a "gap", *i.e.*, a portion of free volume, in the melt. When a gap reaches a segment of chain the corresponding segment may jump to a new position. In the description of the chain dynamics the waiting time between two successive jumps of the same segment plays a fundamental role. The distribution of waiting times is assumed here to be fractal. Such an assumption has turned out to be very fruitful in a similar problem, namely the relaxation of concentrated solutions of electric dipoles [1]. Since the stochastic process is no longer Markovian for such a model, it is not possible to write a Master equation for the time evolution of the system, and thus, the kinematics is investigated by a continuous time simulation.

THE MODEL

Our model is based on a generalization of the Orwoll-Stockmayer model [2]. The elemental motions allowed here for interior links are the same as in the original model, but the end links may jump to a random orientation on the unit sphere. For simplicity we assume that the jump probabilities do not depend on the configurations. If we assume that the gaps are uniformly distributed and undergo a Brownian diffusion, then the distribution of times one must wait for a gap to reach a given point in space is fractal. Thus, the waiting time distribution of bead jumps is fractal. This distribution is generally defined as

$$\psi(t)dt := \text{Prob\{The time between successive jumps for a given bead} \qquad (1)$$
$$\text{is between } t \text{ and } t + dt.\}$$

We use the simplest such fractal distribution

$$\psi(t) = \frac{1}{\beta(1+t)^{\beta+1}}. \qquad (2)$$

For physical reasons we simulate chains with $\beta > 1$.

SIMULATION RESULTS

Two quantities were followed during the run of each simulation: the displacement of the center of mass, $r_c(t)$, and the end-to-end vector, $R(t)$. For the mean-square displacement $\langle r_c(t)^2 \rangle$, we see that 3 regimes exist: Fickian diffusion for short times, fractal non-Fickian behavior for intermediate times, and again Fickian diffusion when the displacement is on the order of one radius of gyration

$$\langle r_c(t)^2 \rangle \sim \begin{cases} t, & \text{First Fickian regime} \\ t^\alpha, & \alpha \text{ dependent on } \beta \\ t, & \text{final Fickian regime, large } t \end{cases} \qquad (3)$$

The dependence of α on β is shown in Table 2. The results for the autocorrelation function of the end-to-end vector $\langle R(t) \bullet R(0) \rangle$ can also be separated into 3 regimes: a complex first regime where all of the normal modes of the chain influence the results, a so-called stretched exponential regime, and the Markovian regime also predicted by the Orwoll-Stockmayer model

$$\langle R(t) \bullet R(0) \rangle \sim \begin{cases} \text{many modes,} & \text{small } t \\ \exp[-(t/\tau_i)^\gamma], & \gamma \text{ is a function of } \beta \text{ and } N \\ \exp[-t/\tau_f], & \text{final regime, large } t \end{cases} \qquad (4)$$

We note that the second, or stretched exponential, regime is much larger for longer chains, and for values of β near 1. The dependence of γ on β and chain length are

Table 1: Values for γ, where γ is the stretched exponential exponent for the intermediate regime of the end-to-end vector autocorrelation function: $\langle \mathbf{R}(t) \bullet \mathbf{R}(0) \rangle \sim \exp[-(t/\tau_i)^\gamma]$

$N\backslash\beta$	1.5	1.3	1.25	1.1
50	0.50	0.38	0.33	024
25	0.51	042	0.34	0.28
12	—	0.44	—	—
9	—	—	0.44	—
5	0.59	0.58	0.47	0.33

Table 2: The second, or fractal, regime of diffusion is described by the relation: $t \sim t^\alpha$, where α is a function of β. The last entry of α is only approximate. The parameter b describes the scaling of the time constant τ_i on number of beads N.

β	α	b
1.5	0.72	2.6
1.3	0.61	3.3
1.25	0.57	4.7
1.1	(0.30)	4.8

shown in Table 1. The dependence of the intermediate time constant τ_i on the number of beads N and β is found from the simulation results to be

$$\tau_i \sim N^b \tag{5}$$

where b is a function of β as shown in Table 2. Finding τ_f is considerably more difficult because of the accurate statistics required for its "measurement".

The diffusion results are consistent with experimental observations made by many groups of workers on many samples using several different techniques (see for example Refs.[3,4,5]). The scaling dependence of the time constant is also consistent with that of the longest relaxation time found from viscoelastic experiments.

REFERENCES

[1] M.F. Shlesinger, *Ann.Rev.Phys.Chem*, 1988, **39**,269–290.

[2] R.A. Orwoll, and W.H. Stockmayer, *Adv.Chem.Phys*, 1969, **15**, 305-324.

[3] G. Fleischer, *Polym.Bull.*, 1983, **9**, 152-158.

[4] I. Zupančič, G. Lahajnar, R. Blinc, D. Reneker, and D Vanderhart, *J.Polym.Sci., Phys.Ed.*, 1985, **23**, 387-404.

[5] H. Kim, T. Chang, J.M. Yohanan, L.X. Wang, and H. Yu, *Macromolecules*, 1986, **19**, 2737-2744.

DISSIPATIVE EFFECTS IN RHEOMETRIC FLOWS AND THE DETERMINATION OF MATERIAL FUNCTIONS OF POLYMER MELTS

J. Schneider

Academy of Sciences of the GDR, Institute of Mechanics,
P.O. Box 408, Karl-Marx-Stadt 9010 GDR

INTRODUCTION AND ABSTRACT

Determining of rheological material functions based on capillary and rotational rheometrical data is always a complicated problem if on the one hand due to environmental dissipation, expansion, reaction, heat and material exchange the rheometer ·flows are superposed by temperature, pressure, or concentration fields and on the other hand if these fields are significantly influencing the material characteristics during the experiment and by that also the measurable rheometric data /1, 6, 7/.

The present paper investigates the effect of the dissipation in both the rheometry of a LDPE melt (according to Meißner's experiments /3, 4, 9/) and a thermally interlacing EPDM caouchouc mix (according to Jepsen /8/).

Modelling and numerical simulation of the thermally .coupled rheometer flows enable computation of the flow field, temperature field, and the chemical reaction. Hence, they can be used to iteratively separate the effects being caused by the temperature field and reaction on the rheological material functions. This separation is realized with either the need of high computational efforts using both a regulation method and an optimization algorithm developed by Hofmann /5/ or it is performed in a heuristic-empiristic way in dialogue with the computer.

The results of separation are material functions of the type $F = F(\dot{\gamma}, T)$ or $F = F(\dot{\gamma}, T, x)$ as well as the parameters for the corresponding rheological constitutive equations and reaction-kinetic equations.

The illustrated examples are based on the viscosity functions of LDPE and EPDM.

METHOD AND MATERIALS

The mathematical description of the nonisothermic rheometer flows characterized by reactions bases on balance equations of mass, pulse, energy, and a corresponding reaction equation. For describing rheometric shear flows these equations can be simplified to a system consisting of onedimensional parabolic and ordinary differential equations under consideration of all significant field couplings and nonlinearities. The couplings between flow field, temperature field, and reaction field have their cause besides the convective heat and material transfer in the dissipation, expansion, temperature dependence of the reaction rate x, and especially on the dependence of the viscosity η , the temperature T, and the conversion ratio x. Other nonlinearities arise from the dependence of the viscosity on the shear rate $\dot{\gamma}$, the temperature dependence on the heat conductibility λ , and the dependence of the reaction rate on the conversion ratio.

The equation system can be completed by initial and boundary conditions as well as the necessary constitutive equations, and it can be numerically solved by means of a predictor corrector difference method (s. /1/). Boundary conditions are wall-stick and temperature conditions of the 3rd kind.

Both the viscosity functions $\eta = \eta(\dot{\gamma}, T)$ of LDPE according to /3/ and $\eta = \eta(\dot{\gamma}, T, x)$ of EPDM according to /8/ can be well described by a Carreau modell of the form

$$\eta(\dot{\gamma}, T) = \frac{C1 \exp\left(C2\left(\frac{1}{T} - \frac{1}{T_B}\right)\right)}{\left[1 + \left(C3 \exp\left(C4\left(\frac{1}{T} - \frac{1}{T_B}\right)\right)\dot{\gamma}\right)^2\right]} \left(C5 + C6 \frac{T - T_B}{T_B}\right)$$

being modified for taking into account temperature dependence which is to be completed by the reaction equation

$$\dot{x} = C7 \exp\left(C8\left(\frac{1}{T} - \frac{1}{T_B}\right)\right) \cdot (1 - x)^{C9}$$

for the interlacing EPDM as well as the relations for changes of viscosity caused by interlacing.

$$\frac{\Delta\eta}{\eta}(\dot{\gamma}, x) = C10 \cdot \dot{\gamma}^{C11} \cdot x^{C12} ,$$

$$\eta(\dot{\gamma}, T, x) = \eta(\dot{\gamma}, T) \cdot \left(1 + \frac{\Delta\eta}{\eta}(\dot{\gamma}, x)\right).$$

Inverse problems are solvable by solving the direct ones with stepwise improring of the constitutive parameters C1 to C6 or C7 to C12.

RESULTS AND DISCUSSIONS

Figures 1 and 2 show selected examples. Figure 1 demonstrates the stationary viscosity function $\eta = \eta(\dot\gamma, T)$ (temperature being parameter) of the LDPE melt measured by Meißner /1/ with a cone-plate and a capillary rheometer. The broken lines represent measuring results by Meißner which are to be approcimated using the Carreau statement. The extraordinarily strong decrease of velocity versus the increase of $\dot\gamma$-values is to be interpreted as the sum of effects of both pseudo-plastic material behaviour and dissipative heating causing decrease of viscosity. The solid curves represent the real material functions $\eta = \eta(\dot\gamma, T)$ gained from measuring results by separating the dissipation effects. The decrease of viscosity of these curves characterizes the pseudo-plasticity.

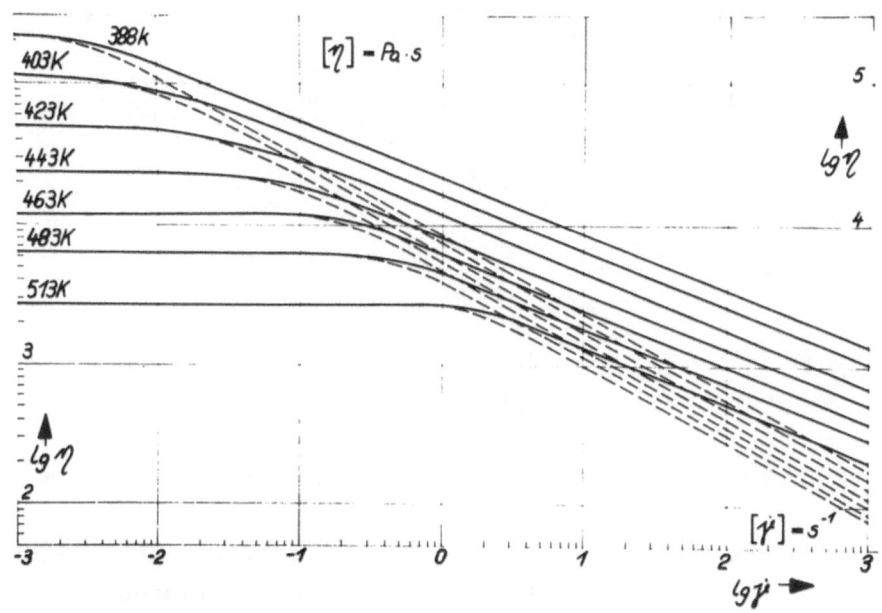

Figure 1: Viscosity functions of LDPE

Figure 2 shows the simulated timed response of temperature and viscosity of a thermally interlacing EPDM mix in a cone-plate rheometer with adiabated wall. The interlacing occurs under the influence of dissipation at shear rates of $\dot{\gamma} = 50\ s^{-1}$ and $\dot{\gamma} = 100\ s^{-1}$ The initial drop of viscosity is a consequence of the dissipative heating only. The thermal interlacing reaction ist initiated by the dissipatively constrained homogeneous temperature increase. The viscosity rises with increasing interlacing per time unit. Interlacing rations of x 0,2 are reached. The lower solid curves represent the viscosity response which, due to dissipation, would appear without interlacing reaction. Consequently, the temperature response has a point of inflection.

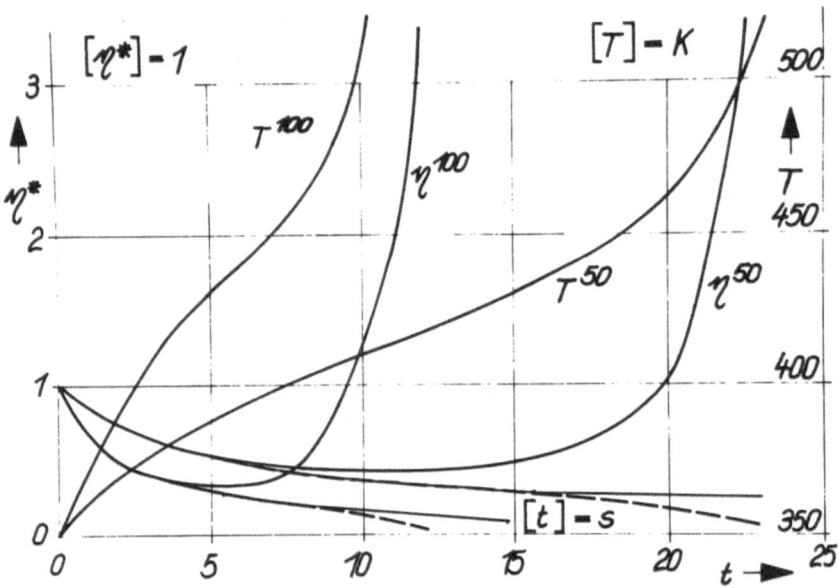

Figure 2: Temperature and viscosity of a interlacing EPDM

Based on the model of a viscous rheometer flow also material data of highly consistent visco-elastic fluids can be thermally corrigated if either the elastically stored energy share is small versus the viscously dissipated one or if these energy shares are separable. Both examples frequently occur in practice.

In comparison with simulation results according to figure 2 corresponding experiments allow to draw conclusions on the parameters C7 to C12.

REFERENCES

1. Schneider J (1975) Dissertation TH Karl-Marx-Stadt
2. Schneider J (1982, 1984) Plaste und Kautschuk 27: 361, 27: 626, 29: 279
3. Meißner J (1971) Kunststoffe 61: 576, 61: 688
4. Meißner J (1972) Journal of applied polymer science 16: 2877
5. Hofmann B (1983) Dissertation TH Karl-Marx-Stadt
6. Schneider J (1985) Report AdW DDR R-MECH 07/85: 115
7. Schneider J (1986, 1988) Progress and Trends in Rheology II, 157
8. Jepsen C (1988) Dissertation Universität Hannover
9. Meißner J (1989) Journal of Rheology 33: 843

DYNAMICS OF NEMATIC SOLUTIONS OF PERSISTENT MACROMOLECULES

A.N.SEMENOV
Physics Department, Moscow State University,
Moscow, 117234 USSR

ABSTRACT

Reptation dynamics of semiflexible persistent macromolecules in athermal nematic solutions is considered. A closed diffusion equation for the orientational distribution function of macromolecular segments is obtained. The Leslie angle and the Leslie viscosities α_2, α_3 are calculated, the result being asymptotically exact in the limit $L \gg l$, $l \gg d$ (L is the total length of a polymer chain, d the effective diameter, and l the Kuhn segment). It is shown that the ratio α_3/α_2 is always positive and tends to zero as the order parameter $s \to 1$.

INTRODUCTION

Let us consider an athermal solution of long rigid-chain persistent macromolecules with contour length L , the diameter d , and the Kuhn segment, l , $L \gg l \gg d$. The solution undergoes a liquid-crystalline transition as the polymer volume fraction φ exceeds some critical value $\varphi^* \simeq 10 \, d/l$ (1,2), where $\varphi = \pi \, Ld^2 c/4$ and c is the number concentration of polymer chains. If $l/d \gtrsim 50$, then the critical volume fraction is rather small; nevertheless the solution is expected to exhibit very interesting and complex equilibrium properties and dynamical behavior near φ^* (see, for example (3)).

It is well known that solutions of rigid-chain polymers form entanglement network already at small φ (4), so that at $\varphi \sim \varphi^*$ a reptation model for macromolecular motion (5) is justified. Moreover, at $\varphi \sim \varphi^*$ the topological constraints are so strong (for $l \gg d$) that a polymer chain can be considered as confined in a very thin "tube" of diameter $a \ll l$ (6).

The aim of this contribution is to calculate the Leslie viscosities α_2 and α_3 (7) of athermal nematic solution of semiflexible persistent chains using the reptation model.

DYNAMICAL EQUATIONS

Orientational steric interactions that give rise to a nematic order in the system originate the mean orientational molecular field:

$$U_{or}(n) = (cT/l) \int B(n,n') \, f(n',x) \, d^2n' dx \tag{1}$$

Here U_{or} is the potential energy of stright Kuhn segment with orientation n, $B(n,n')$ is the excluded volume of two such segments with orientations n and n', $f(n,x)$ is the distribution function for the unit vector n, tangential to a polymer chain at the point x (x is the distance along the chain between one of its ends and the given point, $0 < x < L$).

If the orientational field were absent, the distribution function would be governed by the following equations (5,6):

$$\partial f/\partial t = D \partial^2 f/\partial x^2 \; ; \tag{2}$$

$$\nabla_n^2 f + l\partial f/\partial x = 0, \; x = 0 \; ; \quad \nabla_n^2 f - l\partial f/\partial x = 0, \; x = L. \tag{3}$$

Here D is the diffusion constant for the reptation motion along the "tube".

If U_{or} is nonzero, the equations (2) – (3) must be generalized. Unfortunately, a closed kinetic equation for the function $f_t(n,x)$ does not exist in general case: one has to deal with an infinite system of entangled equations for $f_t(n,x)$ and all other "many-point" distribution functions such as $f_t(n_1,n_2;x_1,x_2)$ etc. (compare with ref.(8)). Nevertheless for semiflexible macromolecules, $L \gg l$, and for $t \gg l^2/D$ the closed equation does exist; it can be written as,(9):

$$\partial f_t/\partial t = D^* \partial^2 [f_t(n,x) - f^{(eq)}(n,x)] /\partial x^2 \; , \tag{4}$$

the boundary conditions being unchanged (see eqns.(3)).
Here $f^{(eq)}(n,x)$ is the equilibrium distribution function corresponding to orientational field $U_{or}(n)$; D^* is the renormalized diffusion constant which depends on orientational field (generally $D^* < D$).

RESULTS AND DISCUSSION

Using eqns.(4),(3) after some analitical work we come to the following final results (9):

$$\alpha_2 = (cL^3/48 \, lD^*) \int_{-1}^{1} dz \, f^{(eq)}(z) \, [2z \, dU_{or}/dz - z^2(1-z^2) \cdot d^2 U_{or}/dz^2] \; ; \tag{5}$$

$$\alpha_3 = (cL^3/48lD^*) \int_{-1}^{1} dz \, f^{(eq)}(z) \, (1-z^2)^2 d^2 U_{or}/d \, z^2 \; , \tag{6}$$

where $z = n \cdot u$ (u is the director of the nematic solution), $f^{(eq)}(z) = (1/L) \int_0^L f^{(eq)}(n,x) \, dx$ and $U_{or} = U_{or}(z)$.

An analysis shows that both viscosities α_2 and α_3 are negative. This implies in particular that in the simple shear flow the director u of the solution would be stationary oriented at Leslie angle to the flow direction, $\tan^2 \theta_L = \alpha_3/\alpha_2$ (7). The angle $\theta_L = 20°$ for the liquid-crystalline phase conjugated to isotropic (i.e. for $s_0 = 0.46$ (2), where s is the order parameter). As s increases, the angle θ_L decreases significantly: $\theta_L \to 0$ in the limit $s \to 1$.

These results are in contrast with that obtained for completely stiff polymers ($l \gg L \gg d$, see refs.10-12). The ratio α_3/α_2 for a nematic solution of hard rods is negative at $s \gtrsim 0.53$, therefore the simple shear flow is unstable here (unless the flow is extremely slow).

Both limits of rigid rods ($l \gg L$) and semiflexible chains ($L \gg l$) are hardly accessible in experiments. An intermediate situation $L \sim l$ is the most common. In order to consider this general case one can use the following approximate equation (compare with eqn. (4)):

$$\partial f_t / \partial t = D^* \partial^2 \psi / \partial x^2 + \hat{L} \qquad , \qquad (7)$$

where $\psi = f - f^{(eq)}$, and $\hat{L}$ is some linear operator which can be rather easily expanded in power series of orientational field U_{or} . It can be proved that retaining only the first, linear term in this expansion we get an equation which is asymptotically exact in both limits $l \gg L$ and $L \gg l$. Thus, one can expect that this equation is a good approximation also in general case, $L \sim l$. The corresponding study of dynamical properties of partially flexible polymers ($L \sim l \gg d$) is now in progress.

REFERENCES

1. Odijk, T., Macromolecules, 1986, 19, 2313 - 28.
2. Semenov, A.N., Khokhlov, A.R., Usp.Fiz.Nauk, 1988, 156, 427 - 476.
3. Lee, S.-D., Meyer, R.B., Phys.Rev.Lett., 1988, 61, 2217-20.
4. Doi, M., Edwards, S.F., J.Chem.Soc.Farad.2, 1978, 74, 560 - 570; 918 - 932.
5. Doi, M., Edwards, S.F., J.Chem.Soc.Farad.2, 1978, 74, 1789 - 1817.
6. Semenov, A.N., J.Chem.Soc.Farad.2, 1986, 82, 317 - 329.
7. De Gennes, P.G., The Physics of Liquid Crystals, Claredon Press, Oxford, 1974.
8. Semenov, A.N., Zh.Eksp.Teor.Fiz., 1989, 96, 194 - 213.
9. Semenov, A.N., Sov.Phys.JETP, 1987, 66, 712 - 716.
10. Semenov, A.N., Sov.Phys.JETP, 1983, 58, 321 - 326.
11. Kuzuu, N., Doi, M., J.Phys.Soc.Jap., 1983, 52, 3486 - 3494.
12. Kuzuu, N., Doi, M., J.Phys.Soc.Jap., 1984, 53, 1031 - 1038.

HELICAL FLOW OF POWER-LAW FLUIDS

J.Sestak,M.Houska,[*]R.Zitny and M.Dostal
Faculty of Mechanical Engineering,Suchbatarova 4,166 07 Prague6
*Food Industry Research Institute,Radiova 7,102 31 Prague 10
Czechoslovakia

ABSTRACT

A numerical solution is presented for the helical flow of a
power-law fluid between coaxial cylinders. The solution is
valid for isothermal steady creeping flow in an annular space
of arbitrary thickness. Results are presented in the form of
relations between dimensionless integral flow characteristics
(e.g. pressure difference, volumetric flow rate and torque
acting on the inner cylinder). Resulting graphs are valid for
arbitrary radii ratio and have the flow behaviour index as a
parameter. Modified rotational rheometer with coaxial cylinders was used for experimental verification of numerical
solution. System of coaxial cylinders having the radii ratio
0.805 was used in experiments. Used model power-law fluids had
the flow behaviour index in the range 0.35-1.0. The agreement
between theoretical and experimental values of dimensionless
pressure difference was found satisfactory. Experimental values
of dimensionless torque were about 20% higher than values calculated for corresponding flowrate and rotational speed.

INTRODUCTION

The helical flow of non-Newtonian fluids in an annular gap is a
common geometry for several industrial processes. Design of
such a process is tedious especially if the gap is not narrow.
    Bird et al.(1) solved this problem for thin annulus.Theory
of the general solution was given by Coleman and Noll (2).
Dierckes and Schowalter (3) applied their theory to the solution of the helical flow of the power-law fluid and together with
Rea and Schowalter (4) verified experimentally the theoretical
prediction.
    The main goal of this work is to present the detailed
numerical solution of the helical flow of power-law fluid for
any cylinder radii ratio and arbitrary flow behaviour index and
experimental verification of the theory.

438

METHODS

Isothermal steady creeping helical flow of power-law fluid was solved. Power-law fluid, characterised by consistency coefficient m (Pa.s^n) and flow behaviour index n (-), flows through the annular gap between coaxial cylinders having radii R_1 and R_2 (m) and length L (m). Velocity of the inner cylinder is v_o (m/s).

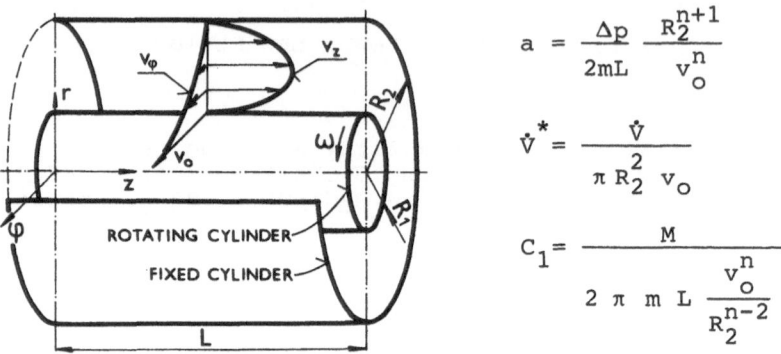

$$a = \frac{\Delta p}{2mL} \frac{R_2^{n+1}}{v_o^n}$$

$$\hat{v}^* = \frac{\hat{v}}{\pi R_2^2 v_o}$$

$$C_1 = \frac{M}{2 \pi m L \frac{v_o^n}{R_2^{n-2}}}$$

Figure 1. Geometry of the helical flow and definition of the dimensionless quantities

The mathematical model of flow consists of two components momentum balance equations in axial and radial directions. Integration of those two nonlinear differential equations provided the tangential and axial velocity distributions. Four integration constants were determined from boundary conditions and the problem was mathematically transfered to the solution of two nonlinear algebraic equations. Using velocity distribution the integral dimensionless quantities were calculated. The results are presented as the graphical dependencies between dimensionless equivalents of the pressure difference Δp (Pa), flowrate $\hat{v}$ (m^3/s) and torque M (N.m) such as a, $\hat{v}^*$ and C_1, respectively, defined in Fig.1. Experiments were made on the set-up consisted of an adapted rotational rheometer with coaxial cylinders Rheotest 2 (GDR). The cup of the rheometer was modified to enable the axial flow of model fluid through the annular gap and measurement of the pressure difference. Flowrate of the fluid was measured by the volumetric method.

TABLE 1
Model fluids characteristics

	Kaolin	Glycerol	Water	ρ(kg/m^3)	m(Pa.s^n)	n (-)	t (^{o}C)
O	-	92.6%	7.4%	1242	0.57	1.00	16.0
△	17.4%	40.3%	42.3%	1224	2.96	0.35	16.6
▽	14.4%	49.3%	36.3%	1232	1.41	0.46	18.6
□	12.3%	55.7%	32.0%	1244	0.91	0.57	18.8

Rheological properties in this table were measured on Haake RV3.

RESULTS

Numerical procedure developed for the helical flow solution provided the dependencies between the dimensionless quantities as mentioned above, see Fig.1. The example of such dependencies is apparent in Fig.2. The example was calculated for radii ratio $R_1/R_2=0.5$ and it is apparent that received dependencies are strongly influenced by the flow behaviour index as a parameter.

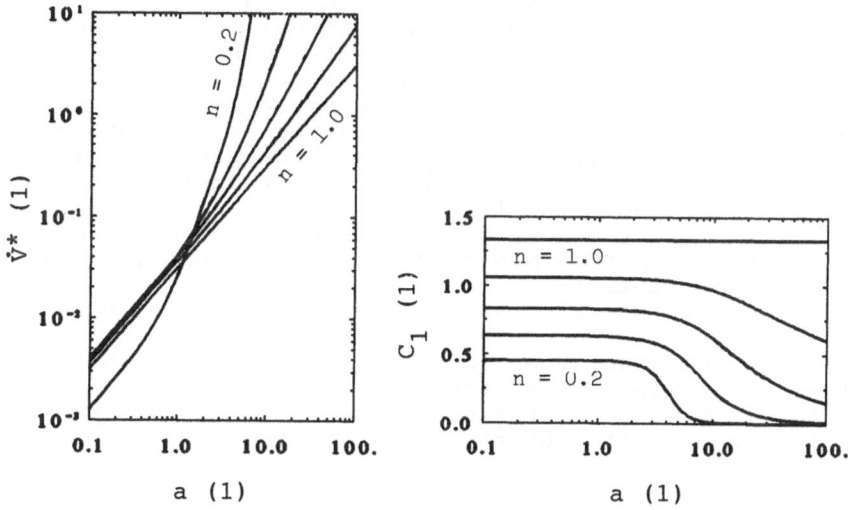

Figure 2. Results of the numerical calculation of the helical flow of power-law fluid, valid for $R_1/R_2=0.5$

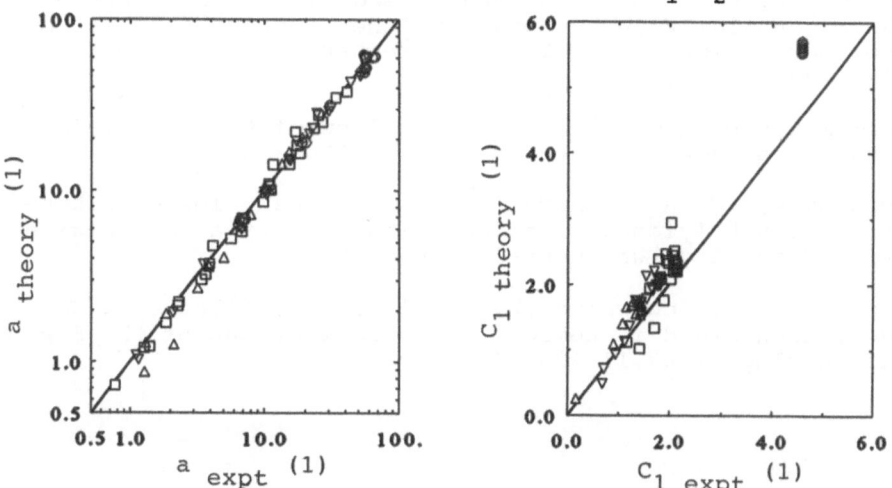

Figure 3. Comparison of the dimensionless pressure difference and torque, calculated for experimental conditions: $R_1/R_2=0.805$ and flow behaviour indexes n=0.35-1.0

Comparison of theoretical and experimental values of the dimensionless quantities is shown in Fig.3. Calculations were made for arbitrary radii ratio, flow behaviour index, rotational speed and flowrate as determined in experiments.

DISCUSSION AND CONCLUSION

Numerical procedure used for solution of the helical flow was shown as accurate and usefull. Calculated graphs can be simply used for determination of integral helical flow characteristics such as pressure difference or flowrate. Good agreement between theoretical and experimental values of the dimensionless pressure difference was found. Experimentally determined values of dimensionless torque are about 20% higher then those determined for the same flowrate by calculation procedure. This result can be partly explained by the fact that fluid covered the inner cylinder and its shaft more than allowed in standard geometry for viscosity measurement. This effect increased the torque about 6-10% as determined by separate experiments and depends on the flowrate.

Experiments provided an evidence that the rotational speed of the cylinder can enhance the axial flowrate as predicted by the theory. The lower the flow behaviour index the higher enhancement of flowrate at the same pressure difference was observed. This effect verified the decrease of the apparent viscosity as predicted by the power-law model for two-dimensional flow.

REFERENCES

1. Bird,R.B.,Armstrong,R.C. and Hassager,O.,Enhancement of Axial Annular Flow by Rotating Inner Annulus,Dynamics of Polymeric Liquids,Vol.I. Fluid Mechanics,Wiley,New York, 1977,226-229

2. Coleman,B.D. and Noll,W.,Helical Flow of General Fluids, Journal of Applied Physics,1959,30,1508-1512

3. Dierckes,A.C. and Schowalter,W.R.,Helical Flow of a Non-Newtonian Polyisobutylene Solution,Industrial and Engineering Chemistry Fundamentals,1966,5,263-271

4. Rea,D.R. and Schowalter,W.R.,Velocity Profiles of a Non-Newtonian Fluid in Helical Flow,Transactions of the Society of Rheology,1967,11,125-143

THE TEST METHOD OF POLYMER MELT BY THE ELONGATIONAL DEFORMATION

VERA SEVRUK
Scientific Production Union "Plastic"
Moscow, USSR

ABSTRACT

In many polymer processing technics including extrusion and orientation of polymer melts both elongation and shear take place. However rheological properties of polymer melts are usually determined by the practical processing under the shear deformation basically by melt index.

In this paper supplementary polymer characteristic are proposed. It's elongation index, measured under uniform elongation at definite constant force and temperature.

INTRODUCTION

The elongational deformation method under a constant force F is analogous to roll's extension from the die with roll's constant angle velocity. In the paper (1) the relation of extrudate dimensions under not uniform extension from the die and sample dimensions under the uniform elongation under F = const is determined. It's shown in the paper (2) that on the base of relation the not uniform extension may be modeled by uniform elongation under F=const, which is easily realized in laboratory conditions. In the paper (3) it's pointed that polymers with analogous shear properties may significantly differ by the elongation behaviour.

METHODS AND MATERIALS

Following polymer grades are researched: polystyrene-PS-168N (Basf) and PS-151 (USSR); high density polyethylene (HDPE)-

vestalen A-6042 (Basf), D6DR-6640 (Union Carbide) and HDPE-274 and 288 (USSR); low density polyethylene (LDPE)-16803-070 and 11503-070 (USSR), Alkaten WNC-71, Bailon 19N430, Peten B-8015 and B-4524; high impact polystyrene (HIPS)-424, 703 and 803 (USSR).

Flow curver were received by capillary viscosimeter Rheograph-1000 (Göttfert).

Elongation deformation was realised in the silicon oil bath, regime F=const is received with loads.

RESULTS AND DISCUSSION

In producing of plane films and bands PS-168N is used. At changing of PS-168N for PS-151 it was impossible to produce good films and bands though the materials have analogical flow curves. It was determine that they significantly differed by their behaviour at elongation (Fig.1).

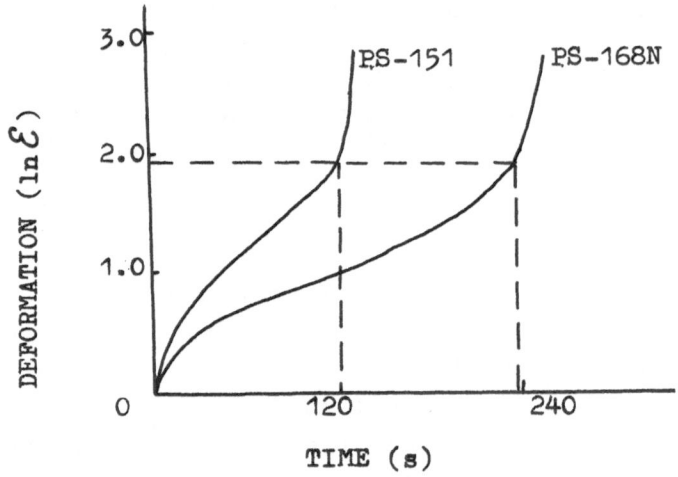

Figure 1. Time dependence of deformation.

as PS-151 has low molecular tail that was shown by gel-chromatografy.

An analogical analysis was made choosing raw HDPE for oriented bands produced on the equipment manufactured by 'Windmüller und Hölscher and LDPE for paper and cardboard lamination. Also various HIPS grades for refrigerators sheets were analysed.

All the researched materials differ at elongation deformation under F=const inspite of analogical behaviour under
shear. These differences well correlate with the polymers behaviour in processing. The test method is very sensitive not
only to the molecular-weight distribution; as well as to the
branching, different additives: stabilizers, plasticisers,
fillers, other polymers, and etc.

For practical use it's convenient to choose one point on
the above curve - the time when the polymer sample obtains
certain deformation. This point may be identified as the melt
elongation index.

REFERENCES

1. Prokunin A.N. Same methods of experimental studies into
   the extension of polymeric liquids. Intern. Polymeric
   Mater., 1980. v.8, pp. 303-321.

2. Prokunin A.N., Sevruk V.D. About the die-swell effect under elongation of the polymeric liquid from the die. Inzhenerno-fizicheskii journal, 1980, V.39, pp. 343-350.

3. Prokunin A.N., Sevruk V.D. Effect of flow slowing under
   different types of elongation. Inzhenerno-fizicheskii journal, 1981, v.41, N1, pp. 74-81.

RHEOKINETICS, FLOW AND CONVECTIVE HEAT TRANSFER OF POLYMERIZING OLIGOMERS

Z.P. SHULMAN, B.M. KHUSID, E.V. IVASHKEVICH, N.O. VLASENKO
Heat and Mass Transfer Institute, BSSR Academy of Sciences
220078 Minsk, USSR

A knowledge of the rheological properties of a composition as a function of conversion degree and temperature as well as macrokinetics of polymerization and heat release processes is necessary to make thermal and hydraulic calculations of a chemical molding process. A space-time separation of thermochemical (reservoir) and thermohydrodynamics (channel) stages is proposed to find rheokinetics and macrokinetics of a reactive fluid. The essence of their separation method is as follows. A test medium is placed into a thermostating cylinder of a charging chamber where it is maintained for a certain period of time to attain a prescribed curing degree. Then the fluid is pumped from the cylinder with an assigned velocity via an independently thermostating capillary. Tests made at different delay times (different curing degree) and at a fixed temperature of the charging chamber enable one to obtain the viscosity of a curing fluid as a function of conversion degree. Tests were made with the following composition: epoxy-diamine resin and curing agents such as metaphenilendiamine. The experimental set-up was composed of the independently thermostated reservoir and channel. The performance parameters were chosen in a special manner. There are: $d_c \ll d_r$ (d_c and d_r are the diameter of the channel and reservoir), $t_d \simeq t_k$ (t_c/t_r)≈ 0.2, $t_r \ll t_k$ where t_d, t_c, t_r, t_k are the delay times in a reservoir, in a channel, discharge time of a reservoir, characteristic curing time, respectively. Optimum experimental regimes (choice of an initial polymerization temperature, reagent ratios and delay time of a curing composition in a reservoir) were determined from the preliminary measurements under the conditions of a considered process similar to the isothermal ones. In our case, the initial curing temperatures were equal to 60, 65 and 67.5°C; the reagent ratios were stochiometric; the delay time of a reactive mixture in a reservoir was varied from 30 to 60 min. The effect of rheokinetics on convective heat transfer from a partially polymerized fluid was studied in the nonisothermal channel fluid, with a channel wall temperature being equal to 40°C. For a cured composition, with regard to chemical and rheological kinetics, the delay times of a reactive fluid in the reservoir were 30 and 35 min. Their conversion degrees were $\beta=0.072$ and $\beta=0.088$, respectively. A fluid pumping velocity range (0.003-0.18 m/s) was chosen. The Pecklet number for the chosen regimes was $(0.6-5)10^3$ and the relative length of the thermal section was 55-110 diameters. Thus, heat transfer occured in the thermal entrance section. The analyzed shear rate range (100-500 s^{-1}) was characterized by the Newtonian nature of the flow curve and by the viscosity independent on a strain rate. Numerical calculations of thermal and kinetic polymerization parameters were made using the macrokinetic model [1]. Local and volume-mean temperature and conversion degree β are calculated in terms of the determined concentration fields of reagents and temperature in the reservoir. The integral characteristics at different regimes of the composition stay in the reservoir are then used as input parameters in the experiments on flow and convective heat transfer in a channel. For $\beta \leq 0.22$, the following relation was obtained for the viscosity of a partially cured composition

$$\eta = a \exp \left\{ \frac{E}{R} \left[\frac{1}{T} - \frac{1}{T_o} \right] + c\beta \right\} ,$$

where $\alpha=10.78$ Pa·s, $c=19.7$ and $T_o=298°K$. The activation energy of viscous flow (E=19.7 kJ/mol) was specified by the resin properties alone. The Nusselt number as a function of the

446

Pecklet number for nonisothermal flow of a partially cured composition was the same as in the case of the newtonian fluid with the nonisothermal Zider -Tayt correction:

$$\overline{Nu} = C \left(\frac{1}{Pe} \frac{L}{d}\right)^{-1/3} \left(\eta_f/\eta_{in}\right)^{-0.14}.$$

Here L is the channel length, η_f- fluid viscosity. Processing of the data on the hydraulic channel resistance under nonisothermal conditions has shown that it is possible to use Poiseuille's formula with the effective viscosity in the form

$$\eta_{eff} = \eta_{in} \exp\left\{\frac{E}{R}\left(\frac{1}{T_f} - \frac{1}{T_{in}}\right)\right\},$$

were η_{in} is the inlet viscosity of a composition. Simultaneously solving the equations for hydraulic resistance and heat transfer, one can calculate the characteristics of nonisothermal flow of a partially cured fluid at a prescribed pressure drop ΔP. The relations are obtained for the Graetz ($\overline{Gz}$) and Stanton ($\overline{St}$) numbers at a relative temperature difference across a channel $\varkappa=(T_{in}-T_w)/T_w$:

$$\overline{Gz} = \overline{Gz}_0 \exp\left[-B\varkappa\left(\frac{1}{\varkappa+1} - \frac{1 - \exp(-\overline{St})}{2\overline{St}+\varkappa(1-\exp(-\overline{St}))}\right)\right]$$

$$\overline{St} = \frac{4C}{\overline{Gz}^{2/3}} \exp\left[-0.07\varkappa \frac{1 - \exp(-\overline{St})}{\overline{St}+0.5\varkappa(1-\exp(-\overline{St}))}\right]$$

The $\overline{Gz}$ and $\overline{St}$ numbers differ from the traditional ones by the factors $4/\pi$ and $d/4\pi$, respectively. The first equation gives the hydraulic resistance under nonisothermal conditions and the second, convective heat transfer.

The authors would like to thank Dr. I.L. Ryklina and V.A. Mansurov for their participance in the work.

REFERENCES

1.Arutyunyan, Kh.A., Davtyan, S.P., Rosenberg, B.A., Enikolopyan, N.S. Curing kinetics of epoxy oligomer ED-5 affected by metaphenilendiamine under adiabatic conditions. Vyssok. Molek. Soed. 1974, 16(A), No. 9, 2115-2122.

UNSTEADY SHEAR FLOWS OF VISCOELASTIC LIQUIDS DRIVEN BY PERIODIC FORCING

D.A. SIGINER
Department of Mechanical Engineering
Auburn University
Auburn, AL 36849 U.S.A.

ABSTRACT

Nearly viscometric flows of viscoelastic liquids represented by integral fluids with fading memory are investigated. Oscillatory shear fields are imposed from the boundary and through pressure gradient oscillations and are superposed onto pure shear. The fluid exhibits flow enhancement effects, independent of the explicit form of the constitutive functions, due to its nonlinear internal structure. Closed form expressions for the change in the mass transport rate are given at the lowest significant order in the perturbation algorithm. It is shown that the interaction of oscillatory shear fields may generate a resonance effect, a sharp increase in the flow rate either longitudinal or orthogonal and independent of the mean pressure gradient, if certain frequency conditions are met. A method to determine the constitutive functions embedded in the internal structure is outlined.

INTRODUCTION

Deviations from Poiseuille flow rate are not observed in both laminar and turbulent flow under moderate periodic forcing when the structure of the liquid is linear. But large deviations from the velocity profile corresponding to that of the Newtonian liquid of the same viscosity and density driven by the same mean gradient may occur when the constitutive structure is nonlinear. Oscillating pressure gradient driven flow has been investigated by several authors with the emphasis placed on differential type constitutive models. Available experimental data and in particular the frequency dependence of the flow enhancement cannot be predicted by any of the popular models such as four constant Oldroyd, Goddard-Miller, fourth order fluid, co-rotational, non-affine network, Wagner, etc. with the possible exception of the generalized Maxwell model. The data shows an increasing enhancement with increasing frequency for moderate frequencies at fixed mean gradient and an increasing and decreasing enhancement for moderate and large mean pressure gradients respectively at fixed frequency. No data is available at high frequencies either at fixed mean gradient or fixed frequency. Experimental data pertaining to the effect of the interaction of a periodic shear field imposed from the boundary with pure shear shows large enhancement effects at small pressure gradients with polymeric liquids. Enhancement increases with increasing frequency and amplitude at fixed pressure gradient and monotonically decreases with increasing pressure gradient at fixed frequency and amplitude. Although existing analytical investigations qualitatively predict the trend displayed by the data for small frequencies quantitative predictions proved to be elusive and the models used, the slightly non-Newtonian and the Rivlin-Ericksen differential models, are open to serious criticisms leveled by recent stability studies.

The contribution of this paper is to study the behavior of an integral fluid, represented by a series of multiple integrals in the strain histories, under the combined effect of pressure gradient fluctuations and longitudinal and transversal boundary oscillations both represented by a finite Fourier series with no restrictions on the magnitude of the strain rate.

ANALYSIS AND RESULTS

Conceptually a framework for a universal constitutive equation has been laid out by Noll and his simple fluid theory has received wide acceptance. The response functional of an incompressible simple fluid

$$\underset{\sim}{S} = \underset{s=0}{\overset{\infty}{\theta}} \left[\underset{\sim}{G}(\underset{\sim}{X},s) \right] = \underset{\sim}{T} + p \underset{\sim}{1} \quad ; \quad s = t-\tau$$

has been developed by Green & Rivlin into a uniform series

$$\underset{s=0}{\overset{\infty}{\theta}} \left[\underset{\sim}{G}(\underset{\sim}{X},s) \right] = \int_{0}^{\infty} \underset{\sim}{K_1}(s)\underset{\sim}{G}(s)ds + \int_{0}^{\infty}\int_{0}^{\infty} \underset{\sim}{K_2}(s_1,s_2)\underset{\sim}{G}(s_1)\underset{\sim}{G}(s_2)ds_1ds_2$$

$$+ \int_{0}^{\infty}\int_{0}^{\infty}\int_{0}^{\infty} \underset{\sim}{K_3}(s_1,s_2,s_3)\underset{\sim}{G}(s_1)\underset{\sim}{G}(s_2)\underset{\sim}{G}(s_3)ds_1ds_2ds_3 + \ldots$$

$\underset{\sim}{G}$, $\underset{\sim}{T}$, $\underset{\sim}{S}$, p, t, and τ are the history of the strain, total stress, extra stress, isotropic stress, present and past time respectively. Velocity and pressure fields are expanded into Taylor series and the extra-stress $\underset{\sim}{S}$ into a Fréchet series. The linear viscoelastic solution is obtained at the first order and the first deviation from linear behavior in the velocity field is observed at the third order. The analysis developed is good for small amplitudes with no restrictions on the frequency. It is shown that in the oscillating pressure gradient case a flow enhancement, dependent on the material parameters of the integral fluid of order three, the frequency and the mean pressure gradient, is possible and a closed form expression for the velocity field and mass transport is given. The enhancement is due to the interaction of the oscillatory shear fields, induced by the components of the quasi-periodic pressure gradient, with the steady shear field and is equal to the sum of the enhancements due to individual pressure gradient waves. In particular we show that if the frequency of any sinusoidal pressure gradient wave is half or double of the frequency of another wave a resonance effect, a mean pressure gradient independent jump in the mass transport, takes place.

Explicit results of the dependence of the discharge on the frequencies of the parallel and orthogonal superposed vibration on the boundary are also presented. A change in discharge is possible both in the case when the direction of vibration is parallel to that of the pressure gradient and when it is transverse to it. In addition we show that if several sinusoidal waves of different frequency and amplitude are superposed either in the longitudinal or transversal direction or simultaneously in both, resonance effects occur for certain ratios of the frequencies, i.e. in addition to a change in the mean discharge which is the sum of the changes which would result from the waves acting independently there is a change in the mean rate of discharge, independent of the pressure gradient. That is to say a mean flow rate is possible in the absence of a constant pressure gradient if the frequencies of the longitudinal and/or orthogonal waves satisfy certain ratios. If the frequency of any sinusoidal boundary wave in either direction is double the frequency of any wave in the other direction a pressure gradient independent mean flow, due to the interaction of the oscillatory fields imposed from the boundary, takes place in the former direction. Further a pressure gradient dependent transversal flow, i.e. secondary flow in pipes, can be generated if the frequency of any transversal wave is equal to that of any longitudinal wave.

If oscillatory shear fields imposed from the boundary and through pressure gradient oscillations are present an additional mean longitudinal flow dependent on the mean pressure gradient may be generated if the frequencies of any pressure gradient wave and any longitudinal boundary vibration are the same. On the other hand if the frequency of any longitudinal wave, either on the boundary or in the pressure gradient, is twice that of any other longitudinal wave a mean pressure gradient independent longitudinal flow may be generated. In the same vein if the frequency of a boundary wave in the transversal direction is equal to that of a longitudinal wave, either on the boundary or generated by pressure gradient oscillations, a pressure gradient dependent transversal mean flow takes place. But if it is double that of a longitudinal frequency the mean transversal flow becomes independent of the mean longitudinal pressure gradient.

In both cases the liquid has to be shear thinning for an increase in discharge to occur. It is also shown that flow driven partially by periodic forcing from the boundary is an inertial phenomenon. In both phenomena the enhancement effects occur independently of the explicit forms of the constitutive functions of the integral fluid of order three. Explicit forms may be introduced as Maxwell functions with multiple relaxation times. Hierarchy equations used in this work are not very popular with rheologists because of the experimental difficulties involved in determining the rather large number of parameters embedded in the constitutive structure. We propose an algorithm which allows the sequential determination of these parameters through a combination of experiments and analytical solutions concerning the two phenomena investigated in this paper combined with established methods of rheometry and show its feasibility. In view of the fact that the highly popular differential and single integral type constitutive structures fail to have universal predictive powers and even fail to make good qualitative predictions in a number of motions important in applications an argument for the possible superiority of the hierarchy equations is made. In particular we take the point of view that it is better to look for an equation as universal as possible for a limited class of fluids than to search for an equation which applies to a restricted class of motions of a large class of fluids.

Up-Scaling In Heterogeneous Drag Reduction Systems

PETRA SITTART AND HANS-WERNER BEWERSDORFF
Department of Chemical Engineering
University of Dortmund, F.R. Germany
P.O. Box 500 500, D-4600 Dortmund 50

ABSTRACT

Heterogeneous drag reduction is obtained when concentrated polymer solutions are injected in turbulent pipe flows and the injected polymer thread stays intact. The friction behaviour of heterogeneous polymer solutions was measured in pipes of different diameters and by using different injector geometries. On the basis of these data an attempt is made to develop an up-scaling method for heterogeneous drag reduction.

INTRODUCTION

Heterogeneous drag reduction obtained by injecting concentrated polymer solutions in turbulent pipe flows of Newtonian fluids differs in many attributes from the well-known drag reduction in homogeneous polymer solutions /1-5/. The polymer thread seems to interact with the large scale structure of turbulence in the core region of the pipe /2,3,6/.
For industrial applications it is necessary to know how to scale-up the friction behaviour from a laboratory scale to an industrial scale.

MATERIALS AND METHODS

The experimental set-up is described in detail in /2,4/. The concentrated polymer solutions (polyacrylamide, Separan AP45) were injected through hollow needles of different diameters in the centre or by annular injectors of different gap widths in the near-wall region of a fully developed turbulent pipe flow of a Newtonian fluid (tap water). Based on the experimental results a new method for up-scaling the friction behaviour by using two characteristic parameters was developed. The first parameter P_1 is the ratio of the actual drag reduction DR to the maximum attainable drag reduction DR_{max} given by Virk /7/. The second parameter takes into account the influence of the rheological properties of the injected polymer solution and is independent of the pipe diameter.

$$P_1 = \frac{DR}{DR_{max}} \qquad \text{relative drag reduction} \qquad (1)$$

$$P_2 = \frac{Re}{d \sqrt{D}} \frac{1}{u_*}^{3/2} \quad \text{flow parameter} \tag{2}$$

where Re is the Reynolds number, d the pipe diameter, D the diameter of the injector for central injection or the gap width for annular injection, u_* the velocity ratio between the injection velocity and the bulk velocity of the pipe flow, and l a missing characteristic length. Using this two parameter model it is possible to predict the heterogeneous drag reduction in pipes of larger diameters.

RESULTS

Fig.1 shows the friction behaviour for heterogeneous polymer solutions injected by annular injectors in pipes of different diameters with different velocity ratios u_*. Using the two parameters described above it is possible to predict the friction behaviour in the larger diameter pipe. The predicted friction factors nearly coincide with the measured ones. In these experiments the gap width and the wall distance of the annular injector were kept constant and the pipe diameter and the velocity ratio u_* were varied.
Fig. 2 shows an up-scaling example for central injection, where the pipe and injector diameter, and the injection ratio were varied. Again the up-scaling method provides a good agreement between the calculated up-scaled and the measured friction factors. Furthermore, by using Prandtl-Kármán coordinates it is even possible to extrapolate the up-scaled friction behaviour.

Illustrations

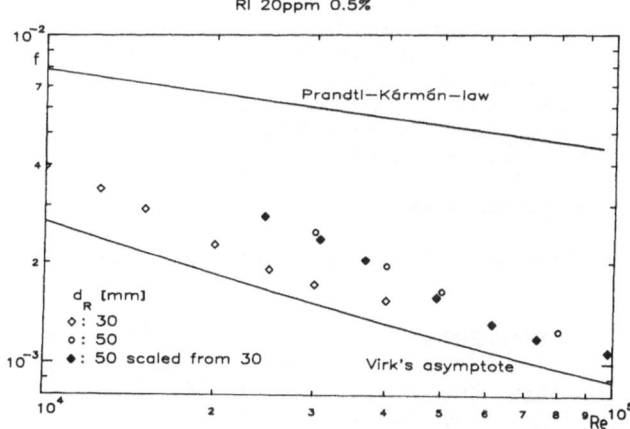

Figure 1: Friction behaviour for annular injection

452

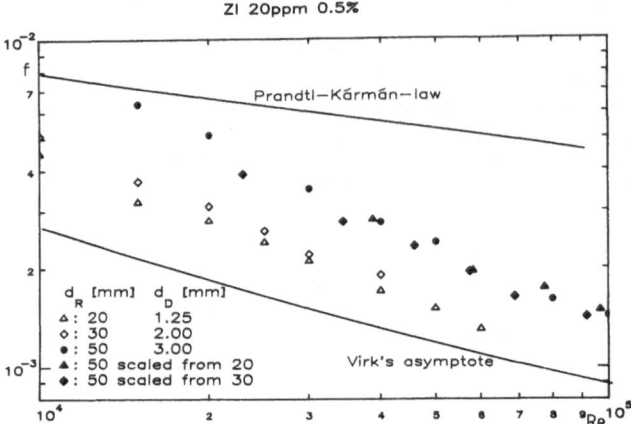

ZI 20ppm 0.5%

Figure 2: Friction behaviour for central injection

CONCLUSIONS

The proposed two parameter model works well for up-scaling the friction behaviour in heterogeneous drag reduction systems. It takes care of the strong influence of the velocity ratio u_* of the injection velocity to the bulk velocity of the pipe flow or the elongation of the injected polymer thread which was found in /8/. Therefore for optimizing heterogeneous drag reduction the elongational properties of the injected polymer solution are very important. By decreasing the injection velocity the polymer thread is elongated by the bulk flow and the drag reduction increases. The model is limited to the same polymer-solvent system which was not varied in the present study. By changing the polymer-solvent system the characteristic length l should change. The model works as long as the injected polymer thread stays intact and is not disrupted into pieces which can occur when the elongation of the polymer thread is too strong /8/. These degradation effects are not considered in the proposed model.

ACKNOWLEDGEMENT

The authors thank the Deutsche Forschungsgemeinschaft (DFG) for supporting this work by Gi 83/9.

REFERENCES

/1/ Vleggaar, J. and Tels, M., Chem. Eng. Sci., 1973, 28, 965

/2/ Bewersdorff, H.W., Rheol. Acta, 1984, 23, 522

/3/ Berman, N.S., Chem. Eng. Commun., 1986, 42, 37

/4/ Frings, B., Rheol. Acta, 1988, 27, 92

/5/ Vleggaar, J., Tels, M., Int. J. Heat and Mass Transfer, 1973, 16, 1629

/6/ Usui, H., Maeguchi, K., Sano, Y., Phys. Fluids, 1988, 31, 2518

/7/ Virk, P.S., AIChE J., 1975, 21, 625

/8/ Bewersdorff, H.W. in: Drag Reduction in Fluid Flows, eds. Sellin, R.H.J. and Moses, R.T., Ellis Horwood, Chichester, 1989, pp. 279

THE EFFECTS OF CO-SOLVENT LEVELS AND NEUTRALISATION ON THE RHEOLOGY OF CARBOPOL GELS

JENNY SLATER
Formulation and Analytical Development Department,
SmithKline Beecham Pharmaceuticals,
Clarendon Road, Worthing.

ABSTRACT

Carbopol (TM, B.F. Goodrich Co.) resins are synthetic, acidic polymers, utilised within the pharmaceutical industry in topical formulations due to their ability to form gels in suitable solvents with the aid of certain neutralising agents.

Systems containing different neutralising agents and co-solvents were prepared to assess the changes in consistency with each formulation. The viscosity in these systems was measured using the flow package on a Carri-Med CS Rheometer, a controlled stress instrument.

This rheometer was found to give comparable results to those obtained with controlled shear instruments. Carbopol gels have a maximum viscosity over a particular pH range when neutralised with sodium hydroxide, and within this range addition of different co-solvents up to a maximum concentration of 40 to 50% increases the system viscosity.

INTRODUCTION

Within the pharmaceutical industry, topical semi-solid formulations and methods for their assessment are constantly under review, to improve drug delivery, stability and patient acceptability.

Rheology using controlled shear instruments is a widely used assessment technique. The use of controlled stress, such as with the Carri-Med CS Rheometer allows the determination of rheological parameters at much lower shear rates enabling a better understanding of the fundamental characteristics of a formulation.

The use of Carbopol resins (acrylic acid copolymerised with polyalkylsucrose) in topical preparations is well documented [1,2,3] and upon neutralisation, the resins in aqueous dispersions form clear gels. Any co-solvent added to improve the solubility of poorly aqueous soluble drugs will affect the gel consistency, as may different neutralisers. These effects were investigated using the Carri-Med Rheometer.

METHOD

One and two percent Carbopol concentrations were dispersed in water, homogenised and neutralised, using 1M sodium hydroxide. Gels were also prepared using propylene glycol, glycerol and Transcutol (TM) Gattefossé as co-solvents with water.

Neutralisation was carried out using triethanolamine, triethylamine, and aqueous ammonia, instead of sodium hydroxide.

Gel pH was measured using a standard electrode and the viscosity of each sample assessed using the Carri-Med Controlled Stress Rheometer in "flow" mode. A 4cm 2° acrylic cone was used for all samples, and shear rates in the region of $200s^{-1}$ were attained.

RESULTS AND DISCUSSION

The gels display plastic characteristics. The 1% gels were less viscous than the 2% formulations. As the pH of the acidic Carbopol rises on addition of sodium hydroxide, the viscosity maximises between about pH 4.5 to 11.0, and drops rapidly above this pH.

The different neutralising agents were used to make gels with a pH corresponding to the maximum gel consistency. With reference to the gel made using sodium hydroxide, gels neutralised using triethanolamine and triethylamine were slightly more viscous at any one shear rate, and slightly less viscous when aqueous ammonia was used.

Similarly, viscosity of gels prepared using the co-solvents glycerol, propylene glycol and Transcutol, increased with increasing co-solvent concentration. However, the order of viscosity of the pure co-solvents (in the ratios glycerol 1011:propylene glycol 26:Transcutol 5:water 1) is not reflected in the corresponding gel viscosities. The differences are not so marked, and in fact propylene glycol gels are slightly thicker than glycerol gels. Gels do not form with these co-solvents at concentrations greater than 40-50%.

Unexpected trends in consistency may be accounted for by changes in system solvation. In non-neutralised dispersions hydrogen bonds are prominent, but stability on neutralisation is due to chemical interactions. In a good solvent, Carbopol molecules are separated and uncoiled to form an open network structure, but in a poor solvent polymer intramolecular forces predominate [4,5]. Hence additives affecting the charge interactions will alter system viscosity; maximum viscosity is achieved with "stretched out" polymers with high intramolecular charge repulsion, the best system being the aqueous propylene glycol gel. As the effective charge decreases, the viscosity decreases.

CONCLUSION

Rheology plays a role in formulation work in assessing the relative
viscosities of different systems. We have looked at the effect of
neutralisation and co-solvents upon the rheology of Carbopol, using
controlled stress rheometry for enhanced definition at low shear rates.
Consistency of Carbopol gels is dependent on three factors: Polymer
concentration, preparation pH and solvent nature. This is useful in the
preparation and investigation of pharmaceutical formulations at molecular
level as well as having practical applications.

REFERENCES

1. Carbopol - water-soluble resins. B.F. Goodrich Company.

2. Martindale Extra Pharmacopoeia. 29th Edition. Pharmaceutical Press.
   p.1433.

3. Handbook of Pharmaceutical Excipients. Pharmaceutical Press, 1986.
   p.41.

4. Bentley's Textbook of Pharmaceutics. E.A. Rawlins. Ballière Tindall,
   8th Edition. p.98.

5. Barry, B.W. and Meyer, M.C., The Rheological properties of Carbopol
   gels. 1. Continuous shear and creep properties of Carbopol Gels.
   Int. J. Pharm., 1979, 2, 1-25.

LINEAR VISCOELASTIC MEASUREMENTS ON LIPID VESICLE DISPERSIONS: HARD-SPHERE BEHAVIOUR AND BILAYER SURFACE DYNAMICS

J. B. A. F. SMEULDERS, C. BLOM and J. MELLEMA
Rheology Group, Faculty of Applied Physics, University of Twente,
P. O. Box 217, 7500 AE Enschede, The Netherlands

ABSTRACT

Linear viscoelastic measurements are carried out on lipid bilayer vesicle
dispersions. The complex viscosity is obtained for frequencies between 70
Hz and 235 kHz with torsion pendula and a nickel tube resonator. Two
relaxation processes are observed. The first pertains to an entropic
relaxation process. As far as this first process is concerned, found
complex viscosities compare very well with those for hard-sphere
dispersions. The second relaxation pertains to vesicle deformation. We
can explain the found relaxation time dependency on vesicle radius and
effective viscosity of the dispersion in terms of the capsule deformation
theory of Oldroyd. We are able to obtain the surface shear viscoelastic
parameters of the bilayer. Furthermore, with some literature values the
Young's modulus and the Poisson ratio of the bilayer were found.

INTRODUCTION

During the last four decades, theoretical studies have been published
into the dynamic properties of fluctuating capsules in which relaxation
times have been expressed in terms of surface mechanical properties. As
yet, no experiments have been carried out to support those theories. We
have carried out linear viscoelastic measurements on monodisperse,
unilamellar lipid vesicle dispersions in search of the relaxation time in
terms of Oldroyd's capsule deformation theory [1] and to obtain the
surface shear viscoelastic parameters of the bilayer. In the experiments
we impose on the vesicle dispersion a harmonic shear flow (70 Hz to 235
kHz) in which the vesicles are deformed at frequencies higher than the
inverse of the deformation relaxation time. We have indeed found a
transition in the complex viscosity due to this capsule deformation. What
is also observed is another transition due to the relaxation of a shear
induced ordering in the dispersion. It is pertaining to the same entropic
elasticity that was recently found for hard sphere silica particles [2].
We have measured 11 vesicle dispersions, varying in radius (36-106 nm)
and volume fraction (24-51%). In figure 1 can be seen a typical result:

two relaxation processes take place in the vesicle dispersion.

HARD-SPHERE BEHAVIOUR

Figure 1 shows that the first relaxation strength is heavily dependent on the volume fraction. Furthermore, the first relaxation time is dependent on the third power of the radius. The same was recently found for hard sphere dispersions [2]. We have observed that the complex viscosities of vesicle dispersions compare very well with those of hard spheres. It is therefore concluded that the first relaxation process pertains to an entropic elasticity of the vesicle dispersion (for which translational diffusion is responsible)

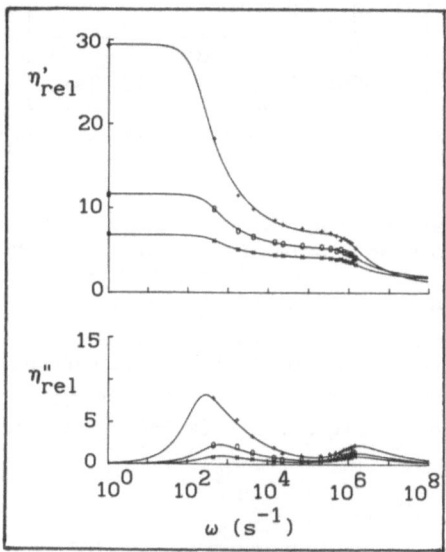

Fig. 1. *Complex viscosity of dispersions of 70 nm vesicles at volumefractions of 39, 44 and 51%.*

DEFORMATION RELAXATION

Oldroyd in his model for fluctuating capsules [1] has found that two relaxation times can occur: one pertaining to the surface shear modulus μ and interfacial tension γ and one pertaining to the dilatational modulus κ. From literature values of κ it follows that we are not able to detect the very small relaxation times pertaining to κ. The expression that applies to γ and μ is

$$\tau = \frac{(23\eta_i + 32\eta_{eff}) \cdot a + 16\zeta}{24(\gamma + \frac{2}{3}\mu)}.$$

(η_i, η_{eff}: internal and external viscosities respectively, ζ: surface shear viscosity, a: radius). Fig.2 shows a plot of τ_2 versus $(23\eta_i + 32\eta_{eff}) \cdot a$. The gradient gives $(\gamma + \frac{2}{3}\mu)$: $(1.9 \pm 0.4) \cdot 10^{-3}$ N/m and the abscissa yields ζ: $(5.9\pm2.0)\cdot10^{-10}$Ns/m (supported by literature). In the literature we find for lipid bilayers: $\gamma \ll 10^{-3}$N/m so μ can be given: $\mu = (2.9\pm0.6)\cdot10^{-3}$N/m.

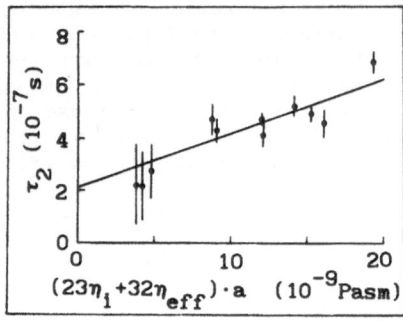

Fig. 2. Second relaxation time τ_2 versus $(23\eta_1+32\eta_{eff})\cdot a$.

Now that we have established μ, it is possible to determine the Young's modulus E and the Poisson ratio ν of the bilayer, conceived as a two-dimensional continuum. E can be expressed as a function of ν and any of the mechanical constants μ, κ and K_c. Various values for κ and K_c can be found in the literature and μ has been obtained as discussed, so that upper and lower limits of the function $E(\nu)$ can be drawn for all three elasticity moduli. The boundaries define a region where E and ν are supported by μ, κ and K_c (fig.3): $E = (2.3\pm0.5)\cdot10^6$ Pa and $\nu = 0.93\pm0.03$.

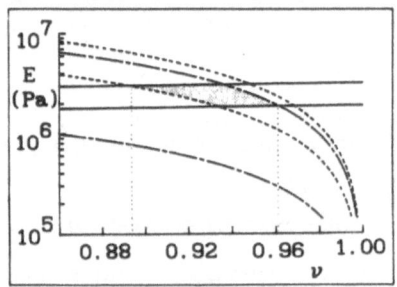

Fig. 3. Three plots of $E(\nu)$ pertaining to μ, κ and K_c with upper and lower limit (——): μ, (– – –): κ and (—— —): K_c.

REFERENCES

1. Oldroyd, J.G., The effect of interfacial stabilising films on the elastic and viscous properties of emulsions. Proc. Roy. Soc. London Ser.A, 1955, 232, 567-77.

2. Van der Werff, J.C., De Kruif, C.G., Blom, C and Mellema, J, Linear viscoelastic behaviour of dense hard-sphere dispersions. Physical Review A, 1989, 39(2), 795-807.

NON-LINEAR FUNCTIONAL RELATIONS AND SUPERPOSITION PRINCIPLES IN THERMORHEOLOGY

ZDENĚK SOBOTKA
Mathematical Institute of Czechoslovak Academy of Sciences
Žitná 25, Praha 1, Czechoslovakia

ABSTRACT

On the basis of tensor expansions, the author derives the non_linear functional relations for the viscoelastic materials which are acted on by the thermal effects and have hereditary characteristics. These relations contain the terms correponding to the first_order as well as to simple and joint second_order effects. The author formulates the non_linear superposition principle for the stress_strain relations of the isotropic materials having the memory of their past states. In contradistinction of the linear Boltzmann superposition principle, the new superposition rule leads to the Stieltjes integral.

CONSTITUTIVE RELATIONS

The non_linear functional relations for the large deformations of viscoelastic bodies define the second Piola_Kirchhoff stress tensor $\pi_{ij}(t,T)$ at the present time t and temperature T as a continuous functional of the Lagrangian strain tensor $E_{ij}(\tau,\theta)$ in the range $t_0 \lesseqgtr \tau \lesseqgtr t$ and $T_0 \lesseqgtr \theta \lesseqgtr T$ and as an ordinary function at the present time t and temperature T. The Lagrangian strain tensor is defined by

$$E_{ij}(\tau,\theta) = \frac{1}{2}\left(\frac{\partial \chi_\alpha}{\partial X_i}\frac{\partial \chi_\alpha}{\partial X_j} - \delta_{ij}\right),\tag{1}$$

where χ_α are the coordinates of a typical particle of a viscoelastic body at the time τ and temperature θ, X_i denotes the material coordinates at the initial time t_0 and tempera_ ture T_0 and δ_{ij} is the Kronecker delta. The stress tensor for an isotropic material having the memory of its past states at N values of time $\tau_0 = t_0$, τ_1, τ_2, ..., τ_{N-1} and temperature $\theta_0 = T_0$, θ_1, θ_2, ..., θ_{N-1} prior to the present state at $\tau_N = t$ and $\theta_N = T$ can be expressed by

$$\pi_{ij}(t,T) = \frac{\varrho_0}{\varrho(t,T)} F_{ij}\left[E_{kl}(t_0,T_0), E_{kl}(\tau_1, \theta_1), E_{kl}(\tau_2, \theta_2 \ldots \right.$$
$$\left. \ldots E_{kl}(t, T)\right], \qquad (2)$$

where ϱ_0 is the initial mass density.

The function on the right-hand side of Eq. (2) can be expanded according to the rules of tensor algebra [1]:

$$\pi_{ij}(t,T) = \frac{\varrho_0}{\varrho(t,T)}\left\{\phi_0\,\delta_{ij} + \sum_{K=0}^{N}\phi_{K+1}\,E_{ij}(\tau_K, \theta_K) + \right.$$
$$+ \sum_{K=0}^{N}\phi_{N+K+2}\,E_{i\lambda}(\tau_K,\theta_K)E_{\lambda j}(\tau_K, \theta_K) +$$
$$\left. + \sum_{K=0}^{N-1}\phi_{2N+K+3}\left[E_{i\lambda}(t,T)E_{\lambda j}(\tau_K,\theta_K) + E_{i\lambda}(\tau_K,\theta_K)E_{\lambda j}(t,T)\right]\right\}, \quad (3)$$

where ϕ_K are scalar functions of invariants.

After determining the invariant functions and introducing them into the foregoing equation, the author has obtained the constitutive deviatoric relation:

$$\pi_{ij}(t,T) - \pi_M(t,T)\delta_{ij} = \frac{\varrho_0}{\varrho(t,T)}\left\{2\sum_{K=0}^{N}\widetilde{G}_K\left[E_I(\tau_K,\theta_K)t,T\right]\left[E_{ij}(\tau_K,\theta_K) - \right.\right.$$
$$\left. - E_M(\tau_K, \theta_K)\,\delta_{ij} + \right.$$
$$+ 4\sum_{K=0}^{N}\widetilde{D}_K\left[E_{II}(\tau_K, \theta_K),t,T\right]\left[E_{i\lambda}(t,T)E_{\lambda j}(\tau_K,\theta_K) - E_S^2(\tau_K,\theta_K)\delta_{ij}\right]$$
$$+ 4\sum_{K=0}^{N-1}\widetilde{H}_K\left[E_{JJ}(t,T,\tau_K,\theta_K)\right]\left\{\frac{1}{2}\left[E_{i\lambda}(t,T)\,E_{\lambda j} + \right.\right.$$
$$\left.\left.\left. + E_{i\lambda}(\tau_K,\theta_K)E_{\lambda j}(t,T)\right] - E_J^2(t,T,\tau_K,\theta_K)\delta_{ij}\right\}\right\}, \qquad (4)$$

where $\widetilde{G}_K$, $\widetilde{D}_K$ and $\widetilde{H}_K$ are the partial modular functions, π_M is the mean stress, E_M is the mean strain

$$E_S = \frac{1}{\sqrt{5}}\sqrt{E_{11}^2 + E_{22}^2 + E_{33}^2 + 2(E_{12}^2 + E_{23}^2 + E_{31}^2)} \tag{5}$$

is the second-order mean strain,

$$E_I = \frac{2}{3}\sqrt{[E_{11}^2 + E_{22}^2 + E_{33}^2 - E_{11}E_{22} - E_{22}E_{33} - E_{33}E_{11} + 3(E_{12}^2 +}$$
$$E_{23}^2 + E_{31}^2)] \,, \tag{6}$$

$$E_{II} = \frac{2}{3\sqrt[4]{2}}\sqrt[4]{E_{ij}E_{j\lambda}E_{\lambda\mu}E_{\mu i} - (E_{ij}E_{ij})^2} \tag{7}$$

is the first-order and second-order intensity of strain tensor,

$$E_J(t,T,\tau_K,\theta_K) = \frac{1}{\sqrt{3}}\sqrt{\{E_{11}(t,T)E_{11}(\tau_K,\theta_K) + E_{22}(t,T)E_{22}(\tau_K,\theta_K) +}$$
$$+ E_{33}(t,T)E_{33}(\tau_K,\theta_K) + 2[E_{12}(t,T)E_{12}(\tau_K,\theta_K) +$$
$$+ E_{23}(t,T)E_{23}(\tau_K,\theta_K) + E_{31}(t,T)E_{31}(\tau_K,\theta_K)]\} \,, \tag{8}$$

$$E_{JJ}(t,T,\tau_K,\theta_K) = \frac{2}{3\sqrt[4]{2}}\sqrt[4]{\{\frac{3}{2}[E_{ij}(t,T)E_{j\lambda}(t,T)E_{\lambda\mu}(\tau_K,\theta_K)E_{\mu i}(\tau_K,\theta_K) +}$$
$$+ E_{ij}(t,T)E_{j\lambda}(\tau_K,\theta_K)E_{\lambda\mu}(t,T)E_{\mu i}(\tau_K,\theta_K) - [E_{ij}(t,T)E_{ij}(\tau_K,\theta_K)]^2\}}$$
$$\tag{9}$$

are respectively the joint mean value and the joint second-order intensity of the tensors $E_{ij}(t,T)$ and $E_{ij}(\tau_K,\theta_K)$.

The relation (4) represents the non-linear superposition principle which leads to the Stieltjes integrals if the number N of the time and temperature intervals tends to infinity:

$$\pi_{ij}(t,T) - \pi_M(t,T)\delta_{ij} = \frac{\varphi_0}{g(t,T)}\left[\left[2\int_{t_0}^{t}[E_{ij}(\tau,\theta) - E_M(\tau,\theta)\delta_{ij}]\,d\tilde{G}[E_I(\tau,\theta),t,T] +\right.\right.$$
$$+ 4\int_{t_0}^{t}[E_{i\lambda}(\tau,\theta)E_{\lambda j}(\tau,\theta) - E_M(\tau,\theta)\delta_{ij}]\,d\tilde{D}[E_{II}(\tau,\theta),t,T] +$$
$$+ 4\int_{t_0}^{t}\{\frac{1}{2}[E_{i\lambda}(t,T)E_{\lambda j}(\tau,\theta) + E_{i\lambda}(\tau,\theta)E_{\lambda j}(t,T)] - E_J^2(\tau,\theta,t,T)\delta_{ij}\} \cdot$$
$$\left.\left.\cdot d\tilde{H}[E_{JJ}(t,T,\tau,\theta),t,T]\right]\right]. \tag{10}$$

REFERENCE

[1]. Sobotka, Z, , Rheology of Materials and Engineering Structures, Elsevier, Amsterdam, Oxford, New York, Tokyo, 1984, pp. 217-294.

THERMOFORMING OF THERMOPLASTICS

W.N. Song, F.A. Mirza and J. Vlachopoulos
CAPPA-D Group
Faculty of Engineering
McMaster University
Hamilton, Ontario, Canada L8S 4L7

ABSTRACT

A finite element analysis of inflation of axisymmetric sheet of finite thickness is presented. Total Lagrangian description and isothermally hyperelastic behavior with incompressibility are used. Both single and multi layer sheets are examined. Comparisons with experimental data for a rubber and PMMA sheet indicated very good agreement.

INTRODUCTION

Thermoforming is the process of raising the temperature of a thermoplastic sheet to a workable level and forming it into the desired shape through any one of several techniques (vacuum forming, draw forming, pressure bubble forming, plug assist forming, etc.). To simulate the process of thermoforming, we developed a finite element algorithm for the inflation of a thick axisymmetric sheet. The difference from previous published work is that the present analysis is for finite thickness while previous publications have dealt with sheets of large aspect ratios and thus applied the membrane approximation. A comprehensive summary of the literature is available elsewhere [1]. Perhaps the most detailed analysis of the thermoforming process using the membrane approximation has been carried out by DeLorenzi and Nied [2]. Experimental data, available in the literature for sheets of finite thickness [3,4,5], have been used for comparison.

FINITE ELEMENT FORMULATION

The finite deformation kinematics based on the total Lagrangian description [6] is employed. We consider the motion of a body, at time $t = 0$, or configuration C_0 and determine the components of the Lagrangian strain tensor in the current configuration C_1 with reference to the initial configuration C_0. The finite element formulation is based on the virtual work principle in the form

$$\int_{0_V} {}^1_0 S^{ij}\ \dot{\gamma}_{ij} d^0V = \int_{1_{A_\sigma}} {}^1 t^* \cdot \delta^1 v\ d^1 A \qquad (1)$$

where ${}^1_0 S^{ij}$ are the contravariant components of the second order Piola-Kirchhoff stress tensor in the configuration C_1, but referred to C_0.

The material behaviour is described by the Mooney hyperelastic model:

$$W = C_{10}(\ I_1 - 3\ ) + C_{01}(\ I_2 - 3\ ) \qquad (2)$$

where I_1 and I_2 are the first and second invariants of the strain tensor, respectively. When $C_{01} = 0$, the model is known as neo-Hookean.

The penalty formulation and the incremental loading method with Newton-Raphson iteration were used to solve the problems presented in this paper.

NUMERICAL RESULTS AND COMPARISONS

Figure 1 shows the inflation profiles and a comparison with Treloar's data [3] is shown in Figure 2. The agreement is very good. More complicated situations were studied and a deformed configuration is presented in Figure 3. Other calculations involved a comparison with Lai and Holt's data [5], and a two layer sheet. These will appear in a future paper [6].

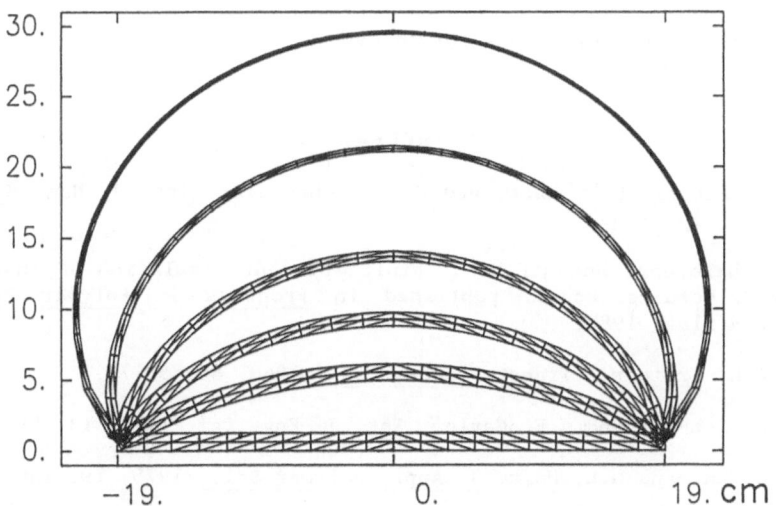

Figure 1. Inflated profiles for the simple supported ends.

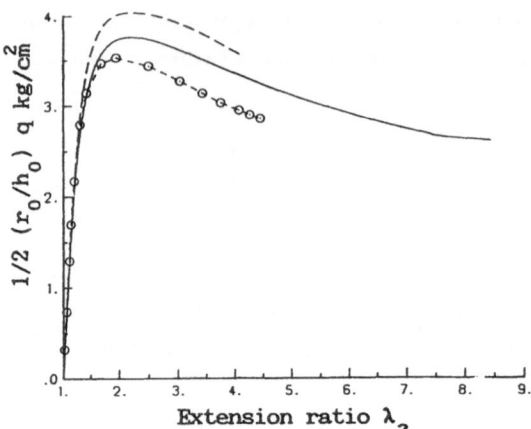

Figure 2. Loading vs. Extension ratio curves. neo-Hookean model, with simple supported(—) or clamped boundary (---) conditions, compared with Treloar's experimental data(-⊖-). $r_0 = 1.25$ cm, and $h_0 = 0.082$ cm.

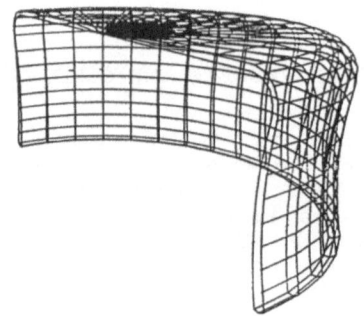

Figure 3. Three dimensional formed cap.

REFERENCES

1. N.G. Zamani, D.F. Watt, and M. Esteghamatian, Int. J. Num. Meth. Eng., 1989, 37, 2681-2693.

2. H.G.DeLorenzi and H.F.Nied, Finite Element Simulation of Thermoforming and Blow Molding, to be published in Progress in Polymer Processing, Hansen-Verlag, 1989.

3. L.R.G. Treloar, Trans. Faraday Soc., 1940, 40, 59-70.

4. L.R. Schmidt and J.F. Carley, Int. J. Eng. Sci. 1975, 13, 563-578.

5. M.O. Lai and D.L. Holt, J. Appl. Polymer Sci., 1975, 19, 1805-1814.

6. W.N. Song, F.A. Mirza, and J. Vlachopoulos, Finite Element Analysis of Inflation of an axisymmetric sheet of Finite Thickness, paper in preparation.

AUTOHESION IN THE PRESENCE AND ABSENCE OF INTERFACIAL MASS TRANSFER

R.G. Stacer

Fraunhofer-Instiut für Chemische Technologie (ICT)
D - 7507 Pfinztal-Berghausen, FRG

ABSTRACT

A study has been conducted to investigate time-dependent autohesion in a series of polybutadiene systems, both with and without diffusing components greater than the entanglement molecular weight M_e. Diffuse interfaces were prepared by joining surfaces of high molecular weight polybutadienes. An equilibrium adhesive fracture energy G_∞ of $2.5 kJ/m^2$ (T=-35°C and c=1mm/s) was measured for these interfaces. Non-diffuse interfaces were produced using a telechelic polybutadiene crosslinked using NCO/OH reactions. Extraction of a sol component was possible in each of these networks, however, subsequent GPC analysis demonstrated that the molecular weight of the mobile sol fraction to be less than M_e. Measured G_∞ values for the non-diffuse interfaces were greater than those of the diffuse interfaces, $G_\infty=8.5 kJ/m^2$ for NCO/OH=1.0, tested at corres-ponding rates. The incorporation of low molecular weight polybutadiene dramatically decreased t_∞, but only slightly influenced G_∞.

INTRODUCTION

Autohesion in polymeric materials is defined as the resistance to separation of two surfaces after they have been joined ogether for a period of time under a given pressure [1]. Currently, two mechanisms have been proposed to account for the time-dependent nature of autohesion [1-3]. The first is bond formation via chain entanglements through polymer diffusion across the interface, while the second is contact area formation (wetting) through viscous flow.

EXPERIMENTAL

Data from three of the materials tested are reported in this preprint. They are a hydroxy terminated polybutadiene (HTPB) cured at an NCO/OH ratio of 1.0 using isophorone diisocyanate, a version of this material designated HTPB/PL containing 25% percent low-molecular weight (4,500 g/mole) polybutadiene, and a high-molecular weight polybutadiene (PB) also produced by free-radical polymerization. Autohesive fracture energies G_a for these materials were measured as functions of contact time t_c and temperature using a T-peel test specimen. More complete experimental details may be found in a previous publication [4].

RESULTS AND DISSCUSSION

An example of the autohesion data measured for each of the materials is presented in Figure 1. Application of the method of reduced variables to these data is illustrated in Figure 2. Use of the method of reduced variables allows the autohesion phenomena to be studied over a much broader experimental time scale than that previously acheived, and enables various proposed mathematical models to be critically examined [4].

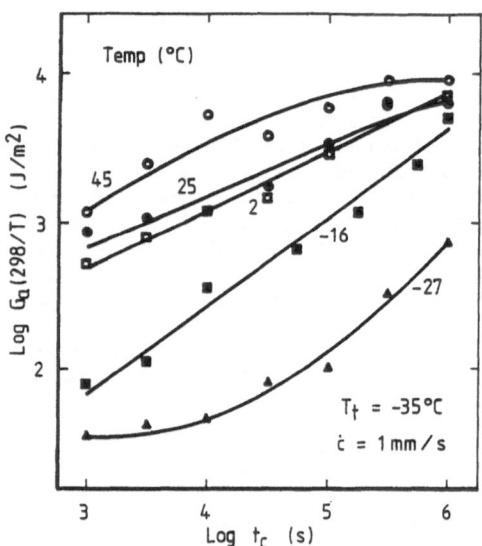

Figure 1. Effect of contact temperature and time on the fracture energy of HTPB.

Also shown in Figure 2 is the corresponding data for the PB material. This material reaches its equilibrium strength approximatley one day sooner, however, its G_∞ value is only a third of that obtained for HTPB even though both materials possess identical microstructures and presumably equivalent segmental mobilities. What is significant in these data is that this crosslinked polymer, without the aid of entanglements across the interface, can acheive a degree of bonding greater than that of the diffuse interface.

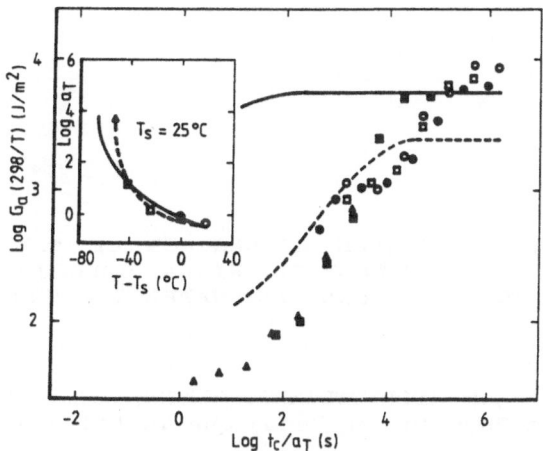

Figure 2. Autohesion master curve for HTPB in comparison with curves fitted to HTPB/PL and PB data.

In an earlier publication [5], it was found that the bulk fracture energy of a similar HTPB formulation was substantially greater than the PB material. This was attributed to the demonstrated presence of hydrogen bonding between the urethane groups. These results suggest that the autohesive strength between HTPB surfaces is related to the degree of physical attraction and/or hydrogen bonding, with the time-dependency originating from the time required for the two surfaces to come into intimate molecular contact (wet). Finally, data are also given in Figure 2 for HTPB/PL. This material acheived equilibrium strength very rapidly when compared to the other two, apparently through better wetting. G_{∞} is, however, 20% lower than the pure HTPB material. This decrease may be related to a reduction in the number of possible bonding sites between the two surfaces.

REFERENCES

1. Hamed, G.R., Tack and Green Strength of Elastomeric Materials, Rubber Chem. Technol., 1981, 54, 576-95.

2. Kausch, H.H. and Tirrell, M., Polymer Interdiffusion, Annu. Rev. Mater. Sci., 1989, 19, 341-77.

3. Wool, R.P., Yuan, B.-L. and McGarel, O.J., Welding of Polymer Interfaces, Polym. Eng. Sci., 1989, 29, 1340-67.

4. Stacer, R.G. and Schreuder-Stacer, H.L., Time-Dependent Autohesion, Inter. J. Fracture, 1989, 39, 201-16.

5. Plazek, D.J., Gu,G.-F.,Stacer, R.G., Su, L.-J., von Meerwall, E.D. and Kelley, F.N., Viscoelastic Dissipation and the Tear Energy of Urethane-Crosslinked Polybutadiene Elastomers, J. Mater. Sci., 1988, 23, 1289-1300.

TIME AND SHEAR STRESSES DEPENDENCE OF THE RHEOLOGICAL PROPERTIES OF HYDROLYZED HIGH MOLECULAR WEIGHT POLYACRYLAMIDE SOLUTIONS IN WATER-GLYCEROL MIXTURES

PHILIPPE STEIERT, CLAUDE WOLFF
Laboratoire de Physique et Mécanique Textiles URA CNRS N°1303
E.N.S.I.T.M.
11, rue Alfred Werner F-68093 MULHOUSE Cedex

ABSTRACT

The viscosity of dilute and semi dilute solutions of high molecular weight hydrolyzed polyacrylamides in water-glycerol-sodium chloride mixtures have been measured as function of the shear stress with a Bohlin Constant Stress rheometer. Shear thinning behaviour is observed up to a critical value of the shear stress, depending on the concentration. At higher values of the shear stress, the viscosity increases up to a maximum value. This behaviour is associated with time effects that have been precisely characterized and which depend also strongly on salt concentration.

INTRODUCTION

We present the results of a study concerning dilute hydrolyzed polyacrylamide solutions. The rheological properties, particularly the influence of shear stress, time, polymer and salt concentration on the viscosity are studied.

EXPERIMENTAL

The molecular weight of the partially hydrolyzed polyacrylamide, determined by low-angle light scattering is 7.2×10^6 and the degree of hydrolysis 32%.The polymer was dissolved in a mixture containing 75% per weight of glycerol and 25% of distilled water. The desired quantity of salt (NaCl) was then added. The polymer concentrations were 125,100,85 and 60 ppm. The viscosity was measured on a Bohlin Constant Stress rheometer (Bohlin Reologi AB), in a cone-plate system and in coaxial cylinders geometry. Measurements were carried out at 20°C.

RESULTS

Figure 1 shows the variation of the viscosity as function of time for various shear stresses. The plot represents the data obtained with the cone-plate system. For shear stresses lower than a critical value σ_c, the viscosity is independent of the time. A sharp increase of the viscosity appears for stresses higher than σ_c, followed by a pseudoequilibrum plateau where the viscosity remains constant up to a critical time t_c. After this time, the viscosity starts to decrease rapidly down to a second plateau. t_c decrease when the shear stress increases. The behaviour is qualitatively the same in the coaxial cylinders geometry.

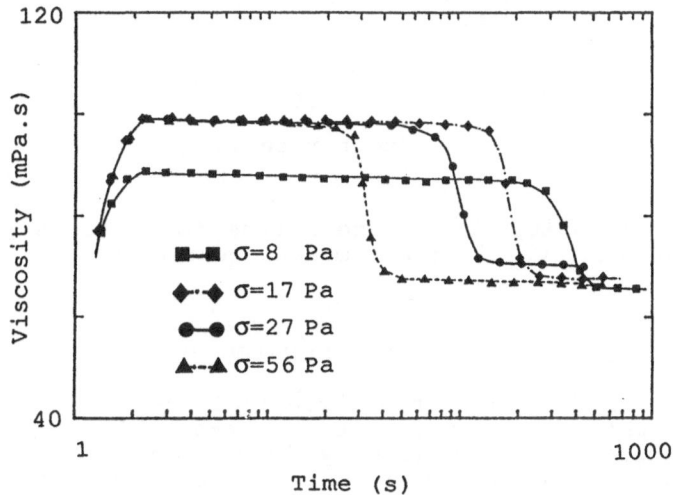

Figure 1. Viscosity vs time for different shear stresses.

Figure 2 shows the viscosity as function of the shear rate for various polymer concentrations and for two different measurement times t_m, respectively lower and higher than t_c.

- For $t_m < t_c$, the fluid exhibits a shear thinning behaviour up to a critical shear rate $\dot{\gamma}_c$. Then, the viscosity increases rapidly up to a plateau. This behaviour disappears progressively with the dilution.

- For $t_m > t_c$, the situation changes drastically. The curves show a weak but clear shear-thickening behaviour and this for all the concentrations studied.

The same behaviour is recorded with the coaxial cylinders. For the same values of the shear stresses, the critical time, t_C, of solutions with salt is higher than without salt.

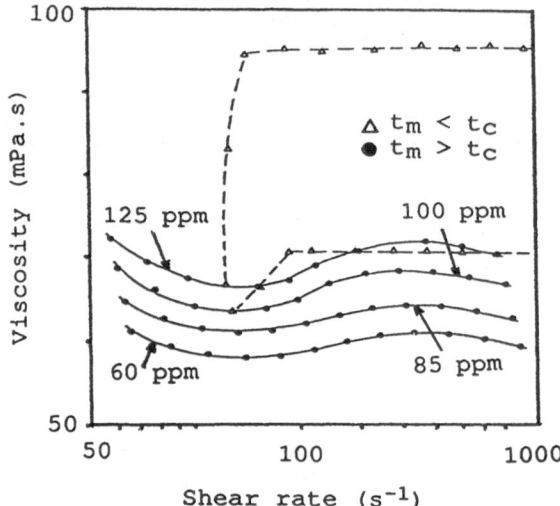

Figure 2. Viscosity-shear rate curves for various polymer concentrations and two different measurement times.

CONCLUSION

The thickening of dilute hydrolyzed polyacrylamide solutions observed in this study, presents the remarkable feature that the minimum position is apparently independent of the concentration. As far as we know such a behaviour has only been predicted in three cases[1,2,3]; but the agreement is not quantitative. An internal viscosity term may perhaps improve the accordance.

REFERENCES

1. PETERLIN A., Gradient dependence of intrinsic viscosity of freely flexible linear macromolecule. J.Chem.Phys.(1960) 33, 1799-1802.

2. DUPUIS D., WOLFF C., Consequences of a conformational change of flexible macromolecules on the non newtonian behaviour of their solutions. Chem.Eng.Comm. (1985) 32, 203-217.

3. KISHBAUGH A.J., McHUGH A.J., A discussion of shear-thickening in Bead-Spring Models. J.Non-Newtonian Fluid Mech. (1990) 34, 181-206.

THE DIGITAL VISCOMETER, A NEW GENERATION OF RHEOMETERS

THOMAS A SUCK
SUCK WISsenschaftliche Geraete ENTwicklung
Grafestrasse 9, D-5900 Siegen, FRG

ABSTRACT

In the V10 Rheometer, whose novel digital operating
principle allows frequency and amplitudinally stable
measurements, either the shear rate or the shear stress can
be taken as the starting factor. As there are only two
conditions, "high" and "low", the requiert motor performance
for maintaining flow is independent of the viscosity of the
solution. This means that a re-adjustment of the speed of
rotation is not necessary.

INTRODUCTION

The development of the digital motor drive, which translates
digital information into mechanical movement, has been
realised in the V10 Viscometer. The setting of the angles
and speed are calculated before measuring commences and fed
to the motor via digital electronics. The system is driven
by TTL impulses. There is no noise, as in analogous wiring,
because there are only two conditions, "high" and "low".
Different speeds can be attained by altering the pulse
frequency. A cosine function, for example, gives the pulse
sequence seen in figure 1, 2. The difficulties of "slow
measuring systems" generally seen in analogous equipment (DC
motors), which are noticeable above all in the starting
phase, are not present in a digital viscometer. If a change
in viscosity occurs during shearing, the motor need not be
adjusted. This is a considerable difference compared to
usual rheometers (where the motor performance for
maintaining the flow field depends on the viscosity of the
solution), which require an additional control circuit. This
means that there are no time differences between shear rate
and shear stress. The digital drive lends itself
particularly to dynamic measurements, i.e. harmonic
oscillation.

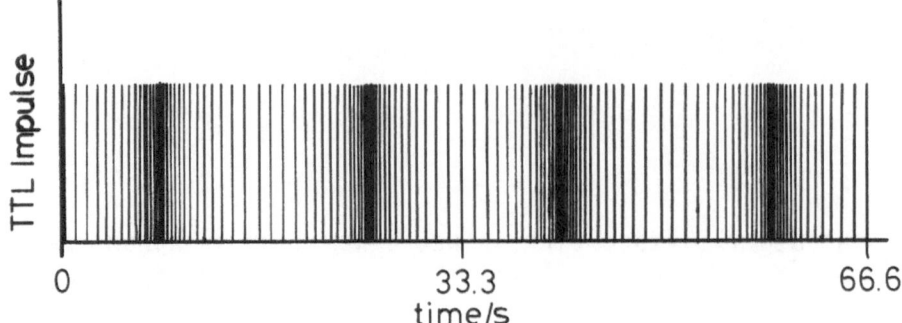

Figure 1. Digital input of a harmonically oscillating wave;
Amplitude: 0.05°, Frequency: 0.03 Hz

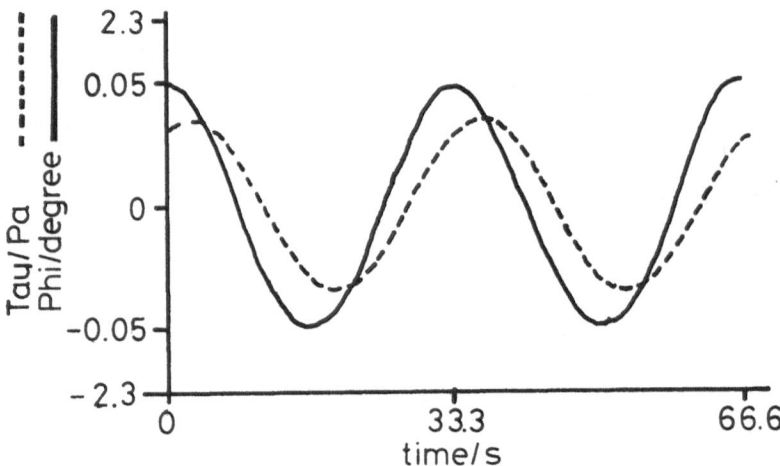

Figure 2. Output of the digital input (see figure 1)

The angle function is reached precisely and need not be
approximated. There is no uncontrolled adjusting around a
time function (which still has to be correlated in order
to obtain useful results), as the desired amplitude and
frequency are strictly adhered to. In order to achieve a

473

laminar flow of the substance in the measuring gap, the V10 rotational viscometer features a coaxial cylinder with radii r_i (inside) and r_q (outside). Furthermore, the condition $r_q/r_i \sim 1$ must be fulfilled. In the couette system used, the outer cylinder moves and the shear rate is obtained from the rotation speed n multiplied by a measuring vessel-related constant ($\dot\gamma = F(r_i, r_q)$). Through the resistance of the liquid to the flow, torque proportional to the viscosity is transferred to the inner cylinder (stator), and is then detected by an inductive compensator circuit. This sensor is contact-free, which means that the stator has no mechanical contact points in the bearing or when torque commences, which guarantees the detection of low torque rates. The main specifications of the rheometer V10 are summarize in table 1.

Table 1

Torque	: $4 \cdot 10^{-8}$-0.046 Nm
Rotation Speed	: $1.2 \cdot 10^{-5}$-50^1/s
Amplitude	: adjustable from 0.05°(0.005°)
Angle Accuracy	: $1.5 \cdot 10^{-3}$°
Frequency Range (Dynamic Measurements)	: adjustable from 10^{-4}Hz

3-D MODELING OF THE INJECTION MOLDING PROCESS

P.A. TANGUY, R. LACROIX and F.H. BERTRAND
CERSIM
Department of Chemical Engineering
Université Laval, Québec, Canada, G1K 7P4

A three dimensional finite element methodology is presented for the modelling of complex injection mold filling, using the Hele-Shaw approximation coupled to a three dimensionsal solution of the thermal equations with phase change.

From a mathematical point of view, the use of the Hele-Shaw assumptions (Tadmor and Gogos, 1979) allows the simplification of the set of governing equations to a simple Poisson problem for the pressure, that can be solved readily by a numerical method. To account for the complex rheological behavior of polymer melts, a rheological equation of state must be introduced into the equations of motion, yielding a non-linear Poisson problem. A complete description of the rheological behavior of polymer melts normally includes the flow curve, normal stress vs shear rate data, and relaxation times. However, from a numerical standpoint, only the non-Newtonian character of the viscosity is generally considered, due to the mathematical complexity of the complete description.

From a practical point of view, the governing equations were solved using the finite element method. In this work, the discretization was achieved using trilinear hexahedral elements.

When the fluid is non-Newtonian, which is the case for polymer melts, the viscosity is no longer a constant and is generally expressed as an explicit function of the rate-of-strain tensor using the Generalized Newtonian Fluid model. In the Hele-Shaw formulation, the viscosity is expressed as a function of the pressure gradients similarly to the other variables in the problem.
Several strategies were used for the resolution of the set of algebraic equations governing filling:

- a straightforward approach based on the PCG method coupled to a fixed-point method with under-relaxation (Lacroix, 1989),

- an augmented Lagrangian formulation to better deal with the non-linearities associated with the rheological model (Tanguy et al., 1984),

– a generalized minimum residual method (GMRES) directly applied to the non-linear equations (Saad et Schultz, 1983).

For the thermal problem, the approach described in (Tanguy and Lacroix, 1987) and which consists of solving a Stefan problem for each advancing front position, was applied.

Illustrations are presented in the context of an industrial mold, and simulation results are validated against experimental data and other numerical data showing a reasonable agreement.

Lacroix R., Ph.D. Dissertation, Laval University, 1989.

Lacroix R. and P.A. Tanguy, A Finite Element Model of a Molten Polymer Solidification Problem Using an Adaptive Mesh Strategy. 1st Int. Conf. on Ind. and Appl. Math., Paris, 1987.

Saad Y. and M.H. Schultz, Research Report, YALEU/DCS/RR-254 Aug. 24, 1983.

Tadmor Z. and C.G. Gogos, Principles of Polymer Processing, Wiley, 1979.

Tanguy P. et al., Int. J. Num. Meth. Fluids, 4:441, 1984.

Three-Dimensional Modeling of Rheologically
Complex Fluid Flow in a LPD Dow-Ross Static Mixer

P.A. Tanguy, R. Lacroix and L. Choplin
Dept. of Chemical Engineering
Laval University, Quebec City, G1K 7P4

Static mixers consist of stationary rigid elements mounted in flow channels or pipes. These elements are used to split the fluid into streams which are then combined and split up again in such a way that a mixing effect occurs by entanglement of the fluid layers [1]. The characterization of parameters governing mixing is generally carried out experimentally. However, the recent progress of advanced numerical technology and the availability of fast computers can provide a powerful tool to speed up the development and screening of new designs, restricting experimental testing to a smaller number of prototypes.

The objective of this work is to demonstrate the capability of the numerical approach in this domain. More specifically, the purpose is to show that accurate predictions on the flow behavior through static mixers can be obtained numerically by using state-of-the-art three-dimensional finite element techniques. The study is carried out in the case of rheologically complex fluids flowing across a LPD Dow-Ross static mixer.

The system considered is composed of a tubular housing in which a series of semielliptical plates has been discriminately positioned. A single mixing element consists of two plates having a 45 degrees angle to the horizontal axis of the flow stream.

We have studied the creeping flow of a shear-thinning fluid obeying the Carreau model of viscosity. Several static mixing configurations and several values of the rheological parameters have been considered.

The numerical work consisted of solving the equations of change using POLY3D finite element software [2].

One difficulty of this work was the generation of the solid model and the finite element grid on which the equations of change are discretized . This was achieved using PATRAN PLUS (PDA Engineering) software. The final mesh included 4016 elements yielding 24503 degrees of freedom for the mixer with one element, and 5536 elements yielding 33644 degrees of freedom for the mixer with two elements.

Numerical tests aiming at the determination of the velocity pattern, pressure field, shear stresses , the distribution of residence time were performed for various flowrates. Results were validated in the Newtonian limit case against experimental data provided by the manufacturer and exhibited a very good agreement (deviation less than 5%). The simulation enabled the study of the mixing mechanism (fluid splitting, layers entanglement, subsequent mixing) across the system and will be presented on video at the conference.

[1] Shintre S.N., Proc. 6th Europ. Conf. on Mixing, 551, Pavia, 1988

[2] P.A. Tanguy, R. Lacroix, F.H. Bertrand, L. Choplin and E. Brito de la Fuente, this proceedings.

THREE-DIMENSIONAL MODELLING OF MIXING
WITH HELICAL RIBBON SCREW IMPELLER

P.A. Tanguy, R. Lacroix, F.H. Bertrand, L. Choplin
and E. Brito de la Fuente

Department of Chemical Engineering
Laval University
Quebec, Canada, G1K 7P4

We present in this work the results of a three-dimensional mixing study in a vessel provided with two kinds of impeller, namely an helical ribbon screw impeller and an helical ribbon impeller. The work includes finite element simulation as well as experimental investigation.

The above impellers are known to be much more efficient than classical impellers like Rushton turbines, marine propellers... in handling highly viscous fluids. If in addition to this viscous character the fluids exhibit elastic properties, actual flow and pumping capacity are considerably reduced resulting in an increase in mixing time and decrease in mixing efficiency. The helical ribbon-type impellers offset these drawbacks by enhancing secondary axial flows [1,2]

In this work, we have considered several model fluids exhibiting different rheological behaviours: Newtonian fluids (glycerol, mixtures of polybutene and kerosene), inelastic shear-thinning fluids (Carbopol, Xanthan solutions), second-order fluids or Boger fluids (mixtures of polybutene and kerosene in which we added a small percentage of polyisobutylene) and viscoelastic fluids (CMC, Separan solutions).

Rheological data of the fluids were measured using a Rheometrics System 4 rheometer in steady and dynamic modes using both cone and plate and parallel plate geometries at room temperature. Then, rheological models were chosen according to the obtained data. For instance, in the case of inelastic shear-thinning fluids, Carbopol solutions exhibited a viscous behaviour which were well represented by the power-law model whereas Xanthan solutions were found to obey either a power-law model or a Carreau model depending on the salt content and the concentration. In the case of the Boger fluids, a simple Criminale-Eriksen-Filbey (CEF) model was selected.

The numerical work was carried out using POLY3D software. This finite element simulator for rheologically complex thermal fluid flow is based on the use of a Galerkin finite element method which allows to transform the governing equations of change into a weak form. The non-linearities associated with the complex rheological behaviour are dealt with using an Augmented Lagrangian Method [3]. The discretization of the weak form uses enriched tetrahedral elements which were proven to provide reliable results for the velocity and the pressure [4]. Finally, the resolution of the matrix system was performed with an Incomplete Uzawa Algorithm in a very cost-effective way [5].

The experimental work includes the determination of power consumption through torque measurements, mixing time and circulation time by thermal pulse techniques [6] as well as flow visualization with various tracers (dyes, small spherical particles...).
A comparison of the numerical predictions with the experimental data will be presented and discussed.

[1] Ulbrecht J.J. and P.J. Carreau, in "Mixing of liquids by mechanical agitation", Vol. 1, Chapter 4, J.J. Ulbrecht and G.K. Patterson eds., Gordon & Breach Publ., N.Y., 1985.

[2] Godfrey J.C., in " Mixing in the process industries", N. Harnby, M.F. Edwards and A.W. Nienow eds., Butterworths, London, 1985.

[3] Tanguy P.A., M. Fortin and L. Choplin, Int. J. Num. Meth. Fluids, 4:441, 1984.

[4] Gadbois M., Ph.D. Thesis Dissertation, Laval University, 1990.

[5] Robichaud M., P.A. Tanguy and M. Fortin, Int. J. Num. Meth. Fluids, 10:429, 1990.

[6] Hoogendoorn C.J. and A.P. den-Hartog, Chem. Eng. Sci., 22:1689, 1967.

NEWTONIAN AND NON-NEWTONIAN FLUID FLOW THROUGH SMALL BIFURCATIONS

MYRIAM TAZI
Institut de Mécanique des Fluides, URA 0005.
Avenue Camille Soula, 31400 Toulouse FRANCE.

ABSTRACT

Our concern is to investigate Newtonian and non-Newtonian steady flow through symmetric bifurcations with cicular cross-section in order to improve the knowledge of atherosclerosis phenomenon. An experimental approach provides axial velocity profiles along the daughter branch. We characterize the influence of the different parameters : Reynolds number, angle of bifurcation, distance from the apex and fluid behaviour on the shape of the axial velocity profiles.

INTRODUCTION

Singularity regions such as bifurcations are generally admitted to be preferential sites for atherosclerosis. It has been seen that haemodynamics factors can occur in the development of this disease. Many experimental and numerical works have been carried out in order to determine the flow characteristics in a bifurcation. Then some authors have measured velocity profiles by different techniques such as Laser Doppler anemometer, Liepsh et al (1), or Ultrasonic Doppler anemometer, Rieu (2), others have been interested in flow visualizations or shear stress measurements.
The purpose of this experimental approach is to analyse laminar steady flow in a symmetric bifurcation with cicular cross-section. We characterize the influence of the Reynolds number, of the bifurcation angle and of the fluid behaviour on the flow shape dowstream the bifurcation. The velocity measurements, using Laser Doppler anemometer are performed in an expanded geometric scale model with respect to Reynolds similarity because of the high difficulties to do measurements inside real vessels. Furthermore blood is supposed to be an homogeneous fluid.

EXPERIMENTAL SET UP

Three different bifurcation models are considered : α = 15, 30, and 45 degrees. They are made of glass. The lengths of the main trunk and of the daughter branches have been correctly chosen in order to obtain a fully developed flow just upstream the test bifurcation and far dowstream it.

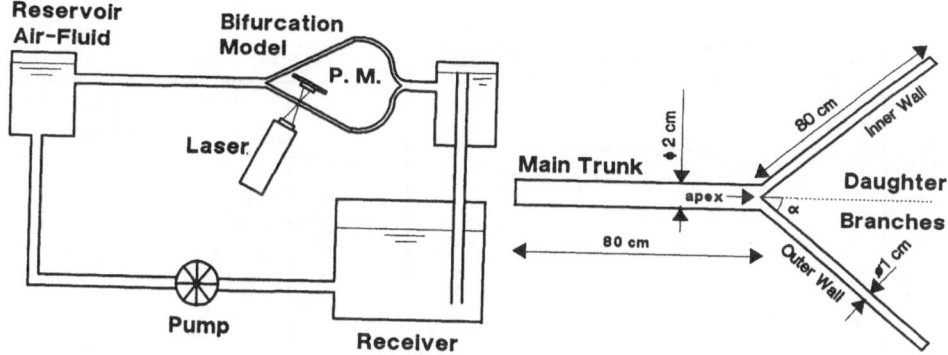

Figure 1 : Diagram of the apparatus and geometry of the bifurcation model.

The hydrodynamic bench in which the model is inserted consists of two reservoirs. The upstream one is an air-fluid reservoir that absorbs the pressure fluctuations coming from the motor pump system. The dowstream one is a constant level reservoir allowing to keep the same pressure drop for all the experiments.
Two different fluids have been used : a water-glycerol mixture, of dynamic viscosity about 60 cP, having a Newtonian behaviour and a CMC solution, having a non-Newtonian behaviour following pseudoplastic power law.
The velocity measurements were performed in the plan of the bifurcation all along the daughter branch of the model.

RESULTS

Because of the sudden change in direction, the flow behaviour in the daughter branch is determined by the opposing influences of the centrifugal force and of the radial pressure gradient. Depending on the balance of forces the deformation of the velocity profiles will be important or not.
A shape factor F, related to Poiseuille profile, is defined to compare all the results. This special parameter allows to characterize the recovery length and the importance of the profile deformation.
The bifurcation angle has an effect on the amplitude of the phenomenon rather than on its nature.

Influence of the Reynolds number
In figure 2 we consider a small distance of 2 cm from the apex because the perturbation is more significant in this region. For high Reynolds number the asymmetry of the profile is evident. The point of maximum velocity is moving towards the inner wall of the bifurcation when the Reynolds number increases. We notice that for Re = 70 the velocity profile is almost parabolic. The recovery length is found to be shorter for lower Reynolds number (see the axial evolution of F in figure 2).

Comparison of Newtonian and non-Newtonian fluid
The axial velocity profiles are quite different. At 10 cm from the apex we can remark in figure 3 that for a Newtonian fluid, we obtain a parabolic Hagen-Poiseuille profile and that for a non-Newtonian fluid it appears slightly flattened.

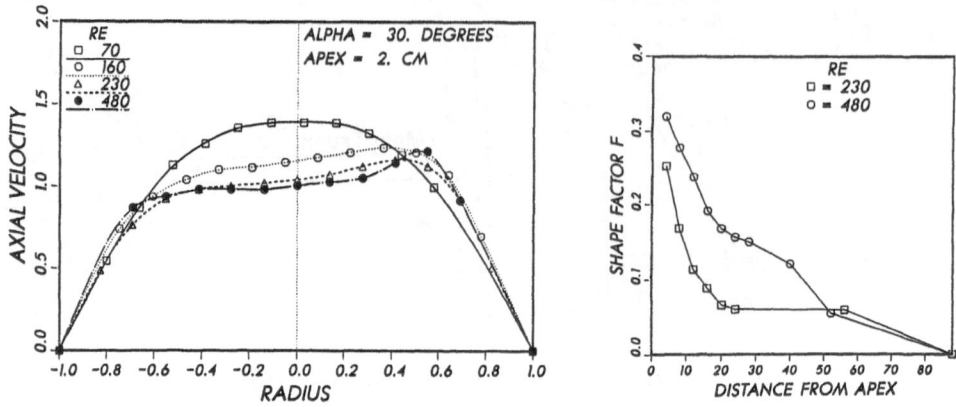

Figure 2 : Influence of the Reynolds number

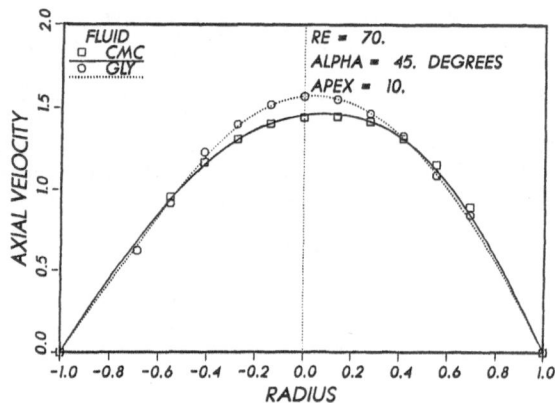

Figure 3 : Comparison of Newtonian and non-Newtonian fluid.

CONCLUSION

We have proved that haemodynamics factors have a significant effect on the flow characteristics through bifurcations. We have introduced a shape factor allowing to compare all the results. This number gives not only the distortion but also the recovery lenght.
Further numerical investigations will be developed in order to confirm these experimental analyses.

REFERENCES

1. Liepsh, D.W., Flow in tubes and arteries. Biorheology, 1986, 23, 395-433.
2. Rieu, R., Pelissier, R. and Farahifar, D., An experimental investigation of flow characteristics in bifurcation models. Eur. J. Mech., B/Fluids, 1989, 8, 1, 73-101.

EQUATION OF THE STATE OF THIXOTROPIC MEDIUM UNDER MECHANICAL INFLUENCES

ANDREW TERENTYEV
Polytechnic Institute of Riga, 226 355 Riga, U.S.S.R.
LECH RUDZIŃSKI
Technical University of Kielce, 25-314 Kielce, Poland

ABSTRACT

For a correct calculation of the stress-strain state of
thixotropic material under the dynamic effect it is necessary
to take into consideration changes in the rheological
properties of material. The above problem always appears in
practice in numerous examples of the manifestation of
thixotropy of disperse systems.

This paper proposes an effective approach which contains
certain general principles, but this approach was applied to a
particular case of taking into account change in the
rheological properties of gas concrete mixture in the shock
technology of material forming. During material forming by
the shock method the mould with swelling mixture falls down
periodically on the rigid foundation.

The purpose of this paper is to find the values of the
rheological parameters of the equation of the state of
thixotropic material as a function of set dynamical actions.

INTRODUCTION

As has been hypothetically assumed the equation of the state of
material obtained as result of quasi-static tests does not
change its form under the dynamic effect with the exception of
the rheological properties of material.

The coefficient of thixotropic thinning $K_\eta = \eta / \eta^*$
(η - viscosity of unaffected medium, η^* - viscosity of thinned
medium) under small heights of the mould fall, insufficient for
limit thinning of gas concrete mixture, is proportional to the
height of the mould fall h or to the specific energy of effect
$W = g \cdot h$. However, in the limit thinning of mixture the
coefficient K_η does not depend on specific energy W (Fig. 1).

Let's introduce the concept of the thixotropic ability of
medium as a function of transmitting of specific energy to the

484

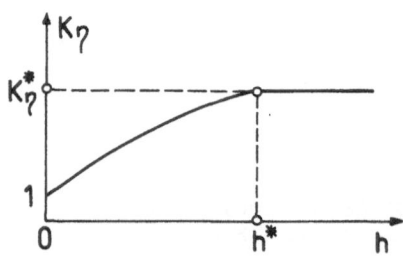

Figure 1. Thixotropic thinning
coefficient vs height of the
form fall

change of the viscosity of
the disperse system using the
following equation:

$$K_\eta = B_\eta(h) \cdot g \cdot h + 1 \qquad (1)$$

where: g — acceleration of
gravity, $B_\eta(h)$ — thixotropic
ability which satisfies the
following conditions (see
Fig. 1):

$$B_\eta(h) \xrightarrow{h > h^*} (K_\eta^* - 1)/g \cdot h \qquad (2)$$

$$B_\eta(h) \xrightarrow{h \to 0} B_\eta = const. \qquad (3)$$

As shock formation of gas concrete blocks occurs at small
energies of the mould fall, which are insufficient to cause a
complete destruction of the material structure and a limit
fall of viscosity, let's consider only the initial segment of
the curve in Fig. 1. Approximating it with a linear
dependence, we obtain $B_\eta(h) = B_\eta = const.$

RESULTS AND DISCUSSION

Behaviour of gas concrete mixture subjected to vibrations is
approximated by the Kelvin-Voigt equation of standard
viscoelastic body. Thixotropic thinning of the medium is
modelled by viscosity decrease under unchanged elastic
parameters of the equation of state. In this connection it
has been established analytically the dependence of the
coefficient of damping of mixture vibrations β on the
coefficient of thixotropic thinning $K\eta$ (Fig. 2). This allowed
to determine the real coefficient of mixture thinning $K\eta$ with
known height h in comparison with experimentally found β_e.

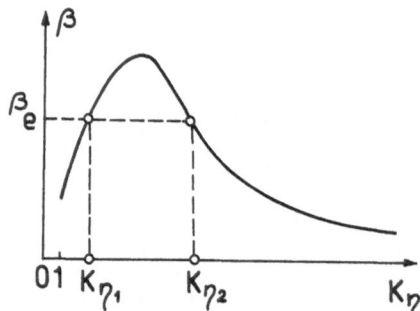

Figure 2. Damping coefficient
of the viscoelastic medium vs
thixotropic thinning factor

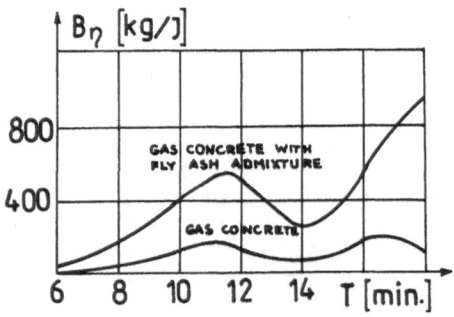

Figure 3. Time dependence of
gas concrete mix thixotropic
ability

Therefore, it was possible to determine the thixotropic ability of mixture Bη from equation (1) taking into consideration (3).

Substituting from equation (1) viscosity η^{*} into the equation of state of cellular concrete mix selected from quasi -statical loading conditions, we obtain the equation of the mixture state which takes into account beside its elastic and viscous parameters thixotropic ability and specific energy of dynamic effect

$$\sigma + \frac{\eta}{E(B_{\eta} \cdot g \cdot h + 1)} \frac{d\sigma}{dt} = H\varepsilon + \frac{\eta}{B_{\eta} \cdot g \cdot h + 1} \frac{d\varepsilon}{dt} \quad (4)$$

where: E,H - moduli of instantaneous and prolonged elasticity respectively.

Equation (4) was used to solve the problem of vibration propagation in the mixture depending on the height of the mould fall on the rigid foundation in shock technology of cellular concrete formation [1].

Values Bη as well as those of all the remaining rheological parameters of the equation of the state of gas concrete mixture of 2 compositions were determined experimentally for the full time of material swelling when the height of block is 0.6 m (see e.g. Fig. 3).

At the outset of the process of swelling, mixture is in a weak gel state, hence disturbance of such a structure causes a slight decrease in viscosity. With swelling, coagulation compounds become reinforced, and thus increase mixture viscosity. After destruction viscosity decreases considerably with increasing coefficient Bη. At the end of swelling the amount of free moisture in the material becomes increasingly small, mixture becomes thick, the coaqulation structure is reinforced and all this limits the possibility of considerable thixotropic thinning. With medium setting the coefficient Bη $\longrightarrow$ 0. The saddle-like cavity shown in Fig. 3 finds experimental confirmation, but so far it has not received a physical interpretation.

It also follows from Fig. 3 that gas concrete mixture on cement binder with admixture of fly ash manifests a considerably higher ability of thixotropic thinning than common gas concrete mixture. It is confirmed by its elevated water-solid ratio W/S = 0.64 in relation to W/S = 0.38 in common gas concrete mixture with identical initial mixture mobility according to Sutthard (S = 13 cm).

Thus, thixotropic ability is a fairly independent and convenient characteristic of the material thixotropic properties.

REFERENCES

1. Terentyev, A.E. and Kunnos, G.J., The rheology of the structurised swelling viscoelastic medium with thixotropy, in Proceedings of the Xth International Congress on Rheology, University of Sydney, Sydney, 1988, pp. 317-19.

THE EFFECT OF CHEMICAL ADDITIVES ON THE DRAG AND SETTLING VELOCITIES OF YIELD STRESS SLURRIES

Q K TRAN* & R R HORSLEY**

*   PhD student, Dept of Mechanical Engineering, Curtin University, Perth, Australia
**   Professor, Dept of Mechanical Engineering, Curtin University, Perth, Australia

ABSTRACT

Terminal settling velocities of single particles, both spherical and cylindrical (with conical ends), were measured in various yield pseudo-plastics slurries. The yield stress of the slurries was reduced by the addition of viscosity modifiers. The results are presented as modified drag coefficients plotted against a modified Reynolds number. For the spherical particles it was found that the drag coefficient is higher for clay/silica slurries than for water. The drag coefficients for the cylinders were found to be higher than those for the spheres. The results are important in understanding industrial processes involving separation of large and small particles.

INTRODUCTION

The suspension of coarse solids in a medium of finer particles and liquid and the separation of coarse or dense particles from slurries depends on the drag exerted by the slurry on the particle. It is known that the yield stress of a yield pseudo-plastic suspension has a large effect on the drag coefficient of falling bodies [1,2] and therefore on the performance of mineral processing separators. Previous work has been concentrated on slurries of very fine particles, usually clay [3,5], whereas slurries of commercial interest are usually mixtures of which clay is just one of the constituents. The other constituents of the slurry are generally of a larger particle size.

Experimental
Mixtures of ball clay and silica were used as models of slurries of commercial interest. 50% of the particles in ball clay/silica 100 g slurries were below 23 μm and in ball clay/silica 350 g slurries were below 10 μm. Their yield stress was reduced by the addition of sodium hexameta-phosphate, as can be seen in Figure 1.

A schematic of the experimental apparatus is shown in Figure 2. Two sets of coils were wound on a clear tube of 54 mm internal diameter at 1000 mm apart. Whenever a particle made of material which affects the magnetic field generated by the primary coil, passes through that field, a secondary voltage is generated and can be recorded. The particles were a steel cylinder (with conical ends) and ball bearings. The temperature of the slurry was taken before its yield stress was measured in the modified viscometer developed by Overend et al [6].

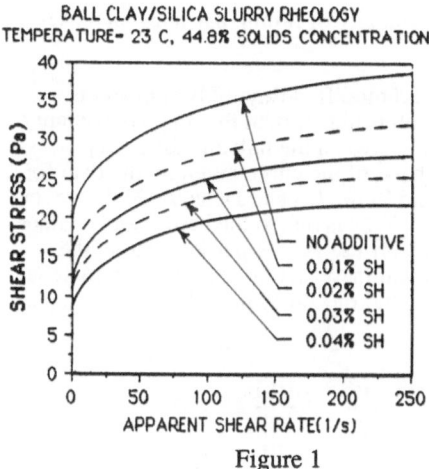

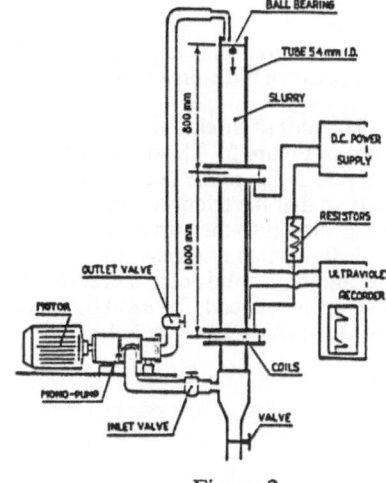

Figure 1   Figure 2

Models

A correlation model is proposed to determine the settling velocities of the cylinder and the sphere based on the experimental data which is shown in Figure 4

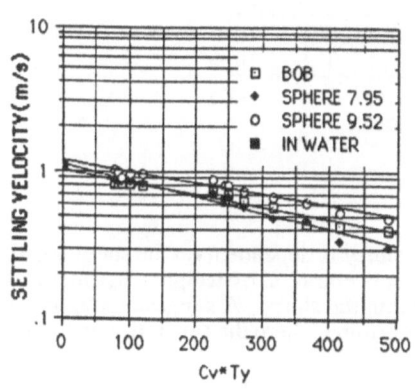

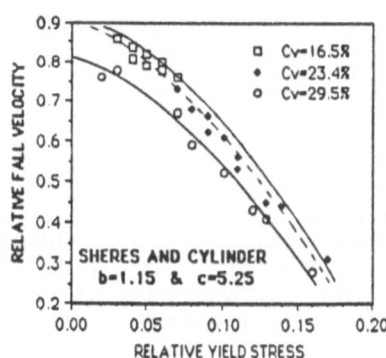

Figure 3   Figure 4

$$\text{viz} \quad U_{s_{rel}} = \left[1 - C_{v_{rel}}^{a}\right]\left[1 - C.\tau_{y_{rel}}^{b}\right]; \quad U_{s_{rel}} = \frac{U_s}{U_{s_{.ter}}}; \quad C_{v_{rel}} = \frac{C_v}{C_{v_{max}}}; \quad \tau_{y_{rel}} = \frac{\tau_y}{\tau_{y_{lim}}} \qquad (1)$$

C_{ymax} is the maximum packing volume fraction, and τ_{ylim} can be determined from Ansley and Smith [4] and Dedegil [5] for the case of no motion. The values of a, b and c in equation (1) depend on the solids concentration and particle size distribution in the slurry. The modified drag coefficient and Reynolds number is determined in [4,5]

$$C_D = \frac{2}{V_s^2 \rho_{sL}}\left[\frac{2}{3}\left(\rho_s - \rho_{sl}\right)d.g - \pi\tau_y\right] \quad \text{and} \quad Re = V_s^2 \rho_{sL}/\tau \qquad (2)$$

Discussion

If solids concentration is kept constant, the settling velocity of falling particles increases as the yield stress in the slurry is reduced.

The plot of modified drag coefficients C_D and modified Reynolds numbers is displayed in Figure 5. The drag coefficient of the bob is higher than the drag coefficient of spheres, this is due to the increase of surface contact between the bob and the slurry particles, so that the friction force increases. The shape of the curve given by the data for spheres is similar to the curves plotted by Ansley and Smith [4] and Dedegil [5], however for Re>150 the drag coefficient is 0.6 instead of 0.4. This is due to the harder and coarser particle, and high solids concentration of the tested slurries. As the particle size in the tested slurry is reduced, for Re>80, the drag coefficient increases as a result of an increase of the surface contact between the falling "particles" and the slurry particles.

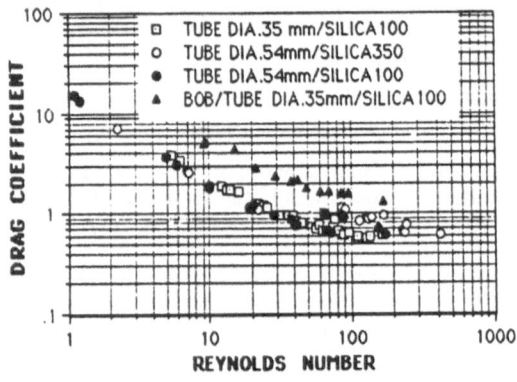

Figure 5

Conclusion

The drag coefficient of particles falling through a slurry is dependent on the rheological properties of the fluid. These are the yield stress, concentration by weight, "hardness" of the slurry particle and particle size distribution within the slurry. A suggested relationship which fits the experimental data can be used to determine where the separation is likely to occur.

REFERENCES

1.　　Pazwash, H. and Robertson, J.M., J. Hyd. Res., Vol 13, N1, pp. 35-55, 1973.

2.　　Reynolds, P.A. and Jones, T.E.R., Int. J. Min. Proc., Vol 25, pp 47-77, 1989.

3.　　Valentik, L. and Whitmore, R.L., Brit. J. Appl. Phys., Vol 16, pp 1197-1203, 1965.

4.　　Ansley, R.W. and Smith, T.N., AlChe Journal, Vol 13, N6, pp. 1193-1196, 1967.

5.　　Dedegil, M.Y., J. Fluids. Eng., Vol 109, pp. 319-323, 1987.

6.　　Overend, I.J., Horsley, R.R., Jones, R.L. and Vinycomb, R.K., Adv. in Rheology, Vol 2, pp. 571, 1984.

BROWNIAN DYNAMICS SIMULATION OF SHEAR THINNING AND SHEAR INDUCED ORDERING IN A COLLOIDAL SUSPENSION

J. M. van der Veer, J. H. J. van Opheusden and R. J. J. Jongschaap.
Department of Applied Physics, Twente University,
P. O. Box 217 7500 AE Enschede, the Netherlands.

ABSTRACT

We studied sheared colloidal suspensions using the Brownian Dynamics (BD) method. We found that the shear thinning behaviour of such a suspension is caused by shear induced ordering of the suspended particles. Two different types of ordering were found, depending on the Péclet number.

INTRODUCTION

Colloidal suspensions are important compounds in modern industry. A good understanding of their rheological behaviour is therefore necessary. Numerical simulation is a convenient tool for gaining insight in the microscopic nature of colloidal suspensions. In the last decade, simulation methods have been developed in order to study colloidal suspensions. Some examples are Molecular Dynamics simulations [1], Stokesian Dynamics simulations [2] and recently BD simulations of suspensions undergoing shear [3].

SIMULATION METHOD

In a BD simulation, one solves the Langevin equation of the suspended particles numerically. This equation for a single particle reads

$$\underline{f} = \underline{f}^H + \underline{f}^I + \underline{f}^B \tag{1}$$

where $\underline{f}^H$ is the hydrodynamic force, $\underline{f}^I$ is the non hydrodynamic systematic force due to interparticle potentials and $\underline{f}^B$ is the Brownian force due to thermal motion of the solvent particles. We followed a procedure as proposed in [3][4], using a repulsive Lennard - Jones potential. Hydrodynamic interactions were limited to Stokes flow and the lubrication approximation at high concentrations [5].

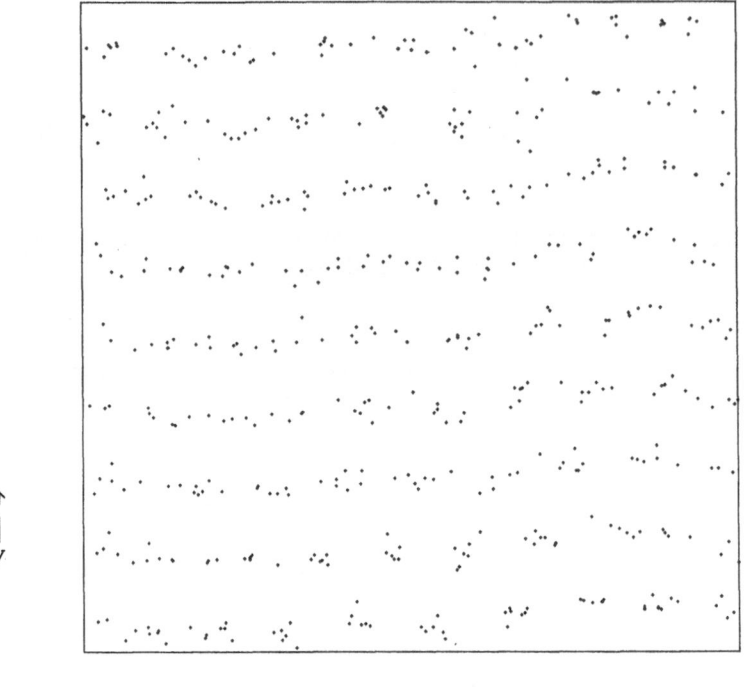

$z \longrightarrow$

Figure 1. *A layered structure induced by shear flow. System parameters were N = 500, ϕ = 0.52 and Pe = 50.*

RESULTS

We chose system parameters $\eta^* = 2868$ and $T^* = 2.5$ and we varied the particle number, the density and the Péclet number. The results were obtained using Stokes flow, work on the lubrication flow is in progress and will presented at the conference. The relative suspension viscosity η^r shows a shear thinning effect as was found by [2]. No dependence of η^r on the particle number could be found. The error in η^r increases quickly when the Péclet number goes to zero. It is therefore not possible to obtain η^o from extrapolation of η^r, so a Green Kubo relation has to be used [3][6]. These results will be presented at the conference. From the behaviour of the stress correlation function it can be concluded that runs may require lengths up to 25,000 Δt, depending on the Péclet number. We found two different types of ordering of the particles, depending on the Péclet number. Figure 1 shows shows a layered structure at Péclet number $Pe = 50$, which breaks up to form a hexagonal structure at $Pe = 100$ (Figure 2). This hexagonal structure develops at volume fractions above $\phi \approx 0.4$. This shear induced ordering causes the shear thinning behaviour of the suspension.

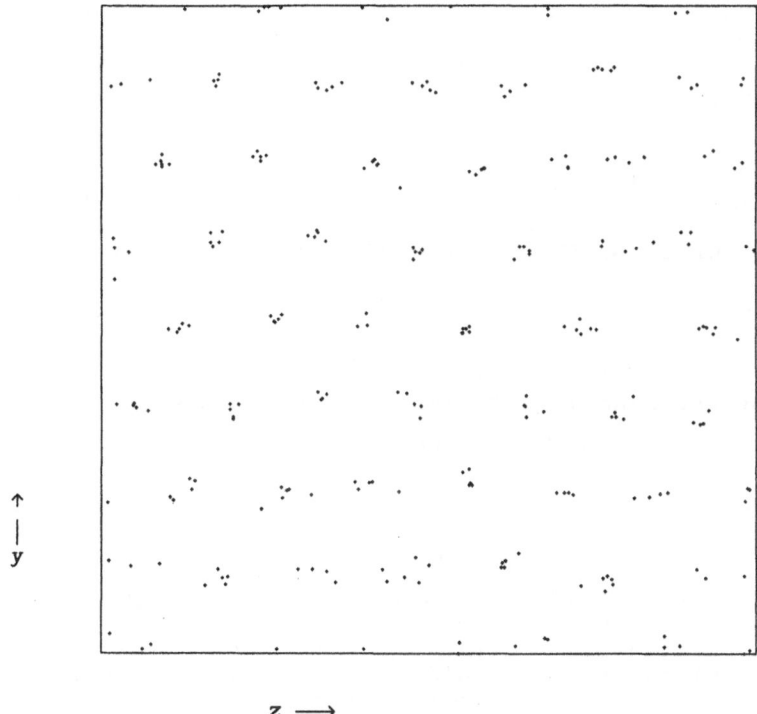

↑
|
y

z ⟶

Figure 2. A hexagonal structure induced by shear flow. System parameters were N = 500, φ = 0.52 and Pe = 50.

ACKNOWLEDGEMENT

We like to thank Prof. Dr. P. F. van der Wallen Mijnlieff for stimulating discussions. This work was supported by AKZO International Research.

REFERENCES

[1] Heyes, D.M., J. Chem. Phys., 1986, **85**, 997.
[2] Bossis, G. and Brady, J.F., J. Chem. Phys, 1988, **88**, 1185.
[3] Heyes, D.M., J. Non Newtonian Fluid Mech., 1988, **27**, 47.
[4] Ronis, D., Phys. Rev. A, 1986, **34**, 1272 .
[5] Frankel, N.A., Acrivos, A., Chem. Eng. Sci., 1967, **22**, 847.
[6] McQuarrie, D.A., Statistical Mechanics., 1976, Harper & Row New York

RHEOLOGY OF ANISOTROPIC VISCOELASTIC FLUIDS

V.S.VOLKOV, V.G.KULICHIKHIN

Institute of Petrochemical Synthesis , USSR Academy of Sciences
Moscow , 117912 , USSR

ABSTRACT

Rheological properties of liquid crystalline polymers are analyzed on the basis of simplest invariant theory of the anisotropic viscoelastic fluids. Oscillatory shear flow of anisotropic viscoelastic fluid with single nondeformable director were examined. The longitudinal and transversal (with respect to director) dynamic moduli are introduced.

INTRODUCTION

The appearance of liquid crystalline (LC) polymers has led to the study of rheological properties of the viscoelastic anisotropic fluids. The theory of anisotropic viscous fluid was formulated by Ericksen [1] and developed by Leslie [2] for a liquid crystals. The Ericksen-Leslie theory describes the main features of dynamic properties of low molecular nematics. In a recent paper, Volkov and Kulichikhin [3] have formulated the simplest constitutive equation for a monodisperse LC polymers:

$$\sigma_{ij} = \sigma^o_{ij} + \sigma'_{ij} \qquad (1)$$

$$\tau_{ijkl} \frac{\Delta \sigma'_{kl}}{\Delta t} + \sigma'_{ij} = \eta_{ijkl}\, \gamma_{kl}$$

This type of LC polymers in a slow relaxation region are regarded as anisotropic viscoelastic fluids with tensorial viscosity η_{ijkl} and relaxation time τ_{ijkl} .The consideration is restricted to a case when LC elasticity can be neglected. It has been assumed that the flow destroys the defects and produces a monodomain anisotropic system. In Eq.(1) σ^o_{ij} and σ'_{ij} are respectively the equilibrium and nonequilibrium parts of the stress tensor σ_{ij}. Also $\Delta/\Delta t$ is the invariant time derivative and γ_{ij} is the strain rate tensor. In this paper we will investigate the linear anisotropic viscoelasticity of nematic polymers, using of the suggested rheological equation of state.

NEMATIC VISCOELASTIC FLUIDS

The viscoelastic fluid with a single director is useful starting point for investigation of the nematic LC polymer flows. In such case, the rheological equation (1) leads to corotational anisotropic fluid defined by the rheologigal equation which involves anisotropic combination of Jaumann derivatives:

$$\tau_\perp \frac{D\sigma'_{ij}}{Dt} + (\tau_\perp - \tau_\parallel) \frac{D_n\sigma'_{ij}}{Dt} + \sigma'_{ij} = 2\eta_\perp \gamma_{ij} + 2(\eta_\perp - \eta_\parallel)\gamma^n_{ij} , \qquad (2)$$

where we introduce the notations : $n_{ij} = n_i n_j$, $n_{ijkl} = n_i n_j n_k n_l$

$$\gamma^n_{ij} = 2n_{ijln}\gamma_{ln} - n_{il}\gamma_{lj} - n_{jl}\gamma_{li}$$

$$\frac{D_n\sigma'_{ij}}{Dt} = 2n_{ijln}\frac{D\sigma'_{ln}}{Dt} - n_{il}\frac{D\sigma'_{lj}}{Dt} - n_{jl}\frac{D\sigma'_{li}}{Dt}$$

The director n_i ($n_i n_i = 1$) is obtained from an additional equation describing the changes in its orientation by the flow

$$\rho \ddot{n}_i = g(\rho, n_i, \dot{n}_i, \nu_{ij}) ,$$

where ρ is the density, ν_{ij} is the velocity gradient. According to the simplest orientational equation [1]

$$\frac{D}{Dt}n_i = \lambda (\gamma_{ij}n_j - \gamma_{km}n_{kmi}) , \qquad (3)$$

where λ is a material constant. Degree of the viscoelastic anisotropy of transversely isotropic (nematic) fluid (2) is defined by two dimensionless parameters

$$\alpha = \tau_\perp/\tau_\parallel , \qquad \beta = \eta_\perp/\eta_\parallel$$

Here $\eta_\parallel, \tau_\parallel$ and $\eta_\perp, \tau_\perp$ are the longitudinal and transversal (with respect to the director) viscosities and relaxation times, respectively.

ANISOTROPY OF VISCOELASTIC MODULI

Analysis of the constitutive equations (2,3) for small-amplitude oscillatory shear flow leads to the following expressions for the components of the apparent complex moduli $G(\omega) = \sigma_{12}(\omega)/\nu_{12}(\omega)$ и $G_1(\omega) = (\sigma_{11}(\omega) - \sigma_{22}(\omega))/\nu_{12}(\omega)$:

$$G' = G'_\perp \sin^2 2\varphi + G'_\| \cos^2 2\varphi \ , \quad G'_1 = (G'_\perp - G'_\|)\sin 4\varphi$$

$$G'' = G''_\perp \sin^2 2\varphi + G''_\| \cos^2 2\varphi \ , \quad G''_1 = (G''_\perp - G''_\|)\sin 4\varphi$$

associated with shear stress σ_{12} and first normall stress difference $\sigma_{11} - \sigma_{22}$ .According to the orientational law (3) φ is the angle between the director and flow direction. Here we define the longitudinal $G'_\|$, $G''_\|$ and transversal $G'_\perp$, $G''_\perp$ storage and loss moduli with respect to the director:

$$G'_\| = G_\| \frac{\omega^2 \tau_\|^2}{1 + \omega^2 \tau_\|^2} \quad , \ G'_\perp = G_\perp \frac{\omega^2 \tau_\perp^2}{1 + \omega^2 \tau_\perp^2}$$

$$G''_\| = G_\| \frac{\omega \tau_\|}{1 + \omega^2 \tau_\|^2} \quad , \ G''_\perp = G_\perp \frac{\omega \tau_\perp}{1 + \omega^2 \tau_\perp^2} \ , \tag{4}$$

where $G_\| = \eta_\|/\tau_\|$, and $G_\perp = \eta_\perp/\tau_\perp$ .The linear dynamic moduli of the viscoelastic nematics depend explicitly on the director orientation. LC polymers differ in viscoelastic behavior from the convencional polymers. The existence of the normal stresses in a linear viscoelasticity region is the main feature of these anisotropic fluids.

REFERENCES

I. Ericksen,J.L.,Transversely Isotropic Fluids. Kolloid. Z., 1960, *173*, 117-122.
2. Leslie,F.M., Some constitutive equation for liquid crystals. Arch. Ration. Mech. Anal., 1968 ,*28*, 265-283.
3. Volkov,V.S. and Kulichikhin,V.G., Anisotropic viscoelasticity of liquid crystalline polymers . J. Rheol., 1990, in press.

THE NONLINEAR STRAIN MEASURE OF POLYISOBUTYLENE MELT IN GENERAL BIAXIAL FLOW AND ITS COMPARISON TO THE DOI-EDWARDS MODEL

M. H. WAGNER

Institut für Kunststofftechnologie, Universität Stuttgart
Böblingerstr. 70, D-7000 Stuttgart 1, Germany

INTRODUCTION

Analysis of nonlinear material behaviour of polymer melts is simplified by the experimentally well-established fact that except for high deformation rates, time and deformation dependence can be separated (1-5). This separability can be expressed in terms of a single integral constitutive equation with a memory functional factorized in a time-dependent linear-viscoelastic memory function and a nonlinear strain measure (6,7). Due to its tensorial nature, this nonlinear strain measure can only be inferred from analysis of experimental data, if widely differing types of flow covering the invariant space are available. Such data sets are scarce, however. Recently, one of the most extensive data sets available to date, namely the multiaxial elongational data measured by Demarmels and Meissner (8,9), has been analysed by Wagner and Demarmels (10).

In this paper, the nonlinear strain measure of PIB as analysed by Wagner and Demarmels (10) is compared to predictions of the molecular model of Doi and Edwards (11), and the corresponding version of the Curtiss-Bird model (12). According to reptation theory, the nonliner strain measure is predicted to be a universal function of strain for linear polymers, independent of molecular weight and molecular weight distribution. By comparing the universal strain measure of the Doi-Edwards model with the experimentally observed strain dependence of a polyisobutylene melt, it is hoped that deeper insights in the nonlinear molecular dynamics of linear polymers can be obtained.

RESULTS AND DISCUSSION

In general incompressible multiaxial flow, two normal stress differences can be measured. These can be converted into two different strain functions $h_{\sigma 1}$ and $h_{\sigma 2}$, which describe the deviation from the rubber-like liquid limit (10). In figs.1 to 3, predictions of the Doi-Edwards model are compared to experimental results for PIB. Note that in planar flow (m = 0), the effective strain function $h_{\sigma 1}$ is equivalent to the damping function in shear flow, and $h_{\sigma 1}$ calculated from the Doi-Edwards model represents the lower limit of the experimental data (fig.1). This is in

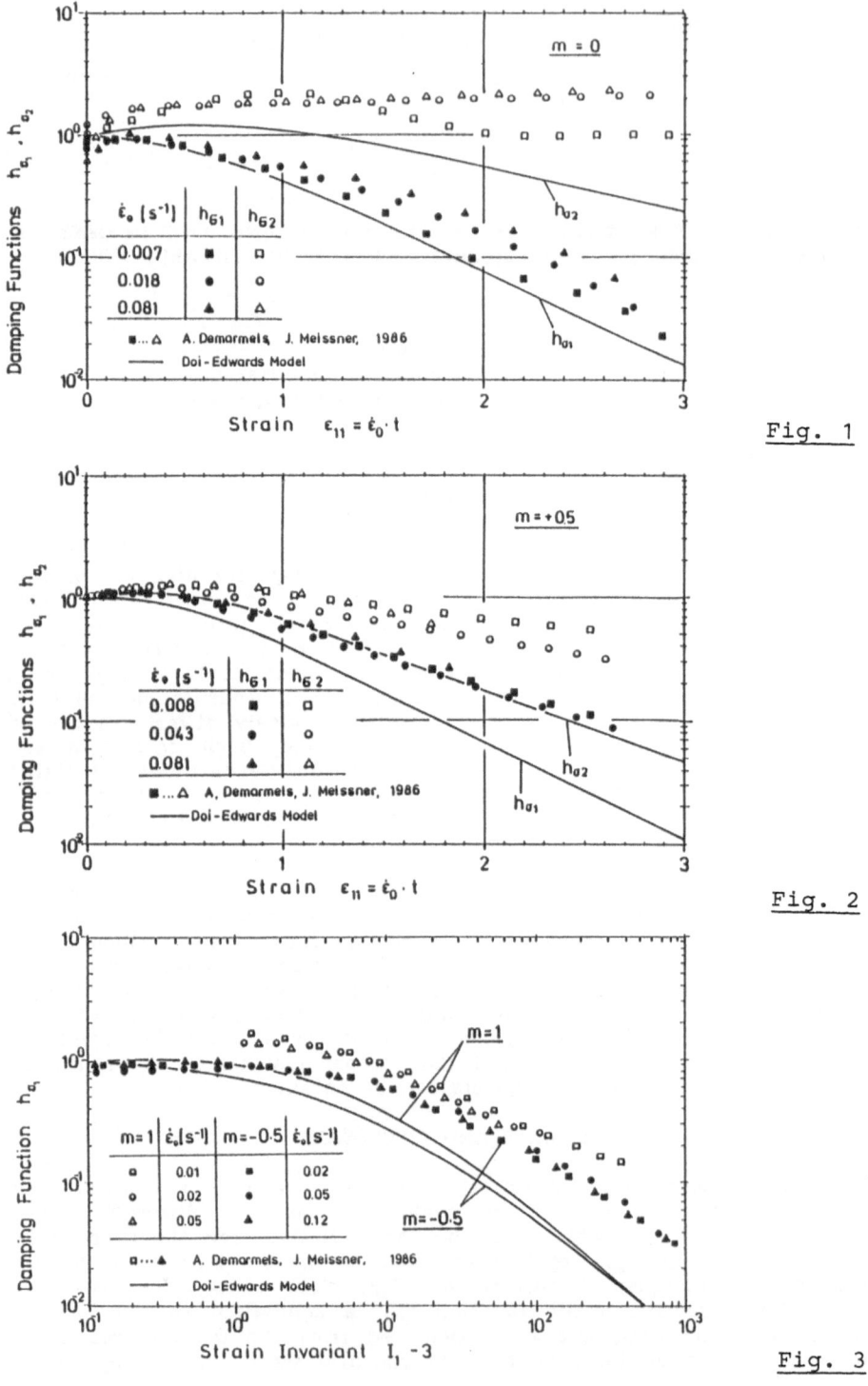

Fig. 1

Fig. 2

Fig. 3

favourable agreement with similar results in shear flow of concentrated polystyrene solutions (11). The coincidence of predictions and experimental results in shear flow was considered as powerful evidence for the correctness of the Doi-Edwards model.

However, this (approximate) agreement between the Doi-Edwards strain measure and experimental data for $h_{\sigma 1}$ in planar flow (m = 0) is the only one found in figs. 1 to 3. There is no agreement between predictions and experimental data for the effective strain function $h_{\sigma 2}$ in planar flow (m = 0), and between predictions and experimental data for $h_{\sigma 1}$ and $h_{\sigma 2}$ in uniaxial (m = - 1/2), ellipsoidal (m = 1/2), and equibiaxial extension (m = 1). Deviations start to appear already at small strains, with predictions being typically a factor of 2 smaller than experimental data at Hencky strains of ε = 1, and deviations increase with increasing strain. At Hencky strains of ε = 3, predictions are up to an order of magnitude smaller than experimental data. The deviations are well outside experimental scatter, and we are therefore compelled to the conclusion, that the strain function of the Doi-Edwards model is unable to predict the nonlinear behaviour of this PIB melt.

The exaggerated strain dependence of the Doi-Edwards strain function results from the assumption regarding the chain retraction process: after equilibration of the chain, the monomer density per unit arc length of chain is assumed to be equal to the equilibrium value, i.e. the primitive path length of the chain is assumed to be independent of deformation. Since the deformation changes the local environment of the individual chain, the validity of this assumption is not obvious. This was already stated by Doi and Edwards (13), and was later discussed by Marrucci et al. (14,15), and Graessley (16). The experimental data presented here suggest a need to reconsider the chain retraction process. Also, it might be necessary to take other constraint release processes into account to achieve quantitative consistency between experimental data and molecular model predictions.

References

1. *M.H. Wagner*, Rheol.Acta **15**, 136 (1976)
2. *M.H. Wagner*, J.Non-Newtonian Fluid Mech. **4**, 39 (1978)
3   *H.M. Laun*, Rheol.Acta **17**, 1 (1978)
4. *M.H. Wagner*, *H.M. Laun*, Rheol.Acta **17**, 138 (1978)
5. *M.H. Wagner*, Rheol.Acta **18**, 33 (1979)
6. *M.H. Wagner*, Proc.8th Int.Congress Rheol., Naples 1980, Vol.2, p.541
7. *M.H. Wagner*, *S.E. Stephenson*, J.Rheol. **23**, 489 (1979)
8. *A. Demarmels*, Ph D dissertation, ETH Zürich No.7345 (1983)
9. *A. Demarmels*, *J. Meissner*, Colloid and Poly.Sci. **264**, 829 (1986)
10. *M.H. Wagner*, *A. Demarmels*, submitted to J.Rheol.
11. *M. Doi*, *S.F. Edwards*, Theory of Polymer Dynamics, Oxford University Press 1986
12. *C.F. Curtiss*, *R.B. Bird*, J.Chem.Phys. **74**, 2026 (1981), and **74**, 2036 (1981)
13. *M. Doi*, *S.F. Edwards*, Faraday Trans.II, **74**, 1802 (1978)
14. *G. Marrucci*, *G. de Cindio*, Rheol.Acta **19**, 68 (1980)
15. *G. Marrucci*, *J.J. Hermans*, Macromolecules **13**, 380 (1980)
16. *W.W. Graessley*, Advances in Polymer Science **47**, 68 (1982)

FINITE ELEMENT METHOD FOR THREE-DIMENSIONAL VISCOELASTIC EXTRUSION

O. WAMBERSIE, M.J. CROCHET
Mécanique Appliquée, Université Catholique de Louvain
2, Place du Levant,1348 Louvain-la-Neuve, Belgium

ABSTRACT

We have developed a time-marching finite element method for solving three-dimensional extrusion which makes use of the conjugate gradient method. The algorithm, initially devoted to generalized Newtonian fluids, has been extended to viscoelastic flow. We briefly explain the method and show some applications.

INTRODUCTION

The numerical simulation of viscoelastic flow has seen important progress over the last few years (see e.g. [1-2] for recent reviews). Available simulations essentially cover two-dimensional problems. In particular, it is now possible to calculate extrudate swelling for fairly high values of the Weissenberg number in planar and circular geometries. It is however agreed that the true predictive value of viscoelastic simulations is expected for three-dimensional extrusion problems through complex dies. Boundary elements have been used for such purpose in [3-4] both for generalized Newtonian and viscoelastic fluids. A finite element technique for solving three-dimensional purely viscous extrusion has been given in [5]; quite recently, three-dimensional viscoelastic extrusion has been calculated in [6] by means of finite elements. A crucial difficulty for solving such problems remains the computer time; boundary elements as applied to three-dimensional extrusion problems require large full matrices, while the cost of frontal elimination with standard finite elements is high if one considers some geometrical complexity.

In a recent paper [7], we have used a time-marching approach coupled with the preconditioned conjugate gradient (CG) method for solving three-dimensional extrusion of generalized Newtonian fluids. For large non-linear systems, we found that the CG method is able to reduce the computational time despite the inherent number of additional iterations. In the present paper, we extend our work to the calculation of viscoelastic extrusion. At the present stage, we use a very simple type of finite element; we concentrate on cost-reducing iterative techniques rather than on the high Weissenberg number. We show the main steps of the algorithm together with its application to dies with circular (for comparison purposes) and square cross-sections. In the present paper, we limit ourselves to Oldroyd-B fluids although the method is available as well for more realistic constitutive equations.

BASIC EQUATIONS AND NUMERICAL ALGORITHM

Let us consider the basic equations for solving the flow of an Oldroyd-B fluid, for which the Cauchy stress tensor is decomposed into the sum of a pressure term $-p\mathbf{I}$, a purely viscous one $2\eta_v\mathbf{d}$ with $\mathbf{d}$ as the rate of deformation tensor, and a viscoelastic contribution $\mathbf{T}$ given by

$$\mathbf{T} + \lambda\,[\partial\mathbf{T}/\partial t + \underline{v}.\underline{\nabla}\,\mathbf{T} - (\underline{\nabla}\underline{v})^T\,\mathbf{T} - \mathbf{T}\,\underline{\nabla}\underline{v}\,] = 2\eta\,\mathbf{d} \ . \tag{1}$$

The momentum equations are given as follows,

$$\rho\,(\partial\underline{v}/\partial t + \underline{v}.\nabla\,\underline{v}\,) = \underline{\nabla}.\,(-p\,\mathbf{I} + 2\eta_v\,\mathbf{d} + \mathbf{T}) + \underline{f} \quad , \tag{2}$$

where $\underline{f}$ is the body force. Incompressibility requires that

$$\underline{\nabla}\,.\,\underline{v} = 0 \quad . \tag{3}$$

We note that, despite our goal to calculate steady-state extrusion, we keep the time derivatives in (1) and (2) because our algorithm is based on a time marching technique. Simultaneously, we need to solve the kinematic equations expressing that the free surface is a material surface.

Let us consider finite element representations for $\underline{v}$, $\mathbf{T}$ and p in terms of nodal vectors $\underline{V}$, $\underline{S}$ and $\underline{P}$. We use eight-node brick elements with P^1 - C^o shape functions for $\underline{v}$ and $\mathbf{T}$ and P^o - C^{-1} shape functions for p. With the standard Galerkin finite element method, it is possible to rewrite (2) and (3) in the following form,

$$\mathbf{M}\,\dot{\underline{V}} + \mathbf{C}\,\underline{P} - (\mathbf{K}\,\underline{V} + \mathbf{C}'\,\underline{S} - \mathbf{N}\,\underline{V} + \underline{F}\,) = \underline{Q} \quad , \tag{4}$$

$$\mathbf{C}^T\,\underline{V} = \underline{Q} \quad . \tag{5}$$

Combining (4) and (5), it is then possible to obtain a consistent discretized equation for the pressure which is given as follows,

$$(\mathbf{C}^T\,\mathbf{M}^{-1}\,\mathbf{C})\,\underline{P} = \mathbf{C}^T\,\mathbf{M}^{-1}\,(\,\mathbf{K}\,\underline{V} + \mathbf{C}'\,\underline{S} - \mathbf{N}\,\underline{V} + \underline{F}\,) - \mathbf{C}^T\,\dot{\underline{V}} \quad . \tag{6}$$

It is clear that the term $\mathbf{C}^T\,\dot{\underline{V}}$ should vanish in view of (5) (although one must save a contribution of earlier time steps [7]).

Let us now briefly explain the time-stepping algorithm. The time-scale is divided into a number of discrete times $t_0, t_1, ... t_n$. Let us assume that, at time t_n, we know the velocity field $\underline{V}^n$, the stress $\underline{S}^n$ and a geometrical vector $\underline{X}^n$ characterizing the free surface. In order to calculate the corresponding quantities at time t^{n+1}, we proceed as follows,

i. Calculate an intermediate pressure field $\underline{P}^*$ by solving (6) on the basis of $\underline{S}^n$, $\underline{V}^n$, $\underline{X}^n$. Use the CG method for solving the system.

ii. Use the discretized form of (1) for calculating $\mathbf{T}^{n+1}$. The matrix is diagonally lumped and thus allows for an explicit integration.

iii. Calculate the new velocity field $\underline{V}^{n+1}$ by solving (4). The method is implicit for the diffusive terms $\mathbf{K}\underline{V}$. The system is decoupled in three sub-systems for each velocity component; each sub-system is solved by means of the CG method.

iv. Calculate the vector $\underline{X}^{n+1}$ characterizing the free surface at time t^{n+1} by means of an implicit integration of the kinematic condition.

We emphasize that our goal is to calculate the steady-state solution. Typically, in problems with a very small Reynolds number, we may neglect the convective term $\mathbf{N}\ \underline{V}$ in (4) and select an arbitrary value for $\mathbf{M}$ (or ρ in (2)). Similarly, we have found that the optimal time-stepping does not necessarily require the same time-step for the fields calculation and the free surface calculation. More details on the algorithm will be given in a forthcoming paper.

NUMERICAL EXAMPLES

We have first tested our algorithm on standard 2D problems, such as the extrusion from a planar die, with a single layer of three-dimensional elements. Next we have considered a true three-dimensional problem with an axisymmetric solution, i.e. the extrusion through a circular die. The ratio of η_v to the total viscosity is 1/9. We have reached a Deborah number of 1.6 with a swelling ratio of 1.4. We have then considered the case of a square die with a finite element mesh given in Fig.1. The extrudate profiles are given in Fig.1 for various values of the Deborah number.

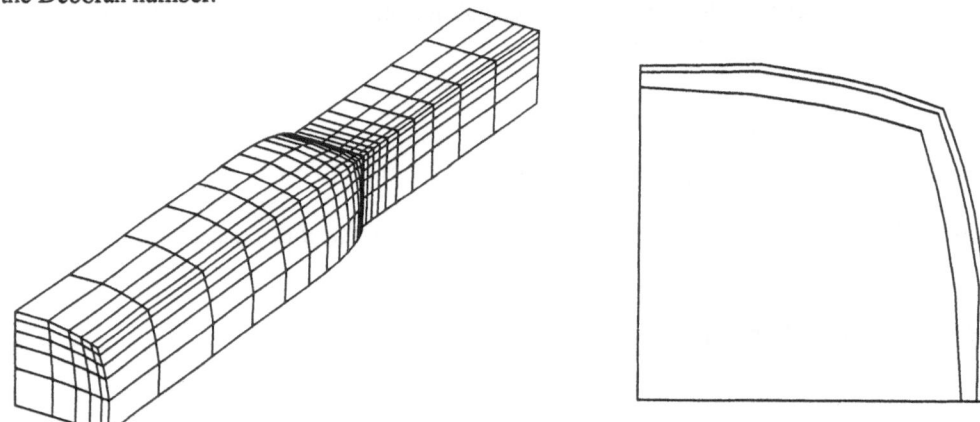

Fig. 1. Finite element mesh for calculating die swell from a square die, and extrudate profiles at De=0, 1., and 1.19.

REFERENCES

1. Keunings R., in *Fundamentals of computer modeling for polymer processing*, ed. Tucker C.L., Hanser Verlag, 1990.
2. Crochet M.J., Rubber Chemistry and Technology, Rubber Reviews 1989, **62**, 426-455.
3. Tran-Cong T., Phan-Thien N., J. non-Newtonian Fluid Mech., 1988, **30**, 37-46.
4. Tran-Cong T., Phan-Thien N., Rheol. Acta, 1988, **27**, 21-30.
5. Karagiannis A., Hrymak A.N., Vlachopoulos J., Rheol. Acta, 1989, **28**, 121-133.
6. Shiojima T., Shimazaki Y., J. non-Newtonian Fluid Mech., 1990, **34**, 269-288.
7. Wambersie O., Crochet M.J., A conjugate gradient method for calculating three-dimensional extrusion, to appear, 1990.

THE RHEOLOGY OF WAXY CRUDE OILS - AN OVERVIEW

L.T. WARDHAUGH AND D.V. BOGER
Department of Chemical Engineering
The University of Melbourne
Parkville, Victoria, 3052,
Australia.

ABSTRACT

The presentation summarizes the work conducted in the Department of
Chemical Engineering at the Unviersity of Melbourne on the flow
characteristics of waxy crude oils. The topic is introduced by
describing a progression which occurred from an extensive consulting
programme on operation and design of an 1800 km pipeline to involvement
with ICI Research Operations Pty Ltd in the successful development of
specific pour point suppressant chemicals for the Jackson crude oil.
The very applied work was then followed by fundamental research on
measuring and interpreting the unique flow properties of waxy crude
oils. The presentation is concluded with an examination of how the
basic flow properties of waxy crudes influence the design process for a
pipeline.

OVERVIEW

The crystallization of wax, which occurs in many crude oils from
Australia and around the world during production, transportation and
storage, results in severe operating difficulties due to the increase
in oil viscosity, oil gelation and deposition. Waxy crude oils are
highly non-Newtonian materials with complex flow properties that
depend, not only on the temperature and shear rate, but also on the
shear and thermal history experienced by the oil. Although
considerable research has gone into the solution of specific industrial
pipelining problems, a study devoted to the understanding of the
rheology of this material has not appeared in the literature. The work
at the University of Melbourne addressed the specific question of how
to obtain material flow property measurements for waxy and viscous
crude oils and how to use the data in steady state pipeline design.
The published thesis also provides an overview of the alternative
methods for pipelining waxy crude oil and of the chemical and physical
nature of the waxes and asphaltenes - the components which give rise to
the non-Newtonian flow properties [1].

A review of the literature [1] has shown that the greatest single hindrance to the development of methods for improving the flow characteristics of waxy crude oils has been the inability to obtain reliable flow property data at temperatures below the pour point of the oil. The first stage of this work has been the development of experimental techniques that are repeatable, and reproducible between different types and geometries of rotational viscometers. Repeatable results are obtained by the control of the shear and thermal history of the sample throughout the measurement, starting from an elevated temperature where the fluid memory has been removed. Continued shear at the target temperature results in a reduction in viscosity of the sample and the almost complete removal of time dependent (hysteresis) effects. This equilibrium position is quite unaffected by time of standing. However, a reduction in temperature in a static sample of as little as 0.2C results in an increase in the viscosity and yield stress and a total change in the nature of the flow properties. Reproducible data are obtained by calibration of the temperature lag that occurs in all instruments during cooling, the correction of contraction effects in the sample and the instrument and the correction of shear rate (concentric cylinder viscometer). The attainment of reproducible equilibrium data (flowcurves) ensures that material data, rather than instrument artifact, are being measured.

The equilibrium flow properties have been shown to be relatively independent of the cooling rate, but strongly influenced by the shear history (the shear rate applied during cooling in a single experiment) to the extent that each shear history results in a unique material with a unique flowcurve. For a given temperature, therefore, waxy crude oils are described, not by a single flowcurve, but by a series of flowcurves each determined by the respective applied shear history. As the shear history is reduced (in separate experiments), the resulting shear stresses go through a minimum then rapidly increase, coinciding with the statically cooled (zero shear) value of shear stress on yielding. This has a number of effects on pipeline operation including the existance of a minimum operating condition, below which flow would cease in a pipeline at a given temperature. The response of flow improver additives (pour point depressants) is also strongly influenced by the shear history effect with the percentage improvement increasing as the shear history is reduced. Since the shear rate varies across the radius of a capillary viscometer, each streamline in laminar flow must represent a unique fluid with the total pressure drop - flowrate being a composite effect of each streamline. The capillary viscometer cannot be expected to give the same result as a rotational viscometer, nor can the results with such an instrument be regarded as unique material properties.

Three quite distinct operating zones have been identified in a pipeline carrying waxy crude oil due to the enormous changes in flow properties which occur as the oil cools, with the division determined by the existance of time dependent and/or shear history dependent flow properties. Modified design techniques are outlined which account for the effect of the shear history. It is shown that existing design procedures for laminar flow can overestimate the flowrate at a fixed pressure gradient in excess of 100%.

Statically cooled samples possess a complex yielding behaviour that cannot be described by existing yield stress models. Three distinct

characteristics of the yielding process, namely, a solid (Hookean)
behaviour; a slow deformation (creep); and a sudden failure of the
sample that closely resembles the brittle or ductile fracture of
solids, have been positively identified by four different techniques -
the vane technique, the cone and plate viscometer (constant rotation);
constant stress viscometry and oscillatory testing. Reasonable
agreement is obtained using the first two techniques in the measurement
of the shear stress at fracture which is the value of most interest to
pipeline designers.

In addition to Reference [1] other papers have been published on this
work [2-10] which may be of interest.

REFERENCES

1. Wardhaugh, L. T., The Rheology of Waxy Crude Oil, Ph.D.
   Thesis, the University of Melbourne, Parkville, Victoria,
   Australia, 1990.

2. Wardhaugh, L. T., Binnington, R. J., and Boger, D. V., The
   Flow characteristics of Australian Waxy Crude Oils,
   Proceedings of the 13th Australian Chemical Engineering
   Conference (CHEMECA 85); Perth, 1985, pp 85-87.

3. Wardhaugh, L. T., and Boger, D. V., The Measurement of
   Time-Independent Flow Properties of Waxy Crude Oil,
   Proceedings of the Fourth National Conference on Rheology,
   Australian Society of Rheology, Adelaide, 1986, pp 237-244.

4. Wardhaugh, L. T. and Boger, D. V., Flow Property Measurement of
   Australian Waxy Crude Oils, Proceedings of the 14th Australiasian
   Chemical Engineering Conference (CHEMECA '86); Adelaide, 1986, pp
   49-53.

5. Wardhaugh, L. T. and Boger, D. V., Measurement of the Unique Flow
   Properties of Waxy Crude Oils, Chem. Eng. Res. Des, 65, 1987, pp
   74-83.

6. Wardhaugh, L. T., Tonner, S. P. and Boger, D. V., Flow Improvers
   for Waxy Crude Oils, Proceedings of the 15th Australiasian
   Chemical Engineering Conference Melbourne, (CHEMECA '87), 2, 1987,
   pp 70.1-70.8.

7. Wardhaugh, L. T., and Boger, D. V., Design Procedures for
   Australian Waxy Crude Oil Pipelines, Proceedings of the 16th
   Australasian Chemical Engineering Conference (CHEMECA '88);
   Sydney, 1988, pp 167-172.

8. Wardhaugh, L. T., Tonner, S. P., and Boger, D. V., Rheology of
   Waxy Crude Oil - Consequences for Pipeline Design and Operation,
   Proceedings of Xth International Congress on Rheology, Sydney,
   1988, pp 360-362.

9. Wardhaugh, L. T., Tonner, S. P. and Boger, D. V., Rheology of Waxy
   Crude Oils, SPE Paper 17625, International Meeting on Petroleum
   Engineering, Tianjin, China, 1988, pp 803-807.

COMPUTER AIDED RHEOMETRY

H.H. Winter, M. Baumgaertel, P. Soskey[†] and S. K. Venkataraman
University of Massachusetts, Department of Chemical Engineering, Amherst,
MA 01003, USA and [†]EniMont Americas, Inc., Princeton, NJ, USA

ABSTRACT

Commercial rheometers have advanced significantly in recent years, but data evaluation remains the most difficult step in the rheometry laboratory. This evaluation involves comparison of data from different instruments and different transient experimental modes, estimation of the range of validity of the data, and determination of the parameters in rheological models. Methods for analyzing rheometry data are well known, however, they involve time consuming and repetitive steps which ideally should be performed by a computer. Over the past years, we have developed software which accepts data from various rheometers and interactively determines the temperature shift factor, the discrete relaxation modes, and the correlation between various transient and steady flows. The software requires a DOS operated PC (IBM or compatible) and is easy to use. Examples will be shown.

INTRODUCTION

In the following we will assume that carefully measured rheological data are available, giving us the components of the complex modulus G', $G''(\omega, Ti)$ over a wide range of frequency ω and at a discrete set of temperatures Ti, which can be shifted into a pair of master curves. This information will be reduced into a discrete relaxation spectrum with which we can simulate arbitrary transient experiments without additional empirical parameters. The simulations are valid in the linear viscoelastic range within the experimental (T, t) window of the initial G', G'' data set. The simulation helps in designing new experiments and allows cross checking between various transient and steady experiments, especially when performed on different instruments (Winter et al., 1990).

DETERMINATION OF THE RELAXATION TIME SPECTRUM

Many different methods have been proposed for determining the relaxation time spectrum (Ferry, 1980; Laun, 1978; Friedrich and Hoffmann, 1985). The measured dynamic moduli G' and G'' may be described with a discrete relaxation spectrum:

$$G'(\omega) = \sum_{i=1}^{N} g_i \frac{(\omega \lambda_i)^2}{1+(\omega \lambda_i)^2} \quad \text{and} \quad G''(\omega) = \sum_{i=1}^{N} g_i \frac{\omega \lambda_i}{1+(\omega \lambda_i)^2}$$

Baumgaertel and Winter (1989) have recently developed a robust numerical method for converting dynamic mechanical data. The method avoids ill-posedness which seems to be inherent in the curve fitting when choosing too many parameters. In the nonlinear regression, g_i and λ_i are treated as freely adjustable parameters to obtain a best fit of G',G". The initial choice $g_{i,0}$, $\lambda_{i,0}$, N is not very important, and is done automatically in the program. The number of relaxation times adjusts during the iterative calculations depending on the needs for an improved fit. No extrapolation of the data (as with a Fourier Transform) and no empirical correlations are necessary. This is especially important when working with complex polymeric materials. The calculated relaxation spectrum also defines the corresponding retardation spectrum, j_i, Λ_i, i=1,2...(N-1). The calculated spectra are, obviously, only valid in a time window which corresponds to the frequency window of the input data, $t_{min} = 1/\omega_{max}$ and $t_{max} = 1/\omega_{min}$. However, if the terminal behavior of the material is known (longest relaxation time for viscoelastic liquids or equilibrium modulus for solids), the discrete spectra are obviously valid for infinitely long times.

SIMULATION OF TRANSIENT EXPERIMENTS

The relaxation modulus, given by the discrete spectrum, is introduced into the general linear viscoelastic equation for the stress (Bird et al., 1987), which is based on Boltzmann's superposition principle. Figures 1 and 2 show simulations of stress growth followed by relaxation and of creep followed by recovery. The agreement between simulation and experiment is a measure of the consistency of the data. In these examples, experiments were kept in the linear viscoelastic region and the agreement is satisfactory.

A similar procedure applies to large strain experiments. However, the constitutive equation will have to be modified (see for instance: Laun, 1978; Larson, 1987). The availability of the spectrum is a prerequisite for selecting and testing new constitutive equations.

CONCLUSIONS

The measurement with a rheometer is only the first step in the rheological characterization. Equally important are the critical evaluation of noisiness, range of validity, and preparation of the data for solving applied problems. A computer program (WBS-Shift, WBS-Relax) has been developed for this purpose. It accepts data files from a large variety of rheometers, superimposes and shifts them interactively on the screen, and allows the comparison of a variety of different transient modes in a computer aided modelling process. The rheological characterization is considered acceptable only after having systematically scrutinized the data and having found consistency within a data set. Obviously, the proposed correlation of various measurements is only applicable within the linear viscoelastic region while deviations from linear viscoelasticity are clearly identifiable in the computer graphs.

An important outcome of the computer aided rheometry is the condensation of the viscoelastic behavior into a small set of material parameters (temperature shift factors and relaxation modes). This set of parameters may be used in computer modelling of viscoelastic fluids or solids, and it can be stored in a database where it is easily accessed during future applications.

REFERENCES

Baumgaertel, M., and Winter, H.H., Rheol. Acta, 1989, **28**, 511.
Bird, R.B., Armstrong, R.C., and Hassager, O.,<u>Dynamics of Polymeric Fluids</u>, Wiley, New York, 1987.
Ferry J. D., <u>Viscoelastic Prop. of Polymers</u>, Wiley, New York, 1980.
Friedrich G., Hoffmann, B., Rheol. Acta, 1983, **22**, 425.
Larson, R. G., <u>Constitutive Equations for Polymer Melts and Solutions</u>.
Laun, H.M., Rheol. Acta, 1978, **17**, 1.
Winter, H.H., Baumgaertel, M., and Soskey, P., to appear in K.T. O'Brien Ed., "Computer-aided Engineering in Polymer Processing: Applications in Extrusion and Other Continous Processes", Hanser, Muenchen, 1990.

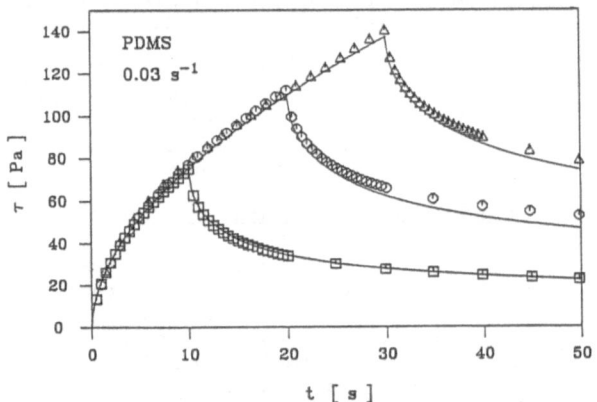

Fig. 1: Shear stress growth at $\dot{\gamma} = 0.03$ s^{-1} for 10, 20 and 30 s, followed by relaxation on cessation of shear; crosslinking PDMS near gel point; 25°C. Solid lines are predicted from dynamic data and points are measured.

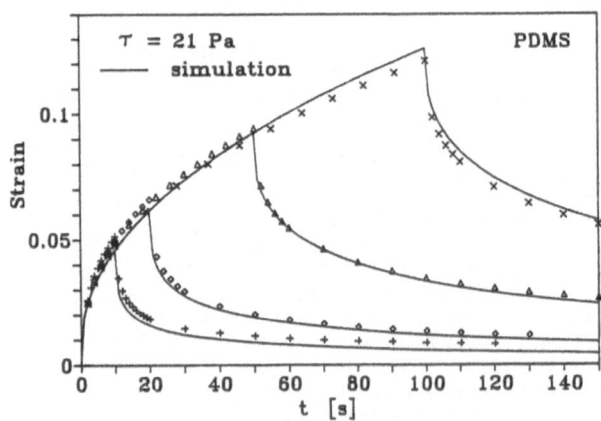

Fig. 2: Creep and recovery response at $\tau=21$ Pa; creep times of 10, 20, 50 and 100 s; same PDMS as in Fig.1.

THE RHEOLOGY OF ADHESIVE HARD SPHERE DISPERSIONS

A.T.J.M. Woutersen and C.G. de Kruif

Van 't Hoff Laboratory, Padualaan 8, De Uithof
3584 CH Utrecht, The Netherlands

ABSTRACT

The steady shear viscosity of a dispersion of sterically stabilized silica spheres in a marginal solvent is investigated. While in a good solvent the particles behave as hard spheres, in a marginal solvent there is an effective attraction between the particles. The particle pair potential is modelled as a hard core repulsion, preceded by an attractive and very short ranged well. The well depth depends on the temperature and can be changed from hard sphere repulsive to strongly attractive which eventually induces phase separation. This so called "adhesive hard sphere" system allowed us to do rheological measurements as a function of volumefraction, shear rate and known potential interaction. The results are compared with theoretical predictions.

RHEOLOGY

The flow properties of a colloidal dispersion are influenced by the size and the shape of the dispersed particles, the nature and the strength of the interparticle forces, the nature of the continuous phase and the stress applied on the dispersion. In a sheared suspension the fluid-motions have a tendency to deform the equilibrium particle distribution. The microstructure is now determined by fluid-mechanical, thermodynamical and potential factors. Since all three affect the particle distribution, the influences of fluid-dynamical interactions, Brownian motion and potential forces cannot be dealt with separately. Unfortunately a complete microscopic description of the stress build up in a concentrated suspension of interacting spheres has not been developed yet. However, in some limiting cases were part of the problem can be reduced, successful theories have been developed.

The simplest problem is that of an infinitely dilute dispersion of hard spheres for which the viscosity was first calculated by Einstein [1]. In this theory each particle is considered as being alone in an unbounced fluid. With increasing volumefraction, particle interactions have to be included. Batchelor [2] developed a rigorous theory for the bulk stress in a suspension of hard spheres. He used exact two-body expressions for the fluid-dynamical and direct interactions. The theory is restricted to low volumefractions and low rates of shear.

An extra difficulty arises if the suspension consists of particles other than hard spheres. The potential interactions produce additional contributions to the bulk stress. A microscopic theory for non-hard-sphere dispersions is developed by Russel [3]. He derived equations for the stress in dispersions where the particles interact either via a short ranged attraction or via a long ranged repulsion by extending Batchelor's theory. Recently Cichocki and Felderhof [4] calculated the stress in a semi-dilute dispersion of adhesive hard-spheres under a weak shear flow. They also based their theory on Batchelor's framework. An important result obtained both by Russel and Cichocki and Felderhof is an equation for the viscosity in an adhesive hard sphere dispersion in the limit of vanishing shear rate.

Other investigators use dimensional analysis methods to approach the problem this giving rise to a variety of semi-empirical expressions for the viscosity as a function of volumefraction and shear rate. In the approach of Krieger [5] and Chaffey [6] electrostatic repulsions and van der Waals attractions between the particles are investigated. The main problem in experimental studies of non-hard sphere dispersions is that the particle interactions are poorly defined and even worse, are poorly controlled.

Silica spheres sterically stabilized by a dense layer of short polymers and dispersed in organic solvents serve as a well characterised model colloidal dispersion. Dispersed in cyclohexane the particles behave as hard spheres. Both the viscoelasic properties and the steady shear rheology were recently investigated [7,8] and are well established now.

When the particles are dispersed in a marginal solvent like benzene there is an effective attraction between the particles. The local interaction between solvent molecules and polymer-chain elements is the most important factor in the interaction process. The strength of the attraction depends on the quality of the solvent and therefore on the temperature. By decreasing the temperature the interaction can be changed from hard sphere repulsive to strongly attractive which then induces phase separation.

Since direct interactions strongly influence the rheology of dispersions, a suitable model description for the interparticle forces is required. The interaction potential is modelled as a hard core repulsion, preceded by an attractive short ranged square well which depth depends on the temperature.

To determine the interaction parameters both equilibrium and transport properties were investigated. In dilute dispersions the static structure factor was measured as a function of temperature using small angle neutron scattering. Dynamic light scattering was used to study the temperature dependence of the diffusion coefficient.

Futhermore, the low shear viscosity was measured as a function of volumefraction and temperature. At very low volumefractions, the relative viscosity of the dispersion is independent of the temperature. In this very dilute region the strength of the direct interaction does not play a role because the attractions are extremely short ranged and the particles never meet. Here Einstein's theory is valid. At higher concentrations, but still in the semi dilute region, interactions between (pairs of) particles must be accounted for. At high temperatures where the particles reach the hard spheres state the viscosity behaves according to Batchelor's theory. At lower temperatures where interparticle forces become important, the viscosity equation for adhesive hard spheres describes the data quite satisfactory. The equation was used to determine the interaction parameters and the results are in good agreement with the results of the other techniques.

At higher volumefractions pair interactions are insufficient to describe the microstructure of the suspension and deviations of the microscopic theories occur. At high temperatures the hard sphere limit is reached and the results can be matched with the other hard sphere results [8]. By decreasing the temperature a tremendous rise in the low shear viscosity is observed. A very weak stress applied on the dispersion at a low temperature (but still above the phase separation point), results in a dramatic decrease in the viscosity. Obviously this is due to the disruption of the microstructure at rest. The high shear viscosity is only moderately influenced by the temperature. At temperatures above the phase separation point no time dependence nor yield stresses are observed.

Dimensional analysis methods are used to study the effect of an interaction potential on the rheology of the dispersion. At high temperatures the effects of shear and Brownian motion on the particle distribution are presented in terms of a scaled viscosity and a known dimensionless number: the Peclet number. A new dimensionless number is introduced, that scales the influence of a non-hard-sphere interaction potential.

LITERATURE

[1] A. Einstein, in "Investigations on the theory of the Brownian movement",R. Furth e.d. Dover, New York (1956).
[2] G.K. Batchelor, J. Fluid Mech. 83(1) (1977) 97.
[3] W.B. Russel, J. Chem. Soc. Faraday Trans. 80 (1984) 31.
[4] B. Cichocki and B.U. Felderhof, private communication.
[5] I.M. Krieger, Trans. Soc. Rheology 7 (1963) 101.
[6] G.E. Chaffey, Colloid & Polymer Sci. 255 (1977) 691.
[7] J.C. van der Werff, C.G. de Kruif, C. Blom and J.Mellema, Phys. Rev. A. 39 (1989) 795.
[8] J.C. van der Werff, C.G. de Kruif, J. Rheol. 33 (1989) 421.

ROLE OF THE DYNAMIC RHEOLOGICAL APPROACH IN KINETORHEOLOGY

YURI G. YANOVSKY
Institute of Petrochemical Synthesis of the USSR Academy
of Science, Leninsky prospect 29, 117912, Moscow, USSR

ABSTRACT

The combination of the two approaches, via the investigation by the dynamic rheological method, vis. the analysis of the varying time-dependent dynamic parameters which are determined, provided $(\omega, T) = \mathrm{const}$, $\gamma_0 = \mathrm{var}$ (ω - frequency, T- temperature, γ_0 - amplitude of deformation) and the analysis of the frequency dependences of the dynamic characteristics (dynamic mechanical spectroscopy) at discrete times, provided $(\gamma_0, T) = \mathrm{const}$, $\omega = \mathrm{var}$ enables one to fully characterize both certain kinetic and the physico-mechanical and structural peculiarities of the system at different time stages of its existence from a low viscous to a practically solid state.

INTRODUCTION

The registration of the rheological properties of the system, whose internal structure changes with time in virtue of reasons for one or another nature (physical, chemical, physico-mechanical, etc.) allows one in a definite manner to characterize the mechanism of the processes occuring within it.

The rheological dynamic method of investigation can substantially complement the information about the mechanism of physico-chemical conversions occuring in the reacting systems and in a number of cases, when, for example, unsoluble three-dimensional products are formed in the course of the reaction, it can successfully replace the traditional classical methods (1-3).

The techniques of investigating the kinetic processes using the forced non-resonance oscillations for systems whose viscoelastic characteristics changes by many decimal orders (tens of millions and even more times) were, for the first time, comprehensively described by us more than ten years ago (1). An important significance in the experiment under consideration proved to be the possibility of varying the deformation frequency spectrum of definite time stages of the occuring processes.

The above-mentioned analyses was the first to use particularly by investigation of the high-conversion radical polymerization of water-soluble monomers of N,N-dimenthyl -N,N-diallyl

ammonium chloride (DMDAAC) in solutions up to the maximum attainable degree of conversion (4). The self-acceleration of the
polymerization, the "gel-effect," proved to be a specific feature of the radical polymerization after a certain degree of
conversion had been attained. In some cases, the reason for
the gel-effect is the formation of a steric molecular network
with either physical or chemical bonds, or both.

RESULTS AND DISCUSSION

Analysis of the body of the rheological and kinetic data
obtained leads us to suppose that the observed self-acceleration of the above mentioned sample polymerization reacter, i.
e. an increase in the process rate and accompanying changes in
the rheological parameters after completing the linear kinetic
or rheological sections is due to the formation of the entanglement network in the reaction solution. It is known that the
appearance of a viscoelasticity plateau at curves $G'(\omega)$ is
a direct experimental evidence of an existing entanglement
network. A series of experiments were carried out to analyze
the frequencies dependences of the G' model PDMDAAC and DMDAAC
solutions. Monomeric polymer mixture solutions were prepared
simulating the reaction system at different high conversions.
By using the widely known viscoelasticity theory expressions,
the molecular mass of the dynamic segment M_e between the entanglement network nodes were estimated from the modulus $G' =
f(\omega)$ values plotted on the plateau.

The rheological data reflect the structure-formation process mechanism in the system in the course of the reaction and
is undoubtedly related qualitatively to the parameters of this
process (Fig. 1) It is important that the above dependences at
various stages of the reaction (initial, gel-effect, slow final
polymerization) are characterized by the linear curve fragments. At low conversion, linear dependences between the G''
and Q values are observed. Besides, the dependence of G' on Q
plays also an important role with increasing conversion. It is
important that there also occurs a definite polymer MM increase alongside with an increase in the entanglement network com-
pactness at the self-acceleration stage. It is evident that
the increasing $\bar{M}$ values and the loss modulus G'' with a rise
in conversion are similar.

The above analysis of the changes in the dynamic rheological parameters during the DMDAAC polymerization has made it
possible to suggest a physical model for the structure-formation process in the course of polymerization in aqueous solutions involving the stage of the increasing molecular mass of
the system, on the one hand, and the formation of the entanglement network on the other. From the data given in Fig. 1, it
is easy to obtain appropriate analytical expressions which relate the rheological and kinetic parameters of the reaction
system during polymerization in the form:

$$G' = K_i Q^{n_i} \; ; \quad G'' = K_j Q^{n_j},$$

where $K_{i,j}$ and $n_{i,j}$ are the approximation parameters.

The expressions obtained in such a manner allow us, to
determine the Q values for any stage polymerization process by
using only the data of the dynamic rheological investigations.

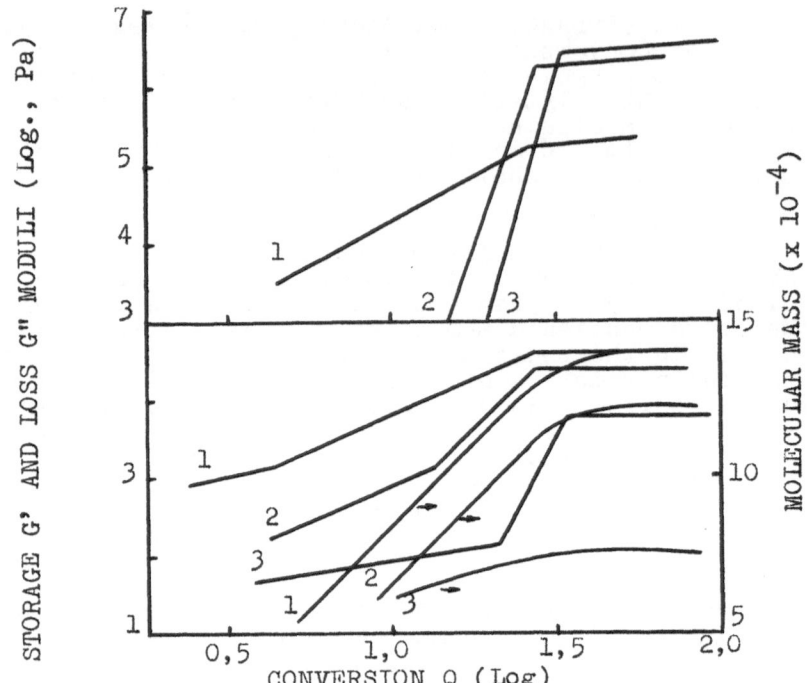

Fig. 1. Dependences of G', G" moduli and mean MM (M) of polymers formed on conversion Q during DMDAAC polymerization in aqueous solutions at 60°C, mol/l. Curves 1-3
correspond to monomer concentrations, mol/l: 4.5, 5.0,
5.5.

CONCLUSIONS

The approach suggested in the present work which is devoted to a rather complete description of the kinetic and structu-
ral-mechanical characteristics of the high-conversion polymerization processes should, in our opinion, be general and used
to investigate other polymerization processes. Moreover, controlling the viscosity and elasticity of the system in the
course of the process paves the way for monitoring the whole
reaction system. This can play an important role in manufacturing polymeric materials with predicted physico -mechanical
and physico - chemical properties.

REFERENCES

1. Andrianov K.A., Vinogradov G.V., Yanovsky Yu.G., Dokl.
   AN SSSR, 1977. **232**, nₒ.4, 810.
2. Yanovsky Yu. G., Zhachernyuk A.B., Burlova E.A., Vysokomol.
   Soyed., 1986, **28A**, №. 4, 829.
3. Winter H.H. and Chambou F.J. Rheol., 1986, **30**, 367.
4. Yanovsky Yu. G., Topchiev D.A., Kabanov V.A., Dokl, AN SSSR,
   1987, **297**, № 2, 428.

THE COMPLEX OF VISCOSIMETERS-DILATOMETERS FOR STUDY OF POLYMERIZATION RHEOKINETICS

DANIIL N. YEMELYANOV
N.I. Lobachevsky State University
Gagarin Avenue 23, Gorky 603600, USSR

ABSTRACT

A description has been made of capillary and rotary viscosi-
meters-dilatometers for simultaneous measurements of conversion depths and rheological properties of polymerizing systems.

In the course of polymer synthesis from liquid monomers
by a radical polymerization the reaction systems can be in the
fluid form as solutions, melts and dispersions. During the
polymerization process the rheological parameters vary usually
in a wide range (1,2). Therefore, an optimal version is the
usage of a complex of devices in these researches.

We have developed a model of a capillary rheodilatometer
and three models of rotary ones. A capillary device is intended
to measure rheological properties at initial conversions and
rotary dilatometers - at medium and high conversions. The
conversion depth of a monomer into a polymer was controlled by
a dilatometric method. Various modifications of rheodilatometers allow us to conduct measurements in varying ranges of
shear rates and shear stresses, under vacuum, at an excess gas
pressure, in the atmosphere of an inert gas and with the
monitoring of the system transparence. Reaction ampoules,
cylinders and dilatometric tubes in devices were made of glass
or steel, inert towards reaction products. The polymerization

was performed under isothermal conditions. Monomers were
carefully purified from impurities and oxygen.

A working unit of a capillary rheodilatometer (CRD) is
a modified suspended-level Ubbelohde viscosimeter. The
modification of the device was achieved by soldering dilatometric tubes between a vessel and measuring capillaries. The
polymerization was carried out in a vessel intended for the
dilution of the solution while estimating intrinsic viscosity. In certain periods of time the viscosity of the polymerizing system was measured. The time for the flow of a
given volume of the polymerizing liquid through the capillary was controlled. The liquid was passing through the
capillary of CRD under the action of a hydrostatic column
of the liquid or an excess pressure of argon.

In rotary rheodilatometers the coaxial cylinders were
used as measuring units: an outward cylinder is a reaction
ampoule and an inward cylinder is suspended by a rigid
magnetic or mechanical dinamometer. The presence of the
dilatometric tube over the ampoule permits to adjust the
conversion depth.

In the rheodilatometer RD-1 most of measurements of the
rheological parameters were made by the mass deformation on
maintaining the constancy of the shear stress. The device
measured the viscosity, elasticity and modulus of elasticity
as well as the kinetics of the strain growth of the polymerizing system at a constant shear stress.

A special feature of the rheodilatometer RD-2 was its
capacity of measuring both viscoelastic and stress-strain
properties of the polymerizing systems.

The device design provided an automatic monitoring of
twist angles of dinamometers and the control of strain
rates.

A sealed glass rheoviscosimeter (SRV) made it possible
to conduct the reaction at a moderate pressure (upto 1.5 MPa)
and concurrently with measuring the rheological parameters
to adjust the transparence of the reaction system.(3). In
device an unit was used to measure the values of torques,

and a magnetic field of a d.c. electromagnet acts as an elastic element in it. The rigidity of such dinamometer was determined by the current value in a winding of the electromagnet. This allowed to adjust shear stresses in a wide range.

The rheodilatometers measured the viscosity of the polymerizing systems in the following ranges: CRD, 10^{-3} - 1 Pa·s; RD-1, 1 - 10^{4} Pa·s; RD-2, 0.1 - 10^{4} Pa·s; SRV, 10^{-3} - 10^{3} Pa·s. The error of the rheological parameter measurements was within 5 - 10 %.

The devices were provided with automatic controls to display the rheological parameters and temperatures.

REFERENCES

1. Yemelyanov, D.N., Smetanina, I.Ye. and Vinogradov, G.V., The rheology of polymerization of vinyl monomers. Rheol. Acta, 1982, 21, 280-87.

2. Yemelyanov, D.N., Smetanina, I.Ye., Barsukov, I.A. and Volkova, N.V., The rheology of polymerization processes followed by phase transition. Progress and Trends in Rheology II. Rheol. Acta, 1988, 26, 300-02.

3. Kamsky, R.A. Rotary viscosimeter with sealed measuring cell. Zavodskaya Laboratoriya, 1988, 3, 45-8.

GELATINATION RHEOLOGY OF WATER-SALT SOLUTIONS OF POLY(VINYL ALCOHOL)

DANIIL N. YEMELYANOV, NADEZHDA K. KIREYEVA
N.I. Lobachevsky State University
Gagarin Avenue 23, Gorky 603600, USSR

ABSTRACT

The relation between the type of structural-rheological
states and the gelatination processes of water-salt
solutions of poly(vinyl alcohol) has been established.

A special feature of aqueous solutions of poly(vinyl
alcohol) (PVA) is their spontaneous gelatination because of
the thermodynamic lability. The application of a rheological
method on studying time changes of solution properties allows
us to establish regularities of the solution-gel transition
and to determine ways of goal-directed control of the gelatination process.

A study has been made on the rheological behavior of
1-20 % aqueous solutions of PVA (molecular weight $6 \cdot 10^{4}$, the
content of residual acetate groups is 1.6 %) in the presence
of salts (NaCl, KCl, KI) the concentration of which did not
exceed 10 mole % relative to PVA.

In a wide range of the polymer concentration C the change
in the greatest Newtonian viscosity η_{o} of the solutions obeys
the general power law which is in force in thermodynamically
stable systems.   Separate portions of a smooth curve
$\lg \eta_{o} = f(\lg C)$ can approach straight lines the boundaries between which are taken as critical concentrations C_{cr1} and C_{cr2}.
The concentrations above permit to distinguish three regions:

$C < C_{cr1}$ (region I), $C_{cr1} < C < C_{cr2}$ (region II) and $C > C_{cr2}$ (region III).

In region I the systems display Newtonian flow. They are characterized by the absence of the relationship between the solution viscosity and shear stress $(0.1-4 \cdot 10^3$ Pa). The solution viscosity does not practically change with time that is testified by the coincidence of the curves $\lg \eta_0 = f(\lg C)$ obtained for the solutions with different storage duration.

In region II the solutions begin to show thixotropic properties. However, the solutions do not exhibit high--elasticity that can be confirmed by a linear character of the dependence of strain γ on time t. In this region of PVA concentrations a slight increase in the viscosity η_0 of the solutions with storage duration is observed.

In region III a pronounced viscosity anomaly is characteristic of PVA solutions, and it becomes greater during the solution storage. The peculiarity of the solution behavior in this region of PVA concentrations is the appearance of flexibility and high-elasticity that manifests itself in the disappearance of the proportionality between γ and t on the strain curves. In the course of time a sharp rise of η_0 values, modulus of elasticity and high-elasticity occurs. The system is gelatinating, its flow being lost completely.

The general character of rheological regularities above remains for all systems studied. In accordance with the solution rheological properties we refer to the state of the systems in regions I, II and III as the viscous-Newtonian state (VNS), structural-viscous state (SVS) and high elastic state (HES). The critical concentrations C_{cr1} and C_{cr2} are boundaries of structural-rheological states. The main factors responsible for the concentration position of the structural-rheological state boundaries for water-salt solutions of PVA are the concentration and nature of salt, duration and temperature of the solution storage as well as molecular weight of PVA. Based on the parameters above we suggested diagrams of system structural-rheological states (fig.1). For each of the systems the curves (1,2) corresponding to the temperature dependence of C_{cr1} and C_{cr2}, respectively, divide the whole area of the diagram into the regions VNS, SVS and HES.

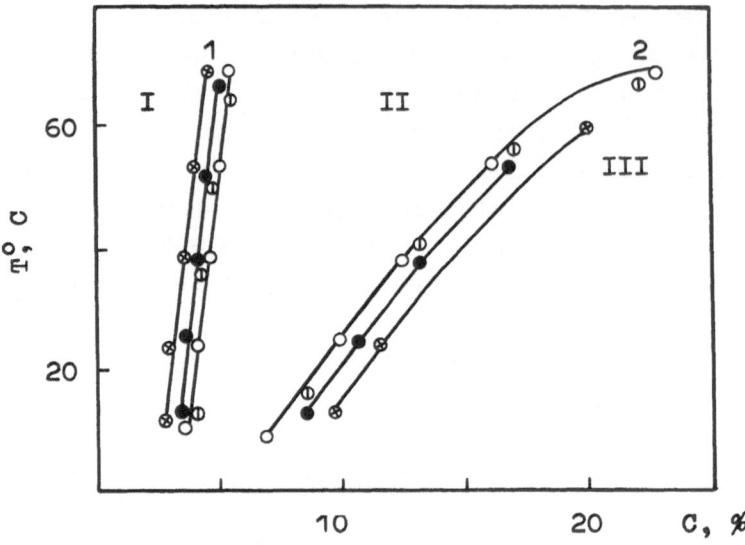

Figure 1. Diagram of structural-rheological states of PVA-H_2O-salt systems: o-NaCl, ⊗-KI, o-KCl, ●-H_2O. The content of salts is 8 mole % (symbols see in the text).

    The variation in the nature and concentration of the salt in the solution leads to a certain shift of the boundaries of the system structural-rheological states without changing the basic diagram. The effect of lyotropic salts on the critical concentrations should be associated with the change in the solvent thermodynamic properties.   The addition of the salts improving the solvent properties (e.g. KI) results in the broadening of the region SVS.

    Thus, the type of structural-rheological states of the system under given conditions determines the conditions for its spontaneous gelatination. The knowledge of the diagrams allows to make a purposeful choice of the method for controlling the gelatination process.

THE ROLE OF MACROMOLECULE HYDROPHOBIZATION IN STRUCTURAL-RHEOLOGICAL PROPERTIES OF WATER SOLUTIONS OF ACRYLIC POLYMERS

D.N. YEMELYANOV, B. V. MYASNIKOV AND L.I. MYASNIKOVA
State University, Gagarin Avenue, 23, Gorky 603600, USSR

ABSTRACT

The increase in fraction of hydrophobic (ester units) in macromolecule copolymers of acrylic acid with its esters causes the increase of viscosity of water solutions and the change in viscosity anomaly nature, that is transition from thixotropy to dilatancy.

The peculiarity of rheological behaviour of polyelectrolytes in aqueous medium is due to the presence of hydrogen bonds of hydrophobic macromolecule interaction alongside those with strong polarity. In this connection it was of particular interest to estimate the role of hydrophylic – hydrophobic equilibrium of forces in formation of acrylic polyelectrolyte solution structure. The equilibrium of hydrophylic – hydrophobic forces was controlled by the change of copolymer composition by acrylic (AA) and methacrylic (MAA) acids with their esters. Hydrophobicity of macromolecules was provided by hydrocarbon copolymer groups and hydrophilicity – by carboxylic ones. The subjects of investigation were 1-10% water solutions of polyacrylic (PAC) and polymethacrylic (PMAc) acids with molecular weight of $1 \cdot 10^5$ – – $6 \cdot 10^5$, as well as copolymer solutions of AA with methylacrylate (MA), ethylacrylate (EA) and those of MAA with methylmethacrylate (MAA). Copolymers and PAA have a close degree of polymerization. Rheological properties of polyelectrolyte solutions were studied by method of rotational viscosimetry.

The availability of the greatest and the lowest Newtonian viscosity and the region of anomalous viscosity decrease is characteristic for PAA solutions with increasing stress or rate of the shear.

The introduction of hydrophobic groups in macromolecules of water soluble polyelectrolytes in transition from PAA to PMAA or from PAA to copolymers of AA with its esters changes the rheological behaviour of solutions. The rheopexic (anti-

thixotropic) anomaly of viscosity occurs. The increase of
ester units in copolymer causes not only the increase of solution viscosity but results in the increase of anomaly and
change of anomalous viscosity behaviour type. In hydrophobi-
zation of PAA molecules the transition from Newtonian to
non-Newtonian reopexic nature of flowing polyelectrolyte solution is the most pronounced. The increase of MA or EA content in the copolymer with AA, promoting the increase of
contribution of hydrophobic interaction in the system structure formation results not only in viscosity increase but
also in the change of flow character (Fig.). In increase of

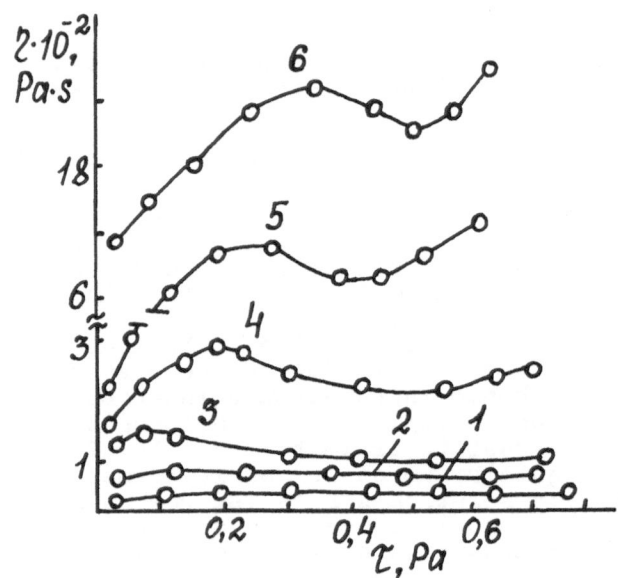

Figure.   The dependence of viscosity (η) upon shear stress
          (τ) in 10% solution of AA-MA copolymer. MA content
          in copolymer, mol.%: I-0; 2-I; 3-5; 4-7; 5-8; 6-10;
          25°, pH=3.

MA content in copolymer to 10 mol.% the solution viscosity
increases by a factor of ten. In this case if the viscosity
of 10% PAA solution is constant within the range of shear
stress studied, with increasing MA content in copolymer by
more than 5 mol.% the solution viscosity becomes a function
of the shear stress. The maximum viscosity appears on the
curves "viscosity - shear stress" with 7-10% MA content in
the copolymer. The further increasing shear stress results
consecutively in viscosity decrease and then in its repeat

increase which is the most pronounced for the copolymer containing 10 mol% of MA. The increase in hydrocarbon radical size of macromolecule ester group (in transition from AA-MA copolymer to AA-EA copolymer) is accompanied with an active growth of equal concentration solution viscosity and with more drastic change of flow character.

The mechanism of the phenomenon studied seems to consist in the following. Under certain conditions of deformation and solution concentration the chain conformation changes as a consequence of which the hydrophobic groups of diphilic macromolecules emerge onto the surface from the coil bulk and this reinforces the chain interaction. The introduction of a small fraction of alkylacrylate units into PAA macromolecules makes it possible not only to increase the solution viscosity but also to control the flow character, imparting Newtonian, thixotropic or rheopexic properties to the system.

522

STRUCTURAL-RHEOLOGICAL STATES AND PROPERTIES OF POLY-MERIZING VINYL MONOMERS

DANIIL N. YEMELYANOV, IRINA YE. SMETANINA
N.I. Lobachevsky State University
Gagarin Avenue 23, Gorky 603600, USSR

ABSTRACT

The scheme which allows to evaluate uniformly the change in the rheological properties of homophase polymerizing vinyl monomers has been suggested.

The polymer formation in the course of homogeneous radical polymerization of vinyl monomers is accompanied with the change in the system rheological properties regularly bonded with molecular weight, molecular-weight distribution and polymer content. For polymerizing methacrylate systems earlier studied with increasing the coversion P the rise in the greatest Newtonian viscosity η_0 in a certain range of conversions was described by the equation $\eta_0 = K P^n$, where K and n are the constants. The portions of conversion curves, where the change of K and n occurred, were called critical conversions P_{cr}. Apart from η_0 , a complex of rheological properties, i.e. coefficient of viscosity anomaly (thixotropy) K_τ, elastic strain γ_{el} and arbitrary-instaneous modulus of flexibility E_{flex}, was used to evaluate properties of polymerizing systems. To systematize the changes in the complex of rheological, kinetic and colloid-chemical proper-ties during the polymerization process the scheme, represent-ing structural-rheological states of the polymerizing systems, has been suggested (fig.1). On developing the scheme the following experimentally found regularities was used. The changes in the system rheological properties with increasing the conversion occur step-by-step.

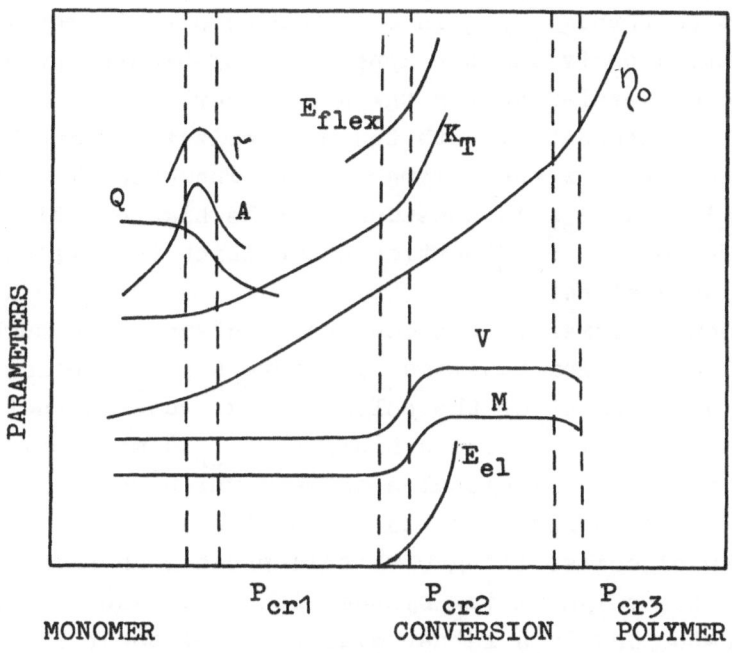

Figure 1. Diagram of conversion change of structural-rheological, kinetic and colloid-chemical parameters (symbols in
the text).

The logariphmic relationship $\eta_o(P)$ represents a curve the
portions of which can approach straight lines. At initial
conversions when $P < P_{cr1}$ the system flow is Newtonian: the
viscosity is not affected by the shear rate of polymerizing
system and the ratio of viscosities measured at different
shear rates is equal to unity, i.e. $K_T = 1$. We refer to this
concentration region as the viscous-Newtonian state. The
upper boundary of this state is the portion of the first critical conversion P_{cr1}. With growing the conversion at $P > P_{cr1}$
the viscosity anomaly increases ($K_T > 1$). Over the portion
P_{cr1} a characteristic change in colloid-chemical properties
of filled polymerizing suspensions is observed: the adsorption A of polymer on fillers approaches the maximum, sedimentation rate of fillers decreases sharply and the degree of
association Γ of macromolecules in monomer solution is the

greatest.   At $P > P_{cr1}$ the polymerizing system is in the
state which we refer to the structural-viscous state. This
state terminates in the portion of critical conversions P_{cr2}.
The transition through P_{cr2} is accompanied with increasing
the viscosity anomaly and the appearance of high-elasticity.
At the same conversion depths the gel effect of the reaction
occurs and, consequently, both the polymerization rate V and
molecular weight M of the polymer formed increase. The conversion region $P > P_{cr2}$ is considered to be high elastic state.
It extends upto the P_{cr3} portion above which the systems pass
into the glassy state.

   The peculiarities of the conversion change of a number of
rheological and colloid-chemical parameters are associated with
the staged development of the polymerizing system structure.
In the region of initial conversions ($P < P_{cr1}$) macromolecules
are in the form of coil-globules between which there is a weak
interaction. The increase in the conversion depth leads to the
accumulation of coils. Along with this process an uniform
increase in the polymerizing system viscosity takes place, and
it obeys Newton law in a rather wide range of shear stresses.
On approaching the critical conversion P_{cr1} there occurs the
overlap of separate macromolecular coils and the origin of
primary structural units which are capable of breaking down
under the influence of shear stress. On deforming this structure the non-Newtonian flow is observed, i.e. the region of
effective viscosity appears on the viscosity versus shear
stress curves. At P_{cr2} fluctuation network spreading over the
whole solution volume arises. The formation of the network
results in the elastic property origin and the viscosity anomaly enhancement. Above P_{cr3} the polymer is a glassy material.

   The critical conversion values depend on the polymer
molecular weight, the reaction temperature and the dissolving
capacity of the medium. The relation between the critical
conversions and the polymer molecular weight allows to construct the diagrams of structural-rheological states for the
homophase polymerizing systems. The diagrams permit to predict the system rheological behavior at different conversions
and polymer molecular weights.

FIBRE SPINNING WITH AXIAL AND RADIAL VISCOSITY DISTRIBUTIONS AS VISCOELASTIC FLOW WITH DOMINATING EXTENSION

STEFAN   ZAHORSKI
Institute of Fundamental   Technological Research
Polish Academy of Sciences
Swietokrzyska 21,   00-049 Warszawa, Poland

ABSTRACT

It is shown that non-isothermal melt spinning with axial and radial viscosity distributions can be considered as a particular case of the flows with dominating extensions discussed elsewhere. The governing equations are solved for weak variability of the extensional viscosity function with respect to the extension rate. Validity of the so-called quasi-elongational approximation, widely used in numerous references, is justified, and possibility of any "skin-effects" excluded.

Various aspects of isothermal and non-isothermal melt-spinning processes were widely discussed in numerous papers and books (for references cf.[1,2]). Almost all contributions were based on the so-called quasi-elongational approximation, i.e. the assumption of uniform velocity profiles across the filament. Only few authors considered some possibilities of radial viscosity distributions trying to justify the assumption of quasi-elongational approximation for various scaling parameters. For instance, Kase [3] presented a series solution in velocity assuming parabolic viscosity distribution across the thread, and proved that under most conceivable spinning conditions the velocity field within the fibre is practically flat and the flow considered - almost purely extensional. A question, however, arises whether under some severe conditions, i.e. for large viscosity gradients across the thread,   real

velocity profiles may be concave, what means that the skin-
-layer of much more viscous liquid is peeled off.

   In the present paper we show that non-isothermal melt-
-spinning with axial and radial viscosity distributions,caused
by the corresponding temperature gradients and crystallization
effects (cf.[4,5]), can be treated as a particular case of the
flows with dominating extensions discussed elsewhere [6]. To
this end the quasi-elongational approximation is assumed for
the fundamental extensional flow, while the additional flow
with shearing effects essentially depends on the extensional
viscosity function as well as on the derivatives of material
functions with respect to the extension rate $q \equiv V'(z)$. The
corresponding linearly perturbed constitutive equations are of
the form:

$$T^* = -p1 + \beta_1 \underset{\sim}{A}_1 + \beta_2 \underset{\sim}{A}_1^2 + \beta_1 \underset{\sim}{A}_1' + \beta_2 (\underset{\sim}{A}_1^2)' + \frac{\partial \beta_1}{\partial q} q' \underset{\sim}{A}_1 + \frac{\partial \beta_2}{\partial q} q' \underset{\sim}{A}_1^2$$

where T^* is the stress tensor, p - the hydrostatic pressure,
β_i and $\underset{\sim}{A}_i$ (i=1,2) - the material functions and the Rivlin-
-Ericksen kinematic tensors, respectively. Primes refer to the
corresponding increments. The non-linear governing equation
for the additional velocity field w, viz.

$$\beta \frac{1}{r} \frac{\partial}{\partial r} (r \frac{\partial w}{\partial r}) + \frac{1}{2} \frac{\partial}{\partial z} [(\frac{\partial \beta_1}{\partial q} + \frac{\partial \beta_2}{\partial q} q)(\frac{\partial w}{\partial r})^2] + \frac{\partial}{\partial z} (3\beta q) = C(z)$$

where $\beta = \beta_1 + \beta_2 q$ is proportional to the extensional viscosity η_E and $C(z)$ is an unknown function of z only, can be
solved with appropriate boundary conditions for weak variability of the material functions with respect to the extension
rate q, i.e. for slightly non-Newtonian fluids for which the
following quantity:

$$k = \frac{1}{\beta} (\frac{\partial \beta_1}{\partial q} + \frac{\partial \beta_2}{\partial q} q) q$$

is small enough. The corresponding results are obtained for
any radial dependence of the extensional viscosity.

   For instance, in the case of purely viscous fluids, we
obtain

$$w_0 = -\frac{3}{4\beta_0}\frac{\partial}{\partial z}(\beta_0 q)(r^2 - \frac{h^2}{2}) - \frac{3}{2}h^2[\int\limits_0^h \frac{rdr}{f}]^{-1} \times$$

$$\times[\frac{q^2}{v} - \frac{1}{\beta_0}\frac{\partial}{\partial z}(\beta_0 q)][\int \frac{1}{r}\int \frac{rdr}{f}dr - \frac{2}{h^2}\int\limits_0^h r\int \frac{1}{r}\int \frac{rdr}{f}drdr],$$

where $f(r)$ is a function responsible for radial viscosity distribution.

It is proved that the quasi-elongational approximation is fully justified for moderate variability of the logarithm of viscosity function along the spinline. An answer to the question posed at the beginning is negative, since it is also shown that velocity profiles for melt-spinning processes are always convex, thus excluding any possibility of the mentioned "skin-effects".

REFERENCES

1. Ziabicki, A., <u>Fundamentals of Fibre Formations</u>, J.Wiley & Sons, London-New York-Sydney-Toronto 1976.

2. Petrie, C.J.S., <u>Elongational Flows</u>, Pitman, London-San Francisco-Melbourne 1979.

3. Kase, S., Studies on melt spinning, III: Velocity field within the thread, <u>J.Appl.Polymer Sci.</u>, 1974, <u>18</u>, 3267-78.

4. Ziabicki, A., The mechanisms of "neck-like" deformation in high-speed melt spinning, 1: Rheological and dynamic factors, <u>J.Non-Newtonian Fluid Mech.</u>, 1988, <u>30</u>, 141-55.

5. Ziabicki, A., The mechanisms of "neck-like" deformation in high-speed melt spinning, 2: Effect of polymer crystallization, <u>J.Non-Newtonian Fluids Mech.</u>, 1988, <u>30</u>, 157-68.

6. Zahorski, S., Viscoelastic flows with dominating extensions: application to squeezing flows, <u>Arch.Mech.</u>, 1986, <u>38</u> 191-207.

SECONDARY RELAXATIONS IN POLYMERS CONTAINING PHENYL RINGS

FRANZ ZAHRADNIK
Institute of Material Science, chair for Polymers
University Erlangen-Nürnberg, Martensstr. 7, D-8520 Erlangen

ABSTRACT

The dynamic mechanical behaviour of bisphenol-A-polycarbonat (PC), polyarylsulfon (PES), polyetherimid (PEI), polyamidimid (PAI) and polyaryletheretherketon (PEEK) has been investigated in the temperature range between -170 and +270 °C and generally two decades around 1 Hz on the frequency axis. Although there are great differences in the chemical structure in regard of their repeating units, all polymers investigated show a maximum in mechanical damping between -100 and -90 °C at an interpolated frequency of 1 Hz. Significant differences in the damping behaviour were found at higher temperatures. While PC, PEEK and PES have shoulders in the neighborhood of 0 °C and a broad minimum around +100 °C, both polyimids are characterised by wide maxima from -20 °C up to their glass rubber transition.

MATERIALS AND METHOD

While bisphenol-A polycarbonates and different types of polyarylsulfones had been object of detailed investigations concerning mechanical properties in connection with molecular structure (1,2,3) polyaryletherketons and thermoplastic polyimides had not been analysed in such great depth. The purpose of this paper is to report the results of dynamic mechanical measurements on five aromatic polymers. Emphasis is placed on the well-known -100 °C (1Hz) transition. Also the damping behaviour between this prominent transition and the glass rubber transition is analysed.

All polymers investigated are commercially obtained materials. The available pellets had been well dried during 24 hours and pressure moulded under vacuum. Thereafter they had been stored in a vacuum desiccator prior to testing. Only PAI was received from the producer in shape of plates. Samples were quickly mounted in the torsion pendulum (4) and heated up to temperatures of $T=Tg+10$ °C for one hour. Subsequently they were quenched to the lowest testing temperature, generally -170 °C by dry nitrogen gas; then measurements were performed in stepwise heating, equilibrating half an hour at each measuring temperature. Chemical structures and glass transition temperatures are given in Table 1.

RESULTS AND DISCUSSION

Dumais et al.(5) (see also the literature cited there) concluded that 180° flips of the phenyl rings are the primary mode of molecular motion of the aromatic rings in

TABLE 1
Chemical structure and glass transition temperature Tg of polymers investigated.

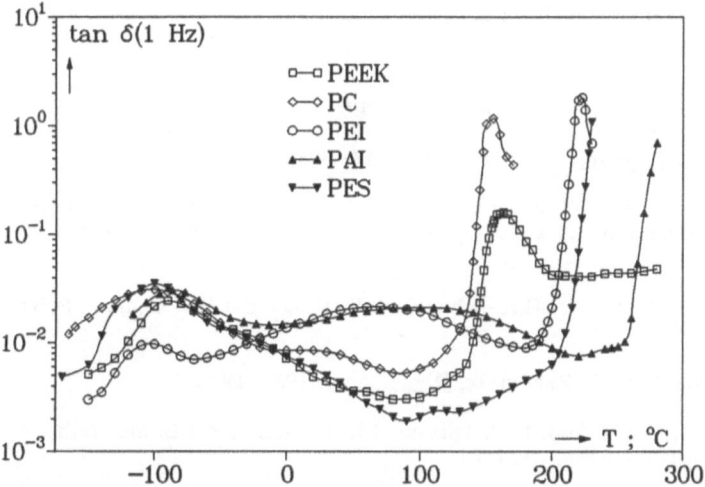

PEI: Polyetherimide $T_g = 217\ °C$

PC: Polycarbonate $T_g = 140\ °C$

PAI: Polyamidimide $T_g = 275\ °C$

PES: Polyethersulfone $T_g = 220\ °C$

PEEK: Polyetheretherketone $T_g = 143\ °C$, $T_m = 334\ °C$

polyarylether (e.g. bisphenol-A polycarbonate, bisphenol-S polyether and related polyarylsulfones) and that these ring flips are the molecular-level process related to their -100°C transition. Assuming this one would predict this prominent transition for all polymers containing molecular moieties like **Ph-X-Ph**.
As shown in Fig.1 all five polymers investigated including the extensively studied and reviewed PC and PES exhibit this prominent secondary relaxation.

Figure 1. Mechanical loss $\tan \delta$ at 1 Hz as a function of temperature for five aromatic polymers investigated.

At higher temperatures our data of PC, PES and PEEK show a shoulder in the temperature region between -40 and +20 °C. Robeson et al. (2) had observed a shoulder for their bisphenol-S polyether without an isopropylidene moiety too. Letton and co-workers (6) presented a mechanical spectrum of bisphenol-A polysulfon (PSF) with a shoulder in this region whose intensity increased by either annealing at 180 °C for 10 hours or by slow cooling from 270 °C. They concluded that this relaxation is associated with the methyl groups in the isopropylidene moiety of their PSF. But which particular motion can be the origin of this transition in polymers containing only $-SO_2-$, $>C=O$ and $-O-$ links beside phenyl rings? The following region in our spectra between +20 °C and the respective Tg's exhibit a characteristic difference between PC, PES and PEEK on one hand and the thermoplastic polyimids PEI and PAI on the other. While data of the first yield a broad minimum in the neighborhood of +90 °C polyimid data exhibit wide maxima at +100 °C. Sasuga et al. (7) observed a similar behaviour in this temperature region for PES which was influenced (and increased in magnitude) by water as well as by electron beam irradiation. Thermal treatement effects this behaviour too. The magnitude of mechanical loss factor decreases with annealing and it reappears with rapid quenching. Gilham et al. (8) presented mechanical spectra of a serie of linear polyimides based on benzophenone tetracarboxylic acid and various diamines which exhibited the same relaxation behaviour as our tested polyimides. Whether the characteristic phthalimid group in all these polyimides including our PEI and PAI may be associated with this transition or not remains to be thoroughly investigated.

CONCLUSIONS

Considering our results as well as prior published investigations we could come to the following conclusions:
- all these polymers containing the Ph-X-Ph moiety exhibit a transition at -100 °C,
- the origin of secondary relaxations between -100 °C and Tg may be found in molecular motions due to molecular packing defects,
- the broad transition in thermoplastic polyimides below their glass transition may be caused by motions involving the phthalimid group.

REFERENCES

1. J. Heijboer, J. Polym. Sci., 1968, Part C 16, 3755.

2. L.M. Robeson, A.G. Farnham and J.E. McGrath in D.J. Meier (Ed.): Molecular Basis of Transitions and Relaxations, Gordon and Breach Sci. Publ., London 1978, page 405.

3. B.C. Auman, V. Percec, H.A. Schneider, W. Jishan and H.J. Cantow, Polymer, 1987, 28, 119.

4. F.R. Schwarzl and F. Zahradnik, Rheol. Acta, 1980, 19, 137.

5. J.J. Dumais, A.L. Cholli, L.W. Jelinski, J.L. Hedrick and J.E. McGrath, Macromolecules, 1986, 19, 1884.

6. A. Letton, J.R. Fried and W.J. Welsh in S.E. Keinath, R.L. Miller and J.K. Rieke (Ed.): Order in the Amorphous "State" of Polymers, Plenum Press, New York 1987, page 359.

7. T. Sasuga, N. Hayakawa and K. Yoshida, Polymer, 1987, 28, 236.

8. J.K. Gilham and H.C. Gilham, Polym. Eng. Sci., 1973, 13, 447.

STUDY ON THE SHEAR STRESS AND NORMAL STRESS OF FOAMS

Zhou Cheng-dong   Chu Jia-ying   Jiang Ti-qian

Chemical Engineering Research Centre
East China University of Chemical Technology
Shanghai 200237, P. R. China

ABSTRACT

Foams became an industrial material at the beginning of this century. The Rheological properties of foam have attracted great interest since the foam fracturing method for enhancement of oil recovery in petroleum production was developed. In the past 20 years, many experimental studies of the rheological properties of foams have been carried out. But the research results often differ because of foams' complex rheological properties. Since the 80s, several theoretical studies of it have become available [1,2,3]. However, the two-dimensional spatial periodic structure they used in the theory is so ideal that the theoretical results differ greatly from the experimental results. This paper makes use of a three-dimensional spatially periodic structure to derive the theoretical expression of the shear stress and normal stress of foams, and checks the expression by experimental datas.

THEORY

First of all, we assure that the foam is a homogeneous monodisperse system. On this basis, a dodecahedron cell——the three-dimensional spatial model of foams has been drawn up. A hexahedron as an element unit has been obtained when the centres of eight adjacent dodecahedron cells are lined up. The elemental unit is the object of investigation. Meanwhile, the stress tensor of foam is given by:

$$\tau = - \frac{1}{V} \int_A \sigma \left(\delta - \vec{n}\,\vec{n} \right) \, dA - \frac{\mu}{V} \int_{V-V_g} \dot{\gamma} \, dV.$$

Where V is the volume of element unit, A is the area of liquid film, δ is the unit tensor, σ is the interfacial tension of foam, $\vec{n}$ is the normal vector of the film, μ is the viscosity of foam and V_g is the gas volume in element unit. Therefore, according to the structure of the unit, the shear and normal stress formula are derived as follows:

$$\tau_{12} = - 0.354 \frac{\sigma}{R^3} \sum_i T_i^1 T_i^2 g_i \left(1 + Ca \frac{R}{2 g_i} \frac{dg_i}{d\gamma}\right) - \frac{0.354 \, \sigma}{R^3} \sum_i \frac{R}{g_{if}} T_i^1 T_i^2 \frac{dg_{if}}{d\gamma} Ca$$

$$\tau_{11} - \tau_{22} = \tau_{12}\dot{\gamma}$$

Where R is the mean radius based on the " surface-volume mean ", T_i^1 and T_i^2 are the vectors in direction 1 and 2 of the i'th film, g_i is the length of the i'th film. φ is the gas volume fraction.

$$Ca = \frac{8}{3} \frac{R \dot{\gamma} \mu}{\sigma} \frac{3}{2} (1 - \sqrt{\varphi}) \sqrt{\varphi}.$$

According to the structure of the unit, we know that the liquid section of the foam exists in pleateau boundry. The deformation may lead the pleateau boundries to combine each other. The characterization of the deformation is quite different before and after combination. Therefore, two deformation models must be established separately. Deformation Model (1) which modifies the foam before combination of its pleateau boundry consists of thirty non-linar transcendental equations:

$$f_j(g_i, T_i^1, T_i^2, \rho, \gamma, \alpha_i) = 0 \quad (j=1, 2, \ldots, 30)$$

Where α_i denotes the areal angle of the dodecahedron. ρ is the radius of pleateau boundry.

Deformation Model (2) which modifies the foam after combination consists of forty non-linar transcendental equations:

$$F_j (g_i, T_i^1, T_i^2, \rho, \gamma, \beta_i, \alpha_i) = 0 \quad (j=1, 2, \ldots, 40)$$

Where β_{i} denotes the acute angle of one plane of the dodecahedron.

For an arbitrary foam system, the deformation process of the unit structure is periodic under the steady shear flow. The effective stress can be obtained by integrating over the period.

$$\langle \tau \rangle = \frac{1}{\gamma} \int_{\gamma} \tau \, d\gamma$$

The theoretical results show that the shear and normal stresses increase when the foam viscosity increase, and decrease when the foam mean radius increase, and increase when the gas volume fraction increases.

EXPERIMENT

In order to check the theoretical results, we measured the rheological properties of foams. The RMS-605 Rheometer, Rotary Viscositymeter-2 and the Capillary Rheometer were used for this purpose. The test material was offered by Beijing Institute of Oil Exploration, to which a few stabilizing agents were added. A rotary mixer of high rotation was chosen as the foam generator for RMS-605 and Rv-2 Rheometer. While, a fixed bed was chosen for Capillary Rheometer. The foam size was determined by the following procedure. First, take the sample in a glass rectangular box, in which the bubble was of approximately spherical segment, then, enlarge the images and take the picture by a camera, and compute the mean radius R based on the surface volume[4]. The interfacial tension of the foam was measured by tensionmeter Jzhy-18. φ was determined by measuring the volume of the liquid before and

after rotation in a rotary mixer, and by controlling the rate of gas and
liquid flow in the fixed bed. Fig.1 and Fig.2 are the results of experiment.
The figures show that the three-dimensional model is better than the two-
dimensional model.

RMS-605, Rv-2 and theory

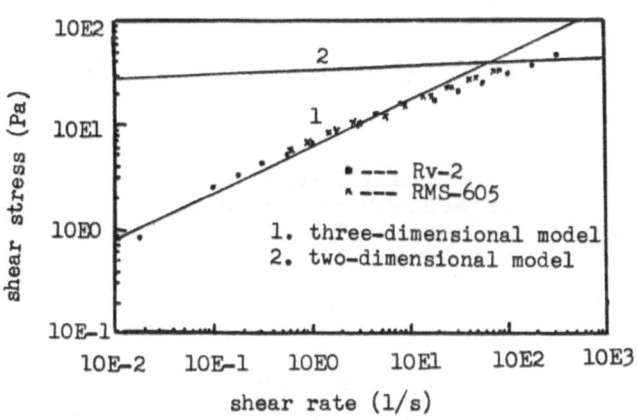

Fig. 1

RMS-605 and theory

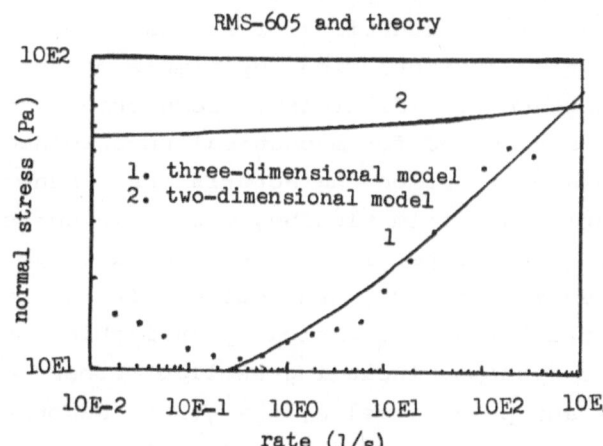

Fig. 2

REFERENCES

1. H. M. Princen, J. Colloid and Interface Sci. , 91, 160 (1983).
2. S. A. Khan & R. C. Armstrong, J. Non-Newtonian Fluid Mech. , 22 (1986).
3. A. M. Kraynic & M. G. Hansen, J. Rheology, 30(3), 409 (1986).
4. H. M. Princen, J. Colloid and Interface Sci. , 105(1), 150 (1985).

534

STRESS-ORIENTATION RELATIONS IN POLYMER FLUIDS

ANDRZEJ ZIABICKI
Polish Academy of Sciences
Institute of Fundamental Technological Research
Warsaw, Poland

Relations between stress and optical anisotropy tensors
in systems subjected to deformation and flow play an important
role in polymer rheology. In systems, where optical anisotropy
is a unique and linear function of stress, complex stress
fields can be easily characterised by simple measurement of
optical birefringence or dichroism. Development of highly
ordered structures required for mechanical performance is controlled by stress-orientation characteristics whenever processing (fibre spinning, film blowing, etc.) is performed in
the fluid state. The steeper is increase of average orientation with increasing stress, the easier it is to produce
highly ordered structures in technically aveilable conditions.

Many polymer systems, including swollen rubbers and gels,
dilute solutions and polymer melts, have been demonstrated to
exhibit linear stress-orientation behaviour in a wide range of
conditions. On the other hand, suspensions of rigid rods,
liquid crystals and some other systems decline from obeying
what is often called "stress-optical law".

We shall discuss several molecular factors affecting
uniqueness of the stress-orientation relations. The macroscopic characteristic of orientation (anisotropy tensor, order
tensor) $\underline{A}$ is defined as a sum of molecular contributions, $\underline{a}$,
dependent on a set of configuration characteristics, $\underline{\xi}$

$$\underline{A} = \int \underline{a}(\underline{\xi}) \; W(\underline{\xi}) \; d\underline{\xi} \qquad (1)$$

where $W(\underline{\xi})$ is configuration distribution.

Orientation, by definition, is additive and controlled by configuration characteristic $\underline{\xi}$.

Let us assume that also stress tensor, $\underline{P}$, is additive and based on configuration-controlled contributions $\underline{p}(\underline{\xi})$

$$\underline{P} = \int \underline{p}(\underline{\xi}) \; W(\underline{\xi}) \; d\underline{\xi} \qquad (2)$$

If this is the case, relation between $\underline{A}$ and $\underline{P}$ can be unique and (in some range of variables) linear. The necessary condition is that stress is purely elastic (i.e. configuration-controlled) and no interactions between particles are present. Examples are provided by ideal crosslinked networks, solutions and melts of flexible chains. Suspensions of elastic, deformable spheres would behave in the same way only if friction effects due to rotation are neglected.

Molecular rigidity introduces new mechanisms into molecular dynamics. "Equilibrium", or "static" rigidity related to non-linear (but purely elastic) relation between the stress contribution $\underline{p}$ and configuration $\underline{\xi}$, does not affect uniqueness of the stress-orientation relation. More important is "dynamic" rigidity, or "internal viscosity" which makes stress contribution dependent not only on configuration, $\underline{\xi}$, but also on configuration rate, $\underline{\dot{\xi}}$

$$\underline{p}(\underline{\xi}) \rightarrow \underline{p}(\underline{\xi}, \; \underline{\dot{\xi}}) \qquad (3)$$

No unique relation between $\underline{P}$ and $\underline{A}$ (still controlled solely by configuration, $\underline{\xi}$) can be expected. Polymer systems composed of viscoelastic molecules do not obey any stress-optical law.

A similar effect results from molecular friction provided by rotation of rigid molecules in a viscous, or viscoelastic continuum. Rotation rate, $\underline{\dot{\xi}}$, contributes to, or even controls stress, while orientation is invariably controlled by configuration alone. Examples are suspensions of rigid particles and solutions of stiff chains in viscous solvents, where elastic (configuration-controlled) effects are accompanied by rotation-controlled energy dissipation.

Intermolecular interactions are an important stress-controlling mechanism in many systems. While orientation remains an additive function of a configuration of a single molecule, and averaging is based on one-molecule distribution, $W_1(\underline{\xi}_1)$

$$\underline{A} = \int \underline{a}(\underline{\xi}_1) \; W_1(\underline{\xi}_1) \; d\underline{\xi}_1 \tag{4}$$

stress includes also an interaction term, $\underline{P}_{int}$ controlled by configurations of interacting particles and pair-distribution function, $W_2(\underline{\xi}_1, \underline{\xi}_2)$

$$\underline{P}_{int} = \iint \underline{p}(\underline{\xi}_1, \underline{\xi}_2) \; W_2(\underline{\xi}_1, \underline{\xi}_2) \; d\underline{\xi}_1 \; d\underline{\xi}_2 \tag{5}$$

This also makes stress-orientation relations non-unique. A typical example is provided by concentrated systems of rigid molecules and liquid crystals, where particle-particle interactions play an determining role in the stress tensor.

VISCOUS FINGERING OF NON-NEWTONIAN FLUIDS IN POROUS MEDIA

E. ALLEN, D.V. BOGER
Department of Chemical Engineering
University of Melbourne
Melbourne, Australia

INTRODUCTION

One of the major restrictions to increased oil recovery from reservoirs is poor sweep efficiency, caused by an interfacial instability phenomenon termed "viscous fingering". The efficiency of miscible two-phase displacement of Newtonian fluids is directly related to the viscosity ratio between the displaced and displacing fluids μ_1/μ_2 where subscript 1 denotes the displaced fluid and subscript 2 denotes the driving (or displacing) fluid. If the viscosity ratio is unity or less, stable displacement occurs and the sweep efficiency is a maximum. If the viscosity ratio is greater than unity, the driving fluid (water) will finger through the displaced fluid (oil) leaving it in place in the reservoir.

Polymers are used in enhanced oil recovery (EOR) to "viscosify" the displacing water and thus increase the displacement efficiency in a reservoir. Because these fluids are relatively expensive, the usual procedure is to inject a small slug of polymer followed by a chasing waterflood. The objective of polymer flooding design is to achieve a stable forward interface between the polymer slug and the displaced oil, and a stable rear interface with the chasing waterflood.

Numerous papers have been written on unstable displacement of Newtonian fluids, in a range of flow geometries. There are very few papers on non-Newtonian displacement. Polymer solutions are non-Newtonian fluids at reservoir conditions, whereas crude oil and water are Newtonian fluids at reservoir conditions. The general approach in the petroleum engineering literature has been to characterise the non-Newtonian fluids with a power-law equation, and then to evaluate the viscosity ratio at the relevant shear rate ([1] and [2]). The assumption is often made that shear rates are significant only in the near well-bore region, where shear thinning behaviour will aid the injection by enabling higher injection rates than would be possible for a Newtonian fluid with the same zero shear viscosity. In the rest of the reservoir it is commonly assumed that the shear rates are so low that the fluid exhibits its low shear Newtonian asymptote, η_o. Thus the Newtonian correlations would appear to be adequate for estimating the displacement efficiency.

However, it is clear from a review of the literature that most of the field applications of polymer flooding have not lived up to the promise of the laboratory tests and theoretical simulations. In fact many polymer floods have been spectacular failures, with <u>very</u> early breakthrough of the chasing waterflood at the producing wells and negligible increase in recovery from that to be expected from a simple waterflood. The problem is generally attributed to a loss of effectiveness of the polymer solutions, due to a variety of unquantifiable causes such as greater heterogeneity than expected in the porous strata, and/or greater polymer degradation, and/or greater polymer retention than expected. Although all of these could be true, there is a simpler rheological explanation which may be crucial to improving the design of future polymer floods. This paper presents the results of a study on the stability of the rear interface of a polymer slug, where a non-Newtonian polymer solution is displaced by water.

DISCUSSION

The shear rate in porous media has been correlated with a range of equations, of which the best appears to be that of Kemblowski and Michniewicz [3]

$$\dot{\gamma}_{K \to M} = [(3n+1)/4n] \quad [\sqrt{3}V_D/(k\phi)^{0.5}] \qquad (1)$$

where V_D is the superficial (Darcy) velocity, k is the permeability, and ϕ is the porosity of the porous medium.

Considering the largest realistic values of porosity and permeability in actual oil reservoirs, using realistic injection rates, and taking the case of radial flow we can obtain a range of minimum probable shear rates experienced in the field. The range is 10 - 1000 s^{-1}, which falls within the power-law region for typical polymer solutions at reservoir conditions, and not in the zero shear region as often assumed. Most reservoirs have porosity and permeability values less than those used here, and therefore the shear rates experienced in the field are likely to be even higher. Therefore fluid behaviour cannot be approximated by Newtonian behaviour at the same zero shear viscosity.

Experiments were conducted using a radial Hele-Shaw cell and three packed radial cells. The fluid was injected using an <u>Instron based</u> positive displacement system, and the developing fingers were photographed from above. The results were evaluated using the previously developed fractal relationship, to characterise the displacement efficiency as a function of the maximum finger radius [4]. The presentation will give the results obtained for polysaccharide solutions (using Shellflo S and Flocon 4800C which are proprietary polymers used by the petroleum industry). The fluids were characterised using a Weissenberg Rheogoniometer in steady shear and oscillatory mode, and have shear thinning inelastic behaviour at the relevant shear rates. The experiments were also conducted at a range of flowrates to cover the relevant shear rates expected in field applications.

The displacement efficiency of Newtonian fluids will be compared with the non-Newtonian fluids, both graphically and visually in the presentation. The displacement data indicates a fractal relationship, in direct conflict with the petroleum engineering literature to date; which will have major implications for the design of enhanced recovery processes in terms of both the optimum well spacing and the polymer slug sizing. The displacement patterns for the non-Newtonian fluids are quite striking, and visually demonstrate a significant reduction in displacement efficiency for the same apparent viscosity ratio. The reason for the markedly different behaviour can be relatively easily explained in terms of the rheological properties. The shear thinning characteristic causes the fluid to be very sensitive to any perturbation in the flow field, and exacerbate any instability.

CONCLUSIONS

1. Two-phase displacement of non-Newtonian fluids cannot be adequately characterised by Newtonian correlations.
2. The displacement efficiency appears to be a function of power-law index n, in addition to the viscosity ratio. Larger values of n correspond to more efficient displacement with reduced viscous fingering.
3. The shear rate under typical production conditions is likely to fall predominantly in the power-law region for the polymer solutions used.
4. The spectacular failure of previous polymer floods may be attributable to loss of polymer slug integrity, due to excessive shear thinning. Future design should look at ways of boosting the viscosity of the displacing fluid without introducing shear thinning.

REFERENCES

1. Lake, L.W., Enhanced Oil Recovery, Prentice-Hall, New Jersey, 1989.

2. Littmann, W., Polymer Flooding,, Elsevier, New York, 1988.

3. Kemblowski, Z. and Michniewicz, M., A new look at the laminar flow of power-law fluids through granular beds. Rheol. Acta, 1979, 18, 730-739.

4. Allen, E. and Boger, D.V., Mobility Control in Polymer-Augmented Waterflooding. SPE 18097, 63rd Annual Technical Conference of the Society of Petroleum Engineers, Houston, Texas, 1988.

540

Wolfgang Mader-Hipp *

Measuring yield points instead of calculating

* Dr. W. Mader-Hipp, PHYSICA Meßtechnik GmbH u. Co. KG, Vor dem Lauch 6,
  D-7000 Stuttgart 80, W. Germany

With conventional rotational rheometers, where the speed is controlled
(shear rate) and the torque is measured (shear stress) it has been
impossible to measure the yield point of a substance. Instead it had to
be determined by means of extrapolation and using mathematical model
equations. Now, with a controlled stress rheometer, the yield point of a
substance can in fact me measured.

Measuring yield points has long been a wish of application technology.
Until now, yield points could only be extrapolated by means of mathematical model calculations from the flow curve for the shear rate value
D = 0. The reason for this is that in conventional rotation-controlled
rheometers only mechanical force transducers (springs, torque rods) are
used and thus a force can only be detected after movement (Hooke's law).
In this case, however, the sample in the measuring gap has already been
sheared before a yield point could be recorded.

Functioning of a controlled stress rheometer

According to the physical bases of the mechanics of fluids, a movement
is induced by a force acting on a body. The two-plate model shown in
Fig. 1 explains this. A force F or shear stress τ acts on the upper,
moving plate A and like this induces a movement v or shear rate D of
this surface against the lower, fixed surface.

Applied to a controlled stress rheometer this means that for instance in
the Searle system the measuring cup represents the fixed and the rotating bob the moving surface.

For the rheological studies explained here, a controlled stress rheometer PHYSICA-RHEOLAB® MC 10 was used (PHYSICA-RHEOLAB ® is a registered
trade-mark of PHYSICA Meßtechnik GmbH u. Co. KG, Stuttgart, West
Germany) with the universal measuring drive system UM, which permits
measurements in cylinder measuring systems as well as in cone/plate and
plate/plate measuring systems.

The working method of the RHEOLAB sensor technique is based on the
foundations of the science of electricity. The law of the dynamic effect
applies, which a current-carrying conductor experiences in a magnetic
field.

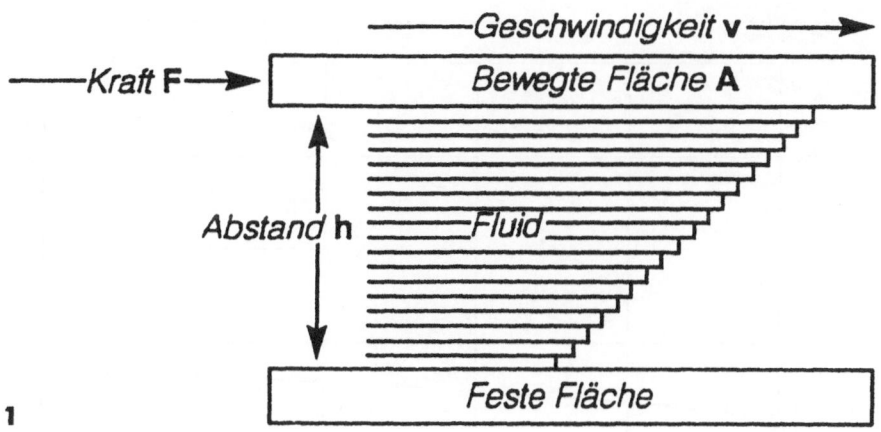

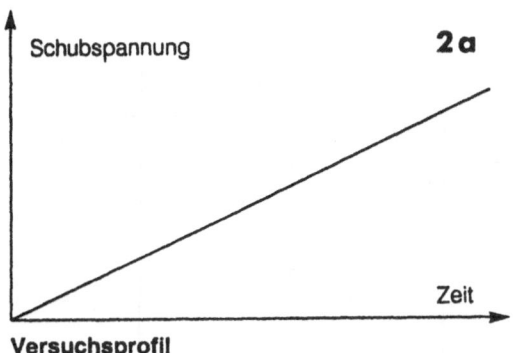

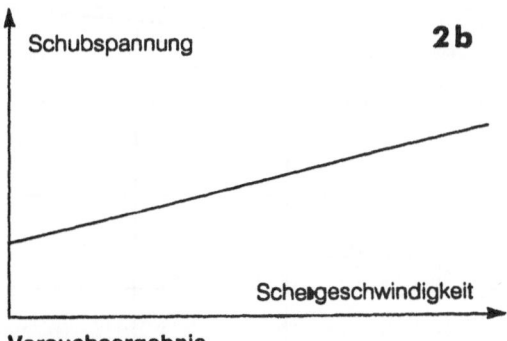

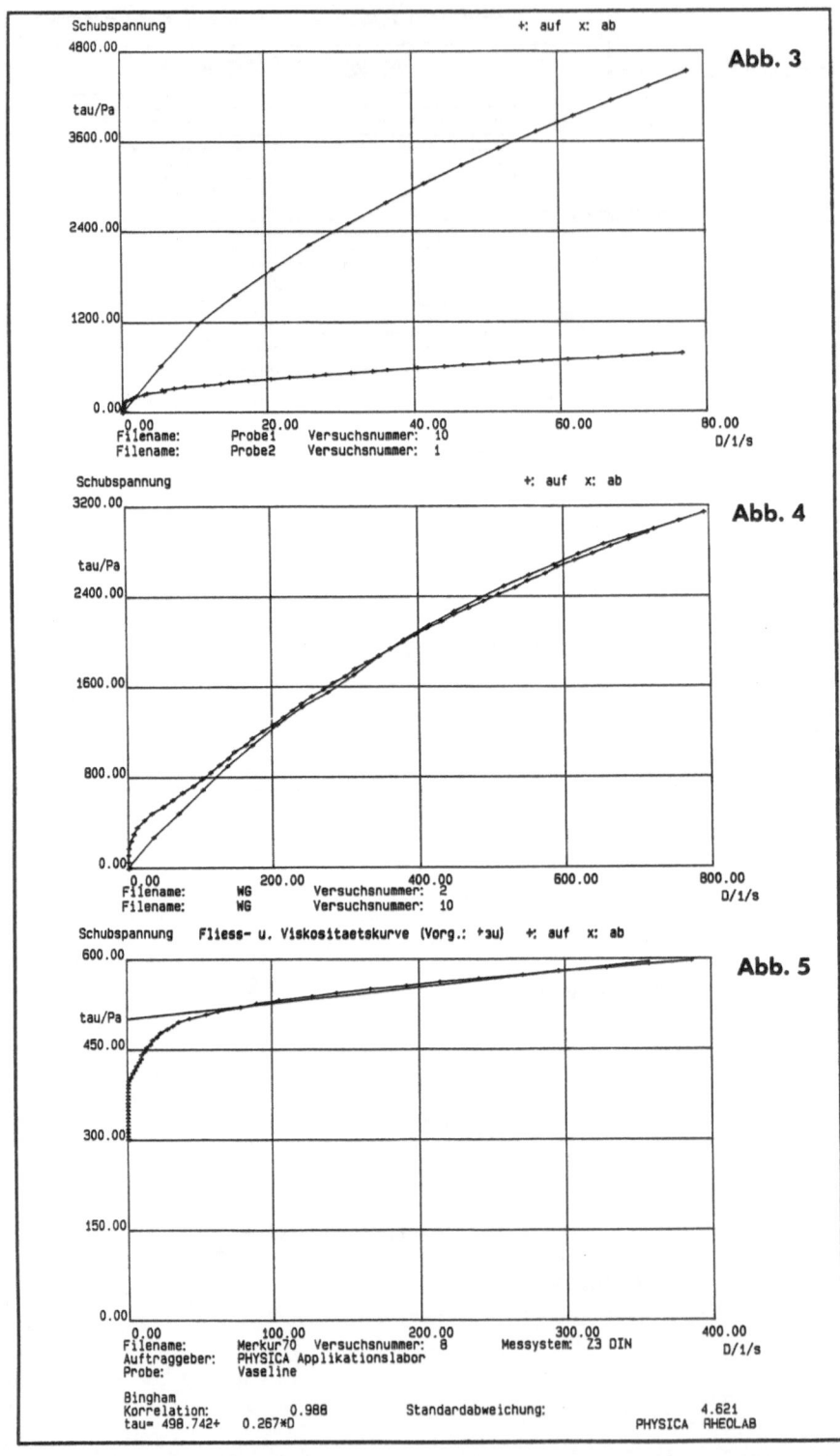

Schubspannung · +: auf x: ab

Abb. 3

tau/Pa

4800.00
3600.00
2400.00
1200.00
0.00

0.00 · · · · 20.00 · · · · 40.00 · · · · 60.00 · · · · 80.00

Filename: Probe1 Versuchsnummer: 10
Filename: Probe2 Versuchsnummer: 1

D/1/s

Schubspannung · +: auf x: ab

Abb. 4

tau/Pa

3200.00
2400.00
1600.00
800.00
0.00

0.00 · · · · 200.00 · · · · 400.00 · · · · 600.00 · · · · 800.00

Filename: WG Versuchsnummer: 2
Filename: WG Versuchsnummer: 10

D/1/s

Schubspannung Fliess- u. Viskositaetskurve (Vorg.: +зu) +: auf x: ab

Abb. 5

tau/Pa

600.00
450.00
300.00
150.00
0.00

0.00 · · · · 100.00 · · · · 200.00 · · · · 300.00 · · · · 400.00

Filename: Merkur70 Versuchsnummer: 8 Messystem: Z3 DIN
Auftraggeber: PHYSICA Applikationslabor
Probe: Vaseline

D/1/s

Bingham
Korrelation: 0.988 Standardabweichung: 4.621
tau= 498.742+ 0.267*D

PHYSICA RHEOLAB

For this dynamic effect the following rule applies:
F = B x L · I,
with B = magnetic flow density
     L = conductor length
     I = current

So there is a direct connection between the acting torque (shear stress) and the electrical values of current and voltage. Thus the acting shear stress can be detected without path (electrically), which, apart from determining the flow and viscosity function, now also permits a direct measurement of the yield point.

Technically, a shear stress increasing over the time (Fig. 2a) is pre-set. As long as the internal holding forces of a substance are greater than the applied shear stress (shear stress = force · surface), no movement, which is directly proportional to the shear rate D, is detected. Only when the applied force becomes greater than the internal holding forces, the substance begins to flow (Fig. 2). Hence, the yield point can be read from the measuring diagram.

Examples of measurements

Fig. 3 shows two production batches of insulating lacquer. Only sample 2 is suited, because it has a yield point and can therefore form a certain layer thickness on the carrier material. The much more viscous sample 1, in contrast, runs down, because it does not have a yield point.

Fig. 4 also shows two different production batches of solder resist, with sample WG2 showing a yield point, WG10 not.

Fig. 5 shows a vaseline substance. If there the yield point is deter-mined according to the conventional mathematical extrapolation method, the result is a value of just under 500 Pa, whereas measured the value is only about 400 pa. In practice this means that the thickness of the layer which can be applied onto a vertical wall is about 20% thinner than assumed according to an extrapolation. Generally it can be noted that the calculated yield points are mostly higher than the measured yield points.

This can have serious consequences in application technology. Following is an example to explain this:

If only the mathematically determined yield point is known of a lacquer and if in reality the yield point is much higher, the lacquer will not sufficiently cover the surface, because the maximum applicable layer thickness is under the expected value.

Other applications for controlled stress rheometers are all fields in
developmental departments and production companies where a defined yield
point setting is required, as for instance in the paint industry, in the
food industry and at producers of coating materials, to name just a few.

Conclusion

By means of controlled stress rheometers it is for the first time
possible to measure exact yield points by presetting a controlled shear
stress. The presetting of the shear stress moreover permits to carry out
other test types which cannot be made with conventional rotational
rheometers, such as creep tests and force-controlled oscillation tests.
However, the present article will not go into these test types. The
modern sensor technology used in the PHYSICA-RHEOLAB ® system also
permits the user to carry out all known speed-controlled test types.

RHEOLOGY OF ASSOCIATING POLYMER SOLUTIONS

M. MOAN, A. OMARI
Laboratoire de Mécanique Physique
I.U.T. de BREST - 29287 BREST CEDEX _ FRANCE

ABSTRACT

We have studied a Polysaccharide-Borax system, in which borate ions serve as crosslinkers between polymer chains. The rheological properties have been investigated in dilute regime as well as in more concentrated regime. We have shown that, under certain conditions, the intrinsic viscosity passes through a maximum and then decreases abruptly. In the semidilute regime, intermolecular associations lead to the formation of a network structure. Oscillatory shear measurements have been used for studying this structure, particularly at the gel point. On the other hand steady shear measurements show a complex and unusual shear thickening behaviour of this structure.

INTRODUCTION

Rheological properties of polymer solutions may be modified by associating interactions between polymers, resulting from the formation of complexes between polymer and crosslinking agents. Such polymer systems can exhibit thickening and gelling properties but also shear thickening behavior. Polysaccharides can be linked by a large number of chemical agents including borax. In this case, the mechanism of formation of crosslinks is governed by complexation equilibria between borate ions and hydroxyl groups of sugar units. The aim of this study is to provide more insight into the effect of associations on the rheological properties of Guar gum-borax system over a large range of polymer-concentration. Guar-borax (Galactomannane) is a neutral flexible polysaccharide of high molecular weight. The study of this system is particularly interesting because the complexation reactions involved and the conditions for gelation are well-established [1,2].

RESULTS

Newtonian viscosity and intrinsic viscosity

Newtonian viscosity of polymer solutions was measured for various polymer and borax concentrations, respectively C and C_B. In the semidilute regime (C > C*, C* being the overlap polymer concentration) the viscosity increases as C_B increases. For a fixed polymer concentration, the viscosity diverges at a critical borax concentration C_B*. Beyond C_B*, a gel phase formation is observed. Indeed, the number of interchain crosslinks increases leading to gelation. In the dilute regime, we observe at a given C a slight decrease of the viscosity as C_B increases. This effect can be explained by measuring the effect of borax concentration on the relative change in intrinsic viscosity $[\eta] / [\eta]_0$, $[\eta]_0$ being the intrinsic viscosity without added borax (Fig. 1). This complex behavior is explained

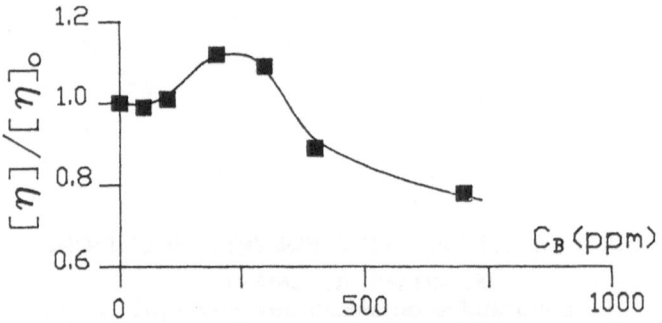

Figure 1. Intrinsic viscosity as a function of borax concentration.

by the competition between various effects which modify the chain conformation [3] : electrostatic repulsions (expansion) between ions fixed on
the chain - contraction due to intrachain crosslinks   (loops) - increase of
ionic strength (contraction). On the other hand, the Huggins coefficient
increases as borax concentration increases, indicating the presence of
molecular associations. This result means that even in dilute regime, some
interchain crosslinks exist in the solution.

Dynamic oscillatory measurements

We have studied the changes in storage modulus $G'(\omega)$ and loss modulus
$G''(\omega)$ induced by the addition of borax to a semidilute polymer solution.
At low borax concentrations, G'' is relatively large while G' is still low.
With   increasing borax concentration, G' increases fastly until, at $C_B \simeq C_B*$,
G' and G'' are congruent and power law dependent, $G'=G'' \sim \omega 0.8$, over
the frequency range used. Thus C_B* could characterized the gel point
[4] . A typical plot obtained for $C_B > C_B*$ is shown in Figure 2. We observe a crossover point followed by a plateau for G'. So, viscous behavior
dominates at low frequencies and elastic behavior dominates at higher frequencies. On the otherhand, the plateau is more extended as borax concentration is high.

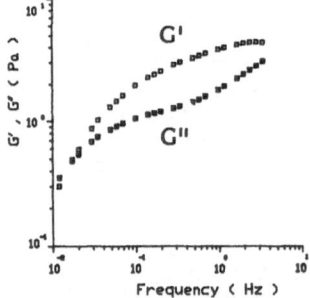

Figure 2. Storage and loss moduli for a polymer solution C = 5000 ppm
and C_B = 200 ppm.

Steady shear viscosity

The effect of shear stress on the viscosity has been studied for semidilute
polymer solutions in which intra and interchain crosslinks are present.
Without added borax and low C_B, an usual flow curve is observed : a Newto-
nian regime followed by a shear thinning behavior at higher shear stresses.
At higher C_B, but lower than C_B*, a small shear thichening peak occurs.
It can be explained by a shift from intra to interchain crosslinks induced

by the shear. At $C_B > C_B{}^*$, starting at a low shear stress and increasing shear stress after a steady state has been reached, we observe, as shown in Figure 3, a high shear thickening peak followed by a smaller peak at higher shear stresses. This second peak can be explained by the fact that in this range of shear stresses, the shearing forces break intrachain crosslinks. This results in an increase of the hydrodynamic volume of individual polymer chains. On the other hand, the flow curve obtained using a fresh solution and decreasing shear stress from a high shear stress (Figure 3) shows a similar shape. However at very low shear stresses, the decreasing shear stress curve returns to a level much higher than that of the increasing shear stress curve and seems reveal the appearance of a yield stress.

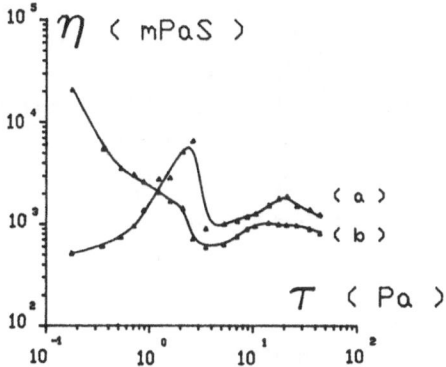

Figure 3. Viscosity as a function of increasing shear stress (a) and decreasing shear stress (b). C = 5000 ppm ; C_B = 150 ppm.

CONCLUSION

This study shows that the addition of complexing ions to a polymer solution modifies the conformation of polymerchains and leads, under certain conditions on the respective polymer and borate ions concentrations, to the formation of a network structure (or gel). In relation to the strength and the lifetime of the crosslinks, the network structure is very sensible to the shear stress in the vicinity of the gel point.

REFERENCES

1. Pezron, E., Ricard, A., Lafuma, F., Audebert, R. Macromolecules, 1988, 21, 1121-1125.

2. Pezron, E., Leibler, L., Ricard, A., Audebert, R. Macromolecules, 1988, 21, 1126-1131.

3. Ochiai, H., Kurita, Y., Murakami, I. Makromol. Chem., 1984, 185, 167-172.

4. Chambon, F., Winter, H. J. Rheol. 1987, 31, 683-697.

Planar Contraction Flow Of FENE Dumbbells

R.A. Keiller

Department of Applied Mathematics and Theoretical Physics,

University of Cambridge,

Cambridge, CB3 9EW,

United Kingdom.

Abstract

Numerical simulations have been performed for the flow of a polymer solution into a planar contraction from an infinite reservoir using a FENE dumbbell model. Finite differencing combined with time-stepping is used on a grid generated by a conformal mapping. This allows the clustering of grid points in the regions of high shear rate and high polymer extension in order to resolve these regions accurately. The mapping also enables the rounding off of the sharp $\pi/2$ corner at the lip of the contraction thereby removing the flow singularity which is a possible cause of numerical problems. Particular attention is given to numerical instabilities which are observed to occur in the downstream shear region and which can be removed by adequate cross-stream grid refinement.

Lip vortices are observed in these simulations but only at sufficiently high Deborah numbers that the finite extensibility of the dumbbells plays a significant role, indicating the importance of additional extensional stresses not present with the Oldroyd-B or Maxwell models. These lip vortices are in qualitative agreement with those observed experimentally by Evans and Walters [1] in the flow of a Boger fluid through an 80:1 contraction.

Reference

Evans, R.E. & Walters, K. (1986) *J. non-Newtonian Fluid. Mech.* **20**, 11–29.

Sudden Strong Strains

E.J. Hinch

Department of Applied Mathematics and Theoretical Physics,
University of Cambridge,
Cambridge, CB3 9EW,
United Kingdom.

Abstract

Computer simulations have been performed of a how a random-coiled polymer unravels when it is placed in a strong straining motion. The computer model has several hundred freely hinged rigid links which are subjected to random Brownian forces and to viscous flow forces (but with no hydrodynamic interactions). The stress is found to build up rapidly in time, to be mainly viscous in nature, and to have a bounded elastic component. These features are different from the behaviour of simple bead-spring models of polymers, which lead to the Oldroyd constitutive relation. The build up in stress is slightly faster than that predicted by Ryskin's *yo-yo* model. That the stress is viscous means that it is proportional to the instantaneous strain-rate, and this produces very much larger stresses than the elastic ones of the bead-spring models in flows such as that through a contraction (where the strain-rate increases rapidly following a fluid element). An investigation of a highly simplified version of the unravelling has revealed that the build up of stress is not related to the overall size of the elongated coil, but is related to the growth of many small sections of the polymer chain which are separately fully stretched.

HIGH DEBORAH NUMBER FLOWS OF DILUTE POLYMER SOLUTIONS

O.G. HARLEN and J.M. RALLISON

Department of Applied Mathematics and Theoretical Physics,
University of Cambridge,
Cambridge CB3 9EW,
United Kingdom.

Abstract

A boundary layer approximation is developed for steady flows of dilute polymer solutions at high Deborah numbers. In these flows polymer molecules are most highly extended within thin elongated regions at or downstream of stagnation points, as only those molecules that pass close to a stagnation point reside in the flow long enough to experience a large strain. This structure can been seen in both optical birefringence experiments (*e.g.* Cressely & Hocquart 1980, Keller & Odell 1985) where the regions of extended polymer appear as bright birefringent lines, and in numerical calculations using a F.E.N.E. dumbbell model (Chilcott & Rallison 1988). For dilute solutions high elastic stresses occur only within these regions of high molecular strain (here called '*birefringent strands*') and outside the fluid behaves as a Newtonian fluid of approximately solvent viscosity.

This structure is used to devise an asymptotic ('*birefringent strand*') technique for flows with stagnation points in which the birefringent strands are treated as line distributions of forces within an otherwise Newtonian fluid. A number of different flow geometries which contain one or more stagnation points are analysed using this technique. These flows contain a variety of different types of stagnation points: isolated stagnation points and separation points on solid boundaries and free surfaces in both two-dimensional and axisymmetric geometries. This method produces results which are in good agreement with both experiment and full numerical solutions, but without requiring large numerical calculations.

References

Chilcott, M.D. & Rallison, J.M. (1988) *J. non-Newtonian Fluid. Mech.* **29**, 381–432.
Cressely, R. & Hocquart, R. (1980) *Optica Acta* **27**, 699–711.
Keller, A. & Odell, J.A. (1985) *Colloid & Polymer Sci.* **263**, 181–201

DYNAMIC MATERIAL PROPERTIES OF A CLOSED—CELL POLYETHYLENE FOAM

A. Geißler
Fraunhofer-Institut f. Chemische Technologie (ICT)
Joseph von Fraunhofer Str.7, D-7507 Pfinztal-Berghausen,
FRG.

H. Weber
Universität Karlsruhe (TH)
Institut für mechanische Verfahrenstechnik u. Mechanik
(MVM) P.O. Box 6980, D-7500 Karlsruhe, FRG

Summary

With today's high standards of technology, transportation stresses can be almost completely described so that certain conclusions for the packaging container can be drawn. But an optimized, costeffective container can only be produced if the material behaviour and the environment of the packaging material used are known /1/. Foams belong to the most important class of packaging materials. Their dynamic behaviour is usually qualified empirically by properties which are not suitable for engineering analysis of packaging con-tainers, but represent only quality control values.

In the context of this paper, it is shown how dynamic mechanical behaviour of a closed-cell PE-foam can be experimentally determined for use in engeneering analysis.

Problem

The dynamic behaviour of foams is complex because of there viscoelastic response. When used as cushioning materials, one can observe geometrical and physical nonlinearities which result from different static deformations.

For the deformation analysis of packaging containers, mathematical procedures such as the Finite Element Method (FEM) become more and more important, because of the availability of more efficient and less costly computing machines /2/. The essential advantage of this approach is the ability to calculate the deformation response of the complex structures used in realistic applications. The precision of FEM-analyses depends on the constitutive equation which describes the response of the material. For foams, this means that the nonlinear viscoelastic material behaviour must be investigated.

Kinematic and Constitutive Equation

In /3/ typical loading conditions for road transportation are listed. The load can be described by a superposition of a large static and a small harmonic strain occuring in the transport unit. Because of the large static deformation, the Green tensor G is used to describe the kinematic situation of the superposed strains $G = G^m + G^d$, with: G^m = static deformatin and G^d = dynamic deformation.

If this strain description is used in a nonlinear constitutive equation /4/, the solution can be divided into static and dynamic components. The result is that the complex moduli of the dynamic part of the solution depend on the excitation frequency and the static deformation. The complex moduli of this dynamic solution can be calculated if the dependence on static deformation for the complex Young's - and shear modulus is known.

Complex Material Description

When a viscoelastic material such as a foam is deformed dynamically, one part of the deformation energy is stored and the other part is dissipated as heat. This material behaviour for a given prestrain $_{xx}$ and an excitation frequency f_o can be described by the complex elastic E* and shear modulus G^*:

$$E^*(\ _{xx}, f_o) = E'(\ _{xx}, f_o) + i\ E''(\ _{xx}, f_o)$$

$$G^*(\ _{xx}, f_o) = G'(\ _{xx}, f_o) + i\ G''(\ _{xx}, f_o)$$

As /4/ shows, they are the basic parameters needed to obtain sufficient material description for dynamic analysis of nonlinear viscoelastic structures.

Results of the Vibration Experiments

The results for the elastic modulus is shown in figure 1. They have the same slope as the shear modulus. The absolut value of the complex moduli increased lightly with increasing frequency but a strong rising was measured increase with prestrain. The damping behaviour has no significant dependence on prestrain.

Application

When the dependence of the prestrain on the shear modulus $G^*(\ _{xx}, f_o)$ and the elastic modulus $E^*(\ _{xx}, f_o)$ is known, a nonlinear constitutive equation can be prepared /4/. The determined functions can then be used in a Finite Element code /5/ to calculate the dynamic deformation of packaging structures with realistic geometries.

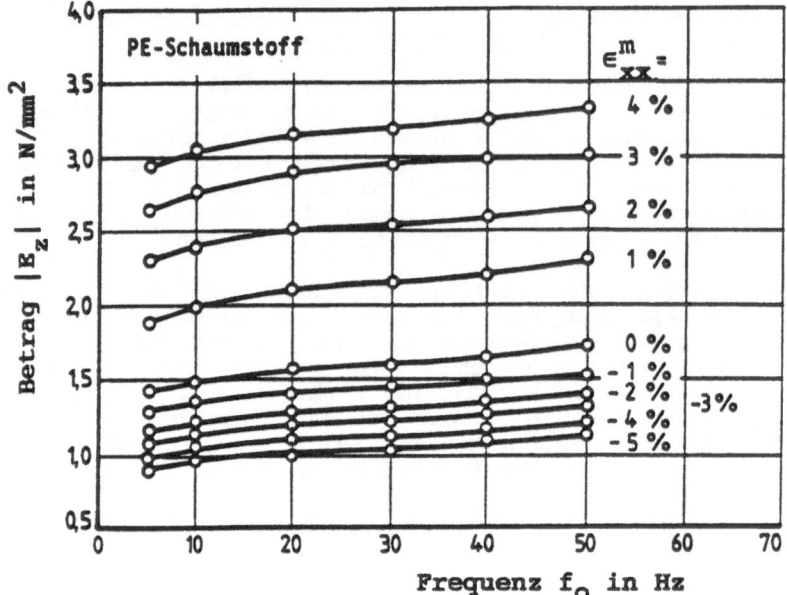

Figure 1: Absolute value of the complex elastic modulus E*

References

/1/ RGV-Handuch Verpackung. Hrsg. von d. Rationalisierungs-Gemeinschaft Verpackung (RGV) im Rationalisierungs-Kuratorium der Deutchen Wirtschaft (RKW) e.V. Berlin: E. Schmidt 1978 ff

/2/ Bathe, K.J.: Finite Elemente Mehtoden. Berlin: Springer 1986

/3/ Ziegahn, K.-F.: Ein Beitrag zur Ermittlung und Beschreibung von mechanischen Transportbelastungen. Fraunhofer-Institut für Treib- und Explosivstoffe, ICT 1987, Pfinztal; (Wissenschaftliche Schrif-tenreihe des ICT) Bd. 1

/4/ Weber, H.; Geißler, A.; Kugler, H.-P.: Einsatz der experimentellen Mechanik in der statischen und dynamischen Prüfung unverstärkter Kunstoffe. In: Experimentelle Mechanik in Forschung und Praxis. Düsseldorf: VDI-Verlag, 1988. VDI-Berichte 679

/5/ Weber, H.; Geißler, E.: Safe Cushion Design for Sensitive Products. Vortrag, 6th IAPRI World Conference on Packaging '89, Hamburg 27.-29. Sept. 1989

INVESTIGATIONS OF A CLOSED-CELL FOAM WITH SCANNING ELECTRON MICROSCOPY

A. Geißler, W. Schmitt

Fraunhofer-Institut für Chemische Technologie (ICT)
Josef-von-Fraunhofer-Sraße 7
D-7507 Pfinztal 1, FRG

Foams belong to the most important class of packaging materials. They are part of a special group of heterogeneous materials that demonstrate nonlinear viscoelastic behaviour. When used as cushioning materials, their damping behaviour is a function of the static load. Microscopic investigations reveal that these non-linearities observed on the macro scale are an attribute of special deformation processes of micromechanical dimensions.

When considering the environmental response of foamed materials, characterization of the macroscopic behaviour is often not sufficient. In time-accelerated tests, the imposed load levels are often icreased, leading to different deformations in the morphological structure of the material. As a result, unrealistic damage can occure. The basic morphological deformation mechanisms should be known for determinating the load levels for time-accelerated testing.

To investigate the deformation mechanisms, a loading device was constructed and inserted into a scanning electron microscope (figure 1). With the aid of this device, we have investigated the response of a closed-cell foam. The micromechanical deformation process reveals unique size distributions within the foam cells. It is important to note that the smaller foam cells produce a stiffer response because of their short rod lenght and push themselves into the larger one's (figure 2). This means that a single cell model cannot completely describe this behaviour.

When determinating of load levels for vibration test, one has to consider that different deformation distributions are to be aspected when changing levels. Very different damage behaviour occure by using an inappropriate increament of the load within the timeaccelerated environmental test. Therefore it is important to understand and to take account for these microscopic deformation processes when designing test matrices to characterize cellular materials.

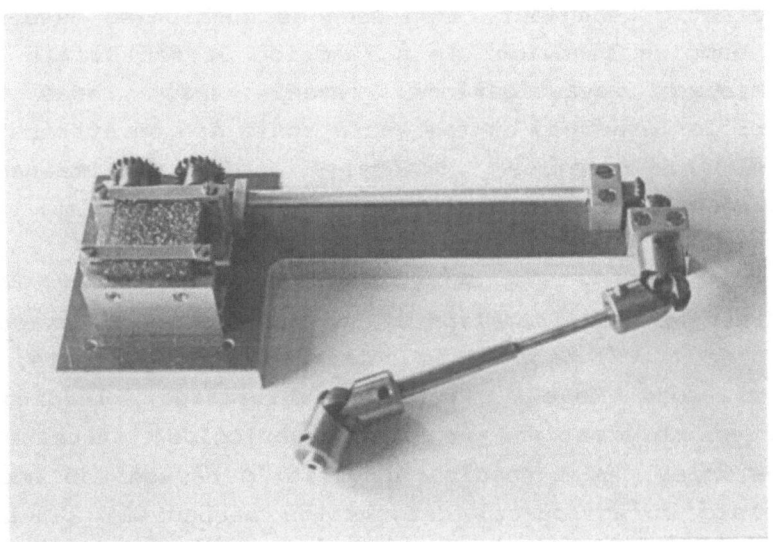

Figure 1: Loading-device for the deformation of a test-sample in a scanning electron microscope

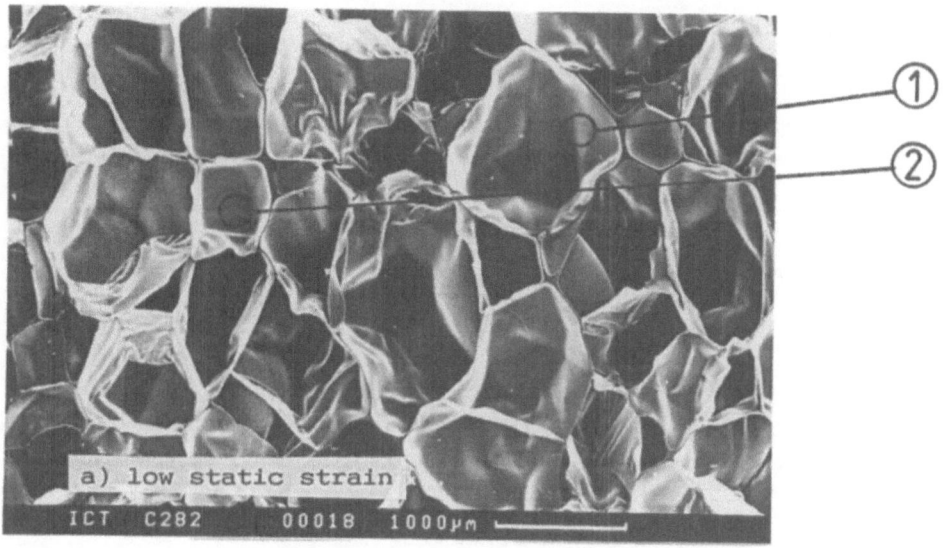

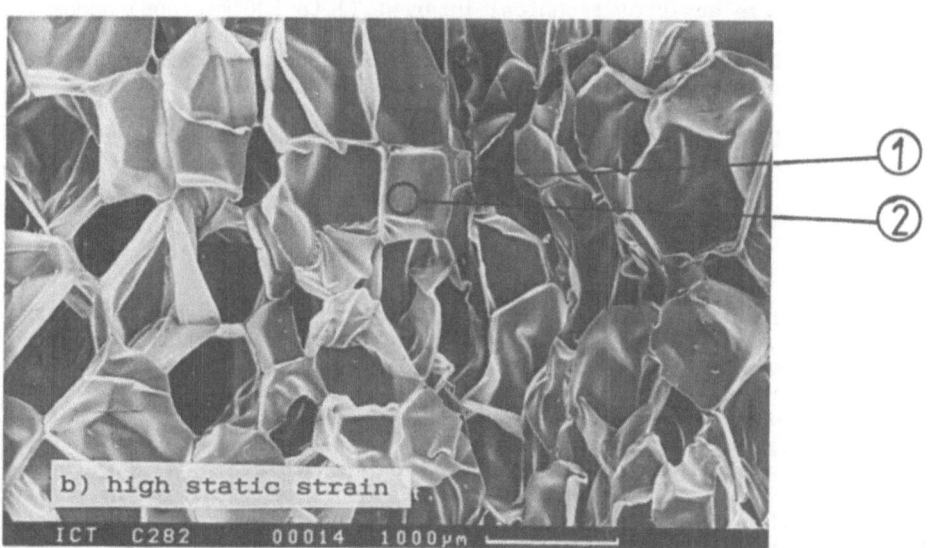

Figure 2: Micromechanic strain distribution of a closed-cell foam at different stress levels
- _location 1 : strong change with local strain_
- _location 2 : slight change with local strain_

STRESS STRAIN RELATIONSHIP OF SMALL DEFORMABLE BODIES UNDER STEADY AND UNSTEADY FLOW CONDITIONS IN NEWTONIAN AND VISCOELASTIC MEDIA

E.WINDHAB[*],B.WOLF[*],V.DENK[**],H.NIERSCHL[**]

* Deutsches Institut für Lebensmitteltechnik;Quakenbrück,Germany

** Technische Universität München,Lehrstuhl für Mechanik Weihenstephan

ABSTRACT

For the experimental evaluation of the deformation behaviour of deformable bodies in the size scale of technical interest (1 to 500 microns) under steady and unsteady shear flow conditions a special opto-rheometric device was built up.With this equipment a "dynamic fixing" of the deformable body in the shear field can be realized.This allows to evaluate the stress-strain relationship under steady and unsteady flow conditions even for fast processes and large deformations.First experimental and modelling results are presented.

INTRODUCTION

Small deformable bodies are present in various fluid systems in the fields of biotechnology (cells,microorganisms), food technology (emulsion droplets, gas bubbles) and chemical engineering.The size scale for small bodies in these studies is about 1 to 500 microns contrary to known studies in the macroscopic range of some milimeters.For practical technological purpose the small bodies have to be treated in different ways with the aim either to destroy them (e.g. emulsifying,homogenizing) or to avoid large deformation and break up (e.g. fermentation process,mixing and stirring of sensitive systems).The forces acting to the small deformable bodies in fluids under different flow conditions are shear and normal forces.It may be supposed that in the micro scale the flow arround the small deformable bodies and following their deformation will be dominated by other forces than in the macroscopic scale.To quantify the "scale down laws" it is useful to compare the results of different size classes at the same Re-numbers.The characteristic length of the "micro scale bodies" leads to Re-numbers lower than 1 even in technical flow processes.This paper presents first results including the newly developed experimental methods,stationary and instationary deformation experiments and modelling of the stress-strain behaviour. Additional studies (not reported here) are carried out to characterize the flow at the surface of the deformable bodies (LDA experiments) to get more information about the interaction mechanism of fluid and small body.

MATERIALS AND METHODS

For the experimental description of the deformation behaviour of deformable
bodies in the size scale of technical interest (1 to 500 microns)under
steady and unsteady shear flow conditions a special opto-rheometric equipment was built up.The transparent measuring cell is a rheometric cone and
plate geometry.With an adapted reversed light microscope the small bodies
can be observed in the shear gap.A video camera connected with an image
analyzing system allows to store digitalized picture data of the deformed
bodies (see fig.1).Cone and plate in the measuring cell are contrarotating
or contraoscillating.This allows especially important for large shear rates
and oscillation amplitudes to realize a "dynamic fixing" of the deformable
bodies in the microscopic observation field (see fig.1)

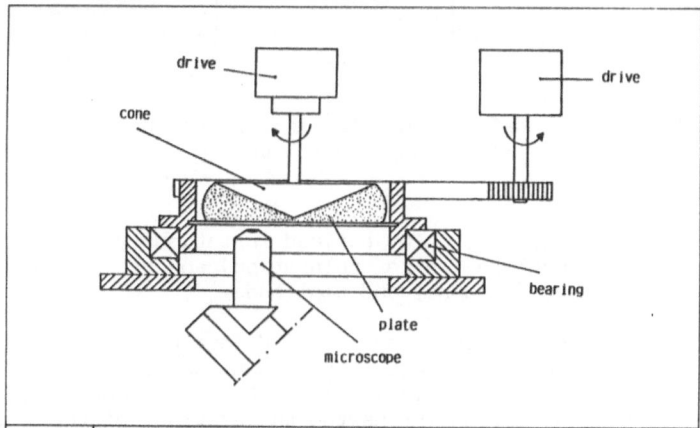

Figure 1.   Principle of opto-rheometric cone and plate device for dynamic
fixing in the microscopic observation field

RESULTS

Stationary Shear Flow
For model experiments droplets of immiscible fluids (preferably silicone
oils and water)with different rheological behaviour and interfacial tension
were used as deformable bodies.In the steady shear experiments carried out
shear stress was varied over a wide range by changing shear rate and viscosity of the surrounding fluid.Figure 2 shows a series of steady state deformations at different shear stresses for a 60 μm water droplet in silicone oil AK 200000.The deformation of the droplet is described by the expression:

$$S(L/B) * L* B / x^2 = f(\tau) \qquad (1)$$

where L and B are the short and long axes of the ellipsoid formed by the
deformable body and x is the diameter of the undeformed spherical body.With
S(L/B) as a definite surface correction function the whole expression (1)
represents a dimensionless surface ratio which is suitable to be introduced.
The droplet deformation depends on the socalled "Weber Number We" which is
proportional to the shear stress τ and $1/\sigma$ (σ = interfacial tension).

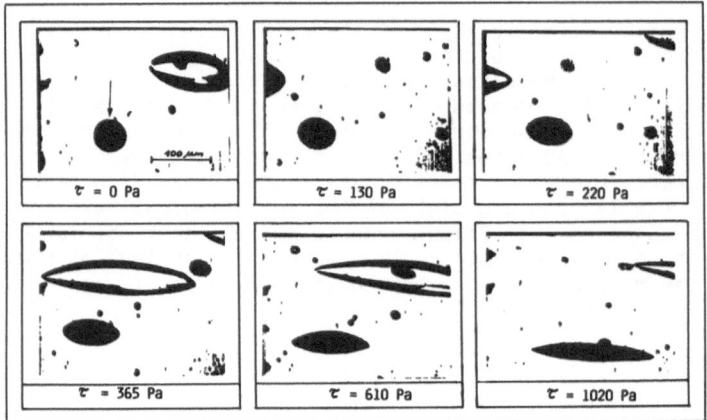

Figure 2. Steady state deformations for increasing shear stress (0 to 1000 Pa) - water droplets (60 µm) in silicone oil AK 2*10^5

Instationary Shear Flow
Two different experimental ways for defined instationary shear flow were performed: definite acceleration (or relaxation) and oscillation of the shear flow with various frequency and amplitude.

Modelling of the Stress-Strain Behaviour of deformable Bodies
First steps in modelling the stress strain behaviour of deformable bodies were carried out including simplifying assumptions. The coupling of energy balance and equation of motion for newtonian behaviour of the inner and outer fluid (η_d, η_c) leads to equation 2 which describes the dependency of dimensionless surface change from the system parameters (K= energy dissipation factor, t= time, η_d = inner viscosity, η_c = viscosity of the continous phase)

$$S(L/B) * \frac{L*B}{x^2} = 1 + \frac{1}{24*K} \left(\frac{\tau * x}{6} \right)^2 * \left[1 - \exp\left\{ - \frac{4}{x * (\eta_d + \eta_c)} * t \right\} \right] \quad (2)$$

The comparison of calculated (eq.2) and experimental results for stationary deformation and instationary change of deformation leads to values in the same order of magnitude. The results are presented.

REFERENCES

1. Taylor, G.J. ,The viscosity of a Fluid containing small drops of another fluid; Proc.Roy.Soc.,London 1934

2. Fewell, M.E. ; Hellums, J.D. ; The secondary flow of newtonian fluids in cone and plate viscometers; Transactions of the Soc. of Rheol.21:4 1977

3. Denk D.,Nirschl H.,Windhab E. ; Interaction between small bodies in enlarged and real scale microfluidmechanics; Proc. CHISA 1990

THE RHEOLOGICAL RESPONSES OF DILUTE POLYMERS AND THE ANALYSIS OF THEIR FLOW INSIDE AND OUTSIDE CHANNELS

I.O.GLOT, N.V.SHAKIROV

Institute of Continuous Media Mechanics Ural Branch –
USSR Academy of Sci. Perm   614061 USSR

The behaviour of Oldroyd and Walters-Fredricson modified models is analysed. To provide the siutable description for shearing and elongational flows it is convinient to use viscosity, relaxation time and retardation time that depend on the second and third invariants of the rate of deformation tensor. The introduction of the third invariant of the rate of deformation tensor in rheological equations is substantiated.

The flow of investigated media inside and outside the channel is considered. The pulsating flow is analysed in details. The authors results are compared with that of another investigators.

The highly elastic deformation stored by the material during its movement through the tube is calculated, and then the swelling ratio is found by using the well-known formulae. The effect of the boundary conditions on the swelling ratio is examined as well.

INDEX OF CONTRIBUTORS